ADVANCES IN INTEGRATED DESIGN AND MANUFACTURING IN MECHANICAL ENGINEERING II

Advances in Integrated Design and Manufacturing in Mechanical Engineering II

Edited by

S. TICHKIEWITCH

INPG, G-SCOP Laboratory, Grenoble, France

M. TOLLENAERE

INPG, G-SCOP Laboratory, Grenoble, France

and

P. RAY

IFMA, Clermond Ferrand, France

Springer

A C.I.P. Catalogue record for this book is available from the Library of Congress.

ISBN 978-1-4020-6760-0 (HB)
ISBN 978-1-4020-6761-7 (e-book)

Published by Springer,
P.O. Box 17, 3300 AA Dordrecht, The Netherlands.

www.springer.com

Printed on acid-free paper

Table of Contents

Mechanical Systems Design and Optimization

Design and Communication

Improved Computational Models

Tolerancing Methods

Metal Cutting and High Speed Machining

Preface

This book presents a selection of papers related to the sixth edition of the International Conference on Integrated Design and Manufacturing in Mechanical Engineering. This Conference has been organized within the framework of the activities of the AIP-PRIMECA network whose main scientific field is Integrated Design applied to both Mechanical Engineering and Production. This conference can be seen as a mirror on international tendencies in Integrated Design in the year 2006. It shows the urgent need for change in product development with the integration during the design activity of the global life cycle of the product.

After Nantes, Compiegne, Montreal, Clermont-Ferrand and Bath, this IDMME Conference has been held in Grenoble, under the responsibility of Professor Serge Tichkiewitch, the Chairman of the Conference and the first General Director of the PRIMECA network. The Scientific Committee members have selected all the lectures from complete papers, which is the guarantee for the Conference of quite outstanding scientific level. After that, a new selection has been carried out to retain the best publications, which establish in a book, a state-of-the-art analysis as regards Integrated Design and Manufacturing in the discipline of Mechanical Engineering.

The 33 papers presented in this book were selected amongst the 97 papers presented during the Conference during 28 sessions. Two keynote papers, one presented by Professor Stephen Lu, from the IMPACT Research Laboratory, University of Southern California, USA, on "Supporting participative joint decisions in integrated design and manufacturing teams", and one written by Professor Stefan Rudolph from Stuttgart University about "Know-How Reuse in the conceptual design phase of complex engineering products or: 'Are you still constructing manually or do you generate already automatically'", introduce the subject of the Conference and are followed by the different themes highlighted during the conference:

The design/manufacturing interface

The unique and important aspect of this conference is to look at the synergies and relationships between the design and manufacture processes. It will also deal with the interrelationships of other technologies in design and manufacture.

Integrated design of manufacturing processes
The manufacturing process or system has to be designed carefully to afford benefit. Within this topic aspects of risk management, IT strategies, design for flexibility and responsiveness are covered. The topic of micro-manufacture will also be covered.

Life cycle design and manufacturing approaches
Life cycle manufacturing is a topic of increasing importance and includes critical aspects dealing with such aspects as design optimisation, cradle to gave management of both products and systems, assembly versus disassembly, flexibility for product up-grade, flexibility for component substitution, variety management, etc.

Agility in design and manufacture
Agility is the key requirements for the 21st century as customers, however defined, require what they want when they want it. In particular the issues of design technologies, computer-aided engineering, product strategy, mass customisation, agility and responsiveness will be considered, particularly in relation to dynamic world markets and globalisation.

Knowledge in engineering
We are actually moving from a resource based society toward knowledge based society. Within this topic, aspects of knowledge management, knowledge modelling, knowledge capitalization, ontologies are covered.

Management in production system
With the notion of extended enterprise, the supply chain management is becoming crucial for industry. Within this topics, logistic, planning, scheduling, control of production system are concerned.

The book is divided into sections reflecting the above themes and will be of interest to academics, students and practitioners specialising in design and manufacturing issues in mechanical engineering. We hope you will find it of the greatest interest to compare your various points of view within the fields broached throughout the Conference.

We should like to highlight the very significant input of the members of the organizing committee for the success of the Conference: Mrs Joëlle Robcis, from the AIP-PRIMECA Grenoble's centre, Professors D. Brissaud, J.F. Boujut, M. Jacomino, Y. Frein, J.C. Léon, F. Noël, H. Paris, H. Tiger and F. Villeneuve, all members of the G-SCOP Laboratory: Laboratory for Sciences of Design, Optimisation and Production of Grenoble, France.

Acknowledgements

The sixth IDMME Conference held in Grenoble in May 2006 received the scientific support of the M.E.S.R. (French Ministry of Higher Education and Research), the C.N.R.S. agency (French National Agency for Scientific Research), the C.I.R.P. (the International Academy for Production Engineering), the A.F.M. (French Association of Mechanics), the VRL-KCiP Network of Excellence (Virtual Research Lab for a Knowledge Community in Production), the Cluster GOSPI (Management and Organization of the Production Systems) and the Group for Research MACS (Modelling, Analysis and Conduct of Dynamical Systems).

The "Région Rhône-Alpes" provided financial support for the publication of this book.

The Editors
Serge Tichkiewitch
Michel Tollenaere
Pascal Ray

Foreword
The French AIP-PRIMECA Network

The 6th edition of the International Conference on Integrated Design and Manufacturing in Mechanical Engineering (IDMME), held from 17th to 19th May 2006 at the "Institut National Polytechnique de Grenoble (France)", was organized under the scientific responsibility of the French AIP-PRIMECA network (Ateliers Inter-Etablissements de Productique/Inter-Institute Productics Workshops and Pôle de Ressources Informatiques pour la Mécanique/Computer Resource Pole for Mechanics). This network is organized along the lines of a joint project: the evolution, in the field of training for Integrated Design in Mechanics and Productics, closely linked to the constantly changing industrial needs of these last 20 years. It has managed to bring together the different participants in the fields of Mechanical Engineering, Industrial Engineering and Productics. The French AIP-PRIMECA network is in charge of promoting both exchanges of experience and know-how capitalization. It has a paramount mission to fulfill in the field of initial and continuous education, technological transfer, international relationship and knowledge dissemination through strong links with research laboratories. The network dynamics resulting from the merger of previous structures has made it possible to generate strong synergies within the regional poles/institutes thus facilitating the medium/long-term consolidation of a development policy for the resource centers many of which are increasingly turned into high-tech and high specificity competence centers. The amalgamation process has allowed the development of an open institute in the field of integrated design as applied to Mechanics and Productics. In addition, it facilitates the already existing programs as regards pedagogical experiments, thus allowing the promotion of research programs being carried out by the various network partners in their own research laboratories. From September 2007, the network will animate an international diploma of master with Tsinghua University on the Product Lifecycle Management.

The French AIP-Primeca network regularly organizes special events such as scientific single day conferences, national colloquia, fall conferences and international congresses. IDMME, as part of this program, has enabled an assessment of the state of the art as regards tool evolution, methodologies, organizations in the field of integrated design and manufacture. The industrial problematic is currently slanted towards the improvement of the design processes now perceived as an inverse approach with the different products to manufacture as the starting point. Needless to say, both methodological improvement, as a blanket term, and model optimization are still topical issues quite in line with the organization of an optimal design. This organization radical change on how to apprehend design is an on-going process, which must not neglect the improvement of already existing tools thus allowing a product validation in the final phase, aiming at being more performing.

A selection of the best papers from this conference has therefore been made which has led to the publication of this book whose ambition is to be considered as a work of reference both for researchers in this particular field and for teaching staff confronted with training methodology in integrated design and productics. It should allow assessment of the scope of development prospects in an extremely wide-ranging field.

Pascal Ray
Director of French AIP-PRIMECA Network

PLENARY TALKS

Supporting Participative Joint Decisions in Integrated Design and Manufacturing Teams

Stephen C-Y. Lu and Amanda Conger

The IMPACT Research Laboratory, University of Southern California, Los Angeles, CA, U.S.A.; E-mail: sclu@usc.edu, amanda.conger@gmail.com

Abstract. Effectively implementing integrated design and manufacturing (IDM) has become a great challenge today as industries try to meet the ever-increasing market demands for low cost, high quality, and short lead-time. Traditionally, the integration between design and manufacturing decisions has been approached mainly as a technical problem of taskwork. However, because most IDM tasks are carried out by teams of engineers, the success of IDM also depends on the effectiveness of teamwork. In order to realize the full potential of IDM, both "taskwork" of individual engineers and "teamwork" of IDM teams must be addressed. This calls for a new socio-technical approach to understand how engineers can, and should, participate in joint decisions collaboratively in a highly dynamic team setting.

There are many different decision-making styles, ranging from a total dictatorship to completely democratic, where a group of stakeholders work together. In the past, studies in this field have been limited to the autocratic individual decisions due to the Arrow's Possibility Theorem, which states that group decision-making is "inherently chaotic and can't be consistent and rational." Consequently, group decisions are often converted to multi-objective, multi-criteria optimization problems whose solutions are available from classical decision science. Such approaches take the collaborative and participative joint decision opportunities away from IDM teams, hence failing to reveal the true benefits and address the real challenges of integrated design and manufacturing. In this paper, we challenge this old myth of group decision-making and propose a new approach to support participative joint decisions in IDM teams. Rather than the traditional deterministic philosophy, our new socio-technical approach is based on a determinism paradigm. We carefully analyze the characteristics of IDM teams and the dynamic social interactions among team members. We present a set of alternative social science models and decision science techniques to support participative joint decisions as a collaborative negotiation process among IDM engineers. This paper opens a new possibility of collaborative engineering by using systemic approaches to support true participative teamwork in IDM teams, henceforth contributing to the ultimate goal of developing joint innovations and collective intelligence in integrated design and manufacturing.

Key words: collaborative engineering, engineering negotiation, engineering team, socio-technical foundation, integrated design and manufacturing.

S. Tichkiewitch et al. (eds.), Advances in Integrated Design and Manufacturing in Mechanical Engineering II, 3–22.

1 Introduction

The never-ending demands for better function, lower cost, higher quality, tighter integration, and shorter lead-time have made the effective integration of design and manufacturing (IDM) an ever more attractive goal in industry today. However, the current approach of integration focuses on only the "technical" issues raised by IDM. This limited viewpoint neglects to consider the important concerns of human social interactions in IDM teams, which are critical to the success of IDM. In order to fully address the challenges of integration between design and manufacturing, as well as other product lifecycle concerns in general, it is essential to develop a better understanding of the socio-technical dynamics among stakeholders in engineering teams. Our research in collaborative engineering has been motivated by this critical need.

Current approaches in IDM and lifecycle engineering often describe engineering teams mainly by their technical functions. This vantage neglects to describe the social interactions of a team and the way in which these human interactions can and do influence engineering decisions. In order to achieve a truly integrated approach, both the social (i.e., teamwork) and technical (i.e., taskwork) components of the group decision-making process must be better understood. Unfortunately, the traditional methods of dealing with group decisions in teams often rely on a "Supra team leader" to make autocratic individual decisions. The support for the current belief in the "Supra team leader" stems from the famous Arrow's Possibility Theorem [1], which suggests that "no social choice procedure could be derived from even simple rules of democratic decision-making in groups." The unfortunate misunderstanding of this Theorem has led to the following Old Myth of Group Decision Making:

> Group decision making is inherently chaotic, because it can't be consistent and rational. Hence, the autocratic decision style by a Supra decision maker is the only consistent way for collaborative engineering to occur.

By expecting one Supra decision-maker to incorporate all team members' views and make a "group" decision, the current approach based on this old myth does not allow a fully collaborative and participative teaming style to support joint decision-making by all members, which is the most exciting aspect of teamwork in IDM and a beneficial feature of collaborative engineering. Furthermore, this "Supra team leader" approach, based on the conventional philosophy of scientific determinism, has led some researchers to insist that there can only be one final correct solution to the technical problems faced by IDM teams [2]. This limiting stance requires that the same unique answer be produced by every IDM team, without regard to the effect of social interactions among the team members. Clearly this conclusion is not reflected in real-life IDM decisions in engineering practice. In our research, we turn to the constructionism paradigm to offer a different intellectual foundation that will allow truly participative and joint decisions to be made by members of IDM teams.

One should note that teams can use many different ways to make group decisions. A key contrast is between a "dictator" style and a "democratic" style of group decision-making. In the former, one person holds the authority to make the final decision for the entire team. Individual team members may give their opinions and discuss the options, but ultimately only one individual makes a decision for everyone. The "Supra team leader," mentioned above, is just such an example of this style. Conversely, democratic group decision-making allows each team member to play a more participative role and have a stake in the final decision. An example of this kind of group decision-making style is called "Participative Joint Decision" (PJD) making. PJD is a managed democratic style of group decision making, in which a consistent and rational group preference is collaboratively established "for all" and "by all" who participate in the team. This PJD process is enabled by the ability to consistently aggregate the preferences of many into a single group preference.

In this paper, we first describe the limitations of current approaches in supporting IDM teams, and introduce the core concepts behind participative joint decisions. Then, we will present the new foundation to support collaborative IDM teams and contrast the previous approaches with our proposed alternative theory for each core concept. Finally, we will describe the new procedure based on our theoretical model.

2 Current Approaches and Their Limitations

Studies of collaborative teamwork in integrated design and manufacturing (IDM) teams require interdisciplinary knowledge and approach, and spans broadly across cognitive, social, organizational, decision, and engineering sciences [3]. With such a broad spectrum of disciplines, this section solely focuses on reviewing those theories and approaches which have given rise to the Old Myth that "a truly democratic group decision-making is never consistent and rational." We should note first that a *myth* does not mean "something that isn't true." In fact, the irrational group decision myth is absolutely true given its specific context [4]. Therefore, we shall carefully examine the specific theoretical contexts and core concepts behind the current approach in order to demystify this Old Myth.

It is important to point out that this Old Myth rises from an underlying philosophy of "scientific determinism," the more popular form of which entails "a unique, mechanical course for the universe, with all future events being caused by the past events" [5]. Based on this scientific determinism philosophy, some researchers have long insisted that the final outcome of a collaborative IDM teaming activity must be a single, predictable, optimized, and stable solution to the IDM problem. The same unique solution is expected from all engineers and every IDM team regardless of the human dynamics and social interactions that take place among them. This viewpoint underscores the current approach to modeling the core components of collaborative engineering teamwork: team behavior, social interaction, group preference and joint decision. Some brief summaries follow:

Team Behavior – Theories that support team behavior attempt to describe how a group of people will, or should, work together toward a common goal. Currently, the neo-classic Economic Theory for Large Open Groups [10] is commonly used to model the behaviors of team members. This theory models an engineer working in an engineering team as existing within an "idealized" environment in which all resources are unlimited, all alternatives are given, all consequences are known, and a complete utility ordering may be constructed.

Social Interaction – Theories that speak to social interaction account for the dynamic effects of social contacts on the individual and group's decision-making process. The currently favored ones, based also on the neo-classic Economic Theory, always assume that individuals work in a large group environment (such as an open market) and have full rationality with a static perspective (i.e., independent of social interaction). In essence, it ignores the important social component of group collaboration.

Group Preference – Theories that address Group Preference endeavor to find a method by which individual preferences can be consistently aggregated into a rational group preference. The most popular ones are those Social Choice Theories [12], which have decision makers rank their preferences toward a set of alternatives within a discrete option space. Following the Arrow's Possibility Theorem, researchers have concluded that a consistent group preference cannot be obtained by any democratic procedure. This leads to the false belief that group decision-making is irrational, inconsistent, and inherently chaotic [6].

Joint Decision – Ideally, theories that attend to Joint Decision seek to allow a decision to be made "by all" and "for all" in teamwork. However, as will be explained more fully, due to the misunderstanding generated by misapplication of Arrow's Possibility Theorem, such fully participative joint decision-making is typically declared impossible. Instead, one group member is assumed to be the "Supra decision-maker" who hands down individual decisions based on his/her interpretations of the different preferences of the group members.

From the brief reviews above, it is clear that the current understandings of, and theories for, supporting collaborative teamwork in IDM teams leave a lot to be desired. The modeling of team behavior as occurring within an idealized environment fails to account for the real-world resource constraints that engineers must face. The assumption of full rationality and static perspective discounts the dynamic effects of social interaction, which cause engineers' preferences to shift quickly in different team settings. Describing the establishment of group preference as "irrational, inconsistent, and inherently chaotic" [6] denies participating engineers the power of rationally and purposefully setting a group preference that fulfills the integrated design and manufacturing requirements. Finally, the reliance on a "Supra decision-maker" denies the engineers contributions to, and true ownership of, their final team results.

In order to overcome the inadequacies of current approaches, we have developed a new foundation that can seamlessly integrate the social (i.e., teamwork) and technical (i.e., taskwork) aspects of engineering teams, as well as an alternative set of social science theories and decision science models that can circumvent the old myth to enable true participative joint decision-making by all engineers in a collaborative IDM team. The rest of this paper presents this new foundation and its supporting approaches.

3 A New Foundation to Support Collaborative IDM Teams

As previously described, current approaches have led to the Old Myth of group decision which hinders the progress of collaborative engineering research and hampers the goal of integrated design and manufacturing. The participative joint decision-making opportunity, which is the most exciting and beneficial feature of integrated design and manufacturing, is taken away by the "virtual detector." The problem is, in effect, conveniently and hastily reduced to a multi-attribute, individual decision-making task within the scope of traditional decision analysis. In other words, this old myth implies that collaborative engineering research is either theoretically impossible, because a true participative joint decision-making by all engineers is unattainable; or practically solved, because classical decision analysis already offers the needed solutions. We believe that such a limiting myth and rigid position must be challenged and changed before any significant progress can be made in collaborative engineering and applied to integrated design and manufacturing teams.

The new foundation developed from our research is a "socio-technical" one which explicitly accounts for the dynamic group interactions taking place in engineering teamwork. In our research, the word "social" is used to refer to the behaviors which take the interests of others into account, as well as the cooperative characteristics between engineers [7]. The term "socio-technical" signifies the mutual consideration of, and the true integration between, the social teamwork and technical taskwork of engineering teams. These definitions acknowledge collaborative engineering as an interface between individual decisions and group interactions, and as an assimilation of social and technical activities operating in parallel over different time, space, and discipline scales in an IDM team. Therefore, a key challenge of collaborative engineering research to support IDM is to understand how designers can effectively participate in group decision-making for technical issues under the dynamic influences of social interactions.

Our socio-technical foundation offers a different view toward engineering teamwork and consists of approaches that are based on an alternative set of social science theories and decision science models. Figure 1 offers a comparative summary of the current and new approaches that will be discussed further in this paper.

For each of the core components of collaborative engineering teamwork – team behavior, social interaction, group preference, and joint decision, Figure 1 identifies the current theoretical approach and contrasts that with the new approach. The differences include the switch from monolithic groups to smaller teams, the focus away

The Core Concept	Current Approach	Proposed Approach	New Paradigm and Procedure
Team Behavior	Neo-Classical Economic Man in Open Large Groups	Modern Organizational Man in Small Teams with Incentives	(WHO) Interaction
Social Interaction	Self-Interested Rationality with Static Perspective	Social Construction Theory with Dynamic Perspective	(WHAT) Understanding
Group Preference	Ordinal Ranking with Discrete Social Choice Models	Cardinal Rating with Spatial Continuous Social Choice Model	(WHY) Preference
Joint Decision	Classic Decision Analysis, and Game Theoretic Approach	Collaborative Win-win Negotiation Framework and Analysis	(HOW) Decision

Fig. 1. The core concepts of current versus new teaming approaches.

from individual to group collaboration, the aggregation of individual preferences to the rational group, as well as replacing the Supra decision-making style with a "by all, for all" method. Through these new approaches, we seek to displace the old myth, and facilitate a truly collaborative teaming experience that will account for the socio-technical nature of collaborative engineering process and allow for consistent and rational joint decisions to integrate design and manufacturing considerations. Both the social-technical foundation and our new approaches are based on the following important assertions of collaborative engineering teamwork:

- Constructionism is a more adequate approach to understand the socio-technical nature, and supports the taskwork-teamwork integration of collaborative engineering teamwork in IDM.
- Social interactions in a collaborative IDM team can be made systematic to achieve a common understanding, based on which all designers' preferences can be rationally expressed and systematically incorporated.
- Group preferences can be consistently achieved if engineers' individual preferences are represented as spatial social-choice models with ordinal ratings of a continuous set of alternatives.
- Participative joint decision-making can be rational if designers are methodically guided through the collaborative negotiation process based on a consistent group preference.

Our new socio-technical foundation (STF) is diagrammatically illustrated in Figure 2.

Figure 2 is based on a typical engineering scenario, where an integrated design and manufacturing team is charged with the task of developing a new project. Mem-

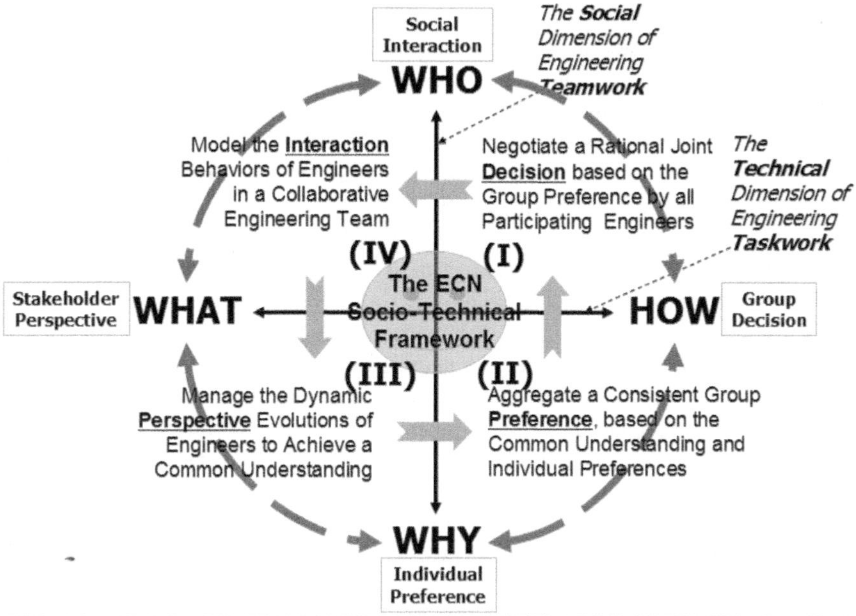

Fig. 2. A new socio-technical foundation for collaborative engineering.

bers of the IDM team must collaboratively address important questions such as "who" (i.e., who are the stakeholders of this product?), "what" (i.e., what should be the functional requirements of the product?), "why" (i.e., why choose to focus on these requirements?), and "how" (how to decide on technical means to satisfy these requirements?). Based on their different expertise, roles, responsibilities and preferences, each engineer has different perspectives to answer these questions. It is also interesting to note that these answers often are not determined by technical facts alone; many social (or non-technical) factors may enter into the considerations. Furthermore, these non-technical factors are always influenced by the social interactions that take place in IDM teams. In other words, social interaction can change individual perspective that, in turn, determines the final results of collaborative teamwork. Effective teamwork depends on members being able to achieve a "shared perspective" during the teaming process. The term "shared perspective" is used here to represent the "common understanding" and "group preference" that are the foundation for making rational and consistent product decisions.

As shown in Figure 2, technical decisions are represented as the "*what* $\leftrightarrow$ *how*" mappings along the x-axis, and social interactions are modeled as the "*who* $\leftrightarrow$ *why*" mappings along the y-axis. The STF seamlessly blends social-technical activities to support a complete "*who* $\leftrightarrow$ *what* $\leftrightarrow$ *why* $\leftrightarrow$ *how*" mapping with feedback to all steps as a continuous and cyclic process during a collaborative IDM engineering endeavor.

This process is initiated by a team of participating engineers [8] and starts from the upper-left (IV) quadrant with the "*who* ↔ *what*" mapping of social *interactions* (i.e., knowledge networks) among engineers whose "perspectives" toward the IDM issue at hand dynamically evolve. Next, the team proceeds to the lower-left (III) quadrant (i.e., *what* ↔ *why* mapping) to socially construct a shared *understanding* of the issue. Anchored in this common team understanding, the "*why* ↔ *how*" mapping (i.e., the lower-right, II, quadrant) establishes a consistent group *preference* for all stakeholders. Finally, supported by the "*who* ↔ *how*" mapping, the process arrives at the upper-right (I) quadrant where a joint *decision* is collaboratively negotiated by all IDM participants. This process continues through the socio-technical cycles until the conclusion of the collaborative IDM engineering endeavor.

It is important to note that, due to the dynamic social interactions that take place at the upstream (i.e., first 3 quadrants), the final outcome from a collaborative IDM task (i.e., "how" in the last quadrant) is intrinsically different from that of a traditional approach. Contrary to the determinism philosophy of the current approaches, the result from collaborative engineering teams, based on our constructionism thinking, is always sufficiently "under-determined" to account for multiple interpretations by different engineers and engineering teams. In other words, whatever decision members of a collaborative IDM team finally agree upon, it could have been different if made by a different IDM team [9].

Our socio-technical foundation is based on the essential primes that "team interaction shapes common understanding which establishes group preference, which in turn influences individual decision." From the teaming point of view, Figure 2 also illustrates an iterative socio-technical mapping process from "who" (the teaming participants) to "what" (the teaming process) to "why" (the evolution of preferences) to "how" (the process of reaching a team agreement). Accordingly, the rest of the paper will examine four core concepts that underlie the IDM team: (1) team behavior, i.e., "Understand Teaming Behavior in the IDM Team (see Section 4)," (2) social interaction, i.e., "Model Social Interaction in the IDM Team (see Section 5)," (3) group preference, i.e., "Construct Group Preference in the IDM Team (see Section 6)," and (4) joint decision, i.e., "Achieve Joint Decision in the IDM Team (see Section 7)." Each section will first present the current approaches being used to instantiate this core concept and then explain our proposed new approach which will enable participative joint decision making in the IDM team. Finally, Section 8 will suggest a new IDM teaming procedure for each core concept.

4 Understand Teaming Behavior in the IDM Team

The prevailing studies of individual behaviors in groups come from economics, the most relevant theories being collective action [10] and neo-classical economics [11]. The former studies the provision of public goods through the collaboration of two or more individuals, and has established different theories for large groups and small groups. The latter has specifically investigated how individuals make rational decisions in large groups (e.g., an open market). Research in collective action [12] has disproved an earlier belief that "groups with common interests will act on these common interests, just like individuals would act on their personal interests." According to the Theory of Large Groups, rational self-interested individuals in large groups will NOT act on their common interests, even when there are unanimous agreements in a group about the common good and the methods of achieving it. Notwithstanding this group paradox, the neo-classical Economic Theory prescriptively models the decision behaviors of an individual in large open groups as an "Economic Man (also called a 'Rational Man' in classical decision science)." The Economic Man has a complete and consistent system of preferences to always choose correctly among the entire set of available alternatives [13]. This idealized behavioral model assumes that, when an individual makes a decision, all the alternatives of choice are given, all of the consequences of each alternative are known in advance, and a complete utility ordering for all consequences is available.

It has been well recognized that engineers engage in integrated design and manufacturing tasks as members of a "purposeful team," rather than individuals in an open social group [14]. Different from an open group, a "purposeful team" consists of a distinguishable set of members who have each been assigned specific roles or functions to perform, interact dynamically and adaptively toward common and valued goals, and have a limited life-space of team memberships [15]. These unique features, plus the fact that an IDM team always operates in a corporate environment that has established corporate procedures and incentive systems, make the teaming behaviors of its members quite different from that of the "Economic Man" in an open market. In fact, the Theory of Small Groups developed from collective action research provides a more adequate model of the teaming behavior of engineers [16]. The theory suggests that separate and selective incentives can stimulate rational individuals in a small latent group to act in a group-oriented way. For example, small groups with members of different interests in the Public Goods (e.g., different engineers assigned to take on different life-cycle responsibilities in an IDM team) will act on their common interests [17]. These conditions and behaviors are all true for collaborative IDM teams within a corporate environment in industry. Therefore, we replace the neo-classical Economic Theory for large open groups with the Organizational Behavior Theory [18] in order to model engineers' teaming behaviors in a collaborative IDM team as "coercion" in small groups.

The Organizational Behavior Theory [18] models the teaming behavior of engineers as the "Organizational Man" functioning in a "small-induced" team. This theory argues that there are practical limits to human rationality in organizations (such as IDM teams), hence, the "bounded rationality" concept [19]. These limits are not static but rather depend on the dynamic social and organizational environments in which individuals' decisions take place. Whereas the neo-classical Economic Man maximizes by selecting the best/optimal alternative from among all, the Organizational Man satisfies by looking for a course of action that is satisfactory or good enough [20]. An alternative is optimal if there exists a set of criteria that permits "all" alternatives to be compared, and the alternative in question is preferred by these criteria to all other alternatives. On the other hand, an alternative is satisfactory if there exists a set of criteria that describes minimally satisfactory alternatives and the alternative in question meets or exceeds all these criteria [21]. These "satisficing" and "bounded rationality" concepts underline our model of the teaming behavior of IDM teams in that, (1) a decision is always exercised with respect to limited resources with an approximate model of the actual situation, and (2) the elements of the situation are not given a priori nor static but are themselves the outcome of sociological processes, including the engineers' own decisions and the activities of other members in a collaborative IDM team [22].

5 Model Social Interactions in the IDM Team

The neoclassic Economic Theory presumes that group members are self-interested individuals who have absolute rationality and are always provided with unlimited resources (e.g., time, information, budget, etc.) to seek optimized decisions [23]. Based on this theory, the traditional approach also supposes that the perspective of individuals (i.e., how they view the world and the problems at hand) is static and unaffected by the dynamic social interactions that take place. However, the traditional approach ignores the fact that social interactions during collaborative teamwork can influence engineers' perspectives, which will alter their preferences in making decisions. It reduces a dynamic, participative joint decision-making task to a static, multi-attribute optimization problem and overlooks the significant benefits of collaboratively constructing a "shared mind" via participative joint decisions by multiple engineers [24].

Cognitive studies [25] have long confirmed that individuals' perspective is always dynamically evolving, especially when they interact with others. In engineering teamwork, engineers often participate in a mutual, reciprocal, social construction process (also called co-construction), whereby each helps to construct the other's perspective (i.e., understanding, interpretation) of the engineering tasks at hand [26]. In essence, an engineer's perspective is a function of her social interactions, and is the basis of her preferences that underline her decision-making in teams [27]. This fact (i.e., decision preferences should be expressed with respect to a common understanding resulting from social interactions) is the centerpiece of constructionism. We must encourage social interactions in collaborative IDM practice and understand the dynamic perspective evolutions in collaborative teamwork in order to benefit from the "two heads are better than one" truism. As will be explained, we forgo the traditional view of self-interested individuals with full rationalities and static perspectives in favor of a means to model the dynamic perspective evolutions that occur during social interactions in a collaborative IDM team.

A collaborative IDM team must simultaneously engage in taskwork and teamwork. Based on the "satisficing" and "bounded rationality" principles, we use the Social Construction Theory [28] to model engineers' teamwork (i.e., social interaction) process. A social construction, or social construct, is an idea that may appear to be obvious to those who accept it, but in reality is just a creation or artifact of a particular culture or society [29]. The implication is that social constructs are human choices rather than laws of the nature. In recent studies [30, 31], this concept referred to "the process of socially shaping an agreement and artifact by interested stakeholders." A typical social construction model has four core components, which are made operational in our new IDM teamwork procedure:

1. *Interpretive Flexibility*: This relativism idea suggests that an IDM task is an "open" process that can produce different outcomes depending on the particular social circumstances of its development [32]. For example, scholars in social construction of technology (SCOT) have applied the interpretive flexibility concept to explain how technological artifacts are, in fact, the product of human negotiations and agreements based on different evolving perspectives [28, 33] hence, they are always "under-determined" to allow for multiple interpretations by members of the IDM team.
2. *Relevant Social Group*: A relevant social group (RSG) is the embodiment of a particular set of interpretations in that its members all share the same interpretations (or perspectives) attached to an IDM issue. An IDM task based on social construction becomes a process in which RSGs, each embodying a specific perspective of a IDM issue, negotiate its interpretation with engineers seeing and constructing different objectives based on different perspectives. This process continues until the negotiation activity brings all engineers to a consensus that their common interpretations work [28]. That is, a collaborative IDM process ceases not because the artifact works in some objective and absolute sense (i.e., according to laws of the nature), but because all involved engineers agree and accept that it works for them.
3. *Closure and Stabilization*: The above social construction process can result in controversies when different interpretations lead to conflicting decisions on an IDM issue. The process must continue until such conflicts are resolved, and the interpretation (or decision) no longer poses a problem for any involved engineer [34]. In short, the collaborative IDM process must achieve a closure (i.e., a final agreement or at least a cessation of further decision) via conflict resolution in negotiation until no further arguments occur and the interpretation or solution stabilizes in its final form (i.e., a joint decision by the team). Such a stable state, which is particular to the said IDM team at the time, becomes the starting point for the next round of social construction as the team, members or time evolves, leading to the cyclic nature of socio-technical collaborative IDM tasks [35].
4. *Wider Context*: In traditional social construction research, it was often assumed that groups are all equal and present in social construction process. Later studies

have recognized that there is always wider socio-cultural-political milieu [31] in which the social construction process takes place in real organizations. These "background conditions" – such as the relations between engineers, the operating procedures ordering the interactions, and differences in the organizational power – are very important to the process and result of collaborative IDM tasks. These wider contexts are all accounted for in our new IDM teaming procedure.

6 Construct Group Preference in the IDM Team

From the decision science viewpoint, the hallmark of rationality is the ability to create a preference by rank-ordering alternatives. Hence, the challenge of rational decision-making is to create a group preference for all when multiple individuals work in groups. This issue has been the subject of study by the Social Choice Theory [36], which investigates procedures to "blend" the preferences of the many into a consistent social ordering of alternatives. Individual preferences are often treated as "given constants" in these studies, which have mainly focused on preference aggregation rather than the more difficult questions of how individual preferences are formed and how they can change due to social interactions over time. The most well-known Social Choice work is the Arrow's Possibility Theorem [1], which suggests that "no social choice procedure could be derived from even simple rules of democratic decision making in groups (e.g., the Voter's Paradox)." More precisely, it proves [37] that no social preference function can satisfy some minimal criteria (i.e., universal domain, Pareto principle, independence from irrelevant alternatives, and non-dictatorship) of democratic decision-making in groups. Some engineering researchers have since used this Arrow theorem to insist on a position that "group decision making by all is not rational, never consistent, and inherently chaotic (hence, Arrow's Impossibility Theorem) [6]." This argument has become the most puzzling roadblock along the pathway of collaborative engineering research.

One must note that Arrow's Theorem is based on the social choice model that ranks a set of discrete preferences [38], which is different from the spatial social choice model that employs a rating of continuous sets of alternatives [39]. Over the past few decades, many studies on the Arrow's dilemma have discovered that instead of additive aggregations of individual preferences expressed as rankings of discrete alternatives, groups can explicitly compare their interpersonal preferences using ratings of continuous alternatives to construct a stable group utility function in a rational manner [40]. The same has been proved analytically by decision scientists [41] who used preference ratings, rather than rankings, to systematically derive a group preference which completely satisfies Arrow's four criteria of democratic decision-making. Therefore, our new approach is based on a spatial social choice model that employs a rating of continuous sets of alternatives. We believe that, especially in the engineering domain where many physics-based numerical simulation models exist, most IDM problems can be formulated using uni- or multidimensional spatial preference models with a continuous set of alternatives. This gives us a great opportunity to overcome Arrow's paradox, and hence open new opportunities to support participative joint decisions in IDM teams.

The ability to consistently aggregate the preferences of many into a single group preference is what makes participative joint decisions possible [42]. As explained above, this ability hinges on how the available alternatives and individual preferences are represented in the social choice model. Rather than using ordinal rankings of discrete alternatives, our approach uses the spatial social-choice models to represent individual engineer's preferences as ordinal ratings of a continuous set of alternatives. With such a spatial model, alternatives are drawn from a mathematically structured space, having the properties of real numbers that can be manipulated with real analysis in mathematics [40]. For example, in a uni-dimensional spatial model the space of alternatives is drawn from the real number line between negative infinity and positive infinity (i.e., $X \subset \mathbf{R}$), and from a high-dimensional space (i.e., $X \subset \mathbf{R}^m$) in a multi-dimensional spatial model. With these continuous alternatives, engineers can express their preferences as in the discrete case. However, to benefit from the mathematical power inherent in the real numbers, the preference relation is represented by an Ordinal Utility Function, Ui, which assigns a numerical value to each alternative (so that the alternative that is more satisfactory has a higher utility value). Note that a Ui is not a unique representation of a preference relation. In fact, any "order-preserving" transformation of Ui can represent a designer's preference.

As proven by social choice researchers [43], the use of Ui with alternatives expressed in a continuous space can circumvent the Arrow's paradox and lead to a consistent group preference that completely satisfies Arrow's four criteria of democratic decision-making. Despite the more complex mathematics required by the spatial social-choice models, they offer a unique opportunity to support participative joint decision-making in collaborative IDM tasks. Unlike many other application domains where continuous alternatives are either impractical or impossible, engineers can use numerical models to generate real-numbered alternatives in a continuous space. Preference relations in Ui can be obtained in our approach via collaborative negotiation, which systematically handles the interpersonal comparisons of engineers' preference strengths. Other than different alternative and preference representations, our approach is fully consistent with the principle of "rational choice theory" [44], which starts with a behavior model of individual rationality that drives the pursuit of personal goals. The differences are that we replace the "absolute rationality" with the "bounded rationality," and set engineers' goals in negotiation as "to satisfy" a set of criteria rather than "to maximize" a single objective in collaborative IDM teams.

7 Achieve Joint Decision in the IDM Team

A direct consequence of the argument based on Arrow's Impossibility Theorem is the conclusion that "rational joint decision making by participating members in a group setting is impossible." To circumvent this dilemma, the current approaches instead focus on how an individual can systematically incorporate the views and interests of others into her own decision-making framework. That is, the preferences of this virtual "Supra decision-maker" depend on the preferences of members in the group [45] – hence, the problem reduces to one of individual decision-making which can be handled by either classical decision analysis (when a single decision is to be made) or theory of non-cooperative games (when multiple interacting decisions are required). In either case, these current approaches do not allow a fully collaborative and participative style to support joint decisions. That is, it does not assume that the preferences of those "others" can also depend on the preferences of other members and the "Supra decision-maker" [46]. A group preference is somehow magically created "for all" by an autocratic leader, and an interactive group task is quickly reduced to a static individual problem. To fully support the collaborative IDM practice, we believe that a group preference must be established "for all" and "by all", and everyone in the IDM team must engage in the participative joint decision-making task collaboratively.

Once the Arrow's paradox is removed, a true joint decision by all engineers becomes possible using the consistent group preference derived from the spatial social-choice models [42]. This possibility frees IDM teams from relying on a Supra decision-maker, and leads to the new possibility of supporting collaborative IDM tasks as true group decision-making. Group decision has been an active area of decision science research, where the interactions and dynamics between multiple unitary decision entities are analyzed [47]. Broadly defined, there are two types of group decisions: interactive decision (by Theory of Games) and joint decision (by Negotiation Theory). Unfortunately, there exists a misunderstanding that treats collaborative engineering, in a mathematical sense, as a "cooperative game" [48].

According to the formal definition from game theory research [49], a cooperative game is one in which players are able to make enforceable contracts. It is not defined as games in which players actually do cooperate, but as games in which any cooperation is enforceable by an outside party. Therefore, strictly speaking, a collaborative IDM practice can be either a cooperative or a non-cooperative game depending on the enforceability of cooperation results by the IDM team rather than the cooperative or non-cooperative nature of the engineers. Furthermore, games alone do not adequately capture all types of collaborative IDM scenarios. For example, while multiple players interact to make separate decisions in games, members in an IDM team must interact to arrive at a single joint decision in collaborative IDM. In fact, we argue that this joint decision scenario is a more important and challenging type of collaborative IDM problem [50].

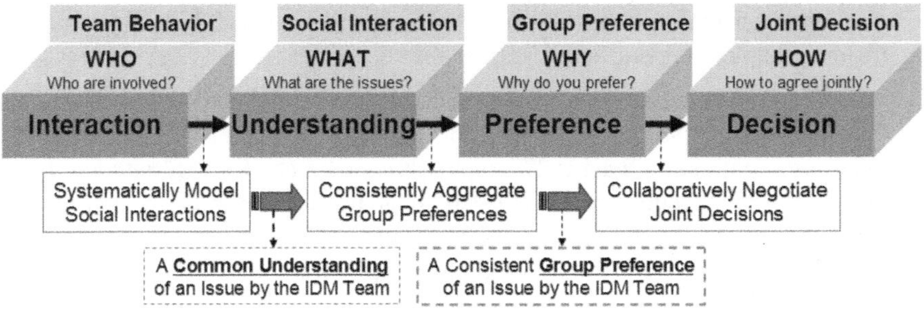

Fig. 3. An overview of the new collaborative IDM teamwork procedure.

Therefore, we employ collaborative negotiation methods to support joint decisions by all engineers in the IDM team. Specifically, we focus on collaborative negotiation analyses [47], which study how "reasonable" individuals (e.g., engineers with bounded rationality) could collaboratively make a joint decision to "satisfy" multiple requirements (i.e., finding good enough alternatives for stated criteria of various IDM concerns) by all members. Engineering Collaboration via Negotiation (ECN) is an emerging research area in collaborative engineering. Many interesting findings and practical methods are now available, which provide prescriptive guidelines for engineers to collaboratively resolve conflicts, bargain for individual/collective advantage, agree upon courses of action, or craft joint decisions that serve their mutual interests [51]. Three major difficulties have always hindered the research progress in this new research area: (1) failure to link the organizational policies and incentives imposed on the team with the negotiation behaviors of engineers in the team, (2) confusion caused by engineers' different understandings and varying interpretations of the same issues during negotiations, and (3) an inability to establish and hold onto a commonly-valued team goal as a base for the needed given-and-take of trade-offs in negotiations. All these difficulties are largely circumvented by our new approaches described in this paper, hence making collaborative negotiation a viable means to support joint decisions by all members in the IDM team.

8 A Collaborative Teamwork Procedure in the IDM Team

Our new approaches described in previous sections provide the theoretical foundation for the assertion that in a collaborative engineering team, interaction alters understanding in which determines preference that underlie decisions. This assertion leads to a new collaborative IDM teamwork procedure for participative joint decisions. Figure 3 shows the four stages of this new procedure with its focuses (WHO, WHAT, WHY, HOW), core concepts (team behavior, social interaction, group preference, joint decision), and the three functional capabilities (model social interaction, aggregate group preference, and negotiate joint decision).

1. *The Interaction Stage*: This first stage focuses on the "WHO" issue with team behavior being the core concept. The behavior of engineers is modeled by the Organizational Behavior Theory. At this preparatory stage, we identify all participating stakeholders, establish IDM goals and expectations, and declare the baseline procedure (i.e., workflows for the IDM taskwork) for team control, ready to carry out the "who to what" mapping operation.

2. *The Understanding Stage*: In the second stage we use the social construction theory to derive a common understanding of the IDM issue by all engineers. For flexible interpretations, we will model the engineers' dynamic perspectives (i.e., changing interpretations of the IDM issue) during social interactions. For relevant social groups (RSGs), engineers who share the same understandings are identified in the team to argue for their perspectives of the given issue. For the closure mechanisms to achieve stable agreements, we analyze engineers' changing perspectives as captured in a Perspective Model State Diagram (PMSD), which represents a collective state of the engineers' current understandings of the issue [52]. Quantitative analyses of "conceptual distance" between perspectives in a PMSD can reveal the "degree of shared understandings" among RSGs. The closure of this Understanding stage becomes a task of minimizing these "distances" to arrive at a PMSD whose interpretations are acceptable by all. Techniques, such as control of dynamic systems, can be used here to guide the convergence toward a common understanding. For the wider contexts, we specify the "roles" and "responsibilities" for each RSG with other contextual information from the baseline process to support PMSD analyses.

3. *The Preference Stage*: The common understanding, socially constructed by the team in the previous stage, now serves as an "anchor" for engineers in order to express their individual preferences toward the said IDM issue to derive a consistent group preference for rational joint decisions. At this stage physics-based numerical simulation models of the technical domain are used to express different engineers' preferences as ordinal ratings of a continuous set of alternatives to mitigate the Arrow's paradox. These spatial social-choice models will capture the relative strengths of individual preferences, obtained by interpersonal comparisons among engineers. This is the hallmark of participative joint decision-making based on a consistent group preference.

4. *The Decision Stage*: With a consistent group preference established "for all," this last stage will allow engineers to negotiate a rational joint design decision "by all." We have developed new collaborative negotiation methods to support the engineers' joint decision-making activity by which the collaborative IDM team can effectively reach a joint decision that can incorporate all life-cycle considerations brought forth by all stakeholders. This stage completes the new collabor-

ative IDM teamwork procedure, leading to a consistent and rational participative joint decision for all the engineers, by all the engineers.

9 Conclusion

The need for efficient taskwork and effective teamwork to accomplish the goal of integrated design and manufacturing demands a fully participative joint decision making process that is unsupported by the current theories and approaches, which are based on the philosophy of scientific determinism. In contrast to the current expectation of a single, optimized solution to each IDM task, we have proposed a new foundation based on social constructionism, which embraces the "under-determinism" thinking that results from the dynamic collaborations among participating engineers.

Our research asks two basic questions relating to the IDM teamwork: what is the true nature of collaborative IDM teams that sets them apart from traditional engineering tasks, and what is an appropriate means to support group decision making in a collaborative IDM team? We are not satisfied with the existing answers given by the current approach, and seek to find alternatives which can both better describe the intrinsic characteristics of collaborative design and manufacturing and effectively facilitate participative joint decision making by allowing all IDM team members to share in the decision making process.

In this paper, we have presented a new foundation for enabling collaborative IDM teamwork. We have looked at the four core components of collaborative engineering teamwork: team behavior, social interaction, group preference, and joint decision. For each component, we have described the previous theory and demonstrated why it is inadequate to truly describe the collaborate nature of engineering teamwork. We then proposed a new theoretical basis that overcomes the drawbacks of the previous methods and allows the IDM team members to enjoy true participative joint decision-making and full ownership of the team results. Finally, we described a new procedure that allows this new theoretical foundation to be put into use in the IDM teaming practice.

Through the use and further development of this new foundation, we believe that collaborative IDM teaming process will be both better understood and more effectively supported [53]. This will allow engineers to more fully engage in the IDM collaboration process and to produce results of a higher quality.

Acknowledgements

Supports of this research have been partially provided by the National Sciences Foundation and the U.S. Army, Corps of Engineers. We also extend our sincere thanks to members of the CIRP (International Academy for Production Engineering

Research) ECN (Engineering Collaboration via Negotiation) Working Group who have actively contributed to the creation and formation of many new ideas presented in this paper.

References

1. K.J. Arrow, *Social Choice and Individual Values*, 1st ed. (1951), 2nd ed. (1963), New York, Wiley Publishers, Inc.
2. G.A. Hazelrigg, On irrationality in engineering design, *ASME Journal of Mechanical Design* **119**, 1997, 194–196.
3. R. Sriram, *Distributed Integrated Collaborative Engineering Design*, Sarven Publishers, 2002.
4. G.A. Hazelrigg, A framework for decision based design, *ASME Journal of Mechanical Design* **120**(4), 1998, 653–658.
5. R.B. Laughlin, Reinventing physics: The search for the real frontier, *The Chronicle of Higher Education* **51**(23), 2005, p. B6. [Laughlin is a 1998 Nobel laureate in physics. This essay is adapted from A Different Universe: Reinventing Physics from the Bottom Down.]
6. J.S. Kelly, *Arrow Impossibility Theorems*, Academic Press, New York, 1978.
7. S.C-Y. Lu and J. Cai, STARS: A socio-technical framework for integrating design knowledge over the internet, *IEEE Internet Computing* **4**(5), 2000, 54–62 [Special Issue].
8. S.C-Y. Lu, Engineering as collaborative negotiation: A new Foundation for collaborative engineering research, The ECN Working Group of the International Institution of Production Engineering Research (CIRP), http://wisdom.usc.edu/ecn, 2001.
9. K.D. Knorr-Cetina, *The Manufacture of Knowledge: An Essay on the Constructivist and Contextual Nature of Science*, Pergamon Press, Oxford, 1981.
10. The economic theory of collective action is concerned with the provision of public goods (and other collective consumption) through the collaboration of two or more individuals, and the impact of externalities on group behavior. The foundational work in collective action was M. Olson, The Logic of Collective Action: Public Goods and the Theory of Groups, *The Harvard Economic Studies* **124**, Harvard University Press, 1965 and 1971.
11. Neoclassical economic theory refers to a general approach to economics based on supply and demand which depends on individuals (or any economic agent) operating rationally, each seeking to maximize their individual utility or profit by making choices based on available information. The foundational work of neoclassical economic theory dates back to Adam Smith, *The Wealth of Nations*, Modern Library, New York, 1937.
12. S. Todd, *Collective Action: Theory and Application*, University of Michigan Press, 1992
13. C.R. McConnell and S.L. Brue, *Economics*, 15th ed., McGraw-Hill Company, New York, 2001.
14. D.M. Wegner, Transactive memory: A contemporary analysis of group mind, in *Theories of Group Behavior*, Brian Mullen and George R. Goethals (eds.), Springer-Verlag, New York, 1987, pp. 185–206.
15. L.L. Thompson, *Making the Team – A Guide for Manger*, 2nd ed., Pearson Education Inc., A Prentice Hall company.

16. M. Olson, R. Zeckhauser, An economic theory of alliances, *Reviews of Economics and Statistics* **XVLIII**, 1966, 266–279.
17. M. Olson, R. Zeckhauser, Collective goods, comparative advantages, and alliance efficiency, in *Issues of Defense Economics*, R. McKean (ed.), New York Bureau of Economic Research, 1967, pp. 25–48.
18. H.A. Simon, *Administrative Behavior: A Study of Decision Making Process in Administrative Organizations*, 4th ed., The Free Press, New York, 1997.
19. H.A. Simon, *Models of Bounded Rationality: Empirically Grounded Economic Reasoning* (Vol. 3), The MIT Press, Cambridge, MA, 1997.
20. J.G. March and H.A. Simon, *Organization*, 2nd ed., Blackwell Publishers, 1993.
21. H.A. Simon, *The Science of the Artificial*, 3rd ed., The MIT Press, Cambridge, MA, 1996.
22. R.M. Cyert and J.G. March, *A Behavioral Theory of the Firm*, 2nd ed., Blackwell Publishers, 1992.
23. J. Von Neumann and O. Morgenstern, *Theory of Games and Economic Behavior*, Princeton University Press, 1944.
24. J. Mathieu, G. Goodwin, T. Heffner, E. Salas, and J. Cannon-Bowers, The influence of shared mental models on team process and performance, *Journal of Applied Psychology* **85**(2), 2000, 273–283.
25. S. Plous, *The Psychology of Judgment and Decision Making*, McGraw-Hill Series in Social Psychology, McGraw-Hill Inc., New York, 1993.
26. P.E. Tetlock, Accountability and complexity of thought, *Journal of Personality and Social Psychology* **45**, 1983, 74–83.
27. D. Liang, R. Moreland, and L. Argote, Group versus individual training and group performance: The mediating role of transactive memory, *Personality and Social Psychology Bulletin* **21**(4), 1995, 384–393.
28. T.J. Pinch and W.E. Bijker, The social construction of facts and artifacts: How the sociology of science and the sociology of technology might benefit each other, in *New Directions in Sociology and History of Technology*, 1984, pp. 17–55.
29. J.R. Searle, *The Construction of Social Reality*, The Free Press, New York, 1995.
30. A. Larsson, Socio-technical aspects of distributed collaborative engineering, Ph.D. Thesis, Department of Applied Physics and Mechanical Engineering, Lulea University of Technology, Sweden, 2002.
31. H.K. Klein and D.L. Kleinman, The social construction of technology: Structural considerations, *Science, Technology, and Human Values* **27**(1), 2002, 28–52.
32. J-F. Boujut and H. Tiger, A socio-technical research method for analyzing and instrumenting the design activity, *The Journal of Design Research* **2**, 2002.
33. W.E. Bijker, P. Hughes, and T.J. Pinch (eds.), *The Social Construction of Technological Systems*, The MIT Press, Cambridge, MA, 1987.
34. S.C-Y. Lu, J. Cai, W. Burkett, and F. Udwadia, A methodology for collaborative design process and conflict analysis, *The Annals of the CIRP* **49**(1), 2000, 69–73.
35. M. Weber, C. Wittich and G. Roth (eds.), *Economy and Society: An Outline of Interpretive Sociology*, University of California Press, 1989.
36. D. Black, *The Theory of Committees and Elections*, Cambridge University Press, Cambridge, 1958.
37. K.J. Arrow and H. Rayuaud, *Social Choice and Multicriterion Decision Making*, MIT Press, Cambridge, MA, 1986.

38. K.J. Arrow, Rational choice functions and orderings, *Economica, New Series* **XXVI**(102), May 1959.
39. A. Downs, *An Economic Theory of Democracy*, Harper and Row Publishers, New York, 1957.
40. P.E. Johnson, Social choice: Theory and research, in *Quantitative Applications in the Social Sciences*, Sage University Press, Thousand Oaks, 1998.
41. R.L. Keeney, A group preference axiomatization with cardinal utility, *Management Science* **23**, 1976, 140–145.
42. R.L. Keeney and H. Raiffa, *Decisions with Multiple Objectives Preference and Value Tradeoffs*, Cambridge University Press, Cambridge, 1993.
43. J.C. Harsanyi, Cardinal welfare, individualistic ethics, and interpersonal comparisons of utility, *Journal of Political Economy* **63**, 1955, 309–321.
44. D. Green and I. Shapiro, *Pathologies of Rational Choice Theory: A Critique of Applications in Social Sciences*, Yale University Press, 1996.
45. P.C. Fishburn, *The Theory of Social Choice*, Princeton University Press, Princeton, NJ, 1973.
46. A.K. Sen, *Collective Choice and Social Welfare*, Holden-Day Publishers, San Francisco, 1970.
47. H. Raiffa with J. Richardson and D. Metcalfe, *Negotiation Analysis – The Science and Art of Collaborative Decision Making*, Harvard University Press, Cambridge, MA, 2002.
48. This statement was quoted directly from the description of Collaborative Design as a Key Issues identified in the Program Description section of the NSF Engineering Design Program. http://www.nsf.gov/funding/pgm_summ.jsp?pims_id=13340&org=DMI&more=Y.
49. J. von Neumann and O. Morgenstern, *Theory of Games and Economic Behavior*, 2nd ed., Princeton University Press, Princeton, NJ, 1947.
50. S.C-Y. Lu, Beyond concurrent engineering: A new foundation for collaborative engineering, *International Journal of Advanced Manufacturing Systems* **9**(2), 27–40.
51. R.J. Lewicki, B. Barry, D.M. Saunders, and J.W. Minton, *Negotiation*, 4th ed., by McGraw-Hill Higher Education Press, 2003.
52. J. Cai, S.C-Y. Lu, F. Grobler, M. Case, and N. Jing, Modeling and managing collaborative processes over the internet, *Journal of Business Process Management* **11**(3), 2005, 255–274.
53. S.C-Y. Lu, A scientific foundation of collaborative engineering, *Annals of the CIRP* **52**(2), 2007.

Know-How Reuse in the Conceptual Design Phase of Complex Engineering Products

Or: "Are you still constructing manually or do you already generate automatically?"

Stephan Rudolph

Institute for Statics and Dynamics of Aerospace Structures, University of Stuttgart, Germany; E-mail: rudolph@isd.uni-stuttgart.de

Abstract. Shortening the design cycle, faster time-to-market, reduction of development costs, more intelligent products and intelligent production are the buzzwords which characterize the current trends in our industries subject to global competition. While in the past many downstream activities in engineering design and mass production have already been subject to automation and are nowadays substituted by robots or software, only little support is currently available for the engineer in the conceptual design phases. Most upstream design activities, i.e. from the initial product idea until the creation of the digital models which represent the blueprint of the final product, still mainly rely on human labor and change management.

Design languages possess the potential to fill this gap by offering the engineer a very flexible way to express the know-how about the design object and the design process. Design object means all aspects proper to the parts of a product, while design process means all aspects of thought processes, compromises and intermediate decision making which occurs during the process. Design languages offer automatic translation into all kinds of digital models, and, if properly designed, automatic compilation under a large variety of topological and parametrical variations of requirements or boundary conditions, thus leading to different product designs and know-how reuse in the conceptual design of complex engineering products.

Keywords: design languages, rule-based design, design patterns, design compiler.

1 Introduction

Engineering design as a research topic has in the last four decades attracted much scientific attention. Due to the heavy impact of the design process on the result as the designed product, researchers have been interested long since in understanding the inner structure of the design process and the art of designing [10, 12]. Germany, with its engineering tradition, has seen the origination of a *systematic engineering*

S. Tichkiewitch et al. (eds.), Advances in Integrated Design and Manufacturing in Mechanical Engineering II, 23–39.

design methodology [13], which structures the design process into the four stages of (1) requirements analysis, (2) functional analysis, (3) principle analysis and (4) form analysis. While these four stages seem to represent a result the design community has agreed on in engineering practice [20], a review of the scientific state-of-the-art of known design thinking and methodologies reveals that in theory the design process is still characterized by a lack of a *unified theory of design* which is generally accepted and based on solid grounds in the philosophy of science [8].

In the absence of such a theoretical foundation for the process of design, several dozens of design approaches (see [7] for an overview) have been developed, and, in absence of a more sound formal foundation, sometimes even been promoted to the status of *design paradigms*. Scientifically speaking however, it should be clear that even with thousands of examples as evidence, inductive reasoning does not offer the same theoretical power than deductive reasoning. As long as a strong theoretical foundation of such a unified theory of design is still missing, any so-called paradigm does only qualify as an educated guess. There exists even a certain danger that these paradigms, if not properly reflected, are perceived as truths, since they are taught and assimilated as a part of our engineering curriculae.

Taking into consideration that the true nature of engineering design is still hidden as long as this unified theory of design is unknown [8], it is worthwhile to take a step back and reconsider what seems to be firmly based on scientific grounds, what qualifies as an assumption and what seems to be an arbitrary choice among many. Going this way back to the roots means *re-thinking design*.

1.1 Re-thinking Design

Thinking about an object in science and engineering means modeling. Modeling is a human thought process of abstraction, where the conception of the human observer about reality is formalized and represented in some sort of a mathematical model. Modeling hereby intends to capture both the relevant distinct and antagonistic aspects of behaviors: *states* and *transitions*. The word states is used here for all static aspects of the design concept, while the word transitions is used for all changes the design concept undergoes from the idea to the final product. The main result of modeling consists therefore in the establishment of an adequate design *representation*.

Concerning the design states firstly, the current status quo of the representations used in the different disciplines of engineering design (systems engineering, design, structures, aerodynamics, production, finance, software, etc., to name a few) is shown in Figure 1. The activity of design includes the creation of the enumerated physical models from some abstract mental model the designer has in mind. The goal of this modeling activity is to create a description precise enough to both represent, simulate, analyze and evaluate the behavior of the future product to be, as well as to represent the properties of the future product to be not only for the purpose of designing, but also to produce the intended product including total product life-cycle considerations with manufacturing, use, recycling and so on.

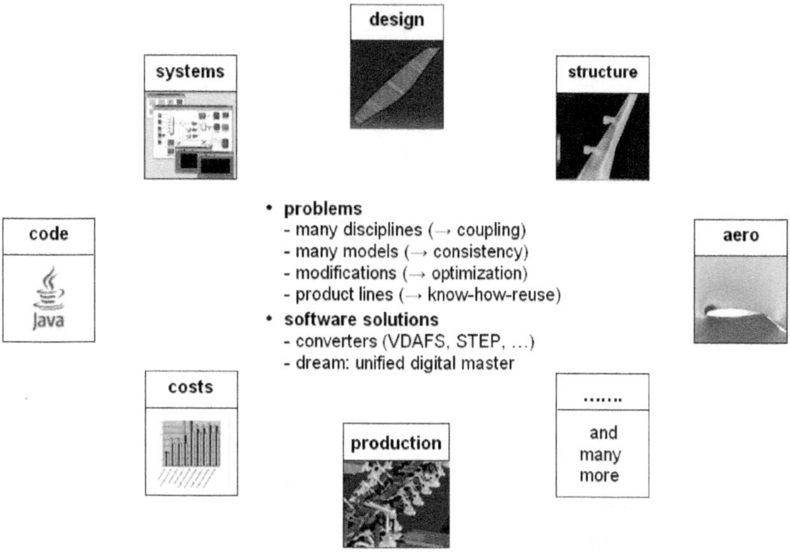

Fig. 1. Design representations in different disciplines.

In a typical industrial environment the landscape of software tools used in the modeling of the above mentioned aspects in Figure 1 is quite diverse. Modeling the design geometry in a modeler like *CATIA*,[1] *PRO-E*[1] or *NX*,[1] the structural mechanics in *NASTRTAN*,[1] *ABAQUS*,[1] or *ANSYS*,[1] the fluid mechanics in *FLUENT*[1] or *STAR-CD*[1] and the controls in *MATLAB/SIMULINK*[1] or *MATRIXX*.[1]

As previously mentioned, design representations undergo many changes during the transition from the idea to the final product. Updating the numerous distributed representations after a design change (called Δx_i hereafter) has occurred means the propagation of one (or several) modified value(s) x_i from one design domain representation to all the others. Otherwise the *model consistency* for subsequent simulations in all domains to identify the modified behavior(s) and possible violation(s) of design requirements is no longer possible. While in the past model consistency has been tried to achieve via more or less powerful standardized interfaces (e.g. *VDAFS, IGES, STEP* for geometry exchange), this problem is currently dealt with by a trend to more integrated software tools (e.g. a geometry kernel with a structural mechanics and a fluid mechanics solver) to allow for automated *process chains*.

[1] *ABAQUS, ANSYS, CATIA, FLUENT, MATLAB/SIMULINK, MATRIXX. NASTRAN, NX, PRO-E* and *STAR-CD* are trademarks of the corresponding companies. Their enumeration emphasizes the occurring combinatorial variety of the software tool landscape, which can easily attend numbers up to 400 different programs for a global player in the automotive or aerospace sector.

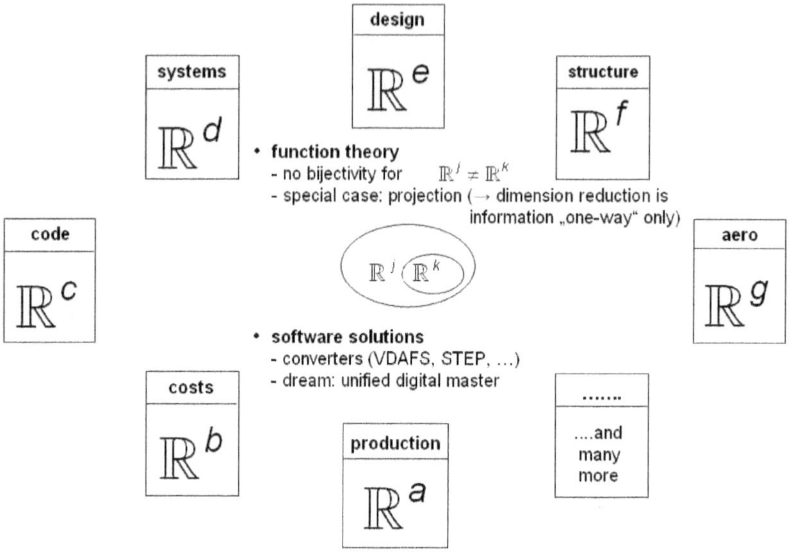

Fig. 2. Parameter representations in $\mathbb{R}^a, \mathbb{R}^b, \ldots, \mathbb{R}^g$ with different dimensionality.

A rigorous mathematical analysis of the situation underlying Figure 1 leads to the following statements based on the dimensionality considerations in Figure 2. It is fact that the dimensionality $\mathbb{R}^a, \mathbb{R}^b, \ldots, \mathbb{R}^g$ of the design spaces used in each of the design domains is different, i.e. in full generality $a \neq b \ldots \neq g$ with $a, b, \ldots, g \in \mathbb{N}$. However, it is also fact that couplings do exist implicitly between the disciplines because different disciplines may share common aspects such as the geometry, one or several individual geometrical design parameters x_i or other non-geometrical design representations among each other partially or entirely such as product function(s).

Mathematically speaking, updating distributed design representations means the propagation of a single or several modified value(s) Δx_i of a design parameter x_i from one design representation to all $(n-1)$ other design representations and back. However, mathematical function theory has proven that *bijective mappings*[2] between two design spaces with dimensionality $\mathbb{R}^j$ and $\mathbb{R}^k$ exist only if $j = k$. As stated, in full generality is $a \neq b \ldots \neq g$. As a consequence, the information flow has an inbuilt flow direction in a one-way sense from the higher dimensional representation space to the lower dimensionality representation space by means of a mathematical projection $\mathbb{R}^j \rightarrow \mathbb{R}^k$ since $j > k$. This is the only possible way to achieve consistent update schemes for representations in design spaces with different dimensionality.

It is this mathematical analysis in Figure 2 which motivates in the following the existence and the use of the so-called *design languages* in the area of engineering

[2] A bijective mapping means that the mapping can be uniquely inverted, i.e. that the value(s) Δx_i of a design parameter x_i can be mapped forward and backward between the domains.

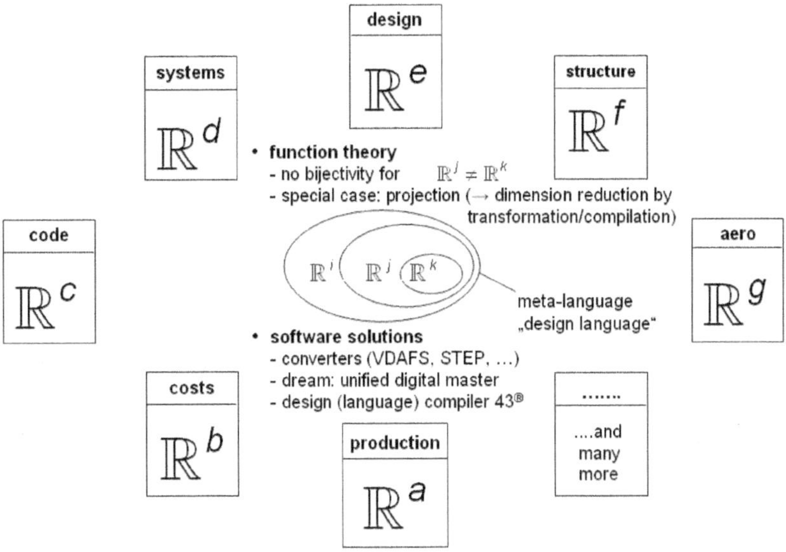

Fig. 3. Parameters representations in different dimensionalities.

design.[3] In Figure 3 the term design language characterizes the design representation with the highest dimensionality, i.e. that "meta-language" design representation level which contains all relevant design information from which all other design model representations can be generated by means of a projection. If the projection is well defined, it can be executed by a machine, which is called *design compiler*.[4]

From the viewpoint of computer science the design language concept offers two big advantages. On the one hand, there is one single place where all the information about a design is stored, and any other model representation needed during design is compiled thereof. On the long run this offers the perspective for the creation of libraries (in the sense of design repositories containing chunks of design solutions) offering the potential of *know-how reuse* in the design of complex products. On the other hand, any formal (computer programming) language exhibits the aspects of *syntax*, *semantic* and *pragmatic*, which are important for the proof of correctness of any problem solution formulated in such a language representation [18].

Finally, the aspect of projection from the high-dimensional meta-level of a design language into the low-dimensional design representations implies the feed-forward aspect in the model generation during the compilation phase of the design language.

[3] The terms engineering *design language* and *design grammar* are used interchangeably in the literature. For a recent overview on formal methods of engineering design synthesis such as engineering design languages and engineering design grammars see [4].

[4] The term mechanical *design compiler* has been introduced by Ward and Seering [21]. The design compiler software tool used here in this work is the *DESIGN COMPILER 43* [9].

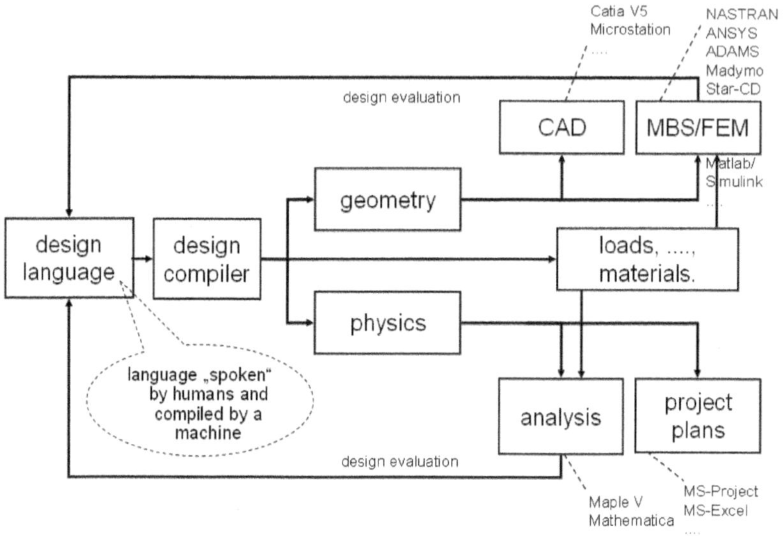

Fig. 4. Design compiler software architecture reflecting the one-way information flow [15].

As shown in Figure 4, the design compiler projects the design language representation on the meta-level into the low-dimensional level of the various input formats defined for the dedicated analysis programs used via interfaces. These interfaces provide appropriate input file formats for simulation and analysis in the different domains such as geometry (here the geometry kernel *CATIA*), for linear static and modal analysis (here the solver *NASTRAN*), for fluid dynamics (here the solver *STAR-CD*), controls (here the program package *MATLAB/SIMULINK*) and so on.

These enumerated interfaces are bi-directional, which means that results written out by the special purpose programs in the results files can be filtered and interpreted, so that specific analysis information can be fed back via the evaluation loops shown in Figure 4. This means that the decision making during the design compilation phase(s) may be influenced by results of partial, intermediate model representations created to decide a certain engineering problem occurring during the design phase.

In order to allow this, a design language offers not only a static representation which characterizes the physical behavior of a component, but disposes also of adequate means to represent the aforementioned state *transitions*. In order to achieve this, a rule-based transformation mechanism is provided in the design language to express the dynamic aspects in the change of the design representation.

As shown in Figure 5, a design language consists of a vocabulary and a set of rules [16, 18]. From the viewpoint of computer science, the vocabulary and the rules define a language grammar, from which it can be decided whether a sentence, i.e. a vocabulary combination generated via a sequence of rules, is syntactically correct.

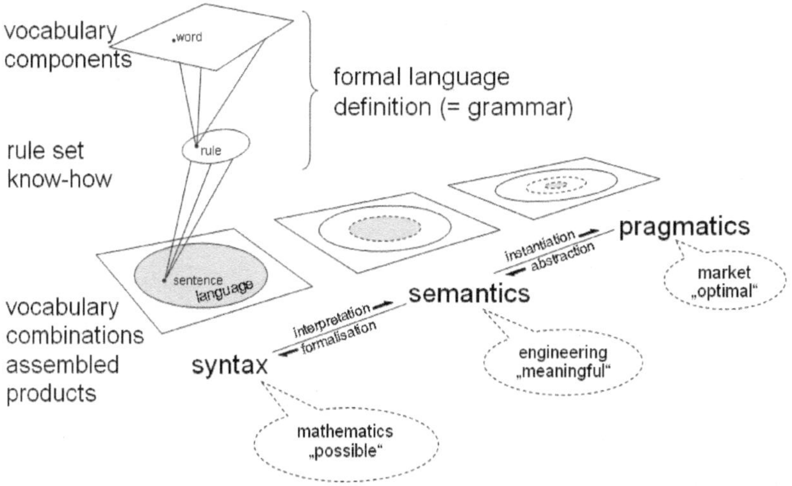

Fig. 5. Syntax, semantic and pragmatic of a design language [16].

The rule set decides as well the kind of sentences which can be generated by the rule set. Since the vocabulary has also a meaning, additional constraints apply to obtain semantically correct sentences. Finally, the requirements and boundary conditions in a certain design situation need to be satisfied, which adds further constraints for the pragmatical correctness of a design representation.

From the viewpoint of engineering, these rather abstract sounding properties are quite useful. The vocabulary represents the design components available as building blocks to the engineer. The rule set stands for the design, build and assembly know-how to create a system from available components. The interesting property of design languages is that both the object behavior and the design know-how can be encoded in a single, consistent, formal and machine-executable form.

The vocabulary and the rules are hereby defined in such a way that the language representation can be read, interpreted, stored, retrieved and executed by both the human (i.e. engineer) and the machine (i.e. compiler). The gray areas in Figure 5 symbolize hereby the potential of a design process by design compilation that the three sets of possible, meaningful and optimal product representations can be generated by the design compiler. This opens up the perspective for full know-how reuse in the conceptual design phase of complex engineering products by automatic model generation instead of tedious manual construction of the various design models.

The benefit of automatic model generation, analysis and update comes of course at a certain price. Libraries of both design vocabulary and rule sets must be established, implemented and tested. This process is almost identical to traditional computer (language) programming, only the format of the (programming) language has changed from a string-based to a graph-based representation [15].

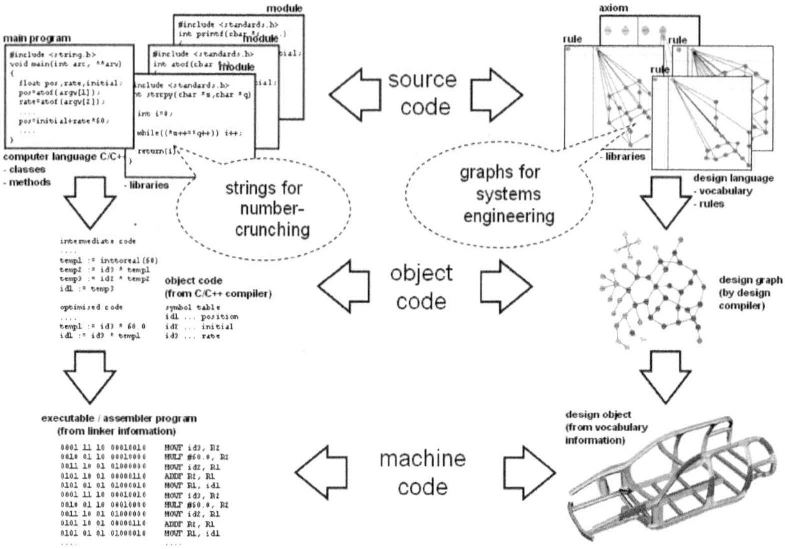

Fig. 6. Graph-based design language versus string-based programming language [16].

This change of language format representation is inspired by a careful analysis of the engineering design process, which requires to manipulate the design topology as easy as the design parameters. Since graphs are ideally suited to express topological existence and connectedness, the graph-based representation offers the more elegant, easier to use and most powerful means to describe multi-disciplinary couplings at a very abstract level of detail. This level of detail can even be chosen accordingly to the specific design task and may change as during the activity of designing the design object description transits from the abstract to the concrete [15].

Writing the code for a graph-based design language for subsequent compilation in a design compiler works in direct analogy to classical programming in string-based programming languages as shown in Figure 6. While this offers the future option to adopt the most modern software technologies and put them into direct productive use in engineering, the string-based program lines are substituted by a transformation rule expressed in a graph-based format. The design compiler creates during the rule execution the so-called design graph, which is automatically converted into the input formats to the dedicated analysis programs via the aforementioned interfaces.

The fact that design is perceived in this view as a sequence of activities is hereby inherited from and shared with earlier design language approaches. Figure 7 shows the L-system [14] representation based on the vocabulary consisting of 5 words $F, +, -, [$ and $]$ and one (combined growing and branching) rule. During repetitive execution of the (growing and branching) rule, a structured stream of symbols is created and interpreted afterwords in a turtle graphic as geometry. This technique

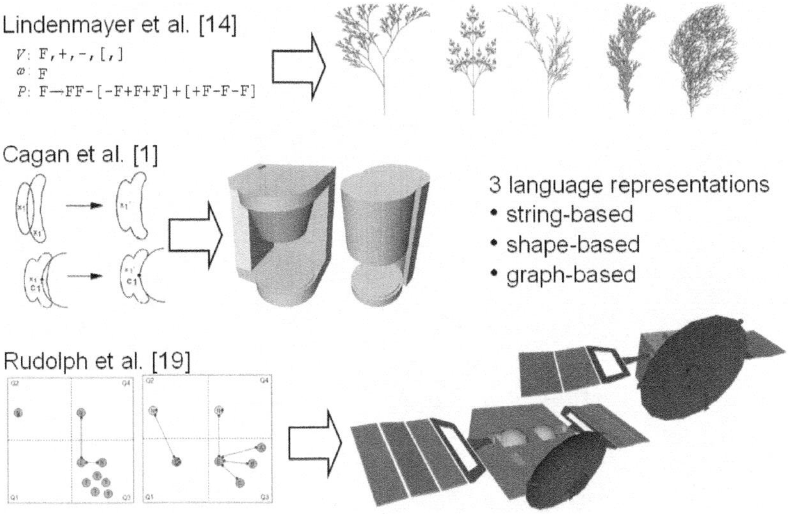

Fig. 7. Three kinds of string-based, shape-based and graph-based design languages.

was refined over the years for the perfect modeling of a large variety of plants, trees and other biological organisms including their growth history [11].

From the large community of shape-based design languages, see [4] for a recent overview, the well-known example of a shape rule for the design of coffee makers [1] is shown in Figure 7 as an example. Both examples of string-based and shape-based rule representations are shown here to give a flavor about the kind of programming effort required to express the design activity.

While a string-based representation is suited for tree-like branching, graph-based rule representations may encode topological closures more easily. Shape-based rule representations follow the view of the dominance of geometry in designing and reflects the CAD paradigm as the central storage for design information. Graph-based rule representations abstract from this and are therefore a priori domain-independent in their knowledge representation and therefore an ideal candidate in the search for a true multi-disciplinary modeling and representation form [2]. This way of expressing the design knowledge in form of a graph-based design transformation is described now in thc following in morc dctail.

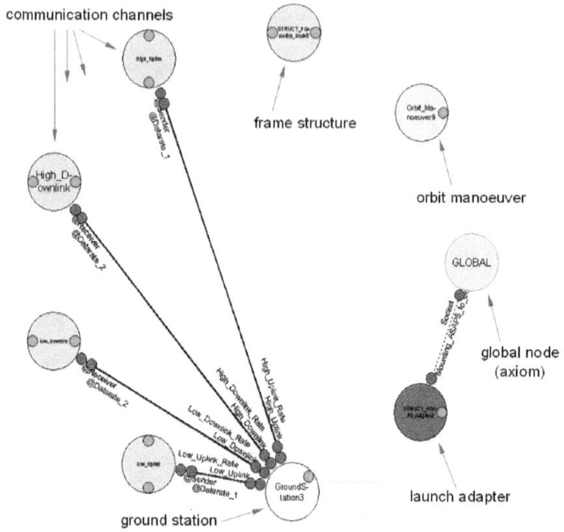

Fig. 8. A complex (satellite design) axiom [19].

2 Graph-based Design Languages

After the execution of each design rule during the design compilation, the design graph is a complete design description at any moment in time.[5] The creation of the design graph starts initially with the axiom and is transformed during the execution sequence of the design rules (i.e. graph transformations, also called *design patterns*), which are described in the following section. Theoretically, the axiom consists of one single graph node (such as the node named *global* node in Figure 8) only, which is expanded to the final design graph during rule execution. However, typically all the customer requirements are already added to the axiom, since they represent the requirements which must be met by any design concept under development.

Figure 8 shows such a complex axiom for a satellite design, where several nodes represent the four required satellite up- and down-links, the need to fly certain orbit maneuvers, the use of a specialized structural adapter to fix the satellite to the rocket and so on. The connections between certain graph nodes signify that some parameter values in the functional or physical description of the nodes are coupled. Links between graph nodes mimic therefore the *call by value* and *call by reference* mechanisms of string-based programming languages for parameter passing.

[5] Of course this statement refers to what is known at a certain moment in time. If little is known, the design graph is small and may contain only little information. With progressing rule execution the design graph may grow and contains more information. More abstract pieces of design representation may hereby even be replaced by more detailed pieces.

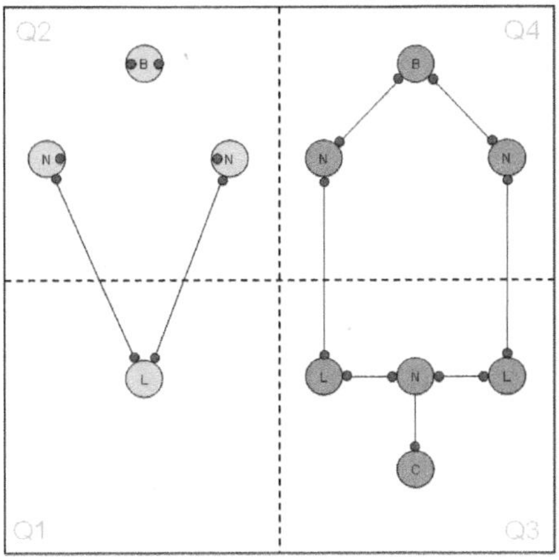

Fig. 9. 4-quadrant scheme for rule definition [2].

2.1 Design Patterns

In Figure 9, the so-called 4-*quadrant scheme* (or modified *x-scheme* [5]) is used to define a design rule in form of a graph transformation. Permitting graph transformations allows much more freedom for the designer to express the context of his design decisions than a simple database access mechanism, which tests for the existence of an item in memory only. Additionally, constraints of any kind (logical, symbolical, numerical, etc.) which can be evaluated at runtime may be used throughout the rule execution process.

The 4 quadrants Q_1, Q_2, Q_3 and Q_4 are used as it is described in the following. The *if*-part of the rule in Figure 9 is specified by Q_1 and Q_2, thus describing the condition of *graph isomorphism* which must be satisfied so that the rule can be executed. Hereby the graph nodes in Q_1 are deleted during the rule execution process, while all nodes in Q_2 are transferred to Q_4.

The *then*-part of the rule in Figure 9 is specified by Q_3 and Q_4, thus describing the creation of new graph nodes in Q_3 and the modification of the context (i.e. of the connection lines between the nodes) in both Q_3 and Q_4. While the syntax is completely described by the above rule, the semantic meaning of the graph nodes B, N, L and C and their corresponding pragmatic values of the parameters do depend on the vocabulary definitions contained in the underlying library.

The graph-based rule mechanism in Figure 9 allows the modification of both the design topology as well as modifications of the design parameters with relative ease in a unified, domain-independent form. Topological modifications as they typically occur during top-down (i.e. a decomposition process) or bottom-up design (i.e. an aggregation process) are therefore possible to model in a straightforward way.

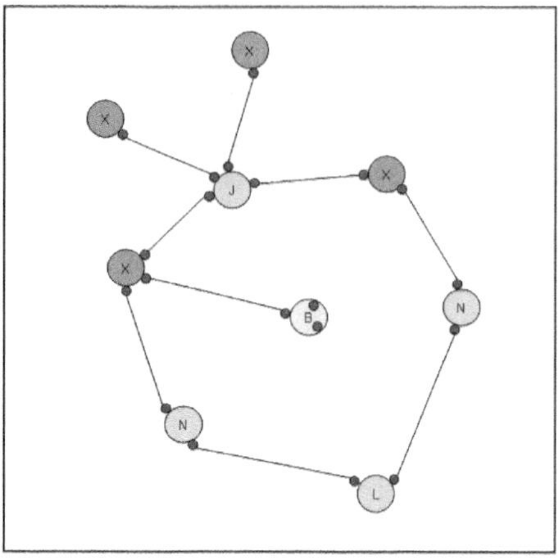

Fig. 10. Design graph at time t (before rule execution) [6].

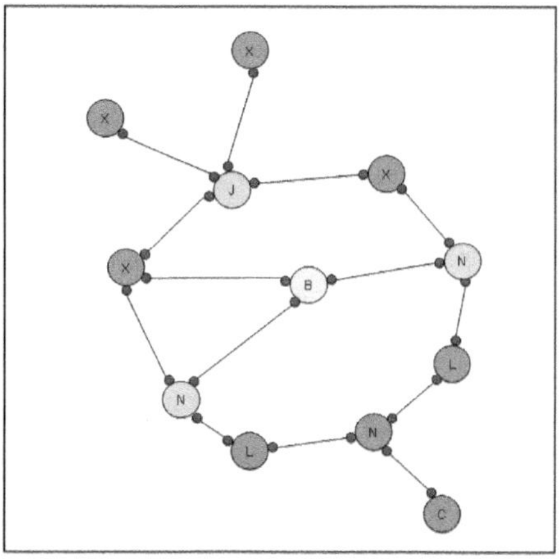

Fig. 11. Design graph at time $t+1$ (after rule execution) [6].

2.2 Design Expansion

Based on the state of the design graph at a certain moment t as shown in Figure 10, the graph transformation shown by the design rule in Figure 9 leads to the design graph at time instant $t+1$ in Figure 11. Since the *if*-part in Figure 9 is isomorphic with the (sub-)graph in Figure 10, the rule can be executed. This means that the former node L is deleted and the four nodes L, N, L and C are newly created. Finally, the modifications of the connecting links lead to the modified design graph in Figure 11.

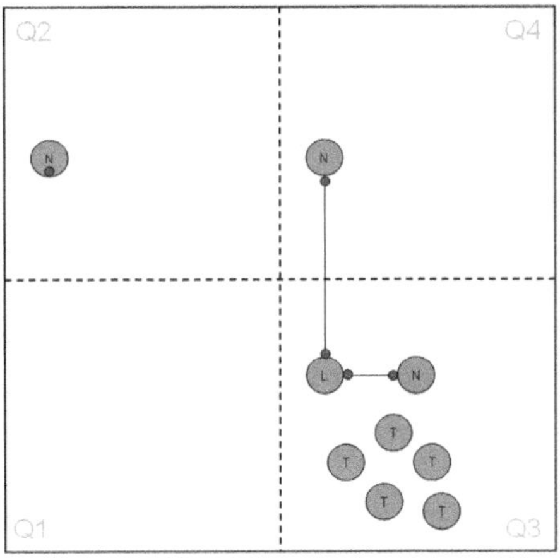

Fig. 12. Satellite propulsion rule [18].

2.3 Satellite Design Sequence

A design language for the early conceptual design phases of satellites [19] has been developed for both interactive and batch mode execution in a design compiler [9]. The satellite design sequence follows a top-down approach from customer mission requirements to the final satellite design embodiment. The overall design sequence consists of more than about 100 rules in the 4-quadrant scheme and of about 100 potential vocabulary elements with additional information about their functionality and physical behavior. Much more detailed information about the satellite design using a design language can be found in [19]. For the reason of space, only the rules presented in Figures 12 and 13 are discussed.

Based on the requirement to fly a certain orbit, the propulsion rule in Figure 12 defines the generation of a chemical rocket motor L, the nozzle N and 5 tanks T. The motor and the nozzle are already explicitly coupled by a node link, which is not yet the case for the tanks. This is done because the propulsion can be finalized only after the mass balance has been determined, thus leading to an increase or decrease in the number of tanks later on, making the manipulation of the connectivity information at this point in time unnecessary.

Figure 13 resolves the overall static energy balance into a predefined number of energy schemes which are resolved in time. These are the three operational modes (e.g. the *safe mode A*, the *nominal mode B* and the *payload mode C*) in the shown energy scheme rule which are used to detail the static energy consumption in the satellite by integrating the different needs over the duty cycle time (e.g. over one or several orbits, depending on the mission requirements). This is beneficial because not every electrical load may be on all the time.

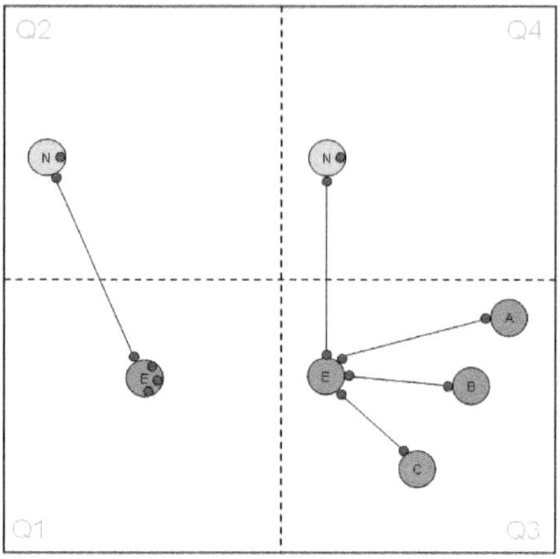

Fig. 13. Satellite energy scheme rule [18].

As a remark, the little bullets at the graph nodes in Figures 8 to 13 are so-called *ports* which allow the coupling of graph nodes based on a syntax check of the underlying data types. Connection lines between the ports of two graph nodes are further used to mimic the call by value and call by reference mechanisms of conventional string-based programming languages for parameter passing. This mechanism is an essential part of the overall effort inside the design compiler to keep all the different disciplinary models which are managed inside the design graph consistent with any ongoing parametrical and topological design modification(s).

The rule scheme is therefore very practical to express design context, since the preconditions can be made as precise and specialized or as general as necessary. Of course, in respect to the desired rule reuse, the most general rule formulations possible are desirable. This allows to decompose complex design activities into smaller, individual actions which are easier to understand and implement in a rule.

In this respect it appears to be of great help that the concept of a vocabulary node ‘rocket motor’ in a graph-based design language is almost the same concept than the semantic hull of the word ‘rocket motor’ in spoken human language. This minimizes the need to learn the semantics of another formal artificial language, facilitates the information acquisition when talking to experienced designers and building the rules from a generalization of the discussion essentials afterwords. It allows even further to establish a formal validation scheme for the semantic correctness of the designs generated by a specific design language [18]. For the upcoming future, this feature may even open a perspective to future design technologies which could be characterized as *semantic engineering*, since design alternatives could be automatically selected from dedicated underlying engineering ontologies defining the domain properties.

Figure 14 finally shows the resulting satellite design geometry after execution of more than 100 design rules. For convenience, the top panel has been removed to al-

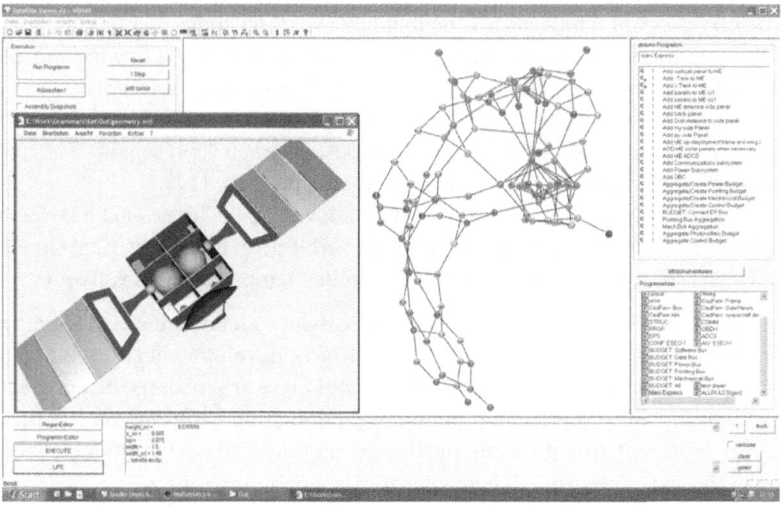

Fig. 14. Screen shot of a satellite design generated in the design compiler 43 [19].

low a look inside the satellite. It should be mentioned that all balances for mass, inertia, energy, data transfer and data processing are satisfied. This can only be achieved, if after inserting the rocket motor, not only the rules for the mass and inertia balances are re-executed, but that also any other balance, which might be affected by a side effect only, is re-run. For all these and any further details see [19].

Once a certain level of detail has been encoded into such a graph-based design language for satellites, modifications of both design topology and/or parameters are relatively easy, fast and convenient to realize. As an example, it is sufficient to change the rule for the selection of the satellite propulsion in Figure 12 from chemical to electrical propulsion, and the energy balance rule will automatically change the solar panels in number and size later on.

3 Conclusions

The graph-based design language paradigm used in all examples in this work offers an intermediate, domain-independent design representation in form of the design graph. This design graph is assembled through a sequence of graph-based design rules. These design rules form a so-called production system and encode the complete design knowledge from the expansion of the design requirements in form of the axiom into the final conceptual design of a satellite.

The design graph as intermediate information storage offers the possibility to incorporate several conceptual advantages in the design information processing architecture [15, 16]. These are, among others, the incorporation of

- a *library concept* which allows the mapping of graph nodes to a specific domain. All domain dependencies can thus be hidden in a library, allowing the design compiler not to be limited to a specific domain or design paradigm;
- a *constraint processing mechanism* for symbolic equations. The design constraints are collected in the intermediate design graph representation at runtime and solved using a symbolic computer algebra package [17];
- a *plug-in programming technique*, which allows the bi-directional interfacing to customer-specific legacy code or any other arbitrarily used numerical simulation and analysis code such as multi-body or finite element analysis software.

Based on these key properties of modern software technologies, shortening the design cycle, faster time-to-market, and reduction of development cost seem to come closer step by step. Of course, there is always still more progress and effort needed to achieve ‘intelligent’ products and ‘intelligent’ production. After the successful solution of the representation problem, implementing powerful design strategies [3] will be one of the most interesting challenges for the upcoming future.

However, the properties of design languages offer already today the potential to model most aspects of thought processes, compromises and intermediate decision making activities which occurs during the design process. Design languages offer automatic translation into all kinds of digital models, and, if properly designed, the automatic compilation under a variety of topological and parametrical variations of requirements or boundary conditions, thus leading to different product designs and know-how reuse in the conceptual design of complex engineering products.

In this respect, the subtitle *“Are you still constructing manually or do you already generate automatically?”* of this paper was hereby intended to motivate the audience to rethink design, reconsider what we usually take for granted and to become aware of the manifold consequences and issues related with the design language approach. Of course many questions still remain unanswered. Despite this it is almost needless to say however that in the opinion of the author graph-based design languages will take for sure a big part in this great future engineering challenge.

Acknowledgments

The invitation to this keynote presentation at the conference for *Integrated Design and Manufacturing in Mechanical Engineering* (IDMME) 2006 in Grenoble was greatly appreciated. The author wants to express in this respect his personal thanks to Professor Michel Tollenaere from the Organizing Committee.

References

1. Agrawal, M. and Cagan, J.: A blend of different tastes: The language of coffeemakers. *Environment and Planning B: Planning and Design* **25**(2), 1998, 205–226.

2. Alber, R. and Rudolph, S.: On a grammar-based design language that supports automated design generation and creativity. In *Proceedings IFIP WG5.2 Workshop on Knowledge Intensive CAD (KIC-5)*, Malta, Malta, July 23–25, 2002.
3. Alber, R., Rudolph, S. and Kröplin, B.: On Formal Languages in Design Generation and Evolution. In *Proceedings 5th World Congress on Computational Mechanics (WCCM V)*, University of Vienna, Vienna, Austria, July 7–12th, 2002.
4. Antonsson, E and Cagan, J (eds): *Formal Engineering Design Synthesis*. Cambridge University Press, Cambridge, 2001.
5. Göttler, H.: *Graphgrammatiken in der Softwaretechnik*. Springer, Berlin, 1988.
6. Haq, M. and Rudolph, S.: *"EWS-Car" – Eine Entwurfssprache für den Fahrzeugkonzeptentwurf.* VDI Bericht 1846, Verein Deutscher Ingenieure VDI, Düsseldorf, 2004.
7. Lindemann, U.: *Methodische Entwicklung technischer Produkte*. Spriner, Berlin, 2006.
8. Lossack, R.-S.: *Wissenschaftstheoretische Grundlagen für die rechnergestützte Konstruktion*. Springer, Berlin, 2006.
9. Ingenieurgesellschaft für Intelligente Lösungen und Systeme (IILS) mbH: The Design Compiler 43. Internet: www.iils.de, Februar 2007.
10. National Science Foundation (NSF): Research opportunities in engineering design – Final Report to NSF. National Science Foundation, USA, April 1996.
11. Noser, H-R., Rudolph, S. and Stucki, P.: Physics-enhanced L-systems. In *Proceedings 9th International Conference in Central Europe on Computer Graphics, Visualization and Computer Vision (WSCG 2001)*, Plzen, Czech Republic, February 5–9, 2001.
12. Oelsner, R.: Geschichte des Konstruierens in Deutschland. Vom künstlerischen Handeln zum formalisierten Wissen. *Konstruktion* **44**, 1992, 387–390.
13. Pahl, G. and Beitz, W.: *Konstruktionslehre*. Springer, Berlin, 1993. Also available as English edition: *Engineering Design*. Springer, Berlin, 1996.
14. Prusinkiewicz, P. and Lindenmayer, A.: *The Algorithmic Beauty of Plants*. Springer, Berlin, 1996.
15. Rudolph, S.: Übertragung von Ähnlichkeitsbegriffen. Habilitationsschrift, Fakultät Luft- und Raumfahrttechnik und Geodäsie, Universität Stuttgart, 2002.
16. Rudolph, S.: Aufbau und Einsatz von Entwurfssprachen für den Ingenieurentwurf. In *3. Forum Knowledge Based-Engineering, CAT-PRO*, Stuttgart, 9 October 2003.
17. Rudolph, S. and Bölling, M.: Constraint-based conceptual design and automated sensitivity analysis for airship concept studies. *Aerospace Science and Technology* **8**, 2004, 333–345.
18. Rudolph, S.: A semantic validation scheme for graph-based engineering design grammars. In: *Conference Proceedings Design Computing and Cognition'06 (DCC'06)*, Gero, J. (ed.). Springer, Dordrecht, 2006, pp. 541–560.
19. Schaefer, J. and Rudolph, S.: Satellite design by design grammars. *Aerospace, Science and Technology (AST)* **9**(1), 2005, 81–91.
20. VDI-Richtlinie 2221: *Methodik zum Entwickeln und Konstruieren technischer Systeme und Produkte*. VDI-Verlag, Düsseldorf, 1986.
21. Ward, A. and Seering, W.: Quantitative inference in mechanical design compiler. In *Proceedings 1st DTM Conference*, Montreal, Canada, 1989.

FORMING TECHNOLOGIES AND DESIGN

Numerical Optimization of an Industrial Multi-Steps Stamping Process by Using the Design of Experiment Method

Yann Ledoux, Eric Pairel and Robert Arrieux

Laboratoire de Mécanique Appliquée (LMécA), Esia, BP 806, 74016 Annecy Cedex, France; E-mail: {yann.ledoux, eric.pairel, robert.arrieux}@univ-savoie.fr

Abstract. The deep drawing process consists in realizing parts with complex shapes like different kind of boxes, cups, etc., from a metal sheet. These parts are obtained through one or several stamping steps. The tool setup motion of the stamping process is difficult to obtain. It requires practice and special knowledge on the process. The design is long and difficult to optimize. It is expensive in machining tool operations because it needs many trials and modifications on the tool before obtaining the right shape and the good working of the tool. The mains reasons of these difficulties come from the strain heterogeneity, the spring back after the tool removing and the decrease of thickness. There are many influent parameters for this kind of process. They modify directly the shape of the part. These parameters can be listed in three different categories: Firstly, the parameters linked to the tool geometry like the die enter radius, the punch diameter and the clearance between die and punch. Secondly, the parameters linked to the manufacturing conditions like the stamping speed, the lubricant and the blank holder force. And thirdly, the parameters linked to the flank geometry.

This paper proposes a systematic method of stamping progression optimization, illustrated on an industrial five part made from five forming steps. By using empirical rules and industrial knowledge, the number of steps, the nominal dimensions and process conditions are defined. From this initial tool definition, the numerical simulation by the finite elements method of the stamping steps is carried out using Abaqus software. Then, the parameters of each stage, which have an influence on the shape of the final stamped part, have to be selected. A range of variation around their nominal values is defined.

Experimental designs are used to test the influence of these parameters in the numerical simulations. This allows to establish a mathematical model between the geometrical variation on the part and selected influent parameters. With this model an optimization of these parameters can be realized to find their best values.

Key words: deep drawing process, design of experiments, finite element method.

S. Tichkiewitch et al. (eds.), Advances in Integrated Design and Manufacturing in Mechanical Engineering II, 43–56.

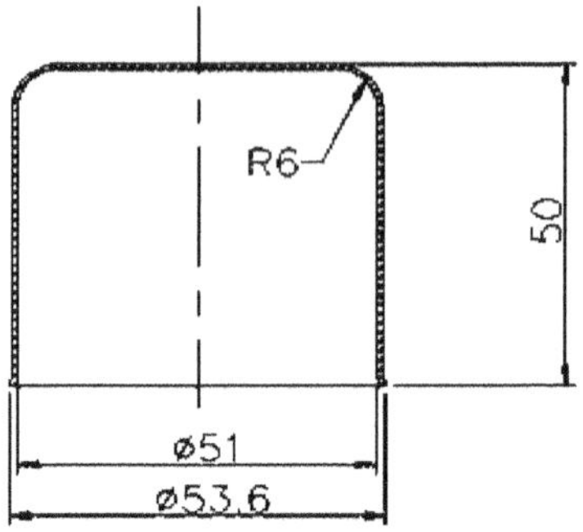

Fig. 1. Drawing of the part.

1 Introduction

In press shop one of the most important problem for engineers is to obtain formed parts with geometrical quality corresponding to the customer specifications. In this paper, we propose the study of a part made from two stamping stages and two additional operations (folding and trimming). Usually, empirical rules combined to "trial and error" method are used to find the right combination of process parameters in order to get the wished part. So this kind of optimization of complex part is not a rigorous procedure and often leads to delay, lack of quality and costing adjustment of tool geometry and process parameters.

In previous works [1, 2], we proposed a method combining both design of experiments and finite element simulations of deep-drawing operation during the optimization phase. It was applied to a simple 2D part made from a single stamping operation. In this paper, we deal with the optimization of a complex part formed from several steps.

These different steps of the proposed method are detailed in this paper and this technique allows to find the optimal configuration of the process conditions.

2 The Studied Part and Initial Plan

The target geometry of the part is presented in Figure 1. It is an axisymmetric part of 50 mm height and 51 mm internal diameter.

A stamping plan is established thanks to some empirical rules used in press shop. After the cutting of the initial blank, two stamping stages (OP1 and OP2) and two additional operations, folding (OP3) and trimming (OP4), are needed, as shown in Figure 2.

In fact, this figure corresponds to a first plan because some parameters such as dies and punch radii, which appear implicitly on the progression, have to be adjusted

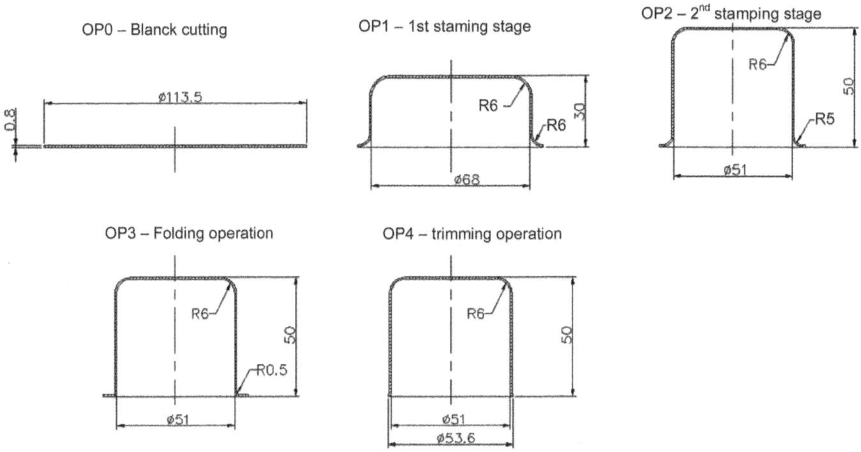

Fig. 2. Initial progression of the stamping part.

to allow the part manufacturing with the required geometry. In addition, we want also to reduce the flange in order to save material.

3 Numerical Simulation of the Process

The blank is discretized by shell elements (S4R) with four nodes. There are seven integration points in the thickness. The stamping tool is modeled by rigid analytical surfaces. The main parameters used for simulation are listed in Table 1. These forming operations are made by explicit Abaqus software.

The material used is a USB steel sheet (DC05) of 0.8 mm thickness. The hardening is determined based on tensile tests using extensometer and image analysis sys-

Table 1. Numerical parameters of the finite element simulation.

Blank discretization:	
Element type	Shell (4 nodes)
Integration points	7
Number of elements	1881
Tool discretization:	
Tool type	Analytical rigid surface
Process parameters:	
Punch speed	3 m/s
Friction coefficent	0.1

Table 2. Main USB material properties.

USB (DC05)	
Young modulus	206.62 GPa
Yield Strengh	168.12 MPa
Poisson modulus	0.298
Density	7800 kg/m^3
	$r0° = 2.09$
Anisotropy coefficients (Lankfort)	$r45° = 1.56$
	$r90° = 2.72$

Table 3. Range variation of the blank diameter.

Parameter	Initial value	Min value (−1)	Max value (+1)
D_0: initial blank diameter (mm)	116.38	115	121

tem for high-strain levels [3]. It is introduced point by point in the FEM code. The used behavior law takes into account the anisotropy according to Hill's 48 criterion [4]. The material characteristics are shown in Table 2.

4 Seek of the Influent Process Parameters for Each Stamping Stage

4.1 Operation 0: Blank Cutting

The possibility to obtain the wished geometry of the part and the minimal flange is directly linked to the blank initial diameter so we have to optimize it. In this aim, we have to estimate its initial diameter D_0 and its variation range. Taking into account the following assumptions:

- H_1: the volume of the blank is constant during the stamping operation (H_1).
- H_2: the thickness is totally constant (the local decrease is compensated by local increase).
- The minimum flange of 1.5 mm around the part is needed to ensure a net final trimming.
- The estimated ears value, due to material anisotropy, is 4 mm.

A geometrical calculation gives an initial diameter $D_0 = 116.38$ mm. From this nominal value, we choose two extreme values to test.

As it is absolutely necessary to have a flange, we cannot decrease this diameter much, so we choose 115 mm as lower value and 121 mm for the upper value. This range is resumed in Table 3.

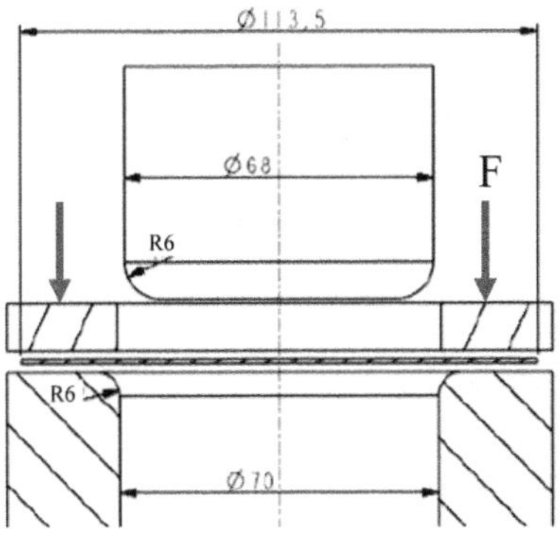

Fig. 3. Op1: first stamping stage (simplified drawing).

Table 4. Range of variation for the selected parameters of the first operation.

Parameter	Initial value	Min value (−1)	Max value (+1)
R_{D1}: die radius (mm)	6	4	8
R_{P1}: punch radius (mm)	6	4	8
F: blank holder force (daN)	1200	800	1600

4.2 Operation 1: First Stamping Tool

Figure 3 shows the stamping tool of the first forming stage. A blank holder is used to avoid the wrinkles. At this stage, there is no flange at the end of this operation. The blank is fully stamped.

For this stage, as we assume that the enter die radius (R_{D1}), the punch radius (R_{P1}) and the blank holder force (F_1) have an influence on the final geometry of the part (thickness), we test their influences. The range of variation of those process parameters are listed in Table 4.

These different variation ranges are chosen around the nominal value. These upper and lower bounds are fixed according to our knowledge in the process. Moreover, it is supposed that there is a least one solution in the considered range.

4.3 Operation2: Second Stamping Stage

This is the last stamping operation. We take the part issued from the previous stage. We use an annular blank holder in order to avoid the wrinkle phenomenon. The stamping tool is shown in Figure 4.

The punch radius is fixed because it is a final radius value of the part. It is possible to modify the die radius (R_{D2}) and the punch travel (T_2) which allow to adjust the final height of the part.

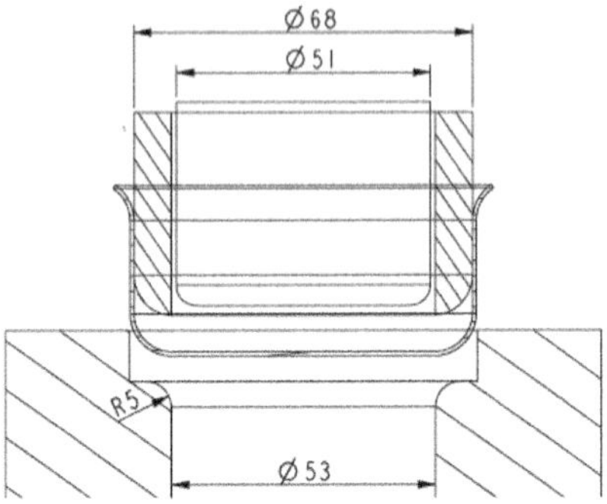

Fig. 4. Op2: Second stamping stage (simplified drawing).

Table 5. Range of variation for the process parameter R_{D2} and T_2.

Parameter	Initial value	Min value (−1)	Max value (+1)
R_{D2}: die radius (mm)	5	4	6
T_2: punch travel (mm)	48	47.5	49

Table 6. Range of variation for the process parameter T_3.

Parameter	Initial value	Min value (−1)	Max value (+1)
T_3: punch travel (mm)	49.2	49	49.5

The range of variation of these process parameters are listed in Table 5.

4.4 Operation 3: Folding Stage

The final height of the part is adjusted at this stage (calibration operation). It depends directly on the punch travel amplitude. It is the only parameter to be adjusted here (Table 6).

4.5 Operation 4: Cutting Stage

This stage consists in cutting out the flange of the part in order to obtain the final part. Here, no parameter can modify the part shape.

4.6 Summary of the Different Process Parameters

Two optimization strategies are possible:
- Optimization of each operation
- Global optimization.

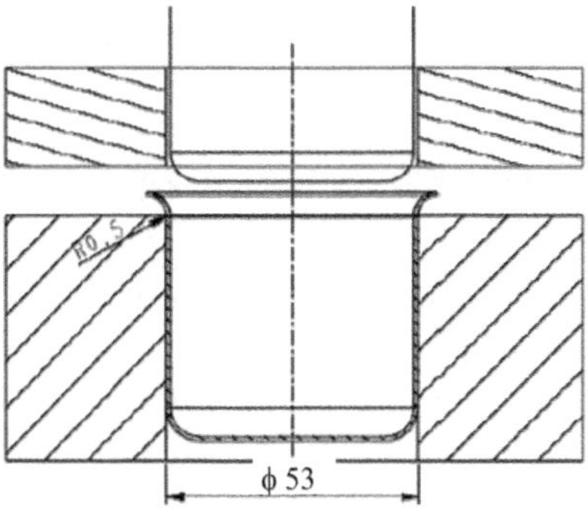

Fig. 5. Op3: folding stage (simplified drawing).

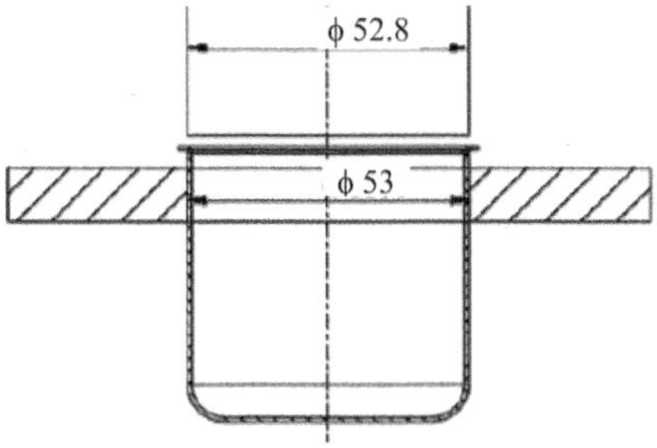

Fig. 6. Op4: trimming (simplified drawing).

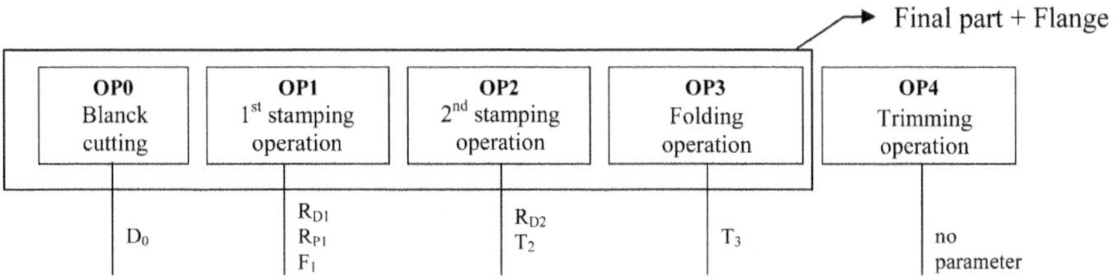

Fig. 7. Synoptic of these operations.

The first strategy needs to arbitrary intermediate objectives while the second one focuses only on final objectives. So the whole of the process steps is considered as a "black box" with seven influent factors. This is represented in Figure 7.

The different variation ranges of the process parameter are summarized in Table 7.

Table 7. Summary of the range of variation for all process parameter tested.

	Parameter	Initial value	Min value (−1)	Max value (+1)
step 0	D_0: initial blank diameter (mm)	116.38	115	121
step 1	R_{D1}: die radius (mm)	6	4	8
	R_{P1}: punch radius (mm)	6	4	8
	F: blank holder force (daN)	1200	800	1600
step 2	R_{D2}: die radius (mm)	5	4	6
	T_2: punch travel (mm)	48	47.5	49
step 3	T_3: punch travel (mm)	49.2	49	49.5

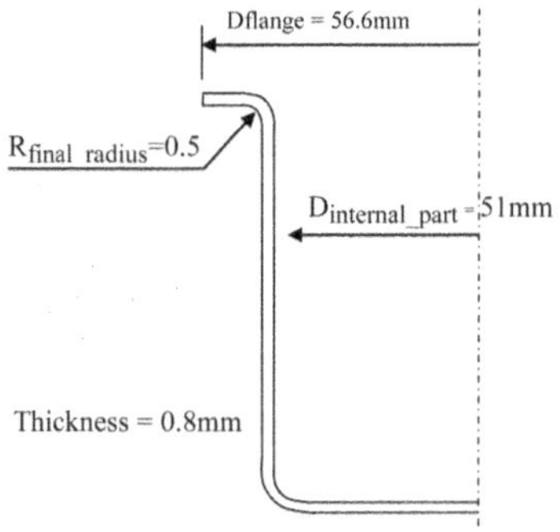

Fig. 8. Calculation of the needed flange diameter.

5 Targets on the Final Part and on the Flange

We have to choose objectives on the part and on the flange. On the final part a 50 mm height and minimum thickness reduction are wanted. On the flange, at least 1.5 mm around the part which corresponds to a 56.6 mm diameter. This diameter is calculated outside the anisotropy ears. These dimensions of the flange are represented in Figure 8.

6 Table of Experiments

For each target on the part and on the flange, a mathematical model has to be identified. In order to minimize the number of numerical trials, the simplest model has been chosen: linear relations without interaction. (In complementary works, we have verified with a factorial design of this operation that the interaction is equal to zero.) So the linear model is presented below.

$$Y = a_0 + a_1 D_0' + a_2 R_{D1}' + \cdots + a_5 T_3', \tag{1}$$

Table 8. Design of experiments L16 of Taguchi.

L16	step 0		step 1			step 2	Step 3	
Nr	$D_0\,(mm)$	$R_{D1}\,(mm)$	$R_{P1}\,(mm)$	$F\,(daN)$	$T_2\,(mm)$	$R_{D2}\,(mm)$	$T_3\,(mm)$	Profile
1	115	4	4	800	47.5	4	49	Profile 1
2	115	4	4	800	49	6	49.5	Profile 2
3	115	4	8	1600	47.5	4	49.5	Profile 3
4	115	4	8	1600	49	6	49	Profile 4
5	115	8	4	1600	47.5	6	49	Profile 5
6	115	8	4	1600	49	4	49.5	Profile 6
7	115	8	8	800	47.5	6	49.5	Profile 7
8	115	8	8	800	49	4	49	Profile 8
9	121	4	4	1600	47.5	6	49.5	Profile 9
10	121	4	4	1600	49	4	49	Profile 10
11	121	4	8	800	47.5	6	49	Profile 11
12	121	4	8	800	49	4	49.5	Profile 12
13	121	8	4	800	47.5	4	49.5	Profile 13
14	121	8	4	800	49	6	49	Profile 14
15	121	8	8	1600	47.5	4	49	Profile 15
16	121	8	8	1600	49	6	49.5	Profile 16

where Y is the value of the geometrical parameters calculated. The coefficients are the effects of the factors. D_0', R_{D1}', ..., are the standardized variables corresponding to the natural variable.

For this, we use table L16 from Taguchi [6]. This table allows the study of seven factors with two levels, in 16 experiments. The different experiments are listed in Table 8.

7 Geometrical Parameters Calculated on the Numerical Part

Each experiment corresponds to a numerical simulation which the result is the "profile i" in Table 8. This profile is a file of points which are the nodes of the simulated formed part mesh. This file is post-treated in order to calculate the final height, the minimal thickness and the flange diameter. The different results are shown in Table 9.

8 Influence of the Process Parameters and Calculation of the Polynomial Models

8.1 Calculation of the Polynomial Equations

From the calculated geometrical parameters, the coefficients of the models are calculated. The following equations are obtained:

Table 9. Results of the different trials of the experimental design.

Profile	Diameter (mm)	Thickness (mm)	Height (mm)
Profile 1	65.61	0.6881	48.47
Profile 2	55.04	0.7501	50.58
Profile 3	58.05	0.7564	49.79
Profile 4	53.75	0.7691	50.63
Profile 5	58.71	0.7377	48.74
Profile 6	53.86	0.7500	50.51
Profile 7	53.86	0.7698	49.52
Profile 8	52.46	0.7695	50.55
Profile 9	70.76	0.7297	49.10
Profile 10	68.78	0.7494	50.01
Profile 11	68.56	0.7677	49.44
Profile 12	66.09	0.7581	51.10
Profile 13	71.08	0.7364	49.02
Profile 14	67.98	0.7385	49.85
Profile 15	68.88	0.7543	49.38
Profile 16	64.69	0.7669	50.87

For minimal thickness:

$$\begin{aligned}\text{Min\_thickness} = 0.749 + 6.38\cdot 10^{-4}\cdot D'_0 + 3.409\cdot 10^{-3}\cdot R'_{D1} \mathbf{+1.45\cdot 10^{-2}\cdot R'_{P1}} + 2.2\cdot 10^{-3}\cdot F' \\ \mathbf{+7\cdot 10^{-3}\cdot T'_2} + 4.2\cdot 10^{-3}\cdot R'_{D2} + 2.7\cdot 10^{-3}\cdot T'_3\end{aligned} \quad (2)$$

For part height:

$$\begin{aligned}\text{Height} = \mathbf{49.85} - 2\cdot 10^{-3}\cdot D'_0 - 4.3\cdot 10^{-2}\cdot R'_{D1} \mathbf{+3.1\cdot 10^{-1}\cdot R'_{P1}} + 3.1\cdot 10^{-2}\cdot F' \\ \mathbf{+6.67\cdot 10^{-1} T'_2} + 6.2\cdot 10^{-3}\cdot R'_{D2} \mathbf{+2.1\cdot 10^{-1}\cdot T'_3}\end{aligned} \quad (3)$$

For the flange diameter:

$$\begin{aligned}\text{Diameter} = \mathbf{62.38 + 5.97\cdot D'_0} - 0.94\cdot R'_{D1} - 1.6\cdot R'_{P1} - 0.2\cdot F' \\ - 2.1\cdot T'_2 - 0.72\cdot R'_{D2} - 0.71\cdot T'_3\end{aligned} \quad (4)$$

In these equations, the most influent process parameters are in bold characters.

8.2 Influence of the Process Parameters on the Part Geometry Variations

8.2.1 Influence on the Minimal Thickness

Figure 9 represents the influence of each parameter of the process on the thickness variation. The most influential parameter is the punch radius R_{P1}. Then, in the decrease order, we have the punch travel T_2, the die radius R_{D2} and R_{D1}.

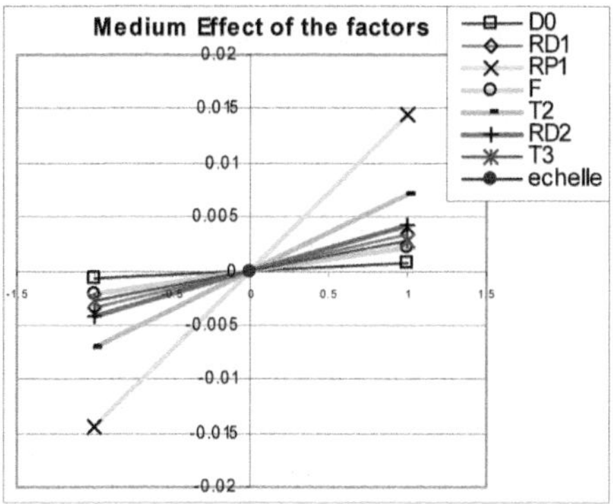

Fig. 9. Influence of each process parameters on the minimal thickness value.

8.2.2 *Influence on the Final Height of the Part*

Figure 10 represents the influence of each parameter on the variation of the part height.

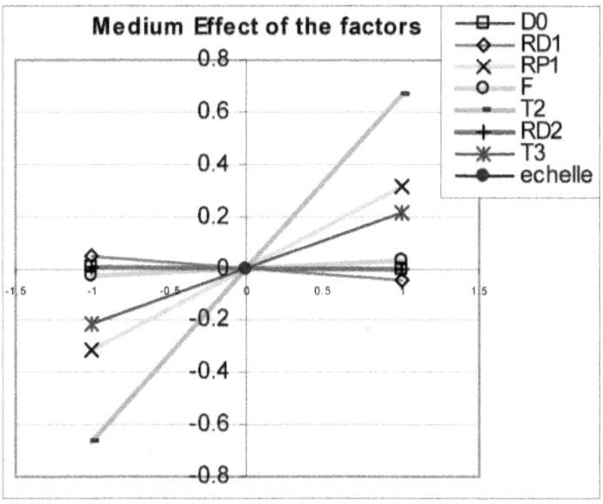

Fig. 10. Influence of process parameters on the final height of the part.

The most influent parameter on the final part height is the punch travel of stage 2 (T_2). The punch travel of the stage 3 is also an influent parameter. This result is logical because the part height depends directly of these travels. The punch radius is the second most important parameter. Indeed, when this radius increases, the length of the profile of the part decreases and allows to obtain a higher part.

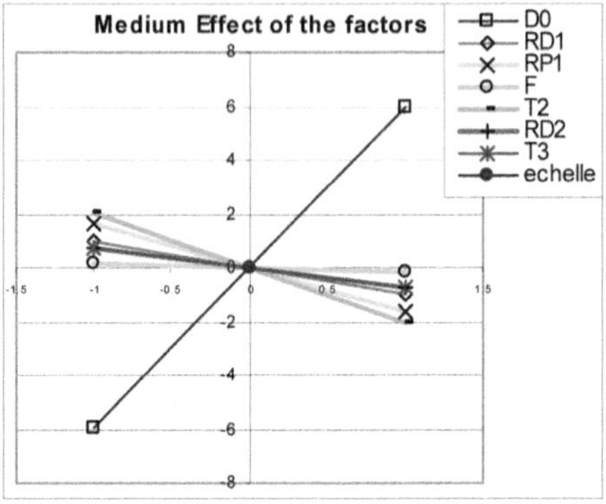

Fig. 11. Influent process parameters on the flange diameter.

8.2.3 Influence on the Flange Diameter

Figure 11 represents the influence of each process parameter in the variation of the flange diameter.

It is noticed that the value of the flange is directly influenced by the value of the initial diameter of the blank D_0. Moreover, an increase of the punch travel T_2 allows a decrease of this value. An increase of the punch radius R_{P1} reduces the dimension of the flange. The other factors have the same tendencies, with lower effects.

9 Optimization of the Process Parameters

The previous mathematical relations are used to determine the values of the process parameters which allow to obtain the best values for the geometrical part parameters, i.e. as close as possible to the target values: Min_thickness = 0.8 mm, Height = 50 mm and flange diameter = 56.6 mm. The principle consists in minimizing a function equal to the sum of the deviations between the theoretical and the wished values. In order to keep this sum positive, the deviations are squared:

$$L = (\text{Min\_thickness} - 0.8\text{ mm})^2 + (\text{Height} - 50\text{ mm})^2 + (\text{diameter} - 56.6\text{ mm})^2$$

The configuration of the process parameters which minimizes the function L is given in Table 10.

Table 10. Optimal combination of the process parameters.

D_0 (mm)	R_{D1} (mm)	R_{P1} (mm)	F (daN)	T_2 (mm)	R_{D2} (mm)	T_3 (mm)	L
116.63	8	8	1600	47.85	6	49.5	0.0012

Table 11. Comparison between calculated values with polynomial models and simulation.

	Geometrical parameters		
	Min_thickness (mm)	Final height (mm)	Flange diameter (mm)
Target	0.8000	50.00	56.60
Polynomial values for parameters	0.7725	50.00	56.60
Calcuated parameter from simulation	0.7680	50.04	55.50
Deviation between target and simulation	0.0320	−0.04	1.10

The optimal configuration of the tools seems to be logical because we find the values of the tools radii at their maximum values (in the aim to minimize to decrease of thickness). The dimension of the D_0 blank is close to the value estimated in Section 4.1 (116.38 mm). The corresponding simulation of this optimum is realized and the results are listed in Table 11.

The proposed optimization enables us to obtain a final part which has a final height very close to the target height (0.04 mm of deviation). The decrease of the thickness is about 0.032 mm. The final geometry of the part is obtained, but, we can notice that there is a bigger deviation between the target and the simulated flange diameter (1.1 mm).

We believe that a quadratic model would be more accurate for the flange diameter.

As the variation of the nominal blank diameter has an influence only on the flange diameter, it could be possible to slightly increase it without effect on the other part parameters.

10 Conclusion

In this paper, we proposed a method that allows the validation and the optimization of the stamping progression defined by industrial knowledge and empirical rules. The method is based on the numerical simulation of the deep drawing operation and design of experiment techniques. The aim of this study is to minimize the decrease of thickness along the different stamping steps and to obtain the final desired geometry.

First, the user has to define the process parameters which have an influence on the geometry of the part and for each of them their range of variation. The different targets of the part have to be clearly identified and their values must be defined. Then, the table of experiments is defined. Geometrical measurement is made on the numerical part and polynomial model for each geometrical parameter is calculated.

The optimal configuration of the process parameters is found and the numerical simulation validates it.

As the results show, the total optimization method proposed here, allows us to obtain the configuration of the process parameters directly, leading to a right part geometry and a minimal flange.

References

1. E. Pairel, Y. Ledoux, L. Tabourot, R. Arrieux: A method to determine relations between process conditions and draw part geometry. In *XVI International Scientific and Technological Conference on Design and Technology of Drawpieces and Die Stamping*, June 14–16, 2004, Poznan, Poland.
2. E. Pairel, Y. Ledoux, R. Arrieux, L. Tabourot, O. Incandela: Geometrical defects correction of stamping parts by numerical simulation and design of experiment. *Archives of Civil and Mechanical Engineering* **4**(4), 2004, 75–85.
3. S. Dumoulin, L. Tabourot, C. Chappuis, P. Vacher, R. Arrieux: Determination of the equivalent stress-equivalent strain relation of a sample of copper in tensile loading. *International Journal of Material Processing Technology* **133**, 2003, 79–83.
4. Hill, R.: A theory of the yielding and plastic flow of anisotropic metals. *Proceedings of the Royal Society of London* **A193**, 1948, 281–297.
5. Hibbit, Karlson and Sorensen: *ABAQUS/Explicit User's Manual*, Version 6.4, 2003.
6. M. Pillet: *Les plans d'experiences par la méthode Taguchi*, Ed les éditions d'organisation, 1997.

Validation Problems of Virtual Prototyping Systems Used in Foundry for Technology Optimization of Ductile Iron Castings

Zenon Ignaszak

CAD/CAE Laboratory of Material Technology (Institute of Material Technology), Poznań University of Technology, 5 Piotrowo street, 60-965 Poznań, Poland; E-mail: zenon.ignaszak@put.poznan.pl

Abstract. The most efficient method of designing and optimization for production processes in the field of material engineering is virtual prototyping (VP). In casting technology can be observed of these methods, in which the coupled physical models are used and their effective solutions are applied for complex technology cases for many years development. Effectiveness of virtual prototyping methods depends on many factors in pre- and post-processing stage. For the efficacy of casting's soundness prognosis, critical solid fraction for mass feeding and critical solid fraction for interdendritic (capillary) feeding are important parameters which describe alloy's feeding abilities. In virtual prototyping/manufacturing simulation systems dedicated to foundry industry, the models based on the physical rules and on corresponding equations, as well as empirical relationships from research works in real conditions are applied. The article concerns the problems of validation of thermal model describing the set of physical behaviours in real casting-mould system. This validation can be done by different ways including two kinds of information sources. The first source consists in observation and thermal parameter measurements in real time of casting process and the other one comes from testing and quality control results out of real casting process time. Chosen examples of validation are presented.

Key words: virtual prototyping/manufacturing systems, foundry, ductile iron castings.

1 Introduction. Description of Problem

Modelling of casting processes is mainly aimed at providing description of motion of the liquid/solid interface in alloy [1, 2] and forecasting quality of virtually formed castings, indicating especially the locations of discontinuities (porosities) [3]. Such a quality of prognosis is affected directly by parameters of calculation stages, so-called pre- and post-processing factors [4].

S. Tichkiewitch et al. (eds.), Advances in Integrated Design and Manufacturing in Mechanical Engineering II, 57–70.

Thermal theory of foundry processes describes the running of phenomena in the casting-mould system. Other hypothesis of individual phenomena and their combination are based on the theory of thermal exchange between casting and mould determining the metal solidification, according to heat transfer. Liquid-solid phase transformation is considered here by phenomenological way. In the known mathematical formulation, it is assumed that casting and mould sub-areas are isobaric; so their non-stationary temperature fields may be formulated using Fourier–Kirchhoff differential equation.

The models, meaning relationships (formulas) between the input and output parameters of object (casting-mould system) in the specific limit conditions (in the time and space domains) are formulated as differential equations. Although the progress of foundry process theory leads to the combination of majority of essential phenomena in the casting-mould system, the existing models are mainly based on a thermal model (heat energy transfer).

The successful creation trials and application of so-called coupled models are more and more frequent. These models connecting energy production and also heat transfer, mass transfer, mechanical phenomena state and their effects, are present in the theory and in the practice of foundry industry. These simulation systems, also called simulation codes (for example Magmasoft, Pamcast, Procast, Calcosoft, Simtec, NovaFlow&Solid), are actually widely used as the optimization tool for technological studies by virtual prototyping method.

One of the reasons that limit the development of these models to be useful in the casting-mould system control, is lack of parallel identification and studies of input and output parameters which are used in the modelling formulas, because these parameters are not always ready for effective application [5].

The simplest models often need the parameter identification to be exactly connected with final experiment. When the essential modelling equation is known, we are looking for the missing coefficients in the differential equations. It is easier if the static characteristics of modelling object are intercepted. It is more difficult if the using of dynamic characteristics is necessary. The Fourier–Kirchhoff equation being the thermal model based on casting solidification in a mould, making possible the phenomenological interpretation of processes, needs just the dynamic identification with the real object conditions [5, 6]. The article presents the problems of validation for selected applied model of shrinkage defects prognoses in the castings made of the ductile cast iron.

2 The Principles of Validation for Foundry Process Model

The validation of model, understood as the physical-mathematical engineering problem solved by means of numerical methods, can concern the preparatory stage, called pre-processing, as well as the procedures which serve for analysis of calculation results called post-processing. This validation is based on the effective adjustment of model components with parameters of process occurring in reality and connected with tests of different variants, which allows determining the margin of error [7].

After simulation calculations are completed (or as partial result are obtained – inter-actively), it is necessary to estimate the state of phenomena in casting. The solidification period requires the interpretation of thermal process effects with reference to temperature fields (local in the time-space coordinates). The final comparison of "calculated" state to experimentally identified "real" state, with possible model correction realized later, is just called the validation. This validation decides about usefulness of simulation system application to solve model and casting technology optimization. Only then it can be said that our simulation tool is ready to optimize the process in real manufacturing conditions (processes).

The condition absolutely required and necessary for the sake of diversity of materials used for mould manufacturing, is energy validation of model i.e. receiving for example the conformity between real and simulated solidification time of chosen casting walls [4]. Only this satisfactory conformity allows to conclude on the virtual solidification and feeding phenomena in the casting. The scheme of classification of information sources is shown in Figure 1.

It is supposed that the uncertainty of thermophysical material data (in preprocessing) is associated with lack of sufficient experimental data (sources of first group, Figure 1). This fact makes energy validation difficult. In this case it is concluded that only one possible estimation of simulation quality results from second group of information (Figure 1). In [1] it was shown that it is possible to obtain the precious I group information, using low cost methods of measurement applied for casting-mould systems.

3 The Validation Example by Inverse Problem Solution

According to the author researches [1, 4] on the sensibility problems of simulation calculations results and on the quality of material data, the importance of mould thermal coefficients is proved and the new sequence of the influence would then be as follows: $\lambda' > (c' \cdot \rho') > c > L > \rho > \lambda \cdot$ (λ, λ'-thermal conductivity, c, c'-specific heat, ρ, ρ'-density, L-latent heat, where $'$ signifies a sand mould). The quality of simulation results (in the example of cast steel casted in sand mould) depends most on the thermophysical parameters of moulding sand, due to its components, its thermal instability and the relations of the moulding sand state with the conditions of its preparation and use, confirmed by another study by the author [6].

In [1] the author proposed a scheme of connections between measuring-calculating procedures related to the obtainment of material data, realized generally in laboratory conditions using procedures with transformed formula of direct methods. In the case of thermal casting processes such a procedure may be called a numerical identification of thermophysical data. In the procedure simplification of physical model, temperature-time conditions, simplifications of geometrical model

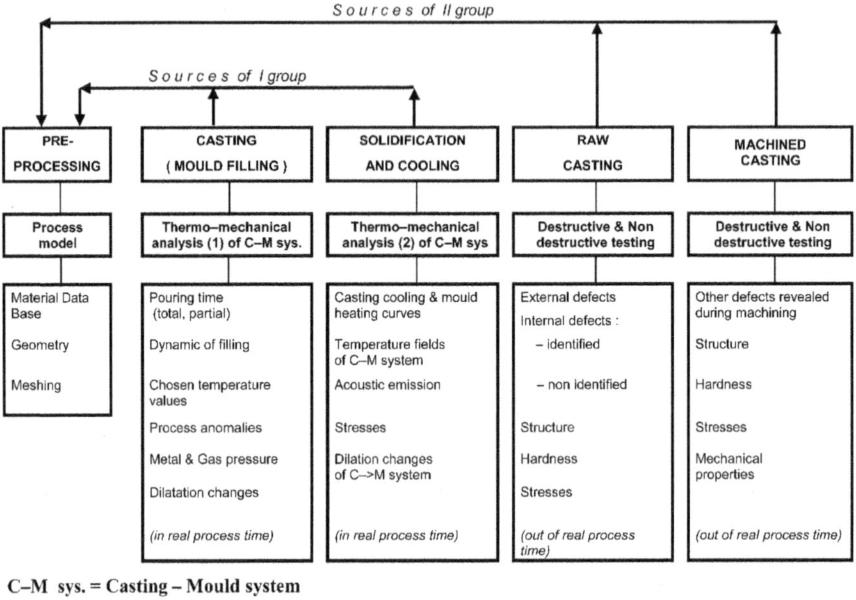

Fig. 1. The scheme of casting model validation. Sources and transfer of information in casting production.

and mesh design are important to be taken into account. The numerical method of identification is an optimization, realized by means of iteration loops, providing minimization of the *objective function*. Validation of material data determined in this way would mean at the same time validation of the whole model (comparison of the results of simulation of real casting). The example of inverse problem solution for cast iron solidification experiment in multi-cavity moulds of furanic sand is presented in Figure 2. The excellent conformity of virtual and real results proves that so-called energy validation is accomplished for this case.

4 The Validation Example of Feeding Parameters during Solidification – Small Castings

The influence of pouring temperature, critical solid fraction parameters and alloy's contraction on casting's soundness prognosis – "shrinkage" parameter have been analyzed. Especially the influence of critical solid fraction parameters for conformity of experiments and simulation results were tested [9]. Classical designing of casting feeding, based on empirical knowledge and realized form many years, consists in defining feeding zones within the whole casting: according to required functions:

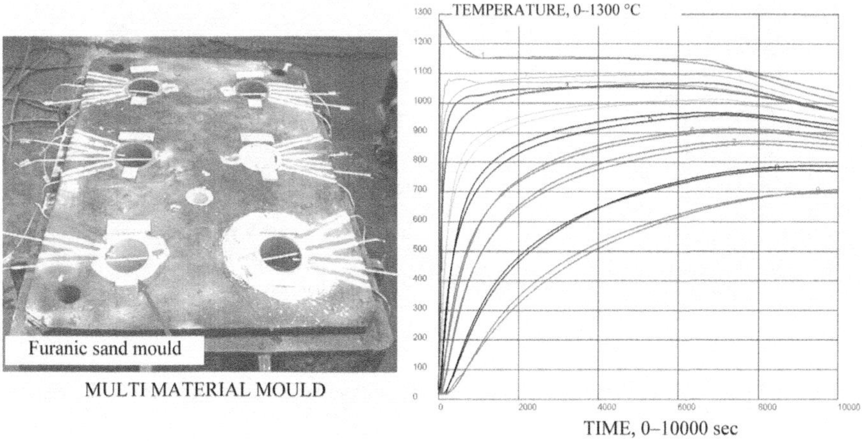

Fig. 2. Experimental investigation of the multi-cavity mould-casting system (single cavity – $\phi 175 \times 900$) by thermal analysis (left) and very satisfactory conformity of temperature T [°C] between measured and virtual (optimized) curves $T = f$ (time) obtained by inverse problem solution using Calcosoft-system. Results for furanic sand – $c \cdot \rho = 0.161E + 07$, $\lambda = 0.106E + 00$, alloy poured: ductile cast iron poured in the cavities (right).

riser–neck–hot spot–connected walls–parasitic walls and balancing analysis of solidification. The analysis includes:

- directional orientation of solidification process in the riser direction, with safe increase in wall modulus of directly feeding wall, about 20%;
- division of the feeding zones of casting among particular risers (e.g. with possible use of chills);
- the distance of under-riser feeding and the presence of natural and artificial boundary zones;
- the balance (for the whole casting) of demand for liquid metal compensating shrinkage of the alloy feeding, in comparison to feeding abilities of the riser (the kind of the riser and FEM coefficient of the risers insulation).

Such a way does not allow to conduct the analysis of the dynamic phenomena of feeding at particular stages of the casting solidification process. It is still a basis for the first technological design – a technological conception of the casting process. It should be noticed that in case of designing the feeding of cast iron castings (especially ductile cast iron) an advantageous compensation of the matrix shrinkage by graphite should be considered, together with strong thermo-mechanical interaction between graphite expansion and mould (with its stiffness).

In this classical procedure, it is impossible to make use of commonly applied parameters characterizing effectiveness of the feeding, and used in the simulation forecasts with regard to the morphology of the alloy solidification and flow (i.e. crit-

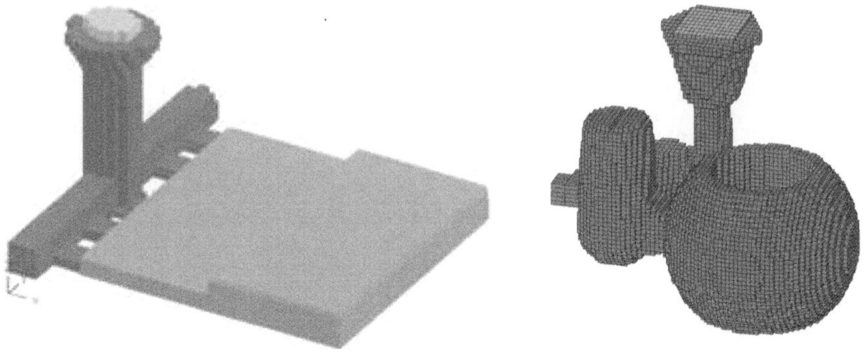

Fig. 3. CAD geometry of the reference plate casting and valve ball casting after FVM meshing.

ical values of solid phase fractions: f_1 and f_2, density variations due to temperature (shrinkage coefficient S), basic thermophysical parameters of the metal and the mould materials) are not considered in classical designing process.

The experimental research has comprised execution of two casting groups: a reference plate casting and a valve ball casting made of a ductile iron EN-GJS-450-10 (Figure 3). In order to assess varying pouring conditions in the industrial casting process of a valve ball, a set of ten castings has been manufactured under varying pouring temperatures (T_{ini} from 1411 to 1305 [°C]).

The aim of simulation calculations was to indicate the conditions of validation for the whole model applied in PamCast 2001 [11], for the sake of compared forecasted and real shrinkage defects. Material data of ductile iron have been adopted in accordance with the PamCast2001 database (ductile cast iron GGG40).

In case of simulation forecasts of the shrinkage defects the following parameters have been a priori defined: alloy shrinkage S, critical solid fraction of mass feeding f_1, critical solid fraction of interdendritic feeding f_2, and initial temperature T_{ini}. The simulation has been made for the following shrinkage coefficient values S: 1, 2 and 3 [%], with values in two of critical solid fractions $f_1 = 0.1$ and $f_2 = 0.3$, $f_1 = 0.4$ and $f_2 = 0.6$, $f_1 = 0.7$ and $f_2 = 0.9$, $f_1 = 0.1$ and $f_2 = 0.5$, $f_1 = 0.3$ and $f_2 = 0.7$, $f_1 = 0.5$ and $f_2 = 0.9$. In this way different cases of connections between solidification morphology and flow abilities in feeding zone were revealed.

Among parameters available in the post-processing stage of calculations in PamCast 2001 particular attention was paid to a classical "shrinkage" parameter and to temperature gradient G, cooling rate R and Niyama criterion N (calculated in selected solidification time points) [9].

Comparison of the distribution of the location and intensity in real and virtual castings (valve ball) has indicated that the values $f_1 = 0.1$ and $f_2 = 0.3$ (Figure 4) describe the properties of considered ductile iron the best. The values are lower than proposed for this ductile iron in PamCast 2001, and in other simulation systems.

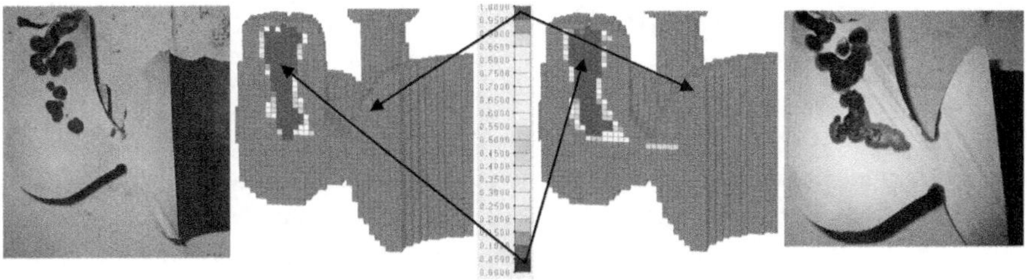

Fig. 4. Shrinkage defect in real and simulated valve ball casting: $T_{\text{ini}} = 1257$, $S = 2\%$, $f_1 = 0.1$, $f_2 = 0.3$ (left) and $T_{\text{ini}} = 1307$, $S = 2\%$, $f_1 = 0.1$, $f_2 = 0.3$ (right).

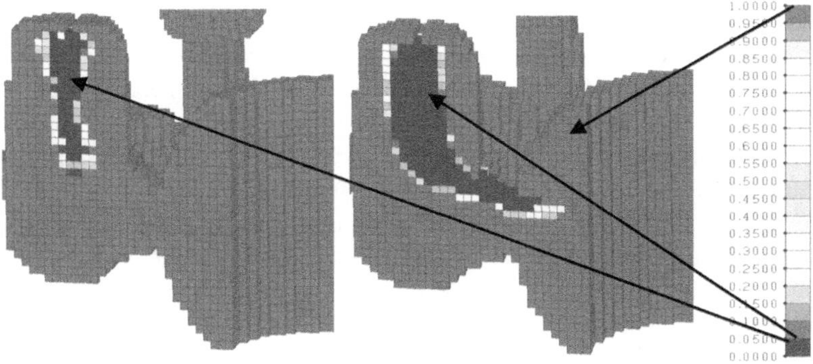

Fig. 5. Shrinkage defect for: $T_{\text{ini}} = 1257$, $S = 1\%$, $f_1 = 0.1$, $f_2 = 0.3$ (left) and $T_{\text{ini}} = 1307$, $S = 3\%$, $f_1 = 0.7$, $f_2 = 0.9$ (right).

Moreover, the investigation showed that the values $S = 1$ and 2% are appropriate, and it has confirmed other opinions on this subject. The comparison of simulation results calculated with different and extreme parameters is shown in Figure 5.

It was proved that the results of simulation forecasts considerably but also diversely affect the parameters that determine the alloy flow abilities during mass (f_1) and interdendritic (f_2) feeding, temperature of the alloy after mold pouring and alloy shrinkage. These factors decide about the location, intensity, and a shape of the forecasted shrinkage defect. In order to perform effective computation aimed at the prognosis of the casting quality (shrinkage defects), previous validation of the parameters affecting the shrinkage defects is necessary. While testing the range of this influence with reference to real castings, it has been affirmed, that border values of critical solid fraction for mass and interdendritic feeding, which gave the best forecasting of shrinkage, are significantly lower than these recommended in materials data bases of simulation systems, also in PamCast 2001. In order to determine these values for another alloys and castings, it is necessary to lead to dynamic temperature validation of thermal model applied in simulation system [1].

5 The Validation Example Concerning Medium Castings Soundness

The compactness of cast products is defined as a feature of their structure consisting in adequate packing of crystallized metallic and non metallic phases obtained by individual foundry technologies [1]. The compactness may concern the whole casting or the analyzed part of casting. It should be emphasized that the compactness of a given alloy depends on the type of technology and the conditions of its crystallization and solidification, i.e. the so-called its morphology. The crystallization and solidification processes thus determine the diversity of distribution and intensity of material lack in the casting volume, called discontinuities. The condition close to perfect compactness characterized by close to zero occurrence of discontinuities in the structure is relatively easier to obtain in the case of castings (or rather ingots) which are subject to routine plastic processing in order to obtain commercial products such as: rods, plates, steel sections, pipes, etc. Among foundry technologies a significantly smaller number of discontinuities is achieved by applying pressure casting (HPDC) and squeeze casting.

As it was underlined in advance, in simulation codes from the Virtual Prototyping/Manufacturing Group, dedicated to the foundry industry, the models based on the physical rules and corresponding equations, as well as empirical relationships from research works in real conditions are applied.

The access of a foundryman technologist into the construction of these algorithms, calculated in a numerical method, is absolutely impossible. During the preparatory stage (pre-processing) he defines the technological geometry of the casting-mould system and selects the necessary parameters from material and technological database, the quality of which will be important in simulating calculations [13].

The simulation systems that enable places in a casting threatened with the occurrence of discontinuities to be indicated are based on the heat transfer model described by the Fourier–Kirchhoff equation. However, a direct interpretation of the results obtained from this model solution (as temperature fields) does not yet provide sufficient basis for formulating the conclusions concerning the localization of shrinkage discontinuities. The practical usefulness of individual systems is evidenced by their conformity with experiments. Hence the validation of the basic model and the interfacing auxiliary models requires experimental tests and their comparison with the prognosis of shrinkage discontinuities location and their dispersal.

Experimental tests consisting in the casting of four plates of ductile cast iron EN-GSJ-400-15, were performed. These plates had different thickness: 15, 30, 60 and 100 mm. The castings were poured through one gating system with the purpose of comparing the impact of wall thickness on the distribution of shrinkage defects in cast iron with the same composition. Schema of castings and its gating system in the form of a CAD drawing and industrial casting with the EN-GSJ-460-12 poured simultaneously, are shown in Figure 6.

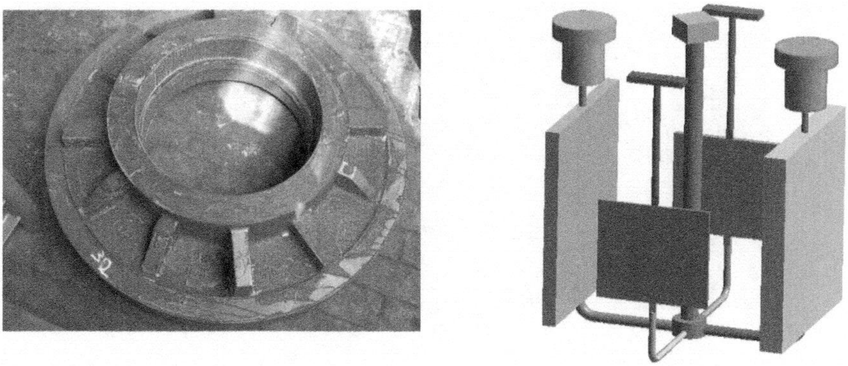

Fig. 6. Ductile cast iron industrial casting (cover, 1300 kg) and CAD drawing of test plate castings used to parallel study.

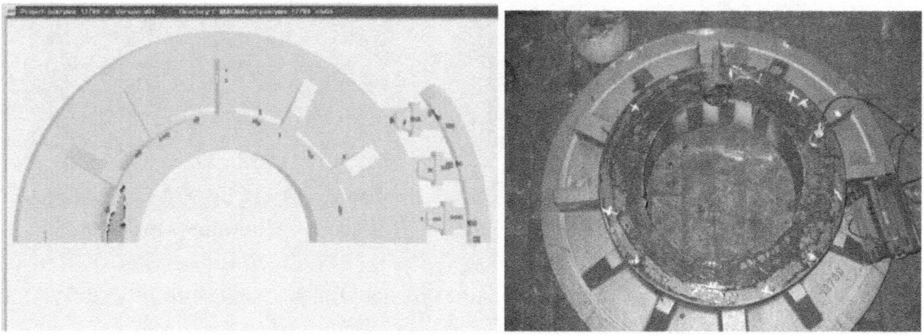

Fig. 7. Comparison of defects location in virtual (from Magmasoft) and real ductile cast iron industrial casting (detection by ultrasonic testing – UT).

The simulation system Magmasoft is used for optimizing the casting technologies by enabling a selection from different versions of technological concepts concerning the same casting to be made. The virtual prognosis of both the location and intensity of shrinkage discontinuities determines the final selection of technology. In Magmasoft system the crystallization heat generation model is based on empirical dependences that describe amount of active nuclei of crystallized phases (austenite and graphite) and nucleation/growth speed of phases as a function of local overcooling (micro-modelization), that is considered as the best virtual prototyping procedure for ductile cast iron by the creators of this VP system.

A relatively good conformity between the distribution of discontinuity defects in industrial casting (cover) according to virtual prognosis by Magmasoft system with real map of these small discontinuities obtained by UT method is shown in Figure 7.

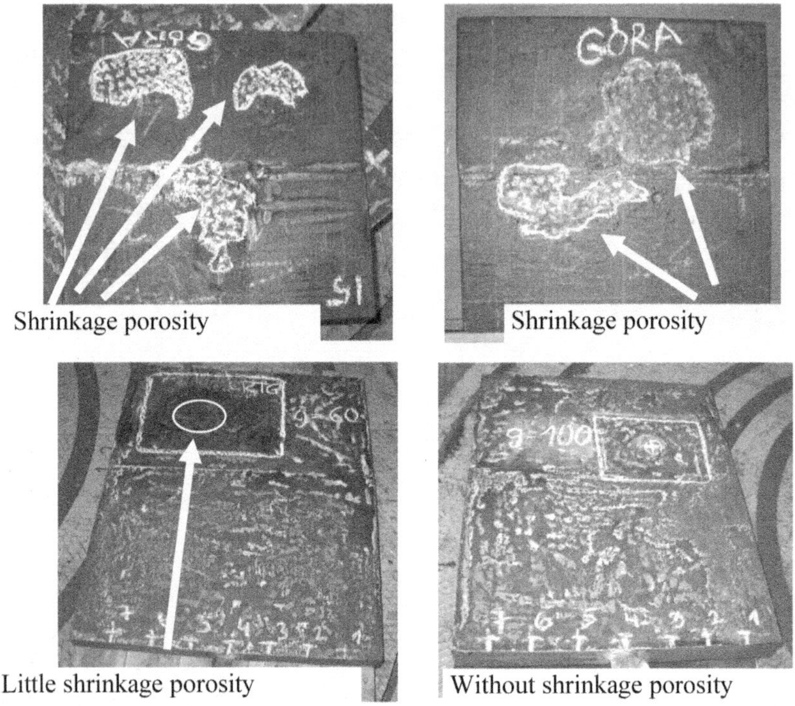

Fig. 8. Real location of shrinkage discontinuities for set of plate castings: from 15 to 100 mm (photos presented in pouring position).

The virtual shrinkage porosities are indicated exactly in regions of casting called hot spots (final solidification sub-volumes).

Other parallel test revealed that the shrinkage defects in plates 15 and 30 mm are located beyond the lower end zone (equal to 2.5 times the thickness of plate). In the case of the 60 mm thick plate, the shrinkage defect only occurred in its upper part (the region over 2/3 of height). In the 100 mm plate no shrinkage defects were detected (Figure 8). The so-called pipe defect in the upper casting area was not found in any of the castings.

The models mentioned above were also applied for simulating the mould filling and solidification (micro-model)of plate castings, and subsequently forecasting shrinkage defects in plates [16]. The results are shown in Figure 9. The comparison of the results obtained in the experiment and simulation tests clearly revealed that maps of shrinkage defects did not overlap.

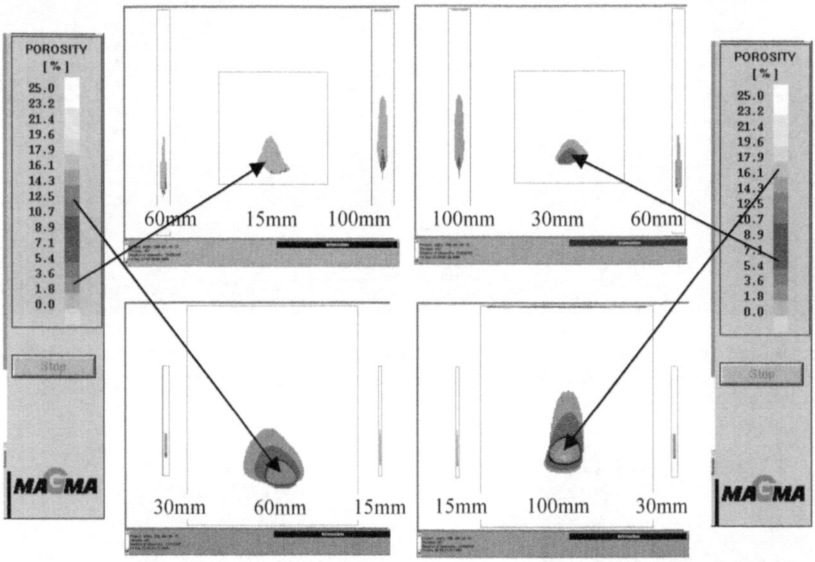

Fig. 9. Prognosis of shrinkage discontinuities by Magmasoft system (with module called Iron) for set of plate castings: from 15 to 100 mm.

6 Conclusions

It is well known that physical grounds of the model for shrinkage defects (discontinuities) determination, coupled with thermal model, are based on temperature distribution during solidification and balance of feeding flow of liquid metal depending on dynamism of shrinkage needs.

Three examples of validation procedures were presented to show the different ways of improvement the effectiveness of VP systems designed for foundry processes, as regards its ability to meet the expectations concerning the accuracy of forecasting shrinkage discontinuity. The first step should always be the heat balance validation called energy validation [1] that conditions next validation procedures. Therefore, the effective methods for the model validation for the sake of critical solid fractions f_1 and f_2 are described in this paper.

The prognosis of these defects in the trial and industrial castings with diversified geometry and weight, as examples, was shown. Dissatisfaction with validation results for trial plate castings (poured as tests) permits to conclude that for industrial ductile cast iron castings (examples in Figure 10) the shrinkage defect prognoses cannot be correct.

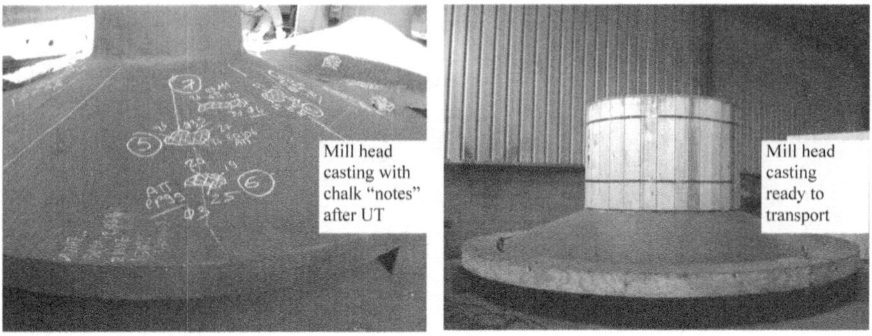

Fig. 10. Example of real discontinuity location in the certain regions and view of complete the big ductile iron casting with wide plate part (mill head, 4 m in external diameter).

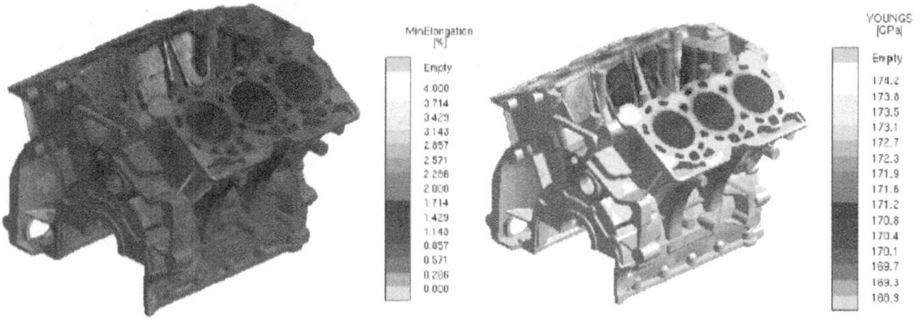

Fig. 11. Distribution of minimum elongation (left) and Young's modulus (right) in the casting (from Magmasoft). The mechanical properties correspond with the microstructure distribution (ferrite; perlite) [17].

The validation methods and their results should be adapted to local specificity of foundry conditions and on the other hand they should take into account the simplifications existing in the mathematical-physical models that the user should be aware.

More validation problems are expected with the introduction of new coupled physical models to simulation practice. The synergy of universities, laboratories and industry workings are required. For examples, the new tendency to introduce the map of local mechanic characteristics and map of residual stresses to simulate and to forecast/optimize the exploitation performances (Figures 11 and 12) are observed.

Moreover, the information about the manufacturing process virtualized by casting simulation, and the transfer of the critical values such as residual stress or an inhomogeneous Young's modulus distribution from casting simulation to FE analysis (as for example: Abaqus or Ansys – FEM systems) is really possible [1, 17, 18].

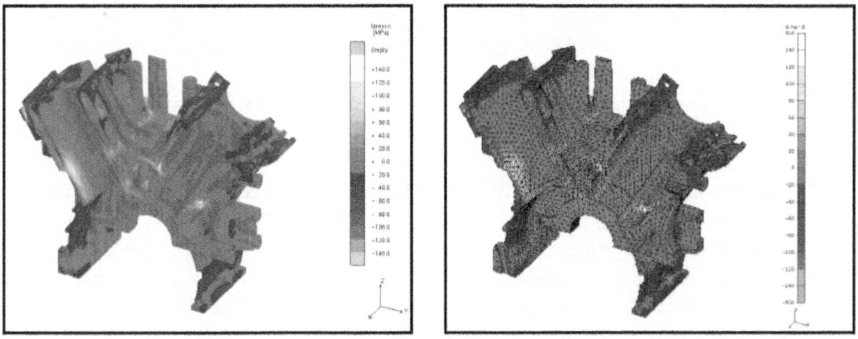

Fig. 12. Residual stress distribution in the chosen section of the raw casting. The values were transferred (mapped) from Magmasoft (left) to an unstructured FE mesh (right) [17].

In the future, the low grade of integration observed in practice for most of the casting design processes will be more caused still by the low level of integration between foundry departments and people, who will be involved in design processes, then by missing links between the used CAE tools [17].

The rationalization of concluding about casting quality, in respect of shrinkage/porosity defects (soundness), and local mechanical characteristics, should take into consideration the quality of the described (based on the author's proper university and industrial experiences [1]) and other validation works.

"*These models are wonderful tools for understanding and learning processes, microstructures, defects, but they are still limited in practical application by a lack of reliable data and validations. The combination of simulation and experiment, done in close collaboration between industrial and industrial and academic partners, should overcome actual limitations and bring these tools to the next step*" (citation from [19]).

References

1. Z. Ignaszak: *Virtual Prototyping in Foundry. Data Bases and Validation*, Politechnika Poznańska, Poznań, 2002 [in Polish].
2. Z. Ignaszak: Energetic and dynamic validation of thermal models in practice of casting technology simulation, in *Proceedings of the International Conference on Advances in Materials and Processing Technologies (AMPT2003)*, Dublin, July 8–11, 2003.
3. J.B. Prunier, Z. Ignaszak: Simulation de remplissage et de solidification. Optimisation d'une traverse de bien d'équipement en EN-GJS-500-7, en collaboration avec le client, par simulation de remplissage et de solidification, *Hommes et Fonderie* **296**, 1999, 34–37.
4. Z. Ignaszak, N. Hueber: Sensibilité du modèle de simulation numérique à la qualité des données thermophysiques. *Homme et Fonderie* **318**, Paris, 2001, 41–45.

5. Z. Ignaszak: La conductivité thermique substitutive du moule. Une nouvelle méthode de mesure pour la simulation de la solidification des pièces, *Fonderie – Fondeur d'Aujourd'hui* **121**, 1993.
6. Z. Ignaszak: *Thermophysical Properties of Mould Materials in Aspect of Solidification Control of Simulation Process*, Poznań University of Technology, Rozprawy nr 211, Poznań, 1989 [in Polish].
7. N. Hueber, Z. Ignaszak: Water effects and thermal behaviour of a sand, thermal properties identification by optimization tools, in *Proceedings of VIII Conference Modelling of Casting, Welding and Advanced Solidification Processes*, San Diego (USA), 1998.
8. Z. Ignaszak: *Etudes expérimentales des matériaux isolants et iso-exothermiques utilisés comme manchons et plaquettes pour le masselottage*, Rapports, Groupe Métallurgique Ferry-Capitain, Joinville , France, 2000–2004.
9. Z. Ignaszak, P. Mikołajczak: Analysis of the influence of alloys feeding parameters during solidification on soundness simulation forecast in ductile iron castings, in *Proceedings of COMMENT – Worldwide Congress of Materials and Manufacturing Engineering and Technology*, 16–19 May 2005, Gliwice-Wisla, Poland.
10. Z. Ignaszak, P. Mikołajczak: *Validation des procedures de post-processing pour la prevision des defauts de retrait utilisees dans PamCast 2001*, Report for ESI Group, PUT, Poznań, 2004.
11. PamCast: 2001, Foundry Simulation Software – ESI Group, www.esi-group.com.
12. Z. Ignaszak, J. Ciesiółka: *Application of Non-Destructive Testing and Virtual Prototyping Systems to Identify the Homogeneity and Compactness of Structure in Castings*, Archiwum Technologii Maszyn i Automatyzacji, Poznań, 2005 [in Polish].
13. Ch. Pequet, M. Gremaud, M. Rappaz: *Modelling of Microporosity, Macroporosity and Pipe Shrinkage Formation during Solidification of Alloys Using a Mushy-Zone Refinement Method*, Manuel Edited by Ecole Polytechnique Fédérale de Lausanne and Calcom SA, Lausanne, Switzerland.
14. Z. Ignaszak, B. Beszterda, J. Ciesiółka: Test of algorithm used by magmasoft system for discontinuity prognosis in ductile iron castings, Paper presented at *II VIDA Conference*, 2005.
15. Magmasoft: Casting Process Simulation, CD (demonstration), Aachen, 2004.
16. B. Beszterda, Z.Ignaszak: Simulation of plate casting pouring and solidification, Internal Report for SREM Foundry, 2005 (not published).
17. A. Egner-Walter, G. Hartmann, M. Kothen: Integration of manufacturing process simulation in to the process chain, in *20th CAD–FEM Users' Meeting 2002*, International Congress on FEM Technology, 9–11 October 2002, Friedrichshafen, Germany.
18. R. Skoczylas, M. Cieplak: Integrierung der Gießimulationsresultate in Programme für Engineering – Analysen, Article in Magmasoft – Casting Process Simulation, CD (demonstration), courtesy of KomOdlew, Krakow, 2005.
19. M. Rappaz: Modelling of microstructure and defect formation in castings, in *Proceedings of 66th World Foundry Congress*, Istanbul, September, 2004.

Simulation of Forming Processes, with Friction, Coupling Finite Elements and Boundary Elements

Dominique Bigot, Hocine Kebir and Jean-Marc Roelandt

Laboratoire Roberval, Université de Technologie de Compiègne, BP 20529, 60205 Compiègne Cedex, France;
E-mail: {dominique.bigot, hocine.kebir, jean-marc.roelandt}@utc.fr

Abstract. Nowadays, the simulation of forming process is rather well integrated in the industrial numerical codes. However, in order to take into account the possible modifications of the tool during rates of working, a dedicated numerical software is currently developed within the Laboratory Roberval. This one more uses coupling finite elements and boundary elements for the respective modelling of the part and the tool. This software will more particularly allow the identification of the fatigue criteria of the tool. The optimal shapes of tools could also be conceived for increase their lifespan and the manufactured part could have the required quality.

Key words: coupling finite elements/boundary elements, penalty, forming process, oriented object programming.

1 Introduction

In the industry, the simulation of the manufacturing processes is now largely used. It became an unavoidable tool in the design of the product to improve the output and quality of the parts.

The quality of the parts obtained by various industrial processes (forming, cutting, etc.) depends on:

- The design of the process itself (e.g., geometry of the tools) whose parameters depend in general on material and the geometry (e.g. thickness of the blank outline, in general.
- The fatigue of the tools during cycles of application of the process. Indeed, this one can generate manufacturing fault (geometry, surface quality, damage, etc.).

The simulation of the processes is rather well integrated in the industrial codes. However, the most current approach consists in regarding the tools of the process as rigid,

S. Tichkiewitch et al. (eds.), Advances in Integrated Design and Manufacturing in Mechanical Engineering II, 71–78.

which does not make it possible to correctly envisage their fatigue and their impact on the finished product.

In order to take into account the possible modifications of the tool during rates of working, a dedicated numerical sofware is currently developed. This one allows more particularly the identification of the fatigue criteria of the tool. From this identification, it will be possible to conceive the optimal shape of the tool in order to increase its lifespan and to ensure a required quality of the part which can also be manufactured.

Our numerical tool allows us to simulate the manufacturing processes (machining, cutting, stamping, etc.). Since the characteristics of each component are different for this manufacturing process, coupling finite elements and boundary elements have been chosen as follows:

- The part is modelled with finite elements. Indeed, the part is subjected to large deformations, like with non-linearity (materials and geometrical).
- The tool is modelled with boundary elements [1] as it is subjected to small deformations. The boundary elements allow to mesh easier or include removal material to optimization of the tool.

Thus the contact [2–5] is done by using these coupling finite elements and boundary elements.

This model is different from those existing since it couples a part, discretized by finite elements (non-linearity: geometrical and material), with a non-rigid tool discretized by boundary elements. This software which couples several independent modules is programmed in Oriented Object, in order to use all the benefits of this language (encapsulation, heritage, integrity of the data, etc.).

One of the aims of the software is to optimize the parameters in order to maximize the lifespan of the tools and to ensure a design quality to the product.

2 Formulations

2.1 Boundary Elements

The principle of the boundary elements method consists in transforming integral equations on contour. In linear elasticity, the study of a part consists in studying his surface behaviour. The discretization transforms the integral equation into a system of linear equations and determines displacements in constraints inside the part using simple integral relations.

In our study, the direct method of the integral equations is used. The formulation into elastostatic is based on the theorem of reciprocity of Maxwell–Betti [6]. In the absence of the forces of volumes and for a point P, the integral equation in displacement is written:

$$C_{ij}(P)u_j(P) + \int_{\partial\Omega} T_{ij}(P,Q)u_j(Q)ds(Q) = \int_{\partial\Omega} u_{ij}(P,Q)t_j(Q)ds(Q), \quad (1)$$

where $u_{ij}(P,Q)$ and $T_{ij}(P,Q)$ are the fundamental solutions of Kelvin and $C_{ij}(P)$ a coefficient which depends on the discontinuity of the normal out of P. $u_j(Q)$ and $t_j(Q)$, called displacement and tension of the point Q pertaining to the contour $\partial\Omega$ of the field, are the unknown factors of the problem.

After discretization of the integral equation (1) in displacement, the following system is obtained:

$$[A]\{u\} = [B]\{t\}, \quad (2)$$

where $\{u\}$ is the field of displacement and $\{t\}$ is the field of tension.

This system (2) can be organized in the form $[K]\{x\} = \{y\}$ where $\{x\}$ (respectively $\{y\}$) is the vector of unknown displacements and tensions (respectively known). The matrix $[K]$ of the preceding system is a full matrix, non-symmetrical and with dominant diagonal, which is a notable difference compared to the finite element methods, for which the matrix of rigidity is symmetrical and bandages.

2.2 Finite Elements

The finite element method is a numerical method used in the calculation of the structures, in particular in nonlinear mechanics. Owing to the principle of virtual work, the integral equation in displacement is:

$$\int_{V_t} \langle D^* \rangle \{\sigma\} dV - \int_{V_t} \langle u^* \rangle \{f_V\} dV - \int_{S_t} \langle u^* \rangle \{f_s\} dS = 0 \quad \forall \{u^*\}, \quad (3)$$

where $\langle D^* \rangle$ is the virtual rate of deformation.

In order to solve the following system, an iterative process of Newton–Raphson is used:

$$[K_T]\{\Delta u_n\} = \{F_{\text{ext}}\} - \{F_{\text{int}}\}. \quad (4)$$

3 Coupling Finite Elements/Boundary Elements

Two two-dimensional examples are treated and for each of them the quadrangle finite elements (respectively the linear boundary elements), in plane strain, were chosen to model the part (respectively the tool). The tool is viewed as "Master" and the part as "Slave". The master segments S_k of the tool and the slaves nodes N_k of the part are considered [7].

The couple, master segments of the tool and slaves nodes of the part, is identified as being into contact. Then, an algorithm (Figure 1) which converges when the conditions of contact are checked, is applied.

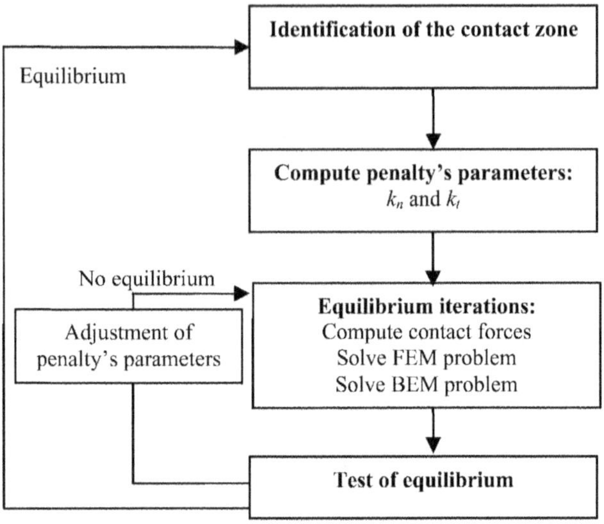

Fig. 1. Contact algorithm.

3.1 Identification of the Contact Zone

For each step, the contact zone is identified and the couple segments and nodes considered in contact are sought. This couple is validated when $X_n \leq 0$, $\overrightarrow{S_1N} \cdot \overrightarrow{S_1S_2} \geq 0$ and $\overrightarrow{S_2N} \cdot \overrightarrow{S_2S_1} \geq 0$.

For each couple Node–Segment, the distance of the penetration X_n and the distance of slip t_1 are calculated (Figure 2).

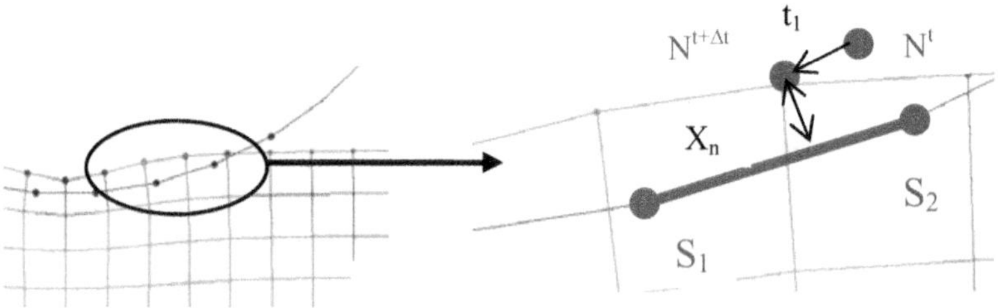

Fig. 2. Identification of the contact zone.

3.2 Compute the Penalty's Parameters

For each node of the part, the normal penalty's parameter k_n is computed. For the first step, teh average of the diagonal coefficients of the stiffness matrix associated to the contact nodes is calculated. For the other steps, the contact forces per node are summed and this sum is divided by the displacement of the last step. The tangential penalty's parameter k_n is computed with $k_t = \mu \cdot k_n$ where μ is the friction coefficient.

3.3 Compute the Contact Force

An iteration procedure is used in order to obtain the equilibrium in displacement with contact forces in coupling the Boundary Elements Method system and Finite Elements Method system. A penalty method [7, 8] is used to compute the normal contact force F_c^n associated with the couple nodes/segments:

$$F_c^n = -\Phi(X_n) \cdot k_n,$$

where X_n is the distance of penetration, k_n is the normal penalty's parameter and Φ is a function of regulation for the normal contact force F_c^n (Figure 3). Thus, the derivative is continuous on all the domain of definition, which improves the change of statute of "no contact-contact". This function Φ is defined by:

$$\Phi(X_n) = \begin{cases} X_n & X_n \leq -2\varepsilon \\ \dfrac{(X_n+\varepsilon)^2}{4\varepsilon} - \dfrac{(X_n+\varepsilon)}{2} + \dfrac{\varepsilon}{4} & -2\varepsilon \leq X_n \leq 0 \\ -\left(\dfrac{(X_n+\varepsilon)^2}{4\varepsilon} - \dfrac{(X_n+\varepsilon)}{2} + \dfrac{\varepsilon}{4}\right) & 2\varepsilon \geq X_n \geq 0 \\ -X_n & X_n \geq 2\varepsilon \end{cases}$$

The Coulomb friction model is taking into account in the computation of the tangential contact force F_c^t:

$$F_c^t = \begin{cases} r_t & \text{(adherent contact)} \\ -\mu |F_c^n| \dfrac{t_1}{|t_1|} \text{ and } k_t = \mu \cdot k_n \cdot \left|\dfrac{x_n}{t_1}\right| \text{ if } |r_t| \geq \mu |F_c^n| & \text{(slipping contact)} \end{cases}$$

where $r_t = -k_t \cdot t_1$, μ is the friction coefficient and k_t is the tangential penalty's parameter.

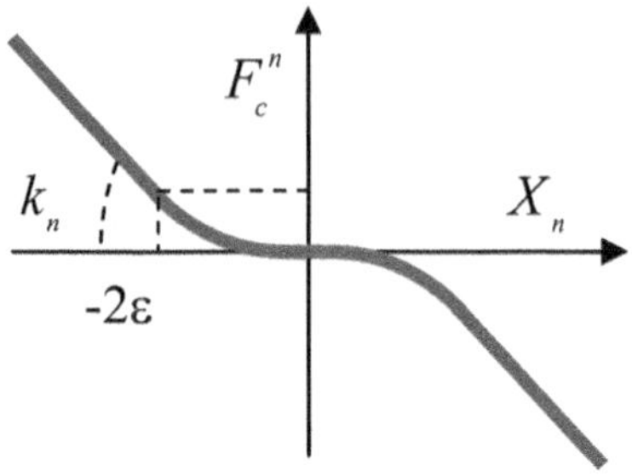

Fig. 3. Function Φ.

3.4 Solve the Systems and Test of Equilibrium

The displacement of the part and the tool is done independently by the resolution of two different systems:

For the part, finite elements lead to solve:

$$[K_T]\{\Delta u_n\} = \{F_{\text{ext}}\} - \{F_{\text{int}}\} + \{F_{\text{contact}}\}$$

and for the tool, the boundary elements lead to solve:

$$[A]\{u\} = [B](\{t\} + \{t_{\text{contact}}\}).$$

After the resolution of these ones, the new distance of the penetration X_n is calculated for testing the equilibrium. If $|X_n| < \varepsilon_1$, the equilibrium is reached and we can pass to the following step.

3.5 Automatic Adjustment of the Normal Penalty's Parameter

If the test of equilibrium is satisfied, the algorithm loops on the next step, otherwise, the normal penalty's parameter is modified for another iteration. The algorithm is inspired on the adjustment of the normal penalty's parameter according to the precision required by the penetration [9].

If the distance of penetration decreased too slowly, for example

$$\left|\frac{X_n^{t+\Delta t} - X_n^t}{X_n^t}\right| < \varepsilon_2,$$

k increased.

If the distance of penetration increased too fast, for example

$$\left|\frac{X_n^{t+\Delta t} - X_n^t}{X_n^t}\right| > \varepsilon_3,$$

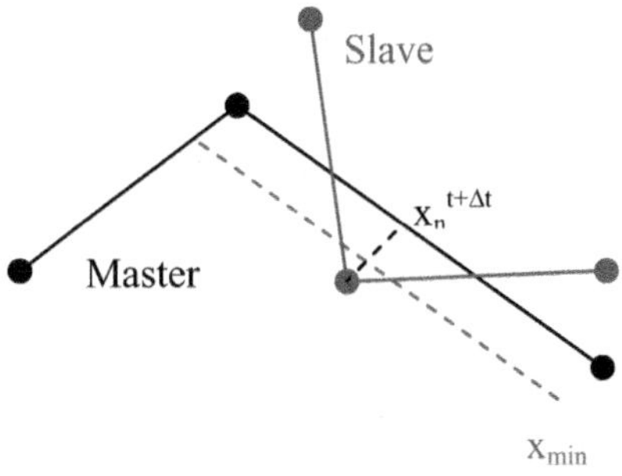

Fig. 4. Adjustment of penalty's parameter.

k decreased. Otherwise, k remained identical.

Numerically, the following penalty's parameters are used (see Figure 4):

$$k_n \rightarrow \Psi(X_n^{t+\Delta t}, x_{\text{min}}) \cdot k_n \quad \text{and} \quad k_t \rightarrow \mu \cdot k_n,$$

where

$$\Psi(X_n^{t+\Delta t}, x_{\text{min}}) = \frac{|X_n^{t+\Delta t}|}{x_{\text{min}}}.$$

4 Results

For the validation of the coupling finite elements/boundary elements with friction, the contact of a part with a cylindrical tool was modelled. Two two-dimensional tests were carried out: the first without friction, and the second with friction (Figure 5).

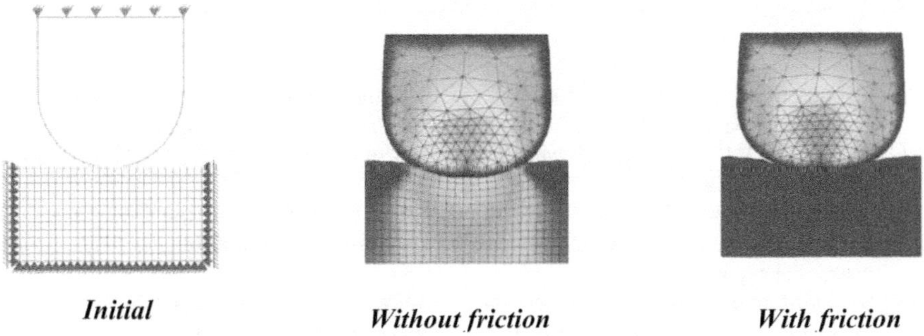

Fig. 5. Example of coupling finite elements/boundary elements.

These first results seem promising. Indeed, in this example, we can see the effect of the friction on the tool: particularly with regard to the contact's surface.

5 Conclusions and Perspectives

Our example represents only one stage towards the three-dimensional (axisymmetric) nearer to the industrial applications.

After having developed the coupling in contact finite elements/boundary elements in the two-dimensional case, our prospects will be as follows:

- To develop the axisymmetric case;
- To develop the optimization of the geometry of the tool;
- To study, in relation with industrial partners, the tools evolution in fatigue and its impact on the manufactured part.

Thereafter, our numerical tool will make it possible to simulate the manufacturing processes (machining, cutting, stamping, etc.).

The coupling finite elements/boundary elements offers new prospects for programming. In our case, the deformation of the tool can be taken into account in the developement of our software. In the future, the aim is the optimization of the geometry of the tool.

References

1. H. Kebir, J.M. Roelandt, J. Foulquier, A new singular boundary element for crack problems, *Engineering Fracture Mechanics* **62**(6), 1998, 497–510.
2. V. Hamel, J.M. Reolandt, J.N. Gacel, F. Schmit, F.E. modeling of clinch forming with automatic remeshing, *Computers & Structures*, 1998.
3. A. Benzegaou, M. Rachik, J.M. Reolandt, P. Duroux, J.N. Gacel, Clinch F.E. Modeling Using an Explicit Method, in *MAT_TEC'96*, A. Niku-Lari (Ed.), 1996, pp. 9–14.
4. A. Benzegaou, M. Rachik, J.M. Roelandt, J.L. Batoz, Modélisation de procédés dŠassemblage par grandes déformations plastiques, in *Acte 12° Congrès Français de Mécanique*, Vol. 2, Strasbourg, 1995, pp. 1–4.
5. K.-Q. Feng, G. De Saxce, G. Touzot, Numerical implementation of the implicit standard material approach for 2D or 3D frictional contact problem, *Int. J. Num. Meth. Eng.*, 1992.
6. C.A. Brebbia, J. Dominguez, *Boundary Elements: An Introductory Course*, McGraw-Hill, first edition, 1989.
7. D. Bigot, H. Kebir, J.-M. Roelandt, Simulation de procédés de mise en forme couplant éléments finis de domaine et de frontière, in *7ième Colloque National en Calcul des Structures*, Giens, 2005.
8. B. Noor-Omid, P. Wriggers, A note on the optimum choice for the penalty parameters, *Communications in Numerical Methods in Engineering* **3**, 1986, 581–585.
9. S.H. Lee, Rudimentary consideration for adaptive gap/friction element based on the penalty method, *Computers & Structures* **47**(6), 1993, 1243–1256.

Experimental and Numerical Study of the Tool in Water Jet Incremental Sheet Metal Forming

Bostjan Jurisevic[1], Viktor Sajn[2], Mihael Junkar[1] and Franc Kosel[2]

[1]*Laboratory for Alternative Technologies, Faculty of Mechanical Engineering, University of Ljubljana, Askerceva 6, SI-1000 Ljubljana, Slovenia; E-mail: {bostjan.jurisevic, miha.junkar}@fs.uni-lj.si*
[2]*Laboratory for Non-Linear Mechanics, Faculty of Mechanical Engineering, University of Ljubljana, Askerceva 6, SI-1000 Ljubljana, Slovenia; E-mail: {viktor.sajn, franc.kosel}@fs.uni-lj.si}*

Abstract. Incremental Sheet Metal Forming (ISMF) is a very flexible technology for fast prototyping and small batch production of sheet metal parts. This contribution deals with an innovative variant of ISMF, where instead of a rigid tool a Water Jet (WJ) is used as the main tool. Such a process can be addressed as Water Jet Incremental Sheet Metal Forming (WJISMF). The main aim of this paper is to present the technological window for WJISMF and characterise the attributes of the WJ used as the main tool. A Finite Element Analysis (FEA) simulation was developed to simulate the impact of a WJ on a rigid surface. The FEA simulation was experimentally validated by measuring the surface pressure at the interface between the WJ and the rigid surface. Numerical results are in good agreement with those obtained experimentally.

Key words: forming, water jetting, technological window, free surface fluid flow, experimental validation.

1 Introduction

The availability of functional prototypes during the development phase of a new product is of great importance. Prototypes should not just be functional and resemble as much as possible the characteristics of the final product, their production has also to be fast, cost efficient and reliable. In the field of sheet metal forming, Incremental Sheet Metal Forming (ISMF) is a very suitable process for fast prototyping and small batch production of complex products. Beside a high degree of flexibility the tooling costs in ISMF are drastically reduced comparing to processes such as deep drawing, where complex and expensive tools are required. This contribution presents an innovative variant of ISMF, in which the main rigid tool is replaced by a Water Jet (WJ).

S. Tichkiewitch et al. (eds.), Advances in Integrated Design and Manufacturing in Mechanical Engineering II, 79–91.

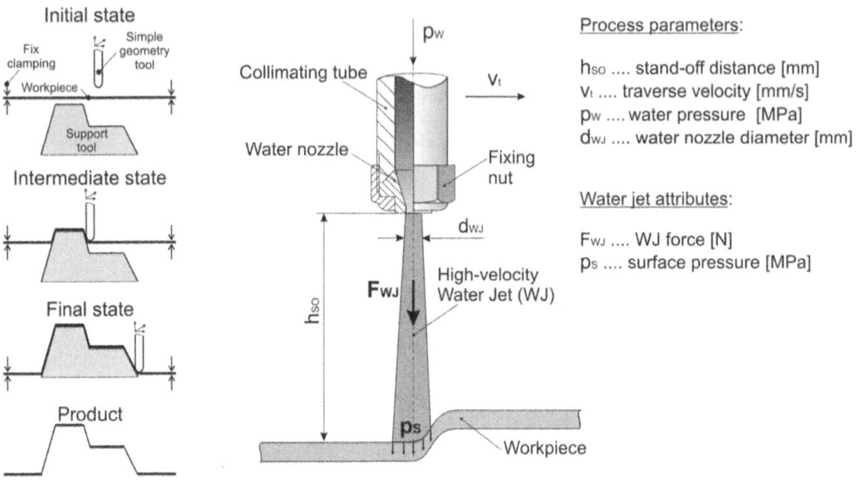

Fig. 1. Principles of ISMF (left) and introduction of a WJ instead of the simple geometry rigid tool (right) [1].

This process is known as Water Jet Incremental Sheet Metal Forming (WJISMF) and the basic principles are shown in Figure 1.

The main aim of this paper is to define the domain of WJISMF by means of technological windows based on the most relevant process parameters such as water pressure (p_W) and stand-off distance (h_{SO}). Since WJISMF is a rather new variant of ISMF, in this contribution the main tool, a high-velocity WJ is analyzed using a Finite Element Analysis (FEA) to simulate the interaction of the WJ with a rigid flat surface, where the surface pressure (p_S) distribution is observed. The numerical simulation results were experimentally validated as described in the sequel.

2 State-of-the-Art in the Field of WJISMF

In the domain of WJISMF very few researches are available. A pioneering work was reported by Iseki [2] in 2001. His work was of paramount importance, since he proved that a WJ can be used as a main tool in ISMF. In 2002 Jurisevic et al. [3] explored the possibilities of converting or upgrading a standard Abrasive Water Jet (AWJ) cutting machine to a WJISMF machine. In this investigation it was realized, that for WJISMF lower water pressures and higher water flows are required comparing to AWJ cutting, what requires a different type of water pump. Also the cutting head used for AWJ is inappropriate for WJISMF and has to be specially developed. However, a WJISMF forming head is much less complex than an AWJ cutting head. In order to have a better overview about those two jet based processes some general differences are presented. The jet diameter in AWJ cutting is normally between 0.5 and 1.2 mm, the water pressure is up to 400 MPa and the water flow around 2 l/min. On the other hand in WJISMF the jet diameter is around 2 mm, the water pressure up to 25 MPa and the water flow is about 50 l/min. Accordingly, it can be observed, that the energy involved in both processes is in the same range. The only difference is that for AWJ it has to be concentrated in order to remove material, while in WJISMF has to be more dispersed to prevent the erosion of the workpiece. A comparison between ISMF and WJISMF was presented by Jurisevic et al. [4], which has shown the advantages and disadvantages of using a WJ as the main tool instead of a simple geometry rigid tool. It showed that the flexibility of the process, workpiece surface integrity, tooling requirements, equipment costs and environmental impact are more favourable in case of WJISMF. On the other hand, the forming accuracy, forming time and energy efficiency are much better in case of ISMF.

From the available literature there are no indications that WJISMF has been observed by means of numerical simulations and this is one of the main aims of the presented contribution.

3 Technological Windows for WJISMF

Forming processes are often characterized by non-dimensional numbers such as the drawing ratio in deep drawing. In case of WJISMF it was found very convenient to define the relative jet diameter (κ), which represents the ratio between the WJ diameter (d_{WJ}) and the blank or workpiece thickness (T_B) as defined in equation (1):

$$\kappa = \frac{d_{WJ}}{T_B}. \tag{1}$$

The definition of the relative jet diameter brings many advantages. Such non-dimensional number is very useful to describe the complexity of products in WJISMF independently on the scale. Smaller κ means that complex features can be formed by WJISMF and consequently the forming accuracy is better. In Figure 2 a technological window of WJISMF depending on the relative jet diameter (κ) and water pressure (p_W) is presented.

The technological window presented in Figure 2 is limited by two lines. On the top it is limited by the water pressure at which erosion occurs on the workpiece surface. Beyond the bottom line the energy concentration of the WJ is too small to induce plastic deformations in the workpiece. The goal of WJISMF is to have the smallest possible κ and according to Figure 2, the only way to achieve that is to be just beyond the water pressure at which erosion occurs on the workpiece surface. Such kind of technological window is rather general, since it takes into account the relative jet diameter κ. A more process related technological window is presented in Figure 3 as function of the water pressure (p_W) and stand-off distance (h_{SO}). Those

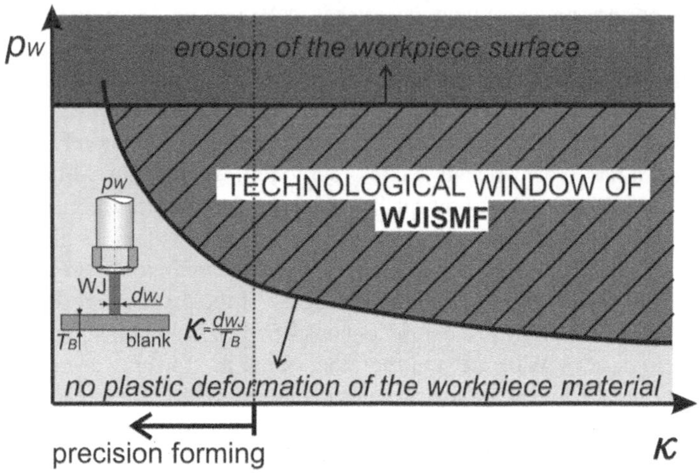

Fig. 2. Technological window of WJISMF.

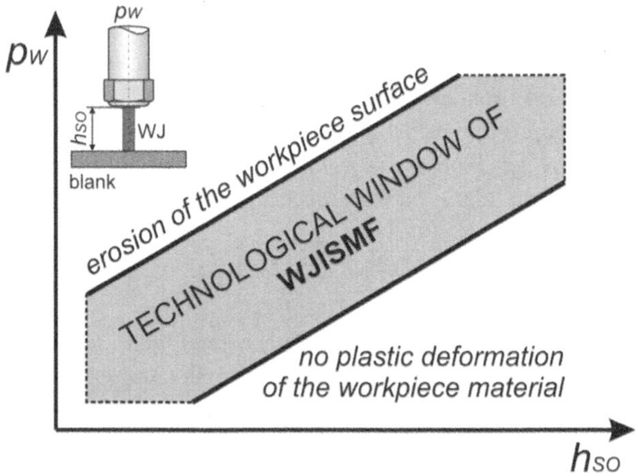

Fig. 3. Technological window of WJISMF as function of the water pressure (p_W) and stand-off distance (h_{SO}).

two process parameters are very important because they can be changed during the process. But what is even more relevant is the fact that their combination represents the energy concentration of the WJ at the interface with the workpiece, what is the main attribute in WJISMF.

Both technological windows presented in Figures 2 and 3 are assumed and not based on any analytical or experimental investigation. Therefore, it is of crucial importance to understand the influence of process parameters on the WJ attributes. In order to achieve this, an experimentally validated FEA simulation was executed as described in the following section.

4 FEA Simulation and Experimental Validation

In order to better understand the process, a high-velocity WJ is observed, both numerically and experimentally, when impinging on a flat rigid surface at impact angle of 90°. The aim of the FEA simulation is to understand the influence of process parameters such as the water pressure (p_W), stand-off distance (h_{SO}), traverse velocity of the WJ (v_t) and water nozzle geometry on the WJ attributes at the interface with the workpiece. Since this investigation is the first of such kind, only the influence of the stand-off distance (h_{SO}) was simulated, while the experimental validation was performed on one set of process parameters only.

4.1 FEA Simulation

As concern the available literature, this is the first numerical simulation in the field of WJISMF. Because of free fluid surfaces in WJISMF the FEA simulation is rather complex. The presented simulation covers a turbulent fluid flow through the nozzle in which the WJ is generated and the WJ impingement on a rigid wall at impact angle of 90°. The FEA simulation in executed in two dimensions and uses a non-stationary turbulent fluid flow model in combination with the standard k–ε turbulence model.

For the FEA simulation the following governing equations of the fluid flow and free fluid surface are used. The continuity equation (2) and the Navier–Stokes equation (3) for uncompressible fluid flow are derived from time averaged Navier–Stokes equations [5, 6]:

$$(\rho \cdot \bar{v}_i)_{,i} = 0, \tag{2}$$

$$\rho_0 \cdot \dot{\bar{v}}_i = \rho_0 \cdot f_i - \bar{p}_{,i} + \mu \cdot v_{i,ij} - \rho_0 \cdot \left(\overline{v_i' \cdot v_j'}\right)_{,j}, \tag{3}$$

where $\bar{v}_i$ and $\bar{p}$ are time averaged velocity and pressure, respectively, v_i' is the fluctuation velocity and μ is the dynamic viscosity of air. The Reynolds tension tensor is defined with the Boussinesque approximation as shown in equation (4):

$$-\rho_0 \cdot \overline{v_i' \cdot v_j'} = \rho_0 \cdot v_T \cdot \left(\overline{v_{i,j}' + v_{j,i}'}\right) - \frac{2}{3} \cdot \delta_{ij} \cdot k_T, \tag{4}$$

where k_T is the average kinetic energy of turbulent fluctuations [7] and v_T is the turbulence viscosity. In the k–ε model the turbulent viscosity is defined in

$$v_T = C_\mu \cdot \frac{k_T^2}{\varepsilon_T}. \tag{5}$$

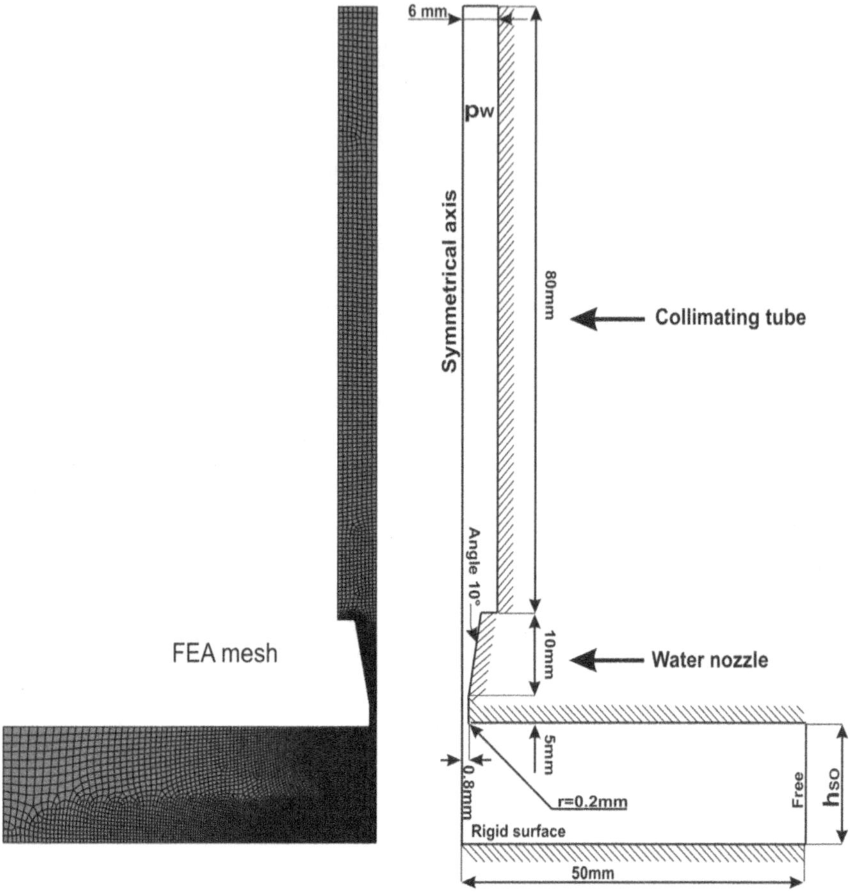

Fig. 4. Finite elements mesh of the numerical model.

Both k_T and ε_T are derived from further equations, which incorporates new empirical constants. The turbulent kinetic energy k_T is derived from

$$\frac{\partial k_T}{\partial t} + \bar{\nu}_i \cdot k_{T,i} = \left[\left(\nu_0 + \frac{\nu_T}{\sigma_k} \right) \cdot k_{T,i} \right]_{,i} + \nu_T \cdot \left(\overline{\nu'_{i,j} + \nu'_{j,i}} \right) \cdot \bar{\nu}_{i,j} - \varepsilon_T. \quad (6)$$

For the dissipation of turbulent kinetic energy ε_T, the following equation has to be fulfilled:

$$\frac{\partial \varepsilon_T}{\partial t} + \bar{\nu}_i \cdot \varepsilon_{T,i} = \left[\left(\nu_0 + \frac{\nu_T}{\sigma_k} \right) \cdot k_{T,i} \right]_{,i} + \nu_T \cdot \left(\overline{\nu'_{i,j} + \nu'_{j,i}} \right) \cdot \bar{\nu}_{i,j} - \varepsilon_T. \quad (7)$$

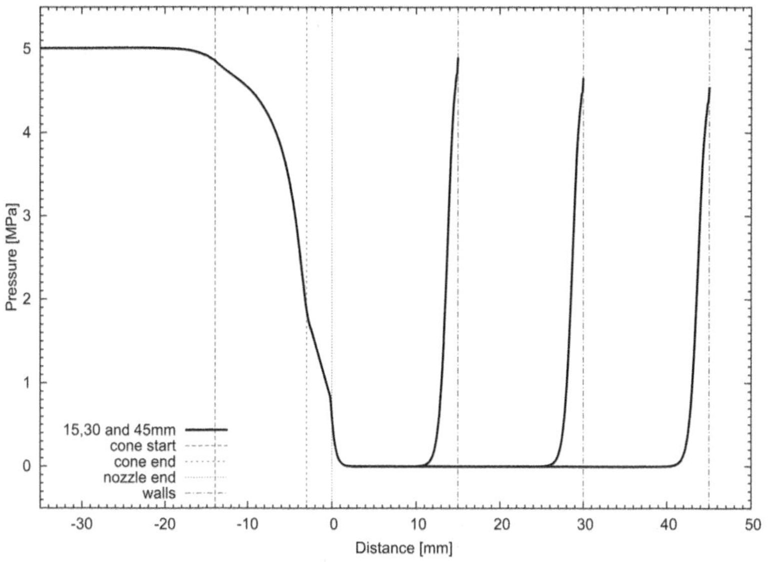

Fig. 5. Pressure profiles along the WJ centreline for stand-off distances 15, 30 and 45 mm.

The values of the empirical constants in equations (5–7) are as follows: $C_\mu = 0.009$, $\sigma_k = 1.0$, $\sigma_{\varepsilon T} = 1.3$, $C_{1\varepsilon T} = 1.44$, $C_{2\varepsilon T} = 1.92$. Equations (6) and (7) are derived based on the closure of time averaged Navier–Stokes equations on the second level [6]. Basic equations are expanded with the addition of equation (8), which describes the material state in a given point [8]:

$$\frac{\partial f}{\partial t} + \bar{v}_i \cdot f_{,i} = 0, \tag{8}$$

where f represents the material state. Equation (8) supplements the mass conservation law by taking into account the free formation of the fluid domain edge. In contrast to the basic equations in fluid mechanics, which are solved with the Euler approach, the material state equation is solved with the Lagrange approach. This facilitates the computation of the fluid flow with the FEA method, where the material state of the element F is defined. The material state is between 0, in which case the element does not contain any fluid, and 1, which means that the observed element is filled with fluid. Elements with $0 < F < 1$ contain a portion of the free fluid surface. After the computation of the basic equation for non-stationary turbulent fluid flow and time increment, the movements of the nodes and the new shape and position of the finite element is defined. Then the new finite elements mesh is moved to the initial position. The new material state of a finite element is calculated with equation (9), which takes into account surface contributions and states of the neighbours elements:

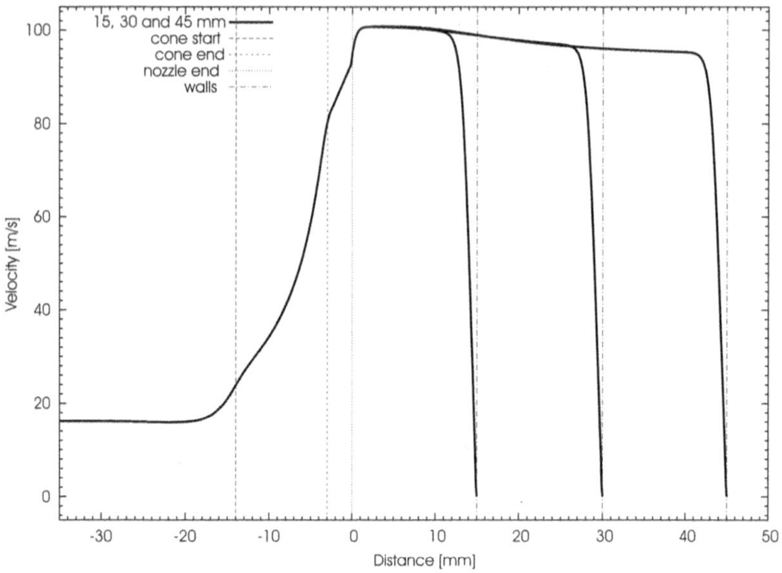

Fig. 6. Velocity profiles along the WJ centreline for stand-off distances 15, 30 and 45 mm.

$$F_{(t+\Delta t)} = \sum_{l} F_{(t)l} \cdot A_{(t+\Delta t)l}, \tag{9}$$

where $A_{(t+\Delta t)l}$ is the surface portion of the finite element in the new mesh, which is superimposed over the currently observed initial mesh. The time increment has to be small enough to prevent a fluid particle crossing any element in a single time increment. The finite element mesh can be observed in Figure 4, where only half of the model was computed due to the symmetry of the problem.

In this investigation three FEA simulation of the WJ generation in the water nozzle and impact on a flat rigid surface were executed. The pressure before the water nozzle is 5 MPa, the nozzle dimension is 1.6 mm and the cone angle of the nozzle 20°. The impacts of the WJ on the rigid surface are observed at three different stand-off distances, what represents the distance between the end of the water nozzle and the impinging surface. These three stand-off distances are 15, 30 and 45 mm, what corresponds to 12765, 19375 and 26543 nodes in the mesh, respectively. The boundary conditions of state inside the nozzles take into account a complete wetness of the inner surfaces, which together with the impingement surface are treated as viscous walls. In the open space between the water nozzle and the impingement surface the wetness is not defined and the fluid edge freely forms. The highest concentration of nodes is on the interface between the WJ and the impingement surface, where the highest gradient of positive static pressure is expected. At the beginning of the simulation the water nozzle is completely empty. During the computation of the non-

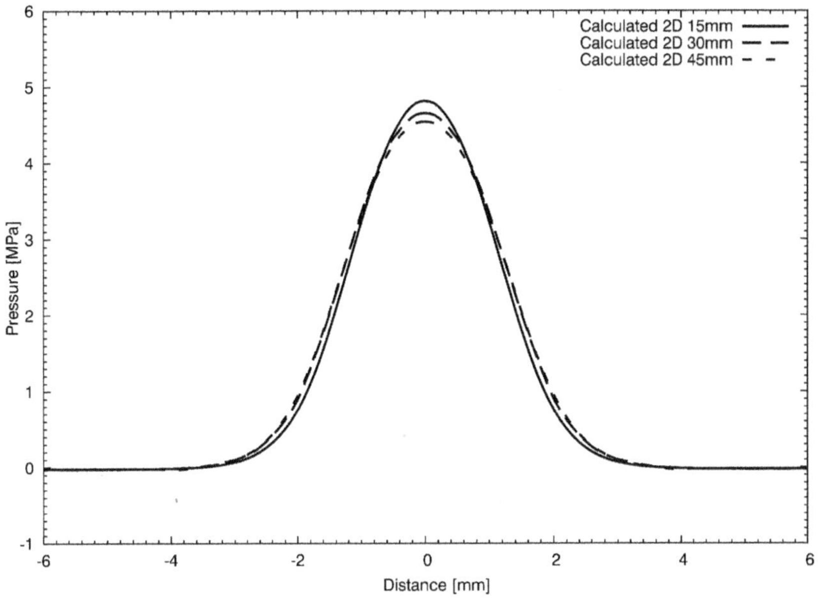

Fig. 7. Pressure profiles at the interface between the WJ and the impingement surface for stand-off distances 15, 30 and 45 mm.

stationary flow the smallest time increment is $12 \cdot 10^{-9}$ s for the transient phenomena and the typical time increment is $250 \cdot 10^{-9}$ s once the flow become stationary.

Results of FEA simulations are given in Figures 5 and 6, in which the pressure profiles and velocity profiles along the WJ centreline are plotted, respectively.

The horizontal axis in Figures 5 and 6 has its origin at the exit point of the water nozzle. The velocity of water inside the collimating tube before the water nozzle is 16 m/s and at the exit from the water nozzle reaches its peak at 100.7 m/s. After that the WJ velocity decreases due to the spreading of the WJ. The highest decrease of the WJ velocity is before the rigid surface on which impacts at an angle of 90°.

At the interface surface between the WJ and the rigid wall the pressure increased, what can be observed in Figure 7, where numerically simulated pressure profiles are plotted for all three stand-off distances (15, 30 and 45 mm). In this case the origin of the horizontal axis represents the WJ centreline.

4.2 Experimental Validation

For the experimental validation of the FEA simulation a measurement system based on a pressure transducer was used as shown in Figure 8.

The experimental validation was carried out at one set of process parameters in which the water pressure was 5 MPa, the stand-off distance 15 mm and the nozzle

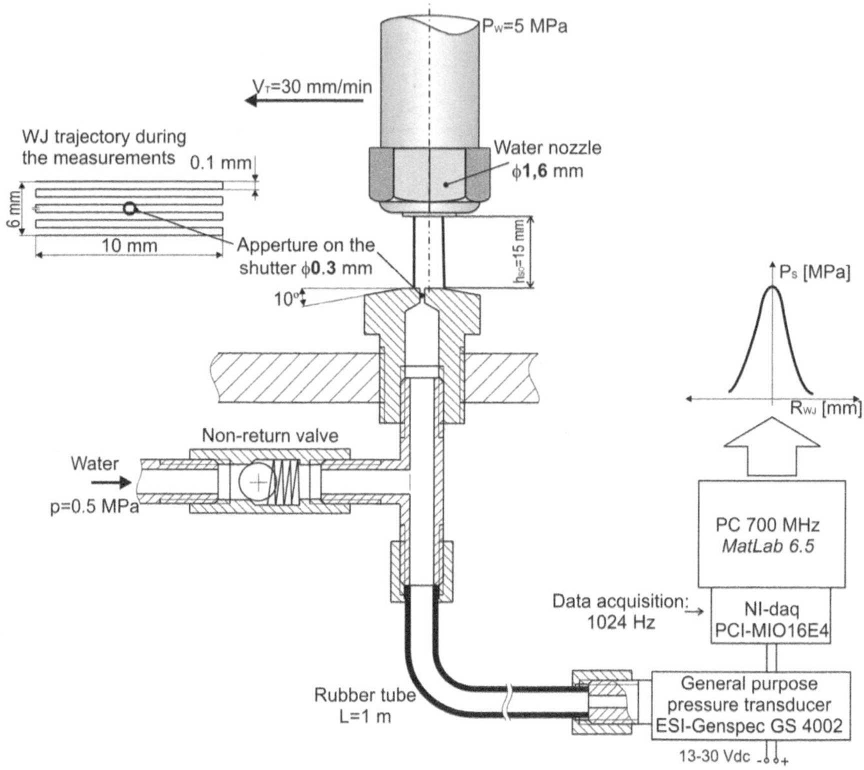

Fig. 8. Measurement system for the pressure profile at the interface between the WJ.

diameter 1.6 mm. The pressure profile of the WJ impinging a rigid surface was obtained by moving the WJ over an aperture of 0.3 mm in diameter on a shutter, which represents a rigid surface in the experiment. During the measurement the WJ moved along the aperture on the shutter with a traverse velocity of 30 mm/min over the trajectory shown in Figure 8. The aperture was connected to a general purpose pressure transducer ESI-Genspec GS 4002 through a flexible rubber tube. The measurement range of the transducer is 0 to 25 MPa, the linearity 0.1% and accuracy 0.25%. In order to prevent the accumulation of air between the shutter and the pressure transducer, the system was all the time flooded with water through a non-return valve. Pressure signals from the transducer were acquired and stored on a PC through a data acquisition card NI-SAQ PCI-MIO 16E4 with a sampling frequency of 1024 Hz. Further data analysis was performed with MatLab 6.5 and the results are shown in Figure 9.

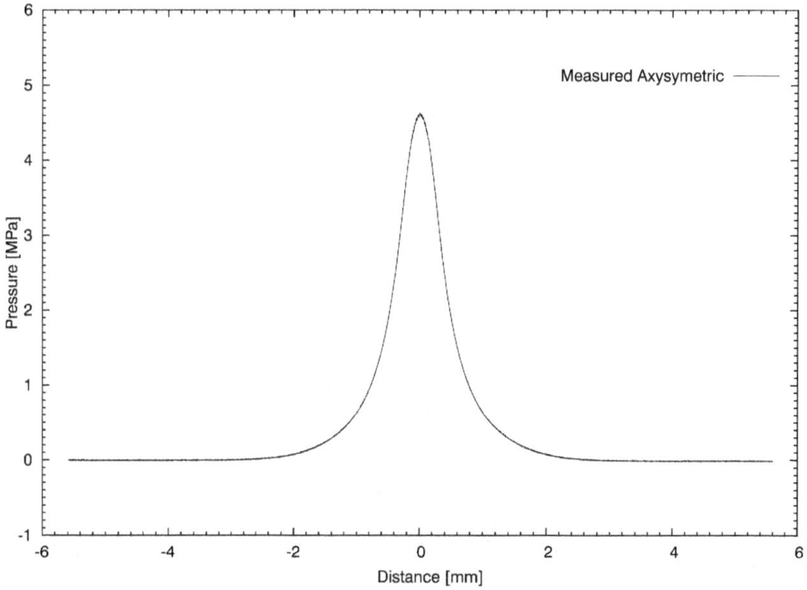

Fig. 9. Measured pressure profile at the interface between the WJ.

During the measurements it was observed that when the WJ passes through the aperture on the shutter the deflected flow disturbs further measurement. For this reason the shutter was partially turned at an angle of 10°, what substantially improved the quality of results.

4.3 Comparison of Numerical and Experimental Results

The comparison between the FEA simulation results and experimental measurements is presented in Figure 10, where the pressure profiles at the interface between the WJ and the impingement rigid surface are compared.

Numerical and experimental results can be compared only qualitatively due to physical difference of the fluid flow in the simulation and in the experiment. Both pressure distributions have a similar shape, but in order to fulfil the continuity equation, the simulated pressure distribution is wider because the problem was treated as a two dimensional flow.

The simulated peak pressure in the centre of the WJ when impacts on the rigid surface is 4.82 MPa, while the measured pressure in this point is 4.64 MPa. The difference between the simulated results and measured value is 3.8%, what leads to the conclusion that the FEA simulation results are in good agreement with the experimental measurements.

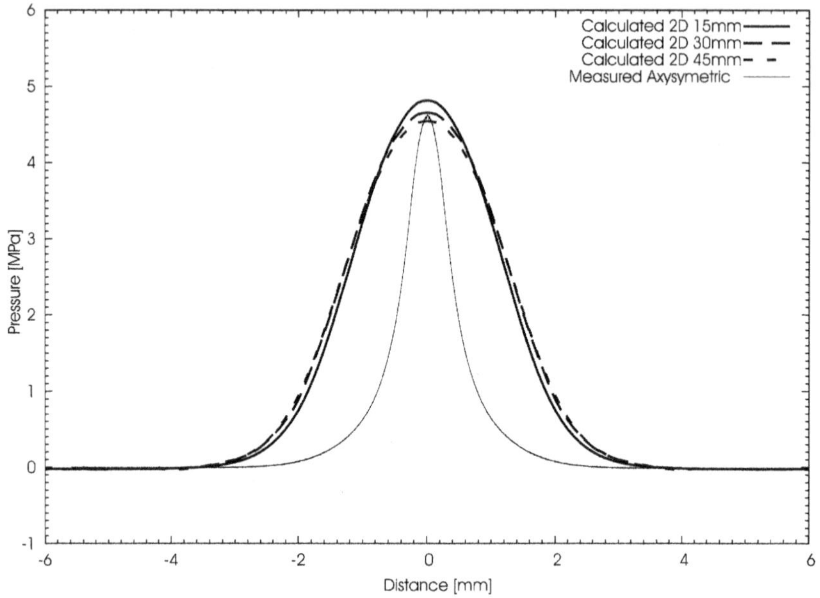

Fig. 10. Comparison of FEA simulation results with experimental measurement.

5 Conclusions and Outlook

The main result of this contribution is the FEA simulation and numerical validation of a high-velocity WJ impinging on a flat rigid surface at an impact angle of 90°, what helps to better understand the influence of process parameters such as the water pressure (p_W) and stand-off distance (h_{SO}) in WJISMF process. Numerical results are in good agreement with those obtained experimentally, even if the FEA simulation was made in two dimensions, while the experimental validation took place in three dimensions. All the information collected in this investigation will be used to better understand and find the limits of WJISMF by means of technological windows.

In the next step a three dimensional FEA simulation will be undertaken and experiments will be extended to several different water pressures (p_W), stand-off distance (h_{SO}), and water nozzle geometries. Further on, numerical simulations and experiments will not be limited to the tool and a rigid surface, but the behaviour of deformable workpiece in WJISMF will be observed.

Acknowledgements

This work is supported by the "Multi-Material Micro Manufacture: Technology and Applications (4M)" Network of Excellence, Contract Number NMP2-CT-2004-500274 and by the "Virtual Research Lab for a Knowledge Community in Production (VRL-KCiP)" Network of Excellence, Contract Number NMP2-CT-2004-507487, both within the EU 6th Framework Program.

References

1. V. Sajn, B. Jurisevic, F. Kosel and M. Junkar. Numerical and experimental analysis of water flow through nozzle, in *Conference "Kuhljevi dnevi 2005"*, J. Korelc and D. Zupan (Eds.), Slovene Society of Mechanics, 2005, pp. 283–290 [in Slovene].
2. H. Iseki. Flexible and incremental bulging of sheet metal using high-speed waterjet, *JSME International Journal*, C, **44**(2), 2001, 486–493.
3. B. Jurisevic, K.C. Heiniger, K. Kuzman and M. Junkar. Incremental sheet metal forming with a high-speed water jet, in *Proceedings of the International Deep Drawing Research Group 2003 Conference*, K. Kuzman et al. (Eds.), 2003, pp. 139–148.
4. B. Jurisevic, M. Junkar, S. Jadhav, M. Kleiner and K. Kuzman. Incremental sheet metal forming process with a water jet and rigid tool, in *Proceedings of the 17th International Conference on Water Jetting*, C. Gee (Ed.), 2004, pp. 71–81.
5. B. Mohammadi and O. Pironneau. *Analysis of the K-Epsilon Turbulence Model*, John Wiley & Sons Ltd., 1994.
6. D.C. Wilcox. *Turbulence Modelling for CFD*, DCW Industries, Inc., La Canada, California, 1993.
7. J.C. Ferziger and M. Peric. *Computational Methods for Fluid Dynamics*, Springer-Verlag, 1999.
8. A. Huerta and W.K. Liu. Viscous flow with large free surface motion, *Computer Methods in Applied Mechanics and Engineering* **69**, 1988, 277–324.

MECHANICAL SYSTEMS DESIGN AND OPTIMIZATION

Hybrid (Bolted/Bonded) Joints Applied to Aeronautic Parts: Analytical One-Dimensional Models of a Single-Lap Joint

Eric Paroissien[1], Marc Sartor[2] and Jacques Huet[1]

[1] *IGM (ENSICA/DGM), 1 place Emile Blouin, 31056 Toulouse Cedex 5, France; E-mail: {eric.paroissien, jacques.huet}@ensica.fr*
[2] *IGM (INSA/GM), 135 avenue de Rangueil, 31077 Toulouse Cedex 4, France; E-mail: marc.sartor@insa-toulouse.fr*

Abstract. The load transfer in hybrid (bolted/bonded) single-lap joint is complex due to the association of two different transfer modes (discrete and continuous) through elements with different stiffness. Analytical methods exist for these two different modes, when considered separately. In this paper two one-dimensional elastic analytical models are presented for the determination of the load transfer in single lap configuration. The first one is developed by using the integration of the local equilibrium equations. From this first method an elastic-plastic approach is presented. The second one uses the Finite Element Method, introducing a new element called "bonded-bars". These models are robust, easy to use and provide the same results. They allow us to analyze the load transfer and to evaluate different geometric and mechanical parameters' influence. Thus they represent the first step for the design of a hybrid joint able to replace its bolted equivalent used on aircraft.

Key words: hybrid (bolted/bonded) joint, single-lap joint, load transfer, analytical analysis, Finite Element Method.

Nomenclature

$E^{(j)}$ Young modulus of the adherend j in MPa
G Coulomb modulus of the adhesive in MPa
$u_i^{(j)}$ Longitudinal displacement in mm of the adherend j in the bay i
b Transversal pitch in mm
d_i Abscissa of the fastener i (d: edge distance in mm; s: longitudinal pitch in mm)
e Thickness of the adhesive in mm
$e^{(j)}$ Thickness of the adherend j
L Length of the lap mm

S. Tichkiewitch et al. (eds.), Advances in Integrated Design and Manufacturing in Mechanical Engineering II, 95–110.

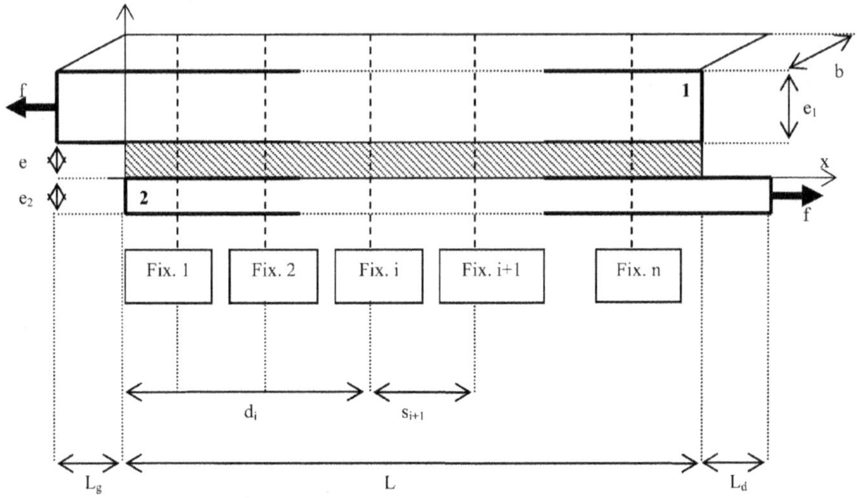

Fig. 1. Nomenclature.

L_g Length of the left bar not bonded in mm
L_d Length of the right bar not bonded in mm
x_p Length of the plastic region in mm
T_p Plastic adhesive shear stress in MPa
N Normal force in N
T Adhesive shear stress in MPa

1 Introduction

The joints under study are joints of civil aircraft. The longitudinal joints of the fuselage are investigated. These longitudinal joints of fuselage are composed of aluminium sheets and titanium bolts. The developed method, which is presented in this paper, has to apply to the other joints on aircraft.

This paper deals with load transfer in hybrid single-lap joints. Hybrid joints are bolted/bonded joints, then associating a discrete transfer mode with a continuous one, each one belonging to its own stiffness. The bolted joint (discrete transfer mode) generates a high overstress around the holes of the fasteners which is prejudicial to the fatigue resistance. The bonded joint (continuous transfer mode) allows a better distribution of the transfer; however it presents a plastic accommodation which is prejudicial to the static strength in the long term. In the domain of aircraft structure assembly, the hybrid joining could be interesting because it could reduce the load

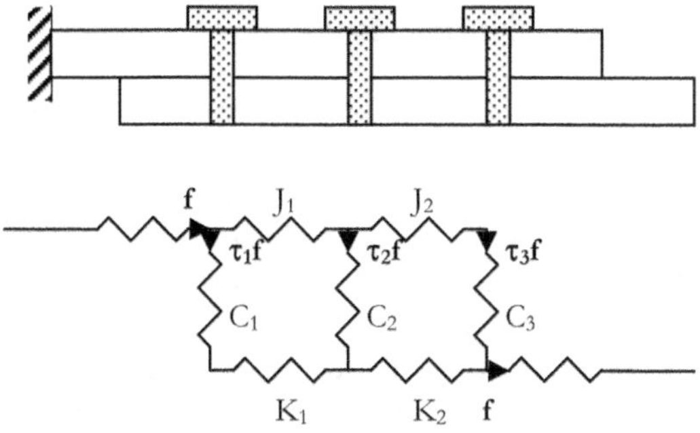

Fig. 2. Electric meshing of bolted joint.

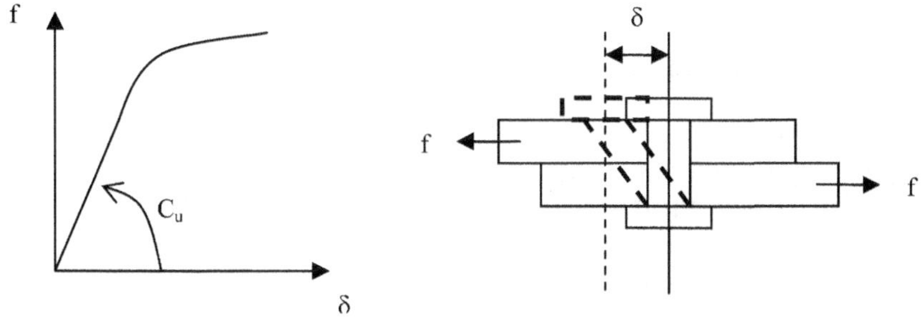

Fig. 3. Behaviour of a fastener.

transferred by the fasteners in order to improve the fatigue life, while ensuring static strength under extreme loads. The idea is to design the hybrid joint in order to share the load between the adhesive and the fasteners in a suitable way. That is why the influence of the joint geometry and the material properties on the load transfer is investigated by means of developing efficient designing tools. Analytical approaches are thus privileged.

Analytical methods exist for the two elementary transfer modes. The second and the third parts of this paper deal respectively with the analytical model of bolted joints design and with the analytical model of the bonded joints design, which will be used again within the analytical approaches of hybrid (bolted/bonded) joints. In the fourth and the fifth part two elastic one-dimensional analytical models are presented for the determination of the load transfer of hybrid joints in single-lap configuration. The results presented in the sixth part concern only the load transfer obtained from these models, although the adhesive shear stress can be deduced from the load transfer. In the seventh part, a perfectly elastic-plastic behaviour of the adhesive is introduced.

2 Analytical Model for Bolted Joints

The load transfer in a bolted joint is a discrete transfer mode. It means that between each bolt (i.e.: on each bay) the transferred load is constant. In [1] the author calculates the load transferred by the fasteners using an analogy with an electric meshing and modelling the fasteners by springs, which work by shearing (Figure 2).The behaviour of a fastener in a joint is a difficult problem and the determination of its flexibility provides numerous studies and formulations [2, 3] (Douglas, Boeing). The behaviour of a fastener can be defined by a curve force-displacement of the joint. The linear part of this curve gives the rigidity of the fastener (Figure 3), quoted C_u.

The idea of electric meshing of the bolted joint model will be used later in this paper. The whole fasteners are assumed to have the same rigidity.

3 Analytical Model for Bonded Joints

In [4], Hart-Smith analyses the stress distribution in a bonded double-lap joint, without taking into account the bending of the adherends and the adhesive peeling stress, since the eccentricity of the load path is not influent in double-lap configuration. The author realized the local equilibrium of an elementary length of the adherend. Considering the case of the single-lap configuration (Figure 4), the equilibrium equations are:

$$\frac{dN^{(2)}(x)}{bdx} = T(x) \tag{1}$$

and

$$\frac{dN^{(1)}(x)}{bdx} = -T(x). \tag{2}$$

The elastic behaviour of the adhesive gives the following equation:

$$T(x) = \frac{G}{e}(u^{(2)}(x) - u^{(1)}(x)), \tag{3}$$

whereas the hypothesis of elastic adherends provides:

$$\frac{du^{(j)}(x)}{dx} = \frac{N^{(j)}(x)}{be^{(j)}\,E^{(j)}} \quad \text{for} \quad j = 1, 2. \tag{4}$$

The author gets then the following differential equation of the second order with constant factors:

$$\frac{d^2T(x)}{dx^2} - \eta^2 T(x) = 0, \tag{5}$$

where

$$\eta^2 = \frac{G}{e}\left(\frac{1}{e^{(1)}\,E^{(1)}} + \frac{1}{e^{(2)}\,E^{(2)}}\right). \tag{6}$$

This bonded joint model will be used later in this paper.

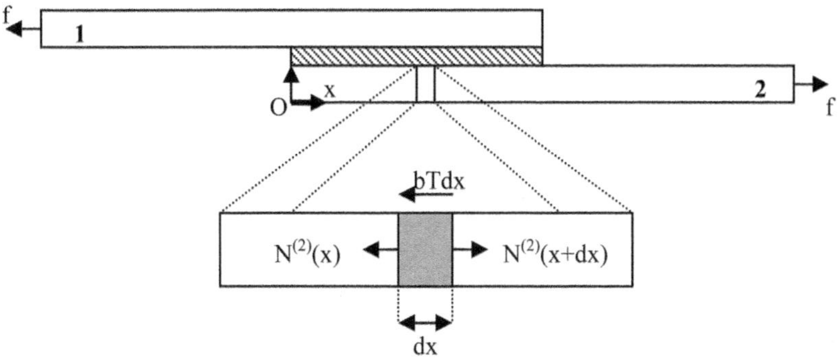

Fig. 4. Bonded single-lap joint and local equilibrium.

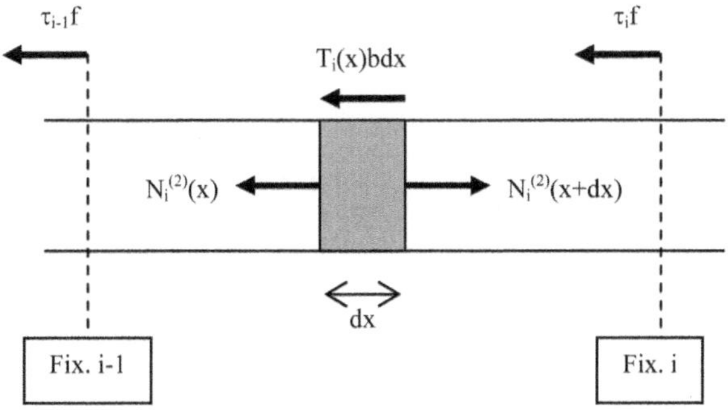

Fig. 5. Bay number *i* of the hybrid joint.

4 First Analytical Model for Hybrid Joints

The goal is to combine both approaches in order to get a load transfer, which is continuous by parts. A similarly approach was developed for the calculation of stepped joints in [5]. The conditions of longitudinal force-equilibrium for a differential element dx in the bay i within the joint are (1) and (2). By differentiating (1) and with (3), it becomes:

$$\frac{d^2 N_i^{(2)}(x)}{b dx^2} = \frac{G}{e}\left(\frac{du_i^{(2)}(x)}{dx} - \frac{du_i^{(1)}(x)}{dx}\right). \tag{7}$$

Using (4) and the equation of the general equilibrium (f is the applied load in N):

$$N_i^{(1)}(x) + N_i^{(2)}(x) = f, \tag{8}$$

it becomes the following differential equation for the bay i (see Figure 5):

$$\frac{d^2N_i^{(2)}(x)}{dx^2} - \eta^2 N_i^{(2)}(x) = \gamma f \quad \text{where} \tag{9}$$

$$\gamma = -\frac{G}{ee^{(1)}E^{(1)}}. \tag{10}$$

Consequently the number of equations is equal to the number of bays i. In order to solve (9) on each bay, it is required to express the boundaries conditions.

The developed approach is based on the following hypotheses:

- an elastic behaviour of materials (adherends, adhesive, fasteners);
- normal stress in the adherends (no bending);
- shear stress in the adhesive (no peeling).

It is assumed that the adhesive thickness is constant along the lap-joint and that the fasteners have the same rigidity. The mechanical and geometric parameters are free.

The solution of Equation (9) is:

$$N_i^{(2)}(x) = A_i e^{-\eta x} + B_i e^{\eta x} - \gamma\eta^{-2} f. \tag{11}$$

For n fasteners (thus $n+1$ bays), there are $2n+2$ unknowns, which are determined thanks to the boundaries conditions.

The first condition is no tensile load at the start of the adherend 2, whereas the second condition corresponds to a complete load transfer at the end of the lap:

$$N_1^{(2)}(0) = 0 \Leftrightarrow A_1 + B_1 = \gamma\eta^{-2} f, \tag{12}$$

$$N_{n+1}^{(2)}(L) = f \Leftrightarrow A_{n+1}e^{-\eta L} + B_{n+1}e^{\eta L} = (1+\gamma\eta^{-2})f. \tag{13}$$

By considering the fasteners, the equation which corresponds to the load transfer ratio τ_i at the fastener i between both bays i and $i+1$ is:

$$N_{i+1}^{(2)}(d_i) = N_i^{(2)}(d_i) + \tau_i f. \tag{14}$$

Moreover the fasteners are simulated by springs, the rigidity of which is C_u (in N·mm^{-1}), so:

$$\tau_i f = C_u(u_i^{(2)}(d_i) - u_i^{(1)}(d_i)). \tag{15}$$

Thus with (1), (3) and (11):

$$\tau_i f = \varphi(-A_i e^{-\eta d_i} + B_i e^{\eta d_i}) \tag{16}$$

where

$$\varphi = \frac{e\eta C_u}{bG}. \tag{17}$$

And with (14) and (16), n additional equations come:

$$re^{-\eta d_i}\ A_i + qe^{\eta d_i}\ B_i - e^{-\eta d_i}\ A_{i+1} - e^{\eta d_i}\ B_{i+1} = 0 \tag{18}$$

with

$$r = 1 - \varphi \tag{19}$$

and

$$q = 1 + \varphi. \tag{20}$$

Finally, the continuity of the adhesive shear stress provides the n last equations:

$$T_i(d_i) = T_{i+1}(d_i) \Leftrightarrow e^{-\eta d_i}\ A_i - e^{\eta d_i}\ B_i - e^{-\eta d_i}\ A_{i+1} + e^{\eta d_i}\ B_{i+1} = 0. \tag{21}$$

Thus, a linear system, the size of which is $2n + 2$, is obtained:

$$\begin{cases} A_1 + B_1 = \gamma\eta^{-2}\ f \\ A_{n+1}\ e^{-\eta L} + B_{n+1}\ e^{\eta L} = (1 + \gamma\eta^{-2}) f \\ re^{-\eta d_i}\ A_i + qe^{\eta d_i}\ B_i - e^{-\eta d_i}\ A_{i+1} - e^{\eta d_i}\ B_{i+1} = 0 \\ e^{-\eta d_i}\ A_i - e^{\eta d_i}\ B_i - e^{-\eta d_i}\ A_{i+1} + e^{\eta d_i}\ B_{i+1} = 0 \end{cases} \quad i \in [1; n]. \tag{22}$$

The resolution of this system gives the parameters A_i and B_i and allows to build the whole functions, which characterize the one-dimensional behaviour of the hybrid joint.

In the particular case of a constant pitch s between the fasteners and a constant distance between the end of the joint and the fastener d (longitudinal edge distance), the linear system (22) may be simplified in a linear system the size of which is 2.

Changing the parameters like:

$$\begin{cases} \alpha_i = e^{-\eta d_i}\ A_i \\ \beta_i = e^{\eta d_i}\ B_i \\ \alpha_{n+1} = e^{-\eta L}\ A_{n+1} \\ \beta_{n+1} = e^{\eta L}\ B_{n+1} \end{cases} \quad i \in [1; n] \tag{23}$$

and setting

$$U_i = \begin{pmatrix} \alpha_i \\ \beta_i \end{pmatrix}, \quad i \in [1; n+1], \tag{24}$$

the linear system (22) becomes:

$$U_i - \frac{1}{2}\Gamma_X\ U_{i+1} = 0, \quad i \in [1; n], \tag{25}$$

where

$$\Gamma_X = \begin{pmatrix} (2+\varphi)e^{\eta X} & -\varphi e^{-\eta X} \\ \varphi e^{\eta X} & (2-\varphi)e^{-\eta X} \end{pmatrix}, \quad \text{and} \quad X = d$$

$$\text{for} \quad i = n + 1 \quad \text{else } X = s. \tag{26}$$

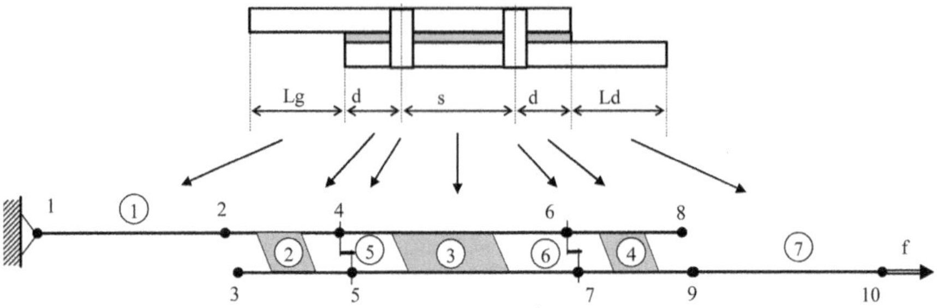

Fig. 6. Structure meshing.

The matrix Γ_X is diagonalisable and $S_{p_{\Gamma_x}} = \{x_1; x_2\}$. In the base of diagonalisation (25) becomes:

$$V_i - \frac{1}{2}\begin{pmatrix} x_1 & 0 \\ 0 & x_2 \end{pmatrix} V_{i+1} = 0, \quad i \in [1, n-1]. \tag{27}$$

A solution is searched like:

$$V_i = \begin{pmatrix} A\tilde{\alpha}^i \\ B\tilde{\beta}^i \end{pmatrix}. \tag{28}$$

Then, it becomes:

$$\begin{cases} \tilde{\alpha} = \frac{2}{x_i}, \\ \tilde{\beta} = \frac{2}{x_2}. \end{cases} \tag{29}$$

Only A and B parameters have to be determined.

5 Second Analytical Model for Hybrid Joints (Bonded-Bars Element)

The Finite Element Method is used in the second approach. The single-lap joint is meshed in 1D-elements (Figure 6). Simple elements are used: bar elements (element 1 and 7 on Figure 6), springs (element 5 and 6) and new elements called "bonded-bars" (element 2, 3 and 4). These new elements are 1D-elements since only the displacements in the direction of the load are taken into account. However they have four nodes (Figure 7) allowing to differentiate the displacement of each adherend.

The rigidity matrix of the bonded-bars element has to be determined. The length of the bonded-bars element is quoted Δ. The subscript i is not useful here.

Thanks to Equations (1) to (4), the following system of differential equations is obtained:

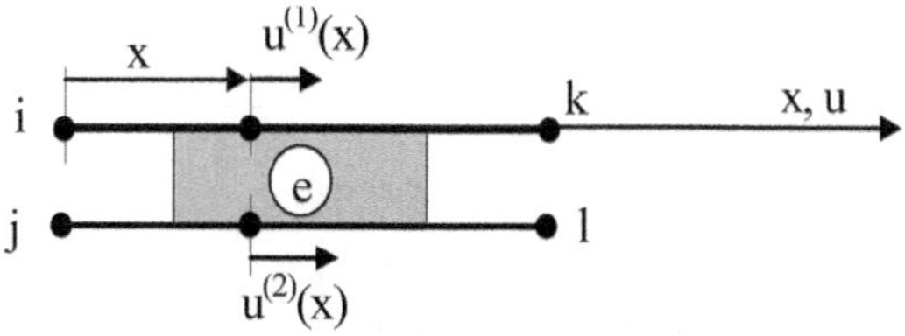

Fig. 7. Bonded-bars element.

$$\begin{cases} \frac{d^2u^{(1)}}{dx^2} + \frac{G}{eE^{(1)}e^{(1)}}(u^{(2)} - u^{(1)}) = 0, \\ \frac{d^2u^{(2)}}{dx^2} - \frac{G}{eE^{(2)}e^{(2)}}(u^{(2)} - u^{(1)}) = 0, \end{cases} \tag{30}$$

and it is solved (by addition and subtraction for example) as:

$$\begin{cases} u^{(1)} = 0.5[-\theta(Ee^{-\eta x} + Fe^{\eta x}) + Cx + D], \\ u^{(2)} = 0.5[\omega(Ee^{-\eta x} + Fe^{\eta x}) + Cx + D], \end{cases} \tag{31}$$

where

$$\psi = \frac{G}{e}\left(\frac{1}{E^{(1)}e^{(1)}} - \frac{1}{E^{(2)}e^{(2)}}\right), \tag{32}$$

$$\theta = 1 + \frac{\psi}{\eta^2}, \tag{33}$$

and

$$\omega = 1 - \frac{\psi}{\eta^2}. \tag{34}$$

The following boundaries conditions:

$$\begin{cases} u^{(1)}(0) = u_i \\ u^{(2)}(0) = u_j \\ u^{(1)}(\Delta) = u_k \\ u^{(2)}(\Delta) = u_l \end{cases} \tag{35}$$

lead to:

$$E = \frac{(u_j - u_i)e^{\eta\Delta} - (u_l - u_k)}{2sh(\eta\Delta)}, \tag{36}$$

$$F = \frac{(u_l - u_k) - (u_j - u_i)e^{-\eta\Delta}}{2sh(\eta\Delta)}, \tag{37}$$

$$C = \frac{u_l\theta + u_k\omega - u_j\theta - u_i\omega}{\Delta}, \tag{38}$$

$$D = u_j\theta + u_i\omega. \tag{39}$$

The rigidity matrix of the bonded-bars element is defined by:

$$\begin{bmatrix} k_{11} & k_{12} & k_{13} & k_{14} \\ k_{21} & k_{22} & k_{23} & k_{24} \\ k_{31} & k_{32} & k_{33} & k_{34} \\ k_{41} & k_{42} & k_{43} & k_{44} \end{bmatrix} \begin{bmatrix} u_i \\ u_j \\ u_k \\ u_l \end{bmatrix} = \begin{bmatrix} Q_i \\ Q_j \\ Q_k \\ Q_l \end{bmatrix}. \tag{40}$$

It follows that:

$$\begin{aligned} \frac{\partial Q_i}{\partial u_i} &= k_{11}, \\ \frac{\partial Q_i}{\partial u_j} &= k_{12}, \\ &\cdots\cdots\cdots \end{aligned} \tag{41}$$

The matrix components can be obtained using the boundaries conditions. Indeed, the normal loads in the adherends are obtained from (4) and (31); the forces Q_i, Q_j, Q_k, and Q_l which the nodes i, j, k and l exert on the element are:

$$\begin{cases} Q_i = -N_1(0), \\ Q_j = -N_2(0), \\ Q_k = N_1(\Delta), \\ Q_l = N_2(\Delta). \end{cases} \tag{42}$$

Finally, the rigidity matrix of the bonded-bars element is:

$$K_{\text{bonded-bars}} = \omega h \begin{pmatrix} ct + \mu & 1 - ct & -cs - \mu & 1 - cs \\ 1 - ct & ct + \mu^{-1} & 1 - cs & -cs - \mu^{-1} \\ -cs - \mu & 1 - cs & ct + \mu & 1 - ct \\ 1 - cs & -cs - \mu^{-1} & 1 - ct & ct + \mu^{-1} \end{pmatrix}, \tag{43}$$

where

$$h = 0.5\frac{E^{(2)}e^{(2)}b}{\Delta}, \tag{44}$$

$$cs = \eta\Delta\cosh(\eta\Delta), \tag{45}$$

$$ct = \eta\Delta\coth(\eta\Delta), \tag{46}$$

and

$$\mu = \frac{E^{(1)}e^{(1)}}{E^{(2)}e^{(2)}} = \frac{\omega}{\theta}. \tag{47}$$

On the other hand, the rigidity matrix of the bar elements, which describe the not bonded portions of the adherends, take the conventional form:

$$K_{\text{bar}} = \begin{pmatrix} K_b & -K_b \\ -K_b & K_b \end{pmatrix}, \tag{48}$$

Table 1. Geometric and mechanical parameters used.

$E^{(j)}$ (Mpa)	G (Mpa)	$e^{(j)}$ (mm)	e (mm)	d (mm)	s (mm)
72000	800	2.4	0.4	9.6	19.2

where

$$K_b = \frac{E^{(1)}e^{(1)}b}{L_g} \quad \text{(element 1)}, \tag{49}$$

or

$$K_b = \frac{E^{(2)}e^{(2)}b}{L_d} \quad \text{(element 7)}. \tag{50}$$

Similarly, the rigidity matrix of the fasteners is:

$$K_{\text{fastener}} = \begin{pmatrix} C_u & -C_u \\ -C_u & C_u \end{pmatrix}. \tag{51}$$

The rigidity matrix of the whole structure can be obtained using the conventional assembly rules of the FEM, and the classical system ($F = Ku$) is solved. For n fasteners, the size of this linear system is $(2n + 5)$.

In order to calculate the load transfer of the fasteners, the nodal forces are determined thanks to (4), (31), (42) and the nodal displacements calculated by the previous system:

$$\begin{cases} Q_i = -0.5E^{(1)}e^{(1)}b[-\theta\eta(F - E) + C], \\ Q_j = -0.5E^{(2)}e^{(2)}b[\omega\eta(F - E) + C], \\ Q_k = 0.5E^{(1)}e^{(1)}b[-\theta\eta(Fe^{\eta\Delta} - Ee^{-\eta\Delta}) + C], \\ Q_l = 0.5E^{(2)}e^{(2)}b[\omega\eta(Fe^{\eta\Delta} - Ee^{-\eta\Delta}) + C]. \end{cases} \tag{52}$$

6 Results

Both previous approaches lead to the same results. For a hybrid joint with 3 fasteners with a Coulomb modulus of the adhesive, the value of which is near from 0Mpa, the models allow to find the results which are given by the analytical model of a bolted joint. In the same way, the models provide the behaviour of a bonded joint, when the rigidity of the fasteners is equal to 0.

From now we will consider examples with two fasteners. Table 1 gives the values of the mechanical and geometric parameters used for the whole following curves. The transversal pitch is taken equal to 1 mm in order to present significantly the effect of the fasteners on the curves, even if it is not realistic.

Figure 8 gives the load transfer along the second adherend in a two fasteners hybrid joint; this load transfer is compared to the load transfer in a simply bonded

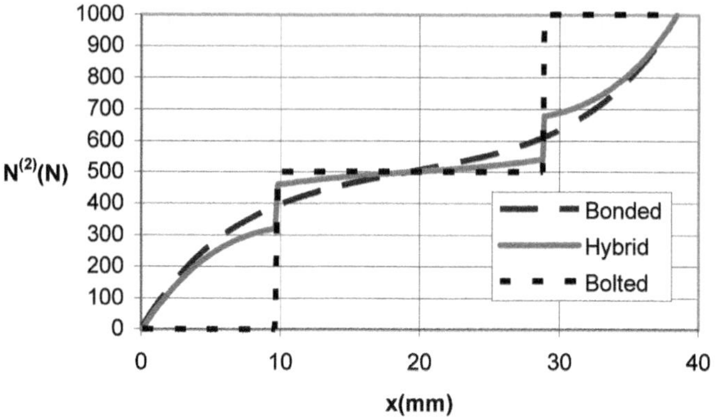

Fig. 8. Load transfer ($b = 1$ mm, $C_u = 40000$ N/mm).

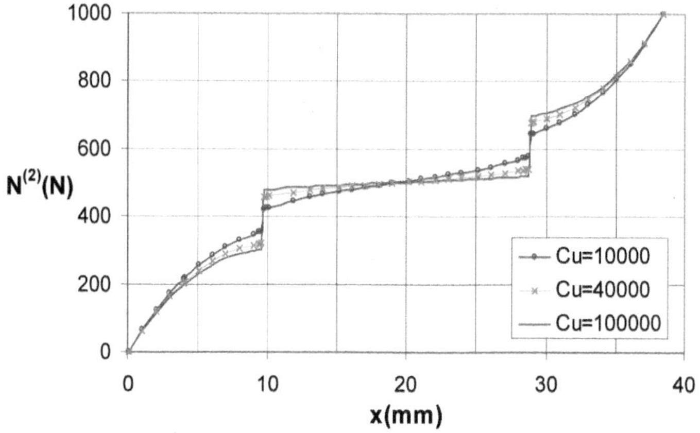

Fig. 9. Load transfer as a function of C_u ($b = 1$ mm).

and a simply bolted joint. This figure shows that the load transfer in a hybrid joint sets between the load transfer in a bonded joint and a bolted joint. Both steps correspond to the transfer by the fasteners and the aspect in *sinh* corresponds to the load transferred by the adhesive.

Figure 9 represents the load transfer for different rigidities of the fasteners. This figure shows that the load transfer is mainly performed at the bonded ends of the joint. Then, compared to a simply bolted joint, the bonding decreases highly the load transferred by the fasteners.

Moreover the parametric study performed thanks to these models fits the trends given in [6] (experimental and numeric study of hybrid joints in single-lap configuration) that is to say the load transferred by the bolts increases when:

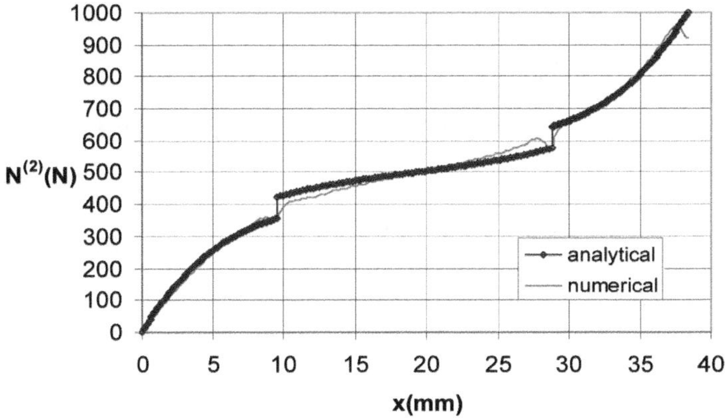

Fig. 10. Comparison between numerical and analytical approaches ($b = 1$ mm, $C_u = 10000$ N/mm).

- the Young modulus of adherends increases;
- the thickness of the adherends increases;
- the length of the lap decreases;
- the Coulomb modulus of the adhesive decreases;
- the longitudinal pitch decreases.

Thanks to these models, it is possible to add that the load transferred by the fastener increases when:

- the rigidity of the fasteners increases;
- the transversal pitch decreases;
- the edge distance decreases.

It is noticeable that the load transferred by the fasteners is nearing 0 when the edge distance increases.

Finally, using the Finite Element code SAMCEF, a two-dimensional model in plane strain is developed. The hypotheses are the same than the one of the analytical approaches. Three superposed layers simulate the joint adherend-adhesive-adherend. These elements are quadrangle elements of degree 2. Each fastener is simulated by quadrangle elements of degree 2. As shown in Figure 10, there is a good correlation between the numerical and analytical approaches.

7 Considering the Elastic-Plastic Behaviour of the Adhesive

In the previous models, the adhesive is perfectly elastic. However, perfectly elastic-plastic behaviour of the adhesive can be computed using the first approach. It is assumed that both edge distances are equal. Moreover, it is assumed that the adhesive has a plastic behaviour only into the edge distances, since the load transfer is more important along both bonded ends. This elastic-plastic approach is inspired by [4].

The length, along which the adhesive has a plastic behaviour, is quoted x_p:

$$0 \leq x_p \leq d. \tag{53}$$

It becomes:

$$\forall x \leq x_p,\ T_1(x) = T_p \tag{54}$$

and

$$\forall x \in [L - x_p;\ L],\ T_{n+1}(x) = T_p. \tag{55}$$

The local equilibrium equations are the same as (1) and (2). Hence, by integrating:

$$\forall x \leq x_p,\ N_1^{(2)}(x) = T_p bx \tag{56}$$

and

$$\forall x \in [L - x_p;\ L],\ N_{n+1}^{(2)}(x) = f - T_p b(L - x). \tag{57}$$

Both following formulae replace (12) and (13):

$$N_1^{(2)}(x_p) = T_p bx_p \tag{58}$$

and

$$N_{n+1}^{(2)}(L - x_p) = f - T_p bx_p. \tag{59}$$

Finally with (11) along the elastic region, the linear system (22) in elastic-plastic behaviour of the adhesive becomes:

$$\begin{cases} A_1 e^{-\eta x_p} + B_1 e^{\eta x_p} = \gamma\eta^{-2} f + T_p\, bx_p \\ A_{n+1}\, e^{-\eta(L-x_p)} + B_{n+1}\, e^{\eta(L-x_p)} = (1+\gamma\eta^{-2}) f - T_p\, bx_p \\ qe^{-\eta d_i} A_i + re^{\eta d_i} B_i - e^{-\eta d_i} A_{i+1} - e^{\eta d_i} B_{i+1} = 0 \\ e^{-\eta d_i} A_i - e^{\eta d_i} B_i - e^{-\eta d_i} A_{i+1} + e^{\eta d_i} B_{i+1} = 0 \end{cases} \quad i \in [1;n]. \tag{60}$$

The method consists in an iterative resolution of the elastic problem. More precisely, for the first iteration x_p is taken equal to 0. Then (60) is solved until the adhesive shear stress in the elastic region is lower to T_p.

Figure 11 represents the ratio of the bolt load transfer with the elastic-plastic adhesive (τ_p) divided by the ratio of the bolt load transfer with the elastic adhesive (τ_e) when the load increase.

This curve shows that the bolt load transfer in elastic-plastic case is around twice as big as the ratio of the bolt load transfer in elastic case, when the adhesive is plasticized all along the edge distance.

In that case, (3) and (15) allow us to write:

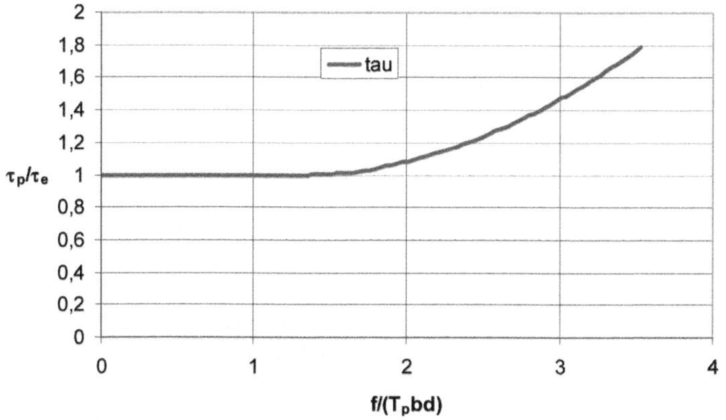

Fig. 11. Bolt load transfer with elastic-plastic adhesive ($b = 19.2$ mm, $C_u = 50000$ N/mm).

$$\tau_p f = C_u \frac{e}{G} T_p. \tag{61}$$

This last equation provides the three most important parameter of the design of a hybrid joint:

- the edge distance (d);
- the rigidity of the fastener (C_u);
- the relative rigidity of the adhesive (G/e).

8 Conclusions

Two one-dimensional elastic analytical models are developed and presented in this paper. They allow us to easily analyse (Figure 12) the influence of the mechanical and geometric parameters on the load transfer in a hybrid joint. These models are robust and simple to use. These models can be regarded as a design tool. Moreover, an original approach, inspired by the Finite Element Method, is presented and validated. Finally, an extension to the perfectly elastic-plastic behaviour of the adhesive is proposed.

Three ways are under consideration to continue. The first way is the development of a two-dimensional model taking into account the bending of the adherend due to the eccentricity of the load path, and the adhesive peeling stress. The second way is the numerical simulation. A three-dimensional model is required to represent accurately and to understand better the behaviour of the hybrid joint. Finally, the last way is the test experience. Static tests using instrumented bolts [6] are launched in order to validate and calibrate these models. In particular, the rigidity of the fastener does not seem to be determined accurately, using to the different existing formulations, which does not take into account the hybrid configuration of joint.

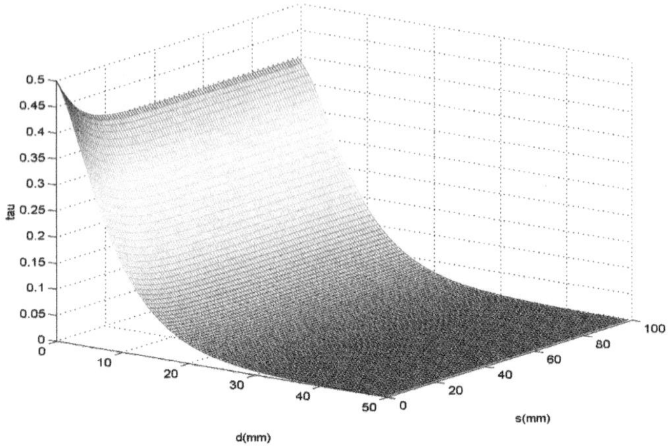

Fig. 12. Influence of d and s on the bolt load transfer ($b = 1$ mm, $C_u = 50000$ N/mm).

Acknowledgements

This study is performed in cooperation with AIRBUS (Toulouse and Saint-Nazaire). The authors acknowledge the industrial partners for their advice and support.

References

1. J. Huet, Du calcul des assemblages par boulons ou rivets travaillant en cisaillement, in *Les assemblages mécaniques: tendances actuelles et perspectives*, CETIM, ISBN 2-85400-328-4, 1995, pp. 133–147.
2. M.B. Tate, S.J. Rosenfeld, Analytical and experimental investigation of bolted joint, NACA, Technical Note 1458, 1947.
3. H. Huth, Influence of fastener flexibility on the prediction of load transfer and fatigue life for multible-row joint, in *ASTM STP* 927, John M. Potter (Ed.), 1986, pp. 221–250.
4. L.J. Hart-Smith, Adhesive-bonded double-lap joints, NASA, Technical Report CR-112235.
5. F. Erdogan, M. Ratwani, Stress distribution in bonded joints, *Journal of Composite Materials* **5**, 1971, 378–393.
6. G. Kelly, Load transfer in hybrid (bonded/bolted) composite single-lap joints, *Composite Structures* **69**(1), 2005, 35–43.

Precision Study of a Decoupled Four Degrees of Freedom Parallel Robot Including Manufacturing and Assembling Errors

R. Rizk[1,2], N. Andreff[1,2], J.C. Fauroux[1], J.M. Lavest[2] and G. Gogu[1]

[1]*Mechanical Engineering Research Group (LaMI), French Institute of Advanced Mechanics, and Blaise Pascal University, Campus de Clermont-Ferrand/Les Cézeaux, BP 265, Aubière Cedex, France; E-mail: {rrizk, fauroux, andreff, gogu}@ifma.fr*
[2]*Laboratory of Sciences and Materials for Electronics and of Automatic, Blaise Pascal University/CNRS, Clermont-Ferrand, France; E-mail: lavest@lasmea.univ-bpclermont.fr*

Abstract. This paper presents a study of geometric errors due to assembly and manufacturing tolerance on the kinematic accuracy of a parallel robot with four degrees of freedom with decoupled motions and fully isotropic in translation – ISOGLIDE4-T3R1.

The effects of geometric errors on the kinematic accuracy are analytically calculated under the hypothesis that the components are rigid bodies by using the ISOGLIDE4-T3R1 forward Jacobian. Analytic calculations give a general form for the pose accuracy in the workspace. They both give a form for the ISOGLIDE4-T3R1 sensitivity to geometric errors, which allows to quantify which defect is more harmful, and how the sensitivity varies in the workspace.

In the second part in this paper, geometric error measurements on the robot prototype are presented. They are carried out using vision-based metrology. This system not only has the required capabilities but also has the advantage of measuring six degrees of freedom in the same coordinate system without contact. The metrology results and analytic calculations allow the hypothesis of rigid bodies to be quantified and open up prospects for the error influences on robot stiffness and accuracy with deformable bodies.

Key words: parallel robot, decoupled motions, accuracy, sensitivity, ISOGLIDE4-T3R1, metrology by vision.

1 Introduction

Parallel mechanisms are emerging in industry (for instance, machine-tools, high-speed pick-and-place robots, flight simulators, medical robots). A parallel mechanism can be defined as a mechanism with a closed kinematic chain, made up of an end-effector with N degrees of freedom and a fixed base connected to each other

S. Tichkiewitch et al. (eds.), Advances in Integrated Design and Manufacturing in Mechanical Engineering II, 111–127.

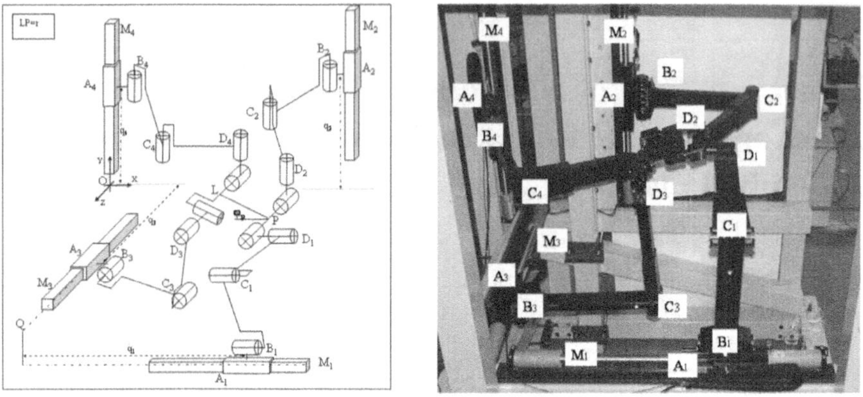

Fig. 1. ISOGLIDE4-T3R1 kinematics diagram and prototype.

by at least two kinematic chains, the motorization being carried out by N actuators [1]. This allows parallel mechanisms to bear higher loads at higher speed and often with a higher repeatability [1]. This paper presents a study of the kinematic accuracy of a parallel robot with four degrees of mobility and decoupled motions called ISOGLIDE4-T3R1, isotropic in translation and have three translations and one rotation for the end-effectors. The mechanism is proposed in [2]. Each of the four kinematic chains of ISOGLIDE4-T3R1 is also called "leg", the body of the leg connected to the actuator is called "arm", and the second body "forearm". An essential property of the parallel mechanisms is the existence of no actuated joints known as "passive joints". For ISOGLIDE4-T3R1, they are revolute joints between the legs and the linear actuators, between the arm and the forearm of each leg, and between the leg and the end-effector formed by a platform (Figure 1).

A parallel robot is characterized by its differential kinematic model which establishes a relation between the infinitesimal variation of articular coordinates $[dq]$ and the infinitesimal variation of the operational coordinates $[dX]$. This relation can be expressed in matrix form as:

$$[A][dX] = [B][dq]. \tag{1}$$

Matrix $[A]$ is known as the parallel matrix and matrix $[B]$ as the serial matrix [3] and this relation can be also expressed in this form if $[A]$ is reversible:

$$[dX] = [{}^{h}J_{P}][dq], \tag{2}$$

where P is the characteristic point of the robot mobile platform, h the reference coordinate system in which the operational variables are expressed. ${}^{h}J_{P}$ is known as the mechanism forward Jacobian matrix. Equation (2) can be considered as the design equation of parallel robotic manipulators and the Jacobian matrix ${}^{h}J_{P}$ as the "design matrix" in terms of axiomatic design [4]. The mechanism is a fully isotropic parallel manipulator if ${}^hJ_P$ is proportional to the identity matrix. It is with uncoupled motions if ${}^hJ_P$ is diagonal and decoupled motions if ${}^hJ_P$ is triangular. In all other cases, the motions of the mechanism are coupled [5]. Several parallel robots of type T3R1 (three translations and one rotation of the end-effector) with coupled motions are proposed in the literature [1, 6–12]. ISOGLIDE4-T3R1 is the first parallel robot with four decoupled motions [20]. The development of the forward Jacobian is based on the closure of the loops which form the various kinematic chains. With the coupling, it is difficult to express the relations between the articular coordinates and the operational coordinates. In a parallel robot with decoupled motions, the forward Jacobian is simpler.

To have motion decoupling, high accuracy of perpendicularity and parallelism between the joints will be imposed. However, the large number of links and passive joints often lead to a lower accuracy [13]. Industrial tolerances can influence decoupling and isotropy and consequently accuracy. The inaccuracies of the kinematic behaviour of the robot as well as the pose (position and orientation) of the moving platform can be calculated by integrating the geometric errors in the forward Jacobian matrix or they can be measured on the robot prototype.

Several methods to perform kinematic identification of parallel mechanisms are proposed in the literature [14–17]. In this paper, measurements are made by artificial vision. This system not only has the required accuracy but also has the advantage of being contactless, measuring six degrees of freedom (position and orientation) in the same coordinate system and finally involving low cost compared to other metrology systems. The goals of this study are to answer the following questions: which defect is most harmful? How does the sensitivity of the end-effector pose vary according to the geometric defects in the workspace? And finally to what extent is the rigid body hypothesis correct?

2 Kinematic Structure and Properties

For the majority of parallel mechanisms, the forward Jacobian does not exist in analytical form. Moreover, its unicity is not assured. This difficulty is due often to the coupling with which, in the majority of the cases, it will be impossible to obtain simple linear relations, since the equations of closure give non-linear equations. In the case of ISOGLIDE4-T3R1 since it is decoupled, the forward Jacobian is simple. In the ideal case (no assembly and manufacturing errors), equation (2) becomes:

$$\begin{bmatrix} dx_P \\ dy_P \\ dz_P \\ d\varphi_P \end{bmatrix} \begin{bmatrix} 1 & 0 & 0 & 0 \\ 0 & 1 & 0 & 0 \\ 0 & 0 & 1 & 0 \\ 0 & -\frac{1}{r\cos\varphi_P} & 0 & \frac{1}{r\cos\varphi_P} \end{bmatrix} \begin{bmatrix} dq_1 \\ dq_2 \\ dq_3 \\ dq_4 \end{bmatrix}. \tag{3}$$

The forward Jacobian matrix expressed in the reference coordinate system $(OXYZ)$ is triangular, therefore ISOGLIDE4-T3R1 has decoupled motions. The submatrix, which connects infinitesimal displacements of the first three actuators, with the end-effector's infinitesimal translations, is the identity matrix, therefore ISOGLIDE4-T3R1 is isotropic in translation. Subsequent calculations will be made in the reference coordinate system $(OXYZ)$ related to the frame, with the Y axis according to the ascending vertical and axes X and Z in the horizontal plane. In the ideal case, the M_1, M_2 and M_3 actuators axes are following axes X, Y and Z respectively, the axis of M_4 is parallel to that of M_2.

3 Assembly Tolerances and Errors

It is known that this type of machine is not infinitely rigid and does not have the same rigidity in all its workspace [18]. Moreover, materials are not perfectly homogeneous and during the realization, tool wear and consequently manufacturing quality are not the same during all the stages of realization. Because of these phenomena, realization is not perfect and errors inevitably exist in the components. Since the paper deals with kinematic accuracy, the study is limited to two particular cases for which the motion is always possible even with infinitely rigid components. The effect of these errors on the kinematic characteristics (isotropy in translation and motion decoupling) was presented in [19] on Version 2 of the ISOGLIDE4-T3R1. In this paper we present the effects on the pose accuracy. The experiment was implemented on the version 1 of ISOGLIDE4-T3R1 (Figure 1). Version 2 has a hyperstaticity degree equal to 4 and version 1 has a hyperstaticity degree equal to 2 [20]. Both versions have the same differential kinematic model so analytical calculations presented in[19] can be applied.

3.1 Error of Perpendicularity between Actuators

Let us consider the case where the axes of actuators, M_1, M_2, M_3 and M_4 form angles α_m, β_m, γ_m and δ_m with axes X, Y, Z and Y respectively (Figure 2). We obtain the following forward Jacobian:

$$\begin{bmatrix} dx_P \\ dy_P \\ dz_P \\ d\varphi_P \end{bmatrix} = \begin{bmatrix} \cos\alpha_m & 0 & 0 & 0 \\ 0 & \cos\beta_m & 0 & 0 \\ 0 & 0 & \cos\gamma_m & 0 \\ 0 & -\dfrac{\cos\beta_m}{r\sqrt{1-\left(\frac{q_4\cos\delta_m - q_2\cos\beta_m}{r}\right)^2}} & 0 & \dfrac{\cos\delta_m}{r\sqrt{1-\left(\frac{q_4\cos\delta_m - q_2\cos\beta_m}{r}\right)^2}} \end{bmatrix} \begin{bmatrix} dq_1 \\ dq_2 \\ dq_3 \\ dq_4 \end{bmatrix}. \tag{4}$$

The inaccuracy can be calculated as the difference with the same articular coordinates between the ideal pose and the pose with orientation errors of actuators

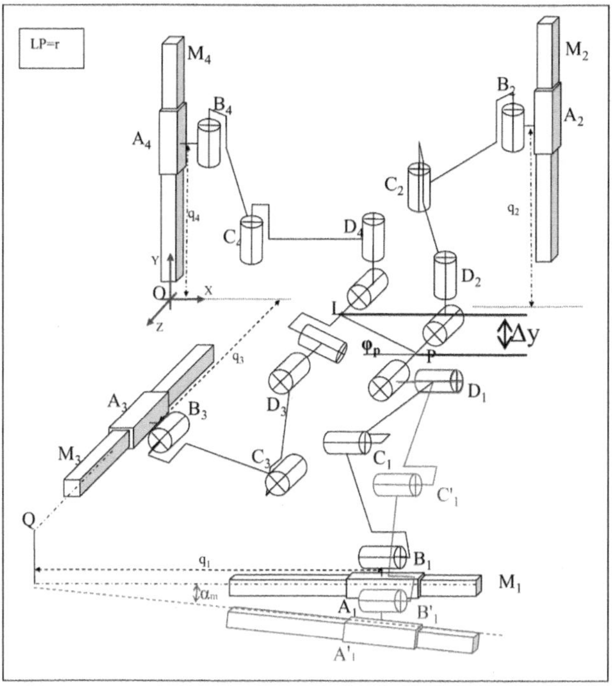

Fig. 2. Kinematic diagram of the ISOGLIDE-T3R1 with orientation defects of the actuators.

$$\begin{bmatrix} e_{dx_P} \\ e_{dy_P} \\ e_{dz_P} \\ e_{d\varphi_P} \end{bmatrix} = \tag{5}$$

$$\begin{bmatrix} 1-\cos\alpha_m & 0 & 0 & 0 \\ 0 & 1-\cos\beta_m & 0 & 0 \\ 0 & 0 & 1-\cos\gamma_m & 0 \\ 0 & -\frac{1-\cos\beta_m}{r\sqrt{1-\left(\frac{q_4\cos\delta_m - q_2\cos\beta_m}{r}\right)^2}} & 0 & \frac{1-\cos\delta_m}{r\sqrt{1-\left(\frac{q_4\cos\delta_m - q_2\cos\beta_m}{r}\right)^2}} \end{bmatrix} \begin{bmatrix} dq_1 \\ dq_2 \\ dq_3 \\ dq_4 \end{bmatrix}.$$

Figure 3 shows the position errors along X caused by M_1 orientation error. This effect is the same for the Y and Z positions. Equation (5) and Figure 3 show that the error does not depend only on the value of the defect but also on infinitesimal displacement dq_1. We have to integrate (5) to obtain the position errors (Figure 3).

Figure 4 and Equation (5) present several important points. The rotation error does not depend linearly on the infinitesimal articular coordinate displacements dq_2 and dq_4. It depends on q_2 and q_4 as well as on β_m and δ_m. The error graphs (Fig-

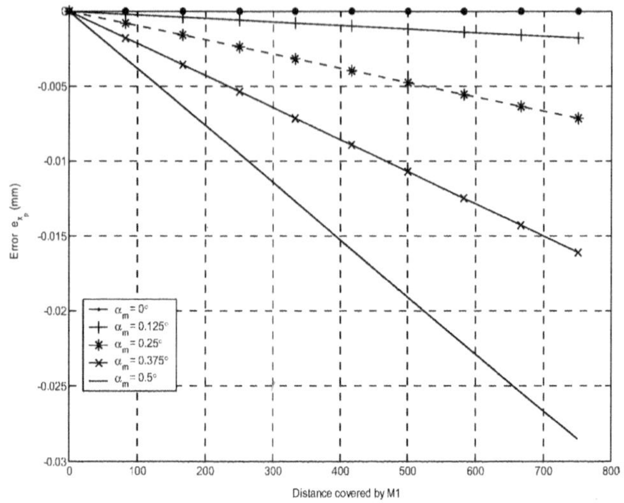

Fig. 3. Position errors along X caused by motor M_1 orientation error (units are in mm).

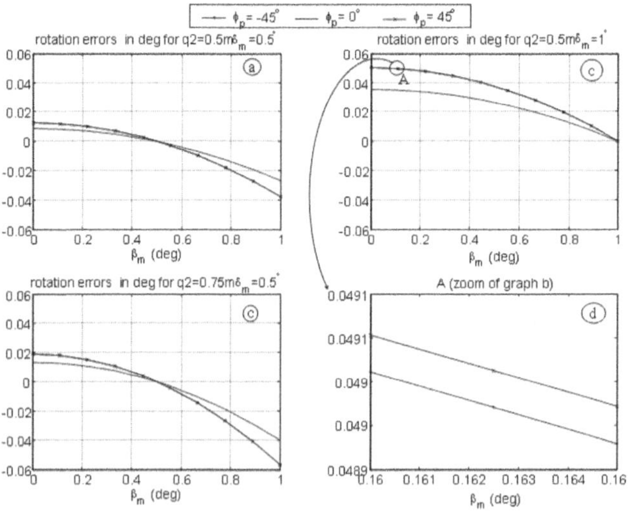

Fig. 4. Rotation errors with actuators orientation errors.

ure 4) show that the rotation error is cancelled when β_m and δ_m are equal. Rotation increases when the difference between the Y coordinate of points L (see the kinematic diagram in Figures 1 and 2) and P increases. When $\beta_m > \delta_m$, Y coordinate of point decreases more than Y coordinate of point L. That leads to a positive sup-

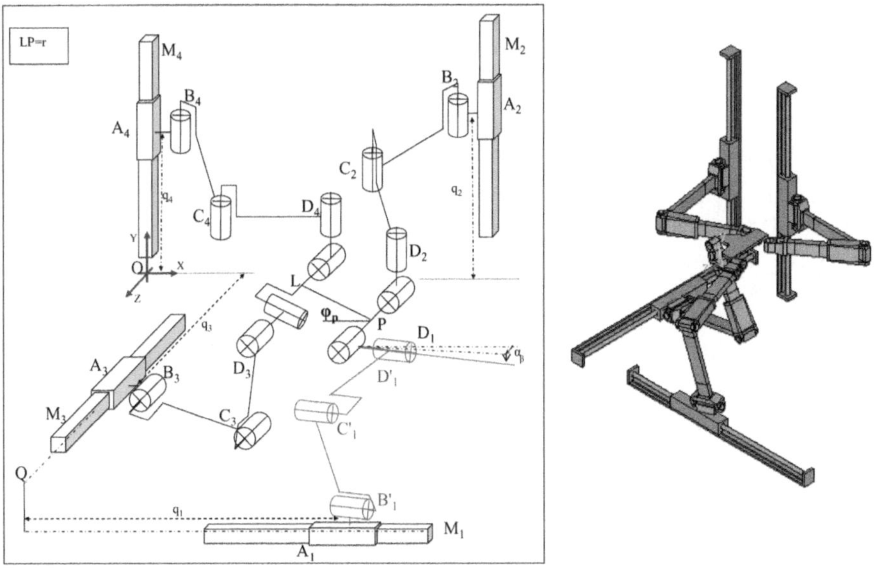

Fig. 5. ISOGLIDE-T3R1 with defects of orientation for leg 1. Kinematic diagram and CAD model.

plement of rotation and therefore a negative error. When both defects are of the same value, both Y coordinate components decrease by keeping the same proportion that preserves the rotation angle. The influence of vertical position q_2 is also significant. In Figure 4, end-effector rotation errors with the same β_m and δ_m defects and desired rotation φ_p are not the same when q_2 changes (see Figures 4a and 4c). When q_2 increases, the end-effector vertical position error increases linearly such as X position error for actuator M_1 in Figure 3, so the difference between the two vertical leg extremities changes and so does the rotation angle. In addition, the rotation error is not the same for orientations φ_p and $-\varphi_p$ (see the zoom in Figure 4d). For φ_p, q_4 is bigger than q_2 for a certain value Y. For $-\varphi_p$, q_4 is smaller than q_2 for the same ΔY. The effect of δ_m error is stronger when q_4 is big such as the effect of α_m on X_P (Figure 3).

3.2 Defect in the Orientation of the Revolute Leg Joints

In this part, the parallelism between revolute joints named B_i, C_i and D_i of leg L_i is assumed (Figure 5). We focused on the orientation defect of the whole leg with respect to the actuator at point B_i. This time, the defect is not the angle between the axis of the leg and the reference axis [19]. If rotation is not around one of the main axes of the reference coordinate system, it is necessary to break up the defect into two consecutive rotations around the principal axes of the reference coordinate system

[21]. All the legs have defects in their orientation. Defect angles will be named α, β, for leg L_1 around axes Y and Z; γ, δ, for leg L_2 around axes X and Z; ε, φ, for leg L_3 around axes X and Y; ζ, η for leg L_4 around axes X and Z. The loop closures give:

$$\begin{bmatrix} 1 & 0 & 0 & 0 \\ 0 & 1 & 0 & 0 \\ 0 & 0 & 1 & 0 \\ 0 & -1 & 0 & 1 \end{bmatrix} \begin{bmatrix} dq_1 \\ dq_2 \\ dq_3 \\ dq_4 \end{bmatrix} = \begin{bmatrix} 1 & \tan\beta & -\tan\alpha & 0 \\ -\tan\delta & 1 & \tan\gamma & 0 \\ \tan\varphi & -\tan\varepsilon & 1 & 0 \\ \tan\delta - \tan\varsigma & 0 & \tan\eta - \tan\gamma & r\cos\varphi_P \end{bmatrix} \begin{bmatrix} dx_P \\ dy_P \\ dz_P \\ d\varphi_P \end{bmatrix}. \tag{6}$$

There,

$$\begin{bmatrix} dx_P \\ dy_P \\ dx_P \\ d\varphi_P \end{bmatrix} = \frac{r\cos\varphi_P}{D}$$

$$\times \begin{bmatrix} 1+\tan\varepsilon\tan\gamma & -\tan\beta - \tan\varepsilon\tan\alpha & \tan\beta\tan\gamma + \tan\alpha & 0 \\ \tan\delta + \tan\varphi\tan\gamma & 1+\tan\alpha\tan\varphi & -\tan\gamma + \tan\alpha\tan\delta & 0 \\ \tan\delta\tan\varepsilon - \tan\varphi & \tan\varepsilon + \tan\beta\tan\varphi & 1+\tan\delta\tan\beta & 0 \\ \frac{A}{r\cos\varphi_P} & \frac{B-1}{r\cos\varphi_P} & \frac{C}{r\cos\varphi_P} & \frac{D}{r\cos\varphi_P} \end{bmatrix} \begin{bmatrix} dq_1 \\ dq_2 \\ dq_3 \\ dq_4 \end{bmatrix}. \tag{7}$$

where D is the determinant of the serial matrix established in (6). A, B and C are functions of the errors and the rotation angle

$$\begin{aligned} D &= r\cos\varphi_P[1+\tan\gamma(\tan\varepsilon + \tan\varphi\tan\beta) \\ &\quad + \tan\delta(\tan\beta - \tan\delta\tan\varsigma) + \tan\varphi\tan\alpha], \\ A &= -\tan\delta\tan\varepsilon\tan\eta + \tan\varphi\tan\eta - \tan\varphi\tan\gamma - \tan\delta + \tan\zeta + \tan\zeta\tan\varepsilon\tan\gamma, \\ B &= \tan\varepsilon\tan\eta + \tan\varepsilon\tan\gamma - \tan\varphi\tan\beta\tan\eta + \tan\varphi\tan\beta\tan\gamma + \tan\delta\tan\beta \\ &\quad - \tan\delta\tan\varepsilon\tan\alpha - \tan\varsigma\tan\beta + \tan\varsigma\tan\varepsilon\tan\alpha, \\ C &= -\tan\eta + \tan\gamma - \tan\delta\tan\beta\tan\eta - \tan\delta\tan\alpha \\ &\quad + \tan\varsigma\tan\beta\tan\gamma + \tan\varsigma\tan\alpha. \end{aligned}$$

Figures 6 and 7 present the error of the end-effector rotation when both vertical actuators are animated for various rotation angles of the platform. Calculations were performed for an example of leg defect orientation (η, φ). Both figures show that the error is increasing with the articular coordinate q_3. Another interesting point is that errors are quite different for two opposite desired rotations as in the previous case. In fact, error is the result of motion coupling is not the angle between the ideal and the real orientation [19]. In spite of the motion decoupling, the pose accuracy depends on x_P, y_P, z_P and φ_P. In Figure 6, errors are null when q_3 is between 200 mm and 300 mm but in Figure 7 when $q_3 = 0$. In fact in Figure 6 for non-zero values, the

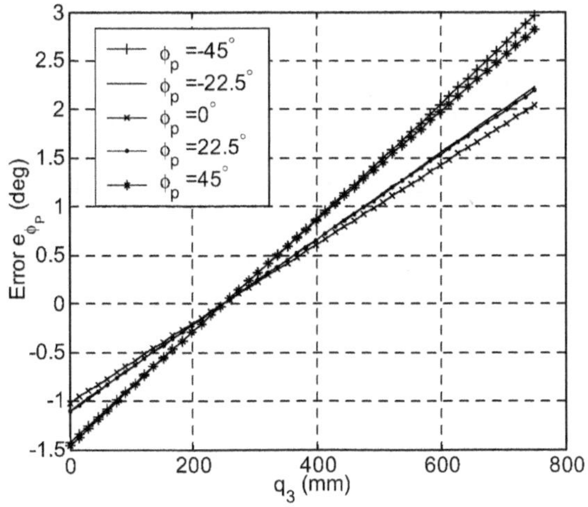

Fig. 6. Errors on rotation according to q_3 for $q_1 = q_2 = 250$ mm, $\eta = \varphi = 1°, \alpha = \beta = \gamma = \delta = \varepsilon = \zeta = 0°$.

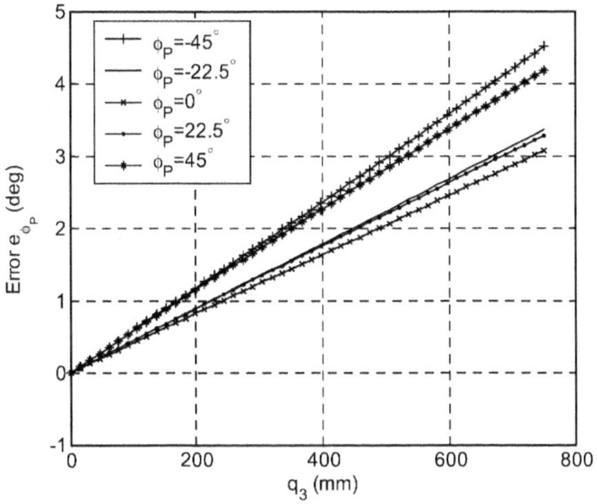

Fig. 7. Errors on rotation according to q_3 for $q_1 = q_2 = 250$mm, $\eta = 1°, \alpha = \beta = \gamma = \delta = \varepsilon = \zeta = 0°$.

defect η and φ are reciprocally compensated. No compensation exists when just one of the two values is non-zero.

4 Sensitivity Analysis

The sensitivity analysis of the ISOGLIDE4-T3R1 to an error is the analysis of the influence of the error variation on each operational coordinate of the robot. Mathematically, the sensitivity of an operational coordinate is its partial derivative with respect to the defect. For actuator orientation errors we obtain:

$$\left|\frac{\partial dx_P}{\partial \alpha_m}\right| = |dq_1 \sin \alpha_m|, \tag{8}$$

$$\left|\frac{\partial dy_P}{\partial \beta_m}\right| = |dq_2 \sin \beta_m|, \tag{9}$$

$$\left|\frac{\partial dz_P}{\partial \gamma_m}\right| = |dq_3 \sin \gamma_m|, \tag{10}$$

$$\left|\frac{\partial d\varphi_P}{\partial \beta_m}\right| = |B(adq_2 + bdq_4)|, \tag{11}$$

$$\left|\frac{\partial d\varphi_P}{\partial \delta_m}\right| = |D[cdq_2 + ddq_4)|, \tag{12}$$

with

$$\begin{aligned}
a &= (-r^2 + q_4 \cos \delta_m (q_4 \cos \delta_m - q_2 \cos \beta)),\\
b &= -q_2 \cos \delta_m (q_4 \cos \delta_m - q \cos \beta_m),\\
c &= -q_4 \cos \beta_m (q_4 \cos \delta_m - q_2 \cos \beta_m),\\
d &= (r^2 - q_2 \cos \beta_m (q_4 \cos \delta_m - q_2 \cos \beta_m)),\\
B &= \frac{\sin \beta_m}{r^3 \left(1 - \frac{(q_4 \cos \delta_m - q_2 \cos \beta_m)^2}{r^2}\right)^{3/2}},\\
D &= \frac{\sin \delta_m}{r^3 \left(1 - \frac{(q_4 \cos \delta_m - q_2 \cos \beta_m)^2}{r^2}\right)^{3/2}}
\end{aligned}$$

The sensitivity of one operational coordinate to an actuator orientation defect depends on the corresponding articular coordinate. Equations (11) and (12) show that orientation sensitivity depends on actuator orientation defects β_m and δ_m and their infinitesimal displacements dq_2 and dq_4. Figure 8a shows the factor of dependence on dq_2 of the sensitivity to β_m. Figure 8b shows the factor of dependence on dq_4 of the sensitivity to β_m. Figure 8c shows the factor of dependence on dq_2 of the sensitivity to δ_m. Figure 8d shows the factor of dependence on dq_4 of the sensitivity to δ_m. If $q_2 \cos \beta = q_4 \cos \delta$. Error on $d\varphi_P$ is null [19], the sensitivity of the rotation to β_m depends only on dq_2 and to δ_m only on dq_4.

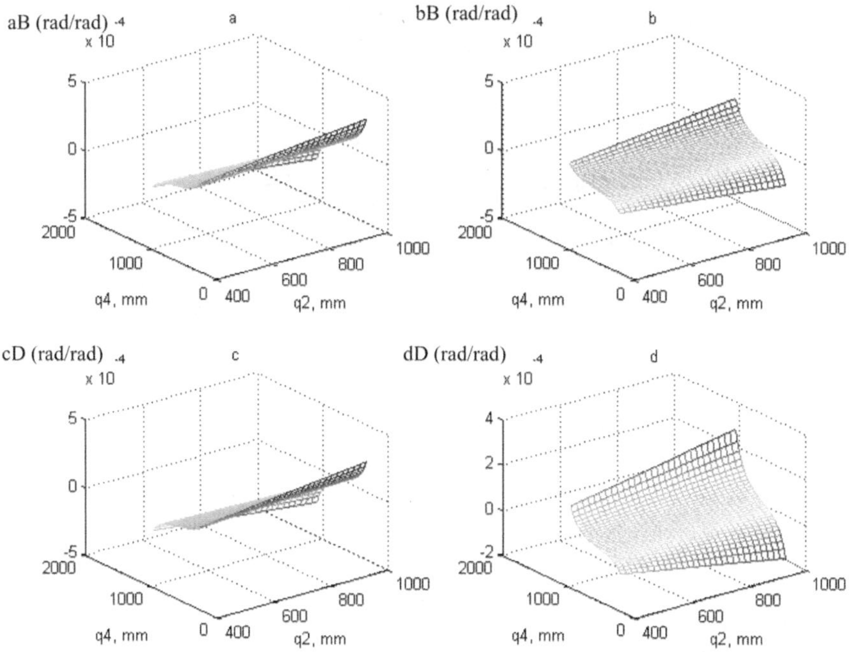

Fig. 8. Sensitivity factors of rotation to M_2 and M_4 actuator orientation $\beta_m = \delta_m = 0.1°$.

5 Discussion

The above results show that the defects of actuator orientation influence linearly according to the articular coordinate and not linearly to the defect amplitude on the ISOGLIDE4-T3R1 operational translations precision. On the other hand, defect effects are much more complicated on rotation, and for two defects of equal values of $\beta_m \delta_m$, inaccuracies are cancelled. This phenomenon can be explained by the coupling of both vertical actuators to define the variation of the rotation angle $(d\varphi_P)$ (Equation (4)). In addition, the inaccuracies due to leg orientation defects are much more significant than the inaccuracies due to actuator orientation. This is due to the coupling induced by the legs perpendicularity defects. Moreover it is seen that the leg orientation defects are coupled (Equation (7)), which also complicates correction by command. Another very interesting point is that the inaccuracies as in an absolute value increased with the articular coordinates.

Fig. 9. Actuator measurement.

6 Experimental Validation

This section presents an experimental validation by artificial vision for the study presented above. Manufacturing and assembly errors are measured. Distances covered by actuators that represent articular coordinates are measured in the same time with the end-effector pose (position and orientation) in the same global coordinate system. The difference between the ideal and the real pose gives errors. In this paper, the parallelism defects between the passive joints of the same leg are ignored.

6.1 Experiment

The experiment is divided into three parts:

- Actuator orientations measurement, which is realized with stereoscopic heads.
- Leg rotation axis orientation measurement is realized with stereoscopic heads.
- End-effector pose with actuator displacement measurement that requires the simultaneous use of stereo heads and a calibrated monocular camera mounted on the end-effector.

6.1.1 Actuators Translational Axis Orientation

On each actuator, a photoreflective mark is stuck (Figure 9). To realize measurements about ten positions are taken. For each position, the stereoscopic head takes 100 images. Results given by the stereoscopic system are stored in a spreadsheet. Using a numerical program, images are filtered and their mean is calculated. Normally, these points are distributed on a perfect line but because of the incertitude due to the actuator profile, an approximate line is fitted in least square sense. The accuracy of this system and the experiment set up has been presented in [22].

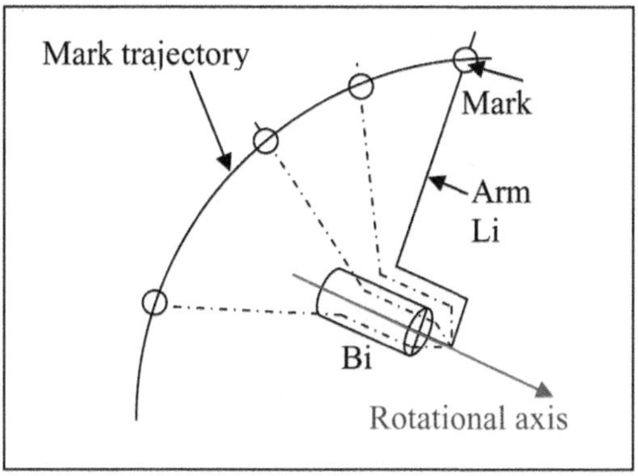

Fig. 10. Leg rotational axis estimation.

6.1.2 Legs Rotational Axis

Considering a mark stuck on an arm. When the arm of leg L_i turns around the revolute joint B_i, the trajectory of every point in the arm, in particular the mark can be seen as a circle located in a plane perpendicular to the revolute joint B_i axis. To calculate the B_i axis, the equation of the plane in which the mark exists should be estimated. Three points are enough to find the plane. In an experiment, we used about 10 points and the best plane is fitted in least square sense, and the normal to the plane can be led to the revolute joint axis B_i (Figure 10).

6.1.3 End-Effector Pose Measurement

A camera was fixed on the end-effector (Figure 11) using a known object (calibration target) (Figure 12). The camera provides for all images the end-effector pose (position and orientation) in the same global coordinate system. Using simulttaneously the stereo heads for actuators and the camera for the end-effector, the real and theoretical end-effector pose errors can be estimated. The comparison of theoretical errors and real errors allows us to quantify the influence of the hypothesis of rigid bodies.

7 Results

Table 1 gives the measured orientation actuator β_m, γ_m and δ_m defects and legs orientation defects α, β, γ, δ, ε, φ, ζ, η. These values are calculated in a coordinate system having axis X parallel to the axis M_1 ($\alpha_m = 0$) and axis Z such that plane XOZ is parallel to the M_1 and M_3 axes. This coordinate system was chosen because the horizontal plane and the vertical direction cannot be easily determinated and allows one defect to be eliminated. (See kinematic diagram in Figure 1 for notation.)

Fig. 11. Monocular camera fixed on the end-effector.

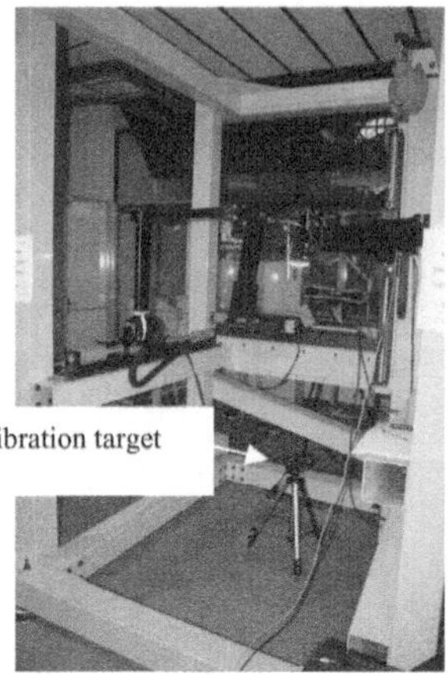

Fig. 12 Experimental device.

Table 1. Geometric errors.

Component	Error	Value (deg)
M_2	β_m	0.215
M_3	γ_m	0.815
M_4	δ_m	0.0917
L_1	α/β	1.760/1.128
L_2	γ/δ	1.985/1.124
L_3	ε/φ	0.510/0.376
L_4	ζ/η	0.821/0.3270

Results presented in Figure 13 have been realized with both vision systems (stereoscopic heads and monocular camera). These results give several informations. The first and most important conclusion is that the end-effector parasite rotations are very small with respect to the position errors and to φ_P fluctuations. Thus, it means that the robot conserves its four degrees of freedom. On the other hand, position errors are bigger in the Y direction. End-effector errors are essentially due to deformations and geometric errors. In this paper, only the self-weight is applied as load. Errors in horizontal plane are due to geometric errors and pre-stress due to the hyperstaticity of the robot. Errors in the Y direction are caused by deflexions of legs L_2 and L_4 and assembly errors. However, deflexion on L_2 and L_4 causes a lumped deformation in the revolute joints between legs and actuators M_2 and M_4, so it introduces a new geometric error that amplifies the vertical position error. The φ_P fluctuation is the result of the difference between both vertical legs deflections. Finally, it is clear that the effect of the self-weight is very important but it can be assumed as an amplification of the vertical legs orientation errors. Note that these measurements were done on an incomplete prototype and before accurate adjustments were made.

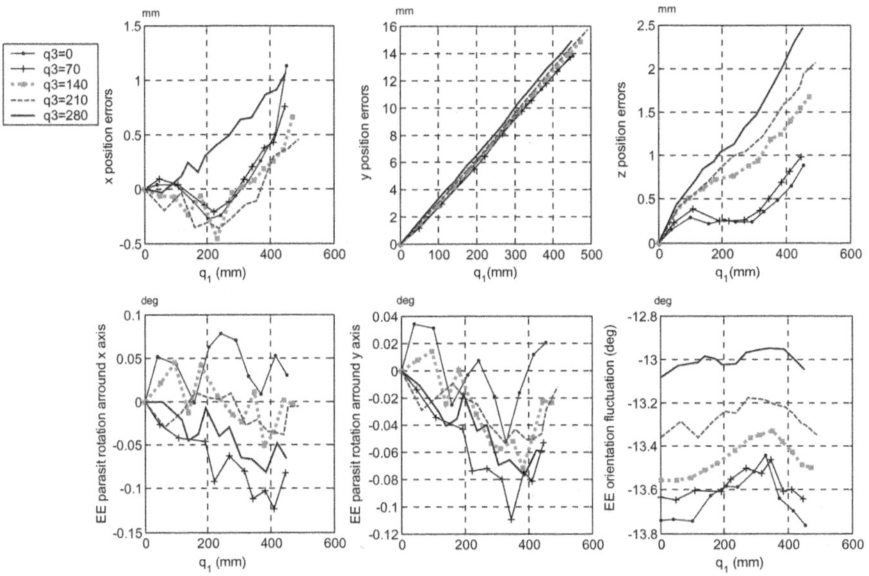

Fig. 13. End-effector pose errors.

8 Conclusions

The method and the results presented in this paper are useful in the design of parallel manipulators with decoupled Schönflies motions. According to previous considerations, we can obtain some answers to the questions addressed in the introduction. The study showed that the robot is more sensitive to leg orientation then to actuator orientation. Moreover, the dominating sensitivity is achieved with respect to orientation errors of actuators M_1 and M_4 and legs L_1 and L_4. In addition, uncertainty increases with the articular coordinates. Finally, in a kinematic study under just self-weight loading, the rigid bodies' hypothesis can be applied and deformations will appear as additional defects for the orientations of leg L_2 and L_4. Manufacturing and assembly errors will create pre-stress in the robot's deformable structure. Their effect analysis in a stiffness study will require a second order finite element analysis which will be presented in our following paper.

Acknowledgment

This work was supported by the TIMS Research federation (FR TIMS/CNRS2586) in the frame of auverfiabilis project of Regional Council of Auvergne.

References

1. J.P. Merlet. *Parallel Robots*. Dordrecht: Kluwer Academic Publishers, 2000.
2. G. Gogu. Structural synthesis of parallel robotic manipulators with decoupled motions, Internal Report ROBEA MAX-CNRS, 2002.
3. C. Gosselin, J. Angeles. Singularity analysis of closed-loop kinematic chains, *IEEE Transactions on Robotics and Automation* **6**(3), 1990, 281–290.
4. Suh. *Principles of Design*, Oxford Series on Advanced Manufacturing, 1990.
5. G. Gogu. Structural synthesis of fully-isotropic translational parallel robots via theory of linear transformations, *European Journal of Mechanics, A/Solids* **23**, 2004, 1021–1039.
6. F. Pierrot, O. Company. H4: A new family of 4-dof parallel robots, in *Proc. IEEE/ASME Int. Conf. on Advanced Intelligent Mechatronics*, 1999, pp. 508–513.
7. S. Briot, P. Maurine, S. Guegan, V. Arakelian. Le PAMINSA: Un nouveau manipulateur d'architecture parallèle aux mouvements découplés, in *Congrès Français de Nécanique*, Troyes, France, 2005.
8. Y. Fang, L.-W. Tsai. Structural synthesis of a class of 4-dof and 5-dof parallel manipulators with identical limb structures, *International Journal of Robotics Research* **21**(9), 2002, 799–810.
9. F. Gao, W. Li, X. Zhao, Z. Jin and H. Zhao. New kinematic structures for 2-, 3-, 4-, and 5-DOF parallel manipulator design, *Mechanism and Machine Theory* **37**(11), 2002, 1395–1411.
10. Z. Huang, Q.C. Li. General methodology for type synthesis of symmetrical lower-mobility parallel manipulators and several novel manipulators, *International Journal of Robotics Research* **21**(2), 2002, 131–145.
11. Z. Huang, Q.C. Li. Type synthesis of symmetrical lower-mobility parallel mechanisms using the constraint-synthesis method, *International Journal of Robotics Research* **22**(1), 2003, 59–79.
12. X. Kong, C.M. Gosselin. Type synthesis of 3T1R parallel manipulators based on screw theory, *IEEE Transactions on Robotics and Automation*, 2004.
13. J. Wang, O. Masory. On the accuracy of a Stewart platform – Part I: The effect of manufacturing tolerances, in *International Conference on Robotics and Automation*, Atlanta, 1993, pp. 114–120.
14. P. Renaud, N. Andref, J.M. Lavest, M. Dhome. Simplifying kinematic identification of parallel mechanism, *IEEE Translations on Robotics*, in press.
15. P. Renaud. Apport de la vision pour l'identification géométrique de mécanismes parallèles. Ph.D. Thesis, University Blaise Pascal, Clermont-Ferrand, 2003.
16. M. Abderrahim, A.R. Whittaker. Kinematic model identification of industrial manipulators, *Robotics and Integrated Manufacturing* **16**, 2000.
17. J. Won Jeong, S.H. Kim, Y. Keun Kwak, C.C. Smith. Development of a parallel wire mechanism for measuring position and orientation of a robot end-effector, *Mechatronics*, 1998, 845–861.
18. B.C. Bouzgarrou, J.C. Fauroux, G. Gogu, Y. Heerah. Rigidity analysis of T3R1 parallel robot with uncoupled kinematics, in *International Symposium of Robotics*, Paris, 2004.
19. R. Rizk, N. Rat, J.C. Fauroux, G. Gogu. Influence des défauts géométriques sur le comportement cinématique d'un robot parallèle, in *Congrés Français de Mécanique*, Troyes, 2005.

20. G. Gogu. Mobility criterion and overconstraints of parallel manipulators, in *Proceedings of CK2005, International Workshop on Computational Kinematics*, Cassino, Italy, May 4–6, 2005.
21. G. Gogu, P. Coiffet. Représentation du mouvement des corps solides, in *Mathématiques pour la Robotique*, Hermes, 1996.
22. T. Cano, N. Andreff, J.M. Lavest, P. Ray. Identification of geometrical defects using a multiple stereovision system, in *IEEE Annual Seminar on Automatic Control, Industrial Electronics and Instrumentation SAAEI05*, Santander, Spain, 2005.

Numerical Simulation of Parallel Robots with Decoupled Motions and Complex Structure in a Modular Design Approach

Zhen Wu[1,2], Rani Rizk[2], Jean-Christophe Fauroux[2] and Grigore Gogu[2]

[1]*Mechanical Engineering Department, Wuhan University, Wuhan 430072, China; E-mail: wuzhenwhu@163.com*
[2]*Mechanical Engineering Research Group (LaMI), French Institute for Advanced Mechanics (IFMA), and Blaise Pascal University, Campus de Clermont-Ferrand/Les Cézeaux, BP 265, 63175, Aubière, France; E-mail: {rrizk, fauroux, gogu}@ifma.fr*

Abstract. In this paper, a modular design methodology is proposed for numerical simulation of parallel robots. Static stiffness is a mechanical characteristic that describes the behaviour of a structure under static force in terms of elastic deflection and can be evaluated for robotic manipulators by means of Finite Element Method numerical simulation. Many parallel robots have multiple identical legs that can be considered as multiple instances of a unique sub-assembly. On this base, we present an efficient approach to perform numerical simulation of these robots. In addition, an application case to the Isoglide family of parallel robots is presented to show the effectiveness of this approach. A new rhombic leg structure is also compared with a classical leg structure. Compliance maps for the Isoglide3-T3 robot with rhombic legs are also provided. Finally, structural symmetry of the geometrical model of the robot is used to find symmetries in the compliance maps and to check calculation correctness.

Key words: parallel robot, stiffness, symmetry, FEM, modular design.

1 Introduction

With the development of advanced robotic technology, mechanical design methods have been extensively studied to create new parallel mechanical systems with specified architecture and number of degrees of freedom (DoF). Original applications for parallel machine tools and parallel kinematics machines have been very recently proposed. They have been applied in various fields such as manufacturing simulators [1, 2], micro robots, industrial high speed pick and place robots [3], medical robots [4]. Parallel mechanisms have become more and more popular because they have better properties, such as high load/weight ratio, velocity, stiffness, precision

S. Tichkiewitch et al. (eds.), Advances in Integrated Design and Manufacturing in Mechanical Engineering II, 129–144.

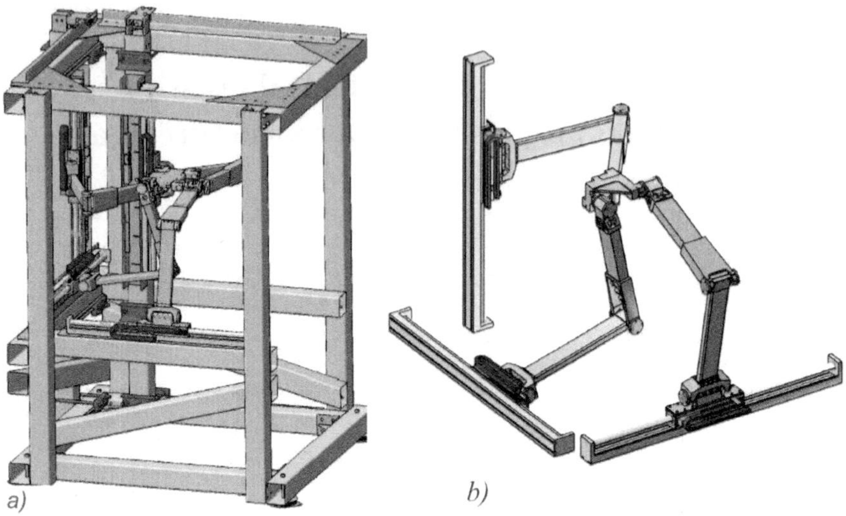

Fig. 1. CAD model of two robots from Isoglide family: (a) Isoglide4-T3R1, (b) Isoglide3-T3.

and low inertia. It is believed that parallel robot mechanisms with few DoFs, usually two to five, are especially prospective because of their simpler structures and lower production costs [5, 6].

Parallel robotic manipulators with decoupled motions and various degrees of mobility have been recently proposed [1, 2]. Figure 1a presents an example of a 4-DoF parallel mechanism whose end effector, called platform, can achieve four independent motions: three orthogonal translational motions and one rotational motion with respect to the fixed base [2].

This manipulator called Isoglide4-T3R1 was designed and implemented by LaMI in a modular approach. The work presented in this paper is applied to translational parallel manipulator Isoglide3-T3 (Figure 1b) but it could be extended to other solutions of the Isoglide robot family including Isoglide4-T3R1.

2 Problem Setting

2.1 Stiffness Analysis

The design of parallel mechanisms is usually based on the use of evaluation criteria involving workspace, dexterity, payload, global conditioning index, and stiffness [7]. Among these properties of mechanical systems, stiffness is a particularly important characteristic for robot specification. In addition, characterizing parallel architectures for practical applications requires evaluating their stiffness. This can be useful for developing analytical design criteria and improving properly prototype performance as proposed for example in [8–10].

A great deal of work has been done on stiffness analysis of parallel mechanical systems and it has direct application in industry. The methods reported in the literature [3, 8–11] can be classified into structural analysis by Finite Element Method (FEM) and Jacobian matrix method. The first method [3, 8–9] is based on an approximation of the original model by a discrete model made of elements and nodes, leading to the stiffness matrix that is dependent on the nature of elements in the structure. The second method is based on the Jacobian matrix that is used to form a generalized stiffness matrix [11]. Significant examples of stiffness analysis on robots can be found in [8–11]. This paper will mainly focus on numerical simulation for the structural analysis method.

2.2 FEM Modelling

Most of the FEM software offer two ways of creating models:

- The first one is the direct *Graphic User Interface (GUI) method*, using a point and click strategy and interactive control on the model.
- The second one is much closer to programming and will be called *scripting method*. Most software include a programming language with basic structures (sequence, test and loop), variable parameters and sub-functions.

Despite its advantage of intuitiveness, the GUI method is not efficient for the numerical simulation of parallel robots. First, for a complex mechanical system, the FEM model is very time consuming. For developing a complex model, it is common to dispatch the developing process on several people that are geographically far one from each other and do not work synchronously. This is what we call the 'distributed method'. The GUI methodology is mainly an individual way of creating models and the distributed method can not be applied easily in this case. Second, the GUI method itself is not a formal, ordered and systematic method. A same model can be created via many different ways. It is based on a point and click strategy. The errors can not be examined and modified easily. Third, in the preliminary design phase of a product, we often need to explore the design space and generate many different alternatives to choose the best one. For parallel robot design, this means building assemblies of standard and reusable components. This is not easily achieved with GUI methodology.

In the process of evaluating the stiffness of Isoglide robot family, we study the stiffness of a parallel robotic manipulator with three isotropic translational motions (Isoglide3-T3) [1]. Figure 2 represents a classical serial leg structure, while Figure 3 represents a rhombic structure, where each leg is a parallel mechanism in itself made of two sub-chains. With the GUI method, it is difficult to make use of the results obtained by solving the classical structure (Figure 2) while we try to study the rhombic

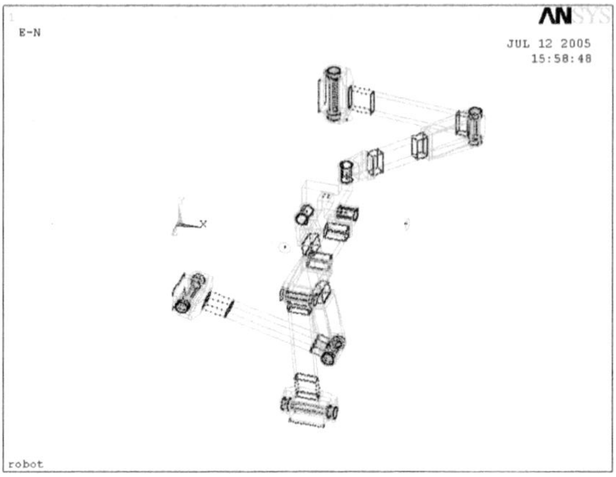

Fig. 2. Leg in classical structure.

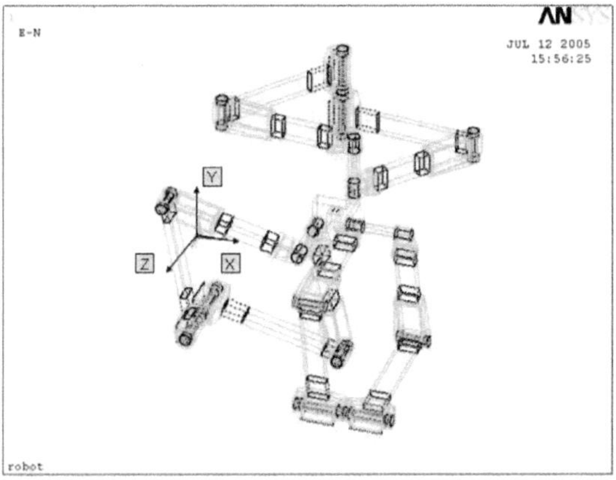

Fig. 3. Leg in rhombic structure.

structure (Figure 3), let alone the comparison of different solutions. Without reusability and exchangeability, which are some of the fundamental concepts first introduced with Object Oriented Programming [12], the GUI method is proved to be unsuitable and cannot meet the demand of numerical simulation.

3 Modular Design for FEM Numerical Simulations

3.1 Modular Design

It is worth noting that almost all parallel robotic manipulators are characterized by their symmetry in structure. The idea of modular design can help us in the FEM numerical simulation of parallel mechanisms. A module is an assembly of parts that can be integrated or repeated several times in the structure of a machine. Modular design takes advantage of repeating patterns and hierarchical relations in the assembly structure of a machine or a mechanism. In fact, modular products are products that fulfill various functions through the combination of distinct modules [13–15]. The modular design of products leads to a large number of different products by creating distinct combinations of modules and components. This can give each product distinctive functionality, features, and performance levels [15–17]. The design of modular products is of considerable importance in enabling companies to respond rapidly to changes in the market environment. Examples of this type of modularity can be found in automobile industry and computer industry. The modular approach promises the benefits of computability, reusability, exchangeability and improved communication.

On the basis of the concepts of modular product design and substructuring, this paper develops a modular design approach to simulate numerically complex parallel mechanical systems.

3.2 Functional Analysis and Assembly Decomposition

A typical FEM simulation is generally divided into three steps, which are respectively model building, load application and solving. With the idea of modular design, we can deal with the three steps in a particular way. Indeed, a FEM model consists of geometrical model and physical model. It is suitable to arrange the building of model into a hierarchical decomposition made of sub-assemblies and components, while leaving behind load application and solving to later steps. In this way, the whole FEM model is modularized functionally. In addition, the FEM model can be modularized physically and geometrically. For example, in a structure with repeated patterns (such as the three legs of a parallel robotic manipulator), we can generate one module to represent the pattern and simply make copies of it at different locations, thereby saving a significant amount of computer time. Considering mainly from the function, computability, reusability and exchangeability of module, we can decompose and disassemble the FEM model into modules. The assembly FEM representation of parallel robot Isoglide3-T3 is shown in Figure 4; the disassembling process is shown in Figure 5. A leg can be disassembled into 10 modules. Considering symmetry of structure, number of modules of Isoglide3-T3 comes to 31 (three legs together with a platform).

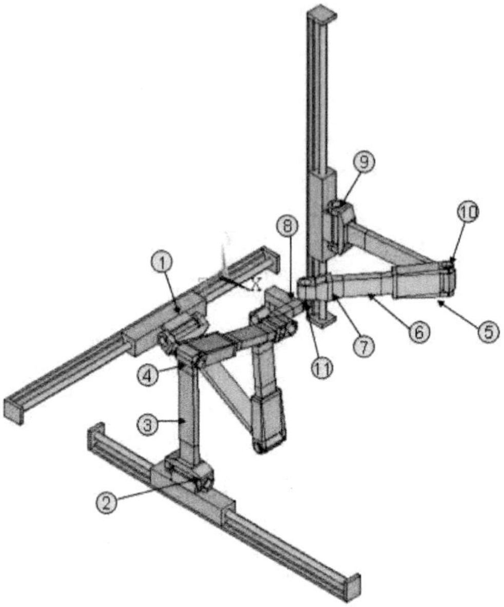

Fig. 4. FEM model of Isoglide3-T3 assembly.

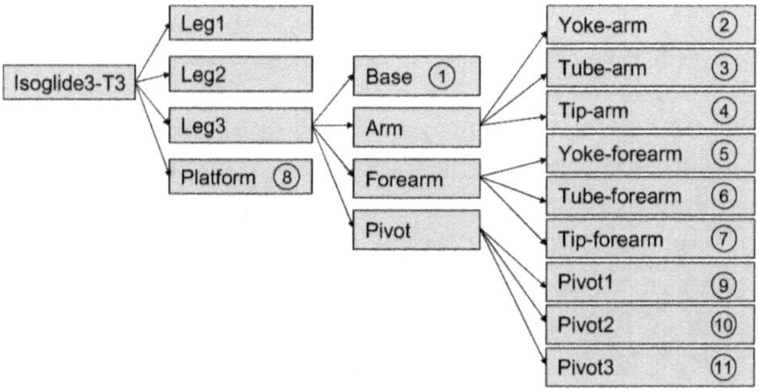

Fig. 5. Disassembling Isoglide3-T3 model into sub-assemblies and components.

3.3 Substructuring

This step concerns mainly the creation of model for module, which is based on the concept of substructure. In ANSYS® software, the substructure analysis is defined as "a procedure that condenses a group of finite elements into one element represented as a matrix. The single-matrix element is called a superelement" [17, 18]. Indeed, the only difference is that the superelement is created first as a module by performing a substructure generation analysis. Modularization and substructuring reduce computing time and allow solving very complex problems with limited computer resources. Nonlinear analyses of structures containing repeated geometrical patterns are typical problems where substructuring can be employed.

APDL, which stands for ANSYS Parametric Design Language, is a suitable candidate for module design to FEM simulation. It is a scripting language that you can use to automate common tasks or even build your model in terms of parameters (variables). APDL also encompasses a wide range of other features such as repeating a command, macros, if-then else branching, do-loops, and scalar, vector or matrix operations.

With APDL, design is completely formal and systematic. A typical module is realized by the following steps:

- Building geometrical model.
- Defining element type, material property and associating element attributes with geometrical model.
- Specifying the analysis type, the type of equation solver, etc.
- Generating superelement equivalent to the considered module (condensing finite elements into one superelement).

The result of modular analysis is a superelement matrix that can be copied at different locations, according to the FEM simulation requirement.

3.4 Assembling Modules and Solving

With all of modules and components available, global model is assembled by importing all superelement matrix files into an assembly file. The assembly file is particular because it concerns only the assembly of modules. It combines all the separate modules into the final global model and generates the solution of model. By modifying this file, numerous combinations of modules can be created. It consists of the following steps:

- Defining a global coordinate system and some local coordinate systems.
- Assembling all modules.
- Applying loads. In ANSYS, this terminology includes boundary conditions and externally or internally applied forces.
- Solving the complete problem.

For instance, the substructure of Tip-forearm and its assembly is shown in Figure 6. With ANSYS, the first step is to create and locate the superelements in global model (SETRAN command). The second step is to import the superelement data that were previously calculated in generation pass (SE command). Loading and solving are encompassed in this step.

By defining the master DoF between interfaces of contacting modules, the main advantage of a modular design-reusability and exchangeability can be guaranteed.

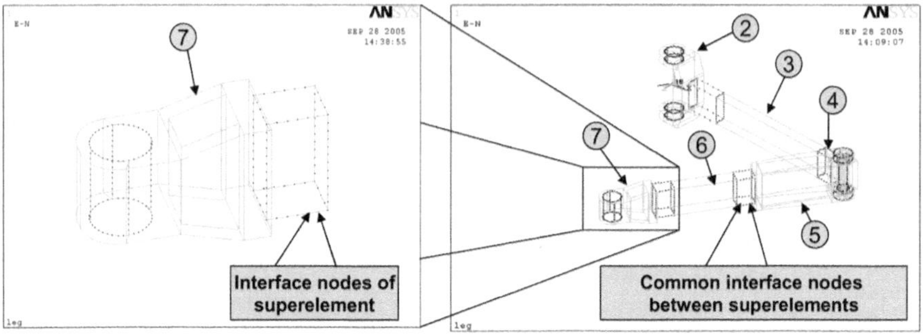

Fig. 6. Sub-structure Tip-forearm and its assembly.

It means that one or more modules can be easily replaced provided the interfaces remain with the same specification.

4 Application to Isoglide Robot Family

4.1 Theoretical Base for Numerical Simulation

Generally, a stiffness evaluation can be represented by stiffness matrix $[K]$ which can be obtained by computing displacements $\mathbf{dp}$ and rotation angles, occurring on the platform at a static configuration when a force $\mathbf{F} = (F_x, F_y, F_z)$ and a moment $\mathbf{M} = (T_x, T_y, T_z)$ act upon it. The stiffness $[K]$ can be formulated as

$$\begin{bmatrix} \mathbf{F} \\ \mathbf{M} \end{bmatrix} = \begin{bmatrix} K_{Fp} & K_{F\theta} \\ K_{Mp} & K_{M\theta} \end{bmatrix} \begin{bmatrix} \mathbf{dp} \\ \mathbf{d\theta} \end{bmatrix}. \quad (1)$$

To calculate the compliance matrix, stiffness equation (1) can be transformed into the following compliance equation:

$$\begin{bmatrix} \mathbf{dp} \\ \mathbf{d\theta} \end{bmatrix} = \begin{bmatrix} S_{pF} & S_{pM} \\ S_{\theta F} & S_{\theta M} \end{bmatrix} \begin{bmatrix} \mathbf{F} \\ \mathbf{M} \end{bmatrix} = \begin{bmatrix} K_{Fp} & K_{F\theta} \\ K_{Mp} & K_{M\theta} \end{bmatrix}^{-1} \begin{bmatrix} \mathbf{F} \\ \mathbf{M} \end{bmatrix}. \quad (2)$$

By numerical simulation, we can determine the compliance matrix. For example, for sub block S_{pF}:

$$S_{pF} = \begin{bmatrix} S_{11} & S_{12} & S_{13} \\ S_{21} & S_{22} & S_{23} \\ S_{31} & S_{32} & S_{33} \end{bmatrix}. \quad (3)$$

The value S_{ij} can be determined and the compliance map can be calculated.

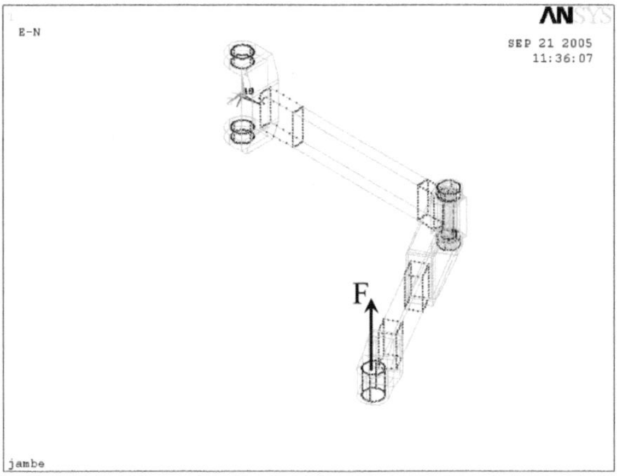

Fig. 7. Leg in classical structure.

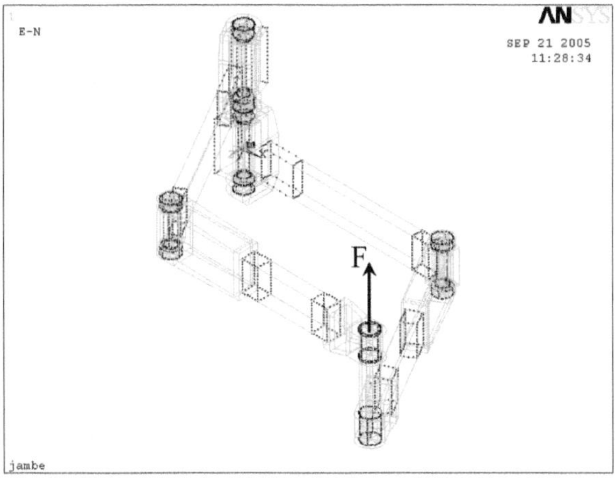

Fig. 8. Leg in rhombic structure.

4.2 Compared Behaviour for Two Types of Legs of the Isoglide Robot Family

First stiffness studies of Isoglide4-T3R1 were performed in [19, 20]. In this work, a modular design approach is used and several solutions are compared. Two possible structures are considered for legs in Isoglide robot family. Figure 7 shows a FEM model of leg in classical structure, while Figure 8 represents a FEM model of leg in rhombic structure. Making use of the substructures available, we have created the leg FEM model as shown in Figure 8. It is interesting to compare the compliance of two

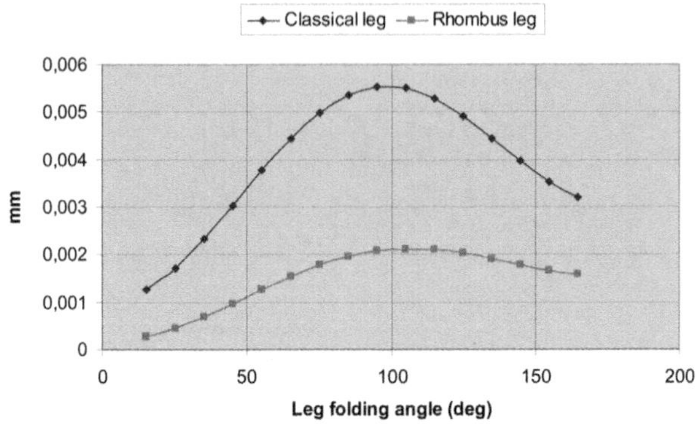

Fig. 9. Graph of compared deflections of classical and rhombic legs.

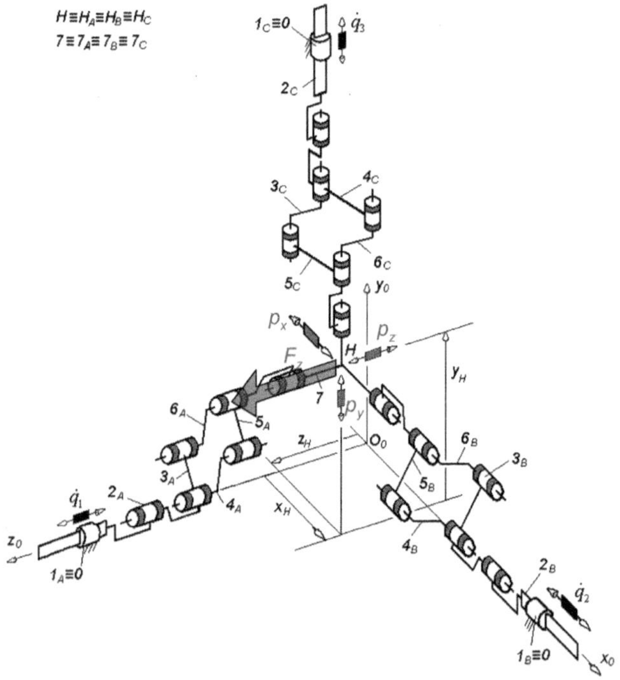

Fig. 10. Kinematic diagram of rhombic Isoglide3-T3.

solutions, which can help us to optimize design. Figure 9 is the graph of compared deflections under an unitary force F.

Several conclusions can be obtained from Figure 9. First, for both types of legs, deflection varies a lot with respect to the folding angle in a 1 to 5 ratio. Deflection is

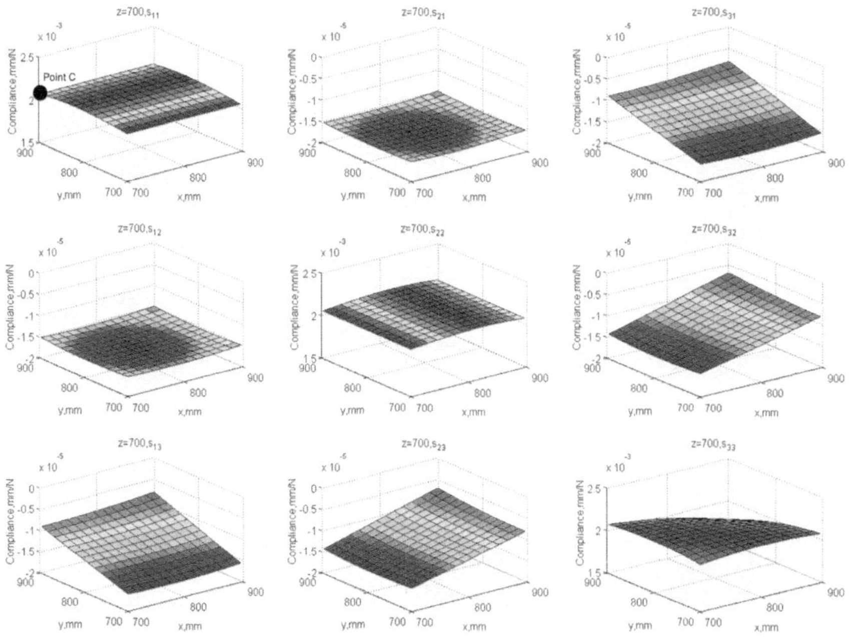

Fig. 11. Compliance map, $Z = 700$ mm.

minimal when the leg is folded and subject mainly to flexion. There is also a local minimum in the unfolded position. Both legs reach their maximum deflection when angle approaches 95 degrees because of torsion moment, which causes the biggest part of deflection. Second, the leg in rhombic structure is greatly reinforced relatively to the classical one. A rhombic leg is geometrically equivalent to two classical legs but is more than two times stiffer.

4.3 Compliance Maps of a Robot of the Isoglide Family

The following results are for Isoglide3-T3 robot with rhombic legs presented in Figure 10. On the base of equation (3), the value S_{ij} can be determined by calculating the displacement p_i and using $S_{ij} = p_i/F_j$ with unitary force F_j. Figure 10 shows the example of the displacement response p_x, p_y, p_z of the robot to a unitary force F_x. The whole series of compliance maps is shown in Figures 11–15. They will help diagnose structural behaviour of this robot around the full workspace.

The compliance maps ranging from vertical coordinate $Z = 700$ mm to $Z = 900$ mm are respectively shown in Figures 11–15. From these figures, we can see that compliance ranges from 2×10^{-5} mm/N to 2.5×10^{-3} mm/N.

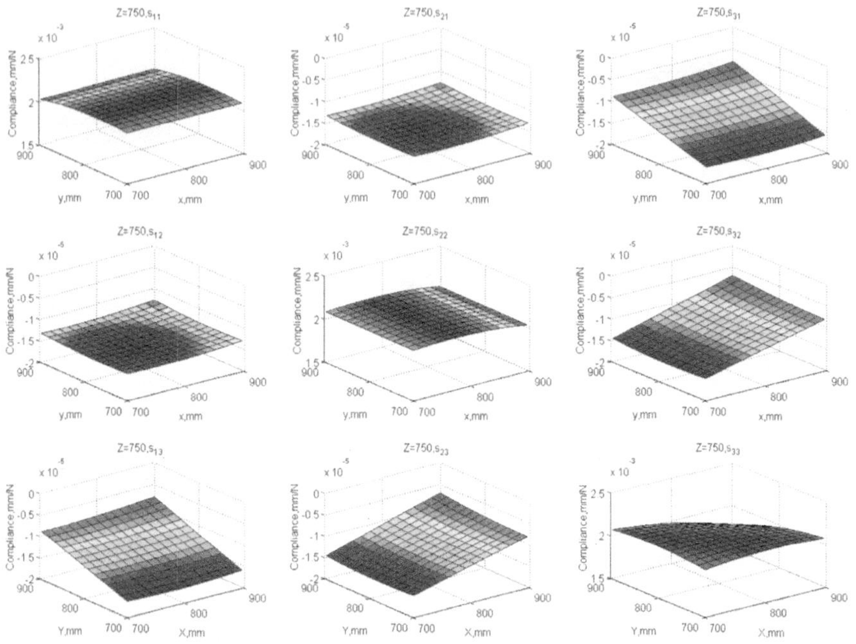

Fig. 12. Compliance map, $Z = 750$ mm.

If vertical coordinate Y is a constant, for example in Figure 11 where $Z = 700$ mm, the values of diagonal terms (S_{11}, S_{22}, S_{33}) are much greater than non-diagonal terms S_{ij} $(i \neq j)$ S_{11} remains constant along the X axis, S_{22} remains constant along the Y axis; S_{33} evolves with the same tendency along axes X and Y.

Comparing Figures 11–15, we can see that side $Y = 900$ of surface S_{11} decreases from 2.1×10^{-3} mm/N to 1.75×10^{-3} mm/N. The opposite side of the S_{11} surface ($Y = 700$) does not change. The evolution of S_{11} is independent on X. It can also be seen in Figure 11 that a unitary force F_x applied in the X direction on the points of axis $Y \approx 800$ mm will generate maximal robot displacement in the X direction. Compliance S_{22} has the same behaviour as S_{11} if the X and Y axes are swapped. Concerning S_{33}, it does not change much from Figures 11–15 and its maximal value is 2.1×10^{-3} mm/N.

The Isoglide3-T3 robot has a very special symmetrical design. It has strong consequences on the shape of compliance maps. Two types of symmetries can be observed. The first one is a triple symmetry that can be observed on every point of the diagonal of the cubic workspace (axis defined by $X = Y = Z$). In Figure 11, when $X = Y = Z = 700$ mm or in Figure 15, when $X = Y = Z = 900$ mm, it can be checked that $S_{11} = S_{22} = S_{33}$. According to the definition of compliance, this

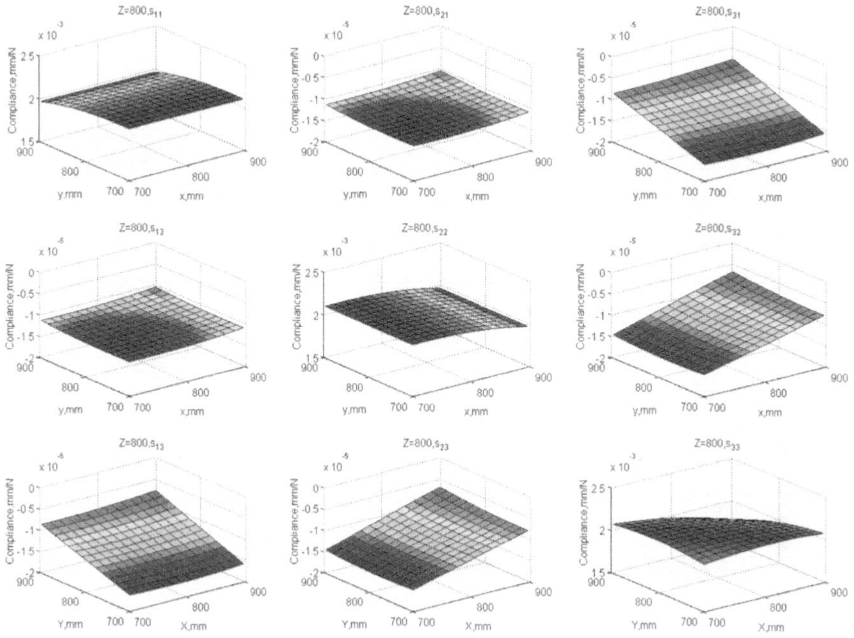

Fig. 13. Compliance map, $Z = 800$ mm.

means that on this point, if we apply a unitary force F in X, Y or Z direction, the robot will generate the same displacements in X, Y and Z direction respectively. As the FEM model was defined without gravity, it is normal to find here a statically isotropic behaviour. The second one is a double symmetry. For example, in Figure 15, we can see that S_{11} on point A (700, 900, 900) and S_{22} on point B (900, 700, 900) are equal to the same value of 1.75×10^{-3} mm/N. Another example would be S_{11} on point C (900, 900, 700) Figure 11 and S_{22} on point D (700, 900, 900) Figure 15: the values are the same and equal to 2.1×10^{-3} mm/N. This is another form of the structural symmetry. From all of these comparisons we can verify the correctness of results.

5 Conclusion

In this article, a modular design approach is presented to simulate the compliance of parallel robots with decoupled motions and complex structure. The use of modular design approach offers several important advantages. First, it is well suited to parallel robots with several identical legs that are modelled as sub-assemblies. Legs are calculated once and used several times. Each FEM model of sub-assembly or module is

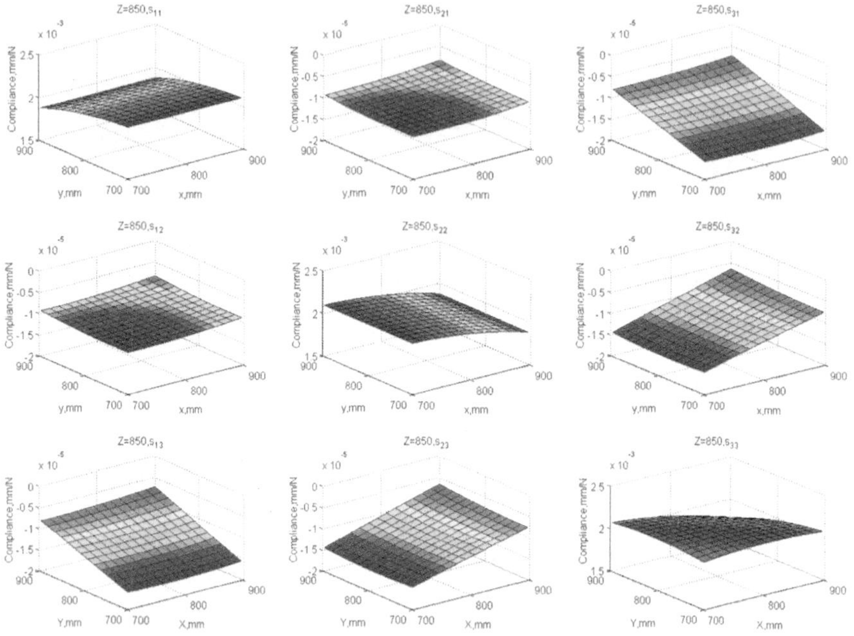

Fig. 14. Compliance map, $Z = 850$ mm.

simple and can be examined, checked, corrected and modified easily. For each module, ANSYS Parametric Design Language is used to develop the program in an object oriented approach. In this way, the distributed policy can be applied and a complex model can be decomposed and solved by several persons. Furthermore, by parameterization and substructuring, the size of the FEM problem is controlled and problem solving is accelerated. With modular design methodology and substructuring, every numerical simulation results are obtained by only re-assembling existing modules. It should be noted that our modular design approach is not limited to parallel robots but is also suitable for every type of machine with a repeating pattern.

The effectiveness of this approach was demonstrated on Isoglide3-T3 robot and some advantages were revealed, especially in the comparative study of different solutions. Two solutions based on classical and rhombic leg structure were tested. The rhombic solution proved to be more than two times stiffer than the classical one. Compliance maps were also computed for the complete rhombic Isoglide3-T3 robot. Triple and double symmetries in compliance maps could be noticed, which is due to the very special symmetrical geometry of Isoglide3-T3. This was used to check calculation correctness. This modular approach can be applied not only in the conceptual design phase but in detailed design as well.

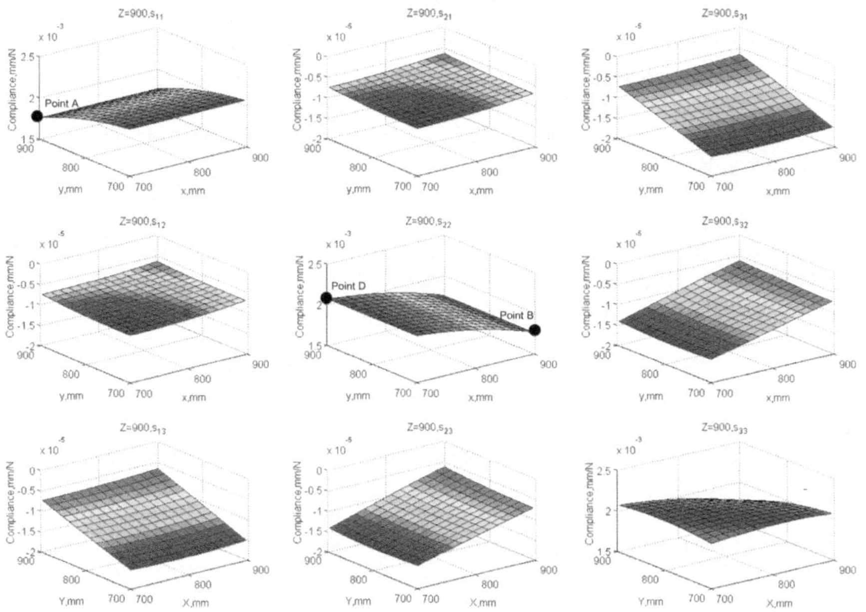

Fig. 15. Compliance map, $Z = 900$ mm.

References

1. G. Gogu. Structural synthesis of fully-isotropic translational parallel robots via theory of linear transformations, *European Journal of Mechanics A/Solids* **23**(6), 2004, 1021–1039.
2. G. Gogu. Mobility criterion and overconstraints of parallel manipulators, in *Proceedings of CK2005, International Workshop on Computational Kinematics*, Cassino, 2005, p. 22.
3. C. Corradini, J.C. Fauroux, S. Krut, O. Company. Evaluation of a 4-degree of freedom parallel manipulator stiffness, in *Proceedings 11th World Congress in Mechanism and Machine Science (IFTOMM'2003)*, Tianjin, China, China Machinery Press, Vol. 4, 2004, pp. 1857–1861.
4. O. Piccin. A parallel 5 DoF positioner for semi-spherical workspaces, in *2004 ASME International Design Engineering Technical Conferences and Computers and Information in Engineering Conference (DETC 2004)*, Salt Lake City, USA, 2004.
5. G. Piras, W.L. Cleghorn, J.K. Mills. Dynamic finite-element analysis of a planar high-speed, high-precision parallel manipulator with flexible links, *Mechanism and Machine Theory* **40**, 2005, 849–862.
6. X.-J. Liu, J. Wang, G. Pritschow. A new family of spatial 3-DoF fully-parallel manipulators with high rotational capability, *Mechanism and Machine Theory* **40**, 2005, 475–494.
7. H. Shen , T. Yang, L-Z. Ma. Synthesis and structure analysis of kinematic structures of 6-dof parallel robotic mechanisms, *Mechanism and Machine Theory* **40**, 2005, 1164–1180.
8. M. Ceccarelli, G. Carbone. A stiffness analysis for CaPaMan, *Mechanism and Machine Theory* **37**, 2002, 427–439.

9. G. Carbone, H. Lim, A. Takanishi, M. Ceccarelli. Stiffness analysis of biped humanoid robot WABIAN-RIV, *Mechanism and Machine Theory* **41**, 2006, 17–40.
10. O. Company, F. Pierrot, J.C. Fauroux. A method for modelling analytical stiffness of a lower mobility parallel manipulator, in *IEEE International Conference on Robotics and Automation (ICRA'2005)*, Barcelona, Spain, 2005, April 18–22, pp. 1195–1121.
11. D. Zhang, F. Xi, C.M. Mechefske, S. lang. Analysis of parallel kinematic machine with kinetostatic modelling method, *Robotics and Computer-Integrated Manufacturing* **20**, 2004, 151–165.
12. G. Booch. *Objected-Oriented Analysis and Design with Applications*, Addison-Wesley Professional, 2nd edition, ISBN 0805353402, 1993.
13. Y. Wang. Modular design process for a micro motion actuator, *Mechatronics* **15**, 2005, 793–806.
14. P. O'Grady, W. Liang. An internet-based search formalism for design with modules, *Computers & Industrial Engineering* **35**, 1998, 13–16.
15. D.W. He, A. Kusiak. Designing an assembly line for modular products, *Computers & Industrial Engineering* **34**, 1998, 37–52.
16. F. Klocke, M. Fallböhmer, A. Kopner, G. Trommer. Methods and tools supporting modular process design, *Robotics and Computer Integrated Manufacturing* **16**, 2000, 411–423.
17. ANSYS Theory reference manual, Chapter 17.6-Substructuring analysis, ANSYS Release 8.0 Documentation help file, 2003.
18. K.J. Bathe. *Finite Element Procedures*, Prentice Hall, ISBN 0133014584, 1995, 1037 pp.
19. B.C. Bouzgarrou, J.C. Fauroux, G. Gogu, Y. Heerah. Rigidity analysis of T3R1 parallel robot with uncoupled kinematics, in *Proceedings of 35th International Symposium on Robotics (ISR'04)*, March, 2004, Paris, France.
20. R. Rizk, J.C. Fauroux, M. Munteanu, G. Gogu. A comparative stiffness analysis of a reconfigurable parallel machine with three or four degrees of mobility, *Journal of Machine Engineering* **6**(2), 2006, 45–55.

Constraints Automatic Relaxation to Design Products with Fully Free Form Features

Jean-Philippe Pernot, Qian Qiao and Philippe Véron

LSIS, IMS, UMR CNRS 6168, CER ENSAM, 2 cours des Arts et Métiers, 13617 Aix-en-Provence Cedex 1, France
E-mail: {jean-philippe.pernot, qian.qiao, philippe.veron}@ensam.fr

Abstract. Feature-based approaches have become popular for the design of products and more widely for the complete product lifecycle management. The geometric model being often used as an intermediate object shared between the different actors of the design process, it is crucial to concentrate on the definition of very powerful tools enabling fast and intuitive definitions/modifications of the product geometry. In this field, we have already proposed the so-called Fully Free Form Deformation Features. They correspond to parts of a free form surface obtained by deformation according to specific constraints and deformation behaviours. Two main control parameters are used: the constraint lines that define both the main directions and the extent of the shape, and the surface behaviours between these lines. Unfortunately, undesirable surface undulations may appear at the end of the constraint lines. This is due to the uncertainty the designer has when defining the blending areas between the different constraint lines. To provide a user-friendly tool, which does not force the user to be very precise, the possibility to relax automatically the end points of the constraint lines is offered. This algorithm is based on an a priori quality criterion which limits the number of inflexion points of auxiliary curves that mimic the shape of the future deformed surface. Once the curvature criterion is satisfied on the curves, the surface deformation process can start while considering the restricted area of the constraint lines.

Key words: design by features, curves and surfaces deformation, shape analysis, curvature, a priori quality criterion.

1 Introduction

Nowadays, the CAD tools tend to become some real PLM solutions enabling the creation and simulation of the whole product lifecycle. They do not simply regroup several modules but they really integrate the product models around the concept of feature. The features are used to give a meaning to a set of entities manipulated by an application and are also a good way to convey semantic information. A hole inside

S. Tichkiewitch et al. (eds.), Advances in Integrated Design and Manufacturing in Mechanical Engineering II, 145–160.

a geometric model will be considered as a bearing housing for the designer and as a bore with a surface roughness and a tolerance on its diameter for the manufacturing process. Thus, each actor can enrich the product definition and can rapidly react to the modifications performed by the other actors. Among the different types of feature, this work focuses on those relative to the geometry of the product. Since the geometric model is often used as an intermediate object shared between the different actors of the design process, it is crucial to be able to both structure it and give the designers very powerful tools to manipulate this structured model. This corresponds to the definition of a feature-based approach. In a more advanced semantic-based system, this geometric model should also support application dependent information [1, 2], e.g. a material for some parts, some boundary conditions directly expressed on the geometric elements, the tolerances associated to each face or the type of tool used to manufacture the different parts.

Form features, i.e. geometric features defined by analytic surfaces such as planes, cylinders and so on, have become popular in mechanical engineering [3] and are now accessible in most of CAD software. Hole, rib, slot or blending features are so many high level entities the designers can use to define step by step the shape of a mechanical part. They do not anymore manipulate directly the surfaces but the features themselves which can be parameterised by numerical parameters such as length, width, height and so on. The modelling sequences are stored within a feature model defining in some sense the skeleton of the geometric model and ensuring the coherency between the different entities. It is also used as an entry to any modification. Figure 1a shows how Form Features are used to define some parts of a commands bloc of a car (Figure 1b). Unfortunately, form features do not support all the needs in terms of shape definition during the design process. Effectively, analytic surfaces are unable to represent free form shapes widely used in aesthetic and engineering designs (Figure 1c). To overcome these limits, the various concepts proper to form features have to be extended to free form surfaces that are often composed of several B-Spline and NURBS patches connected together. Since the manipulation of such surface mathematical models requires a deep knowledge and a great skill in the manipulation of the appropriate parameters, the need of higher level entities is even more justified than in the case of analytic surfaces. This work addresses such complex problem. To the best of our knowledge, the use of a free form feature-based approach to design complex shapes (Figure 1c) is not yet possible in the actual CAD system, the free form surfaces being built from a succession of classical operations (e.g. sweep, loft or global deformation).

Nevertheless, several techniques for the definition and manipulation of Free Form Features have been proposed and can be classified in three categories according to the level of control the user has when defining its shape (Figure 2). *Semi Free Form Features* are defined by free form surfaces obtained by classical rules such as loft/sweep operations, interpolation rules or specific relationships directly expressed between the control points. The user often has few possibilities to control the resulting shapes

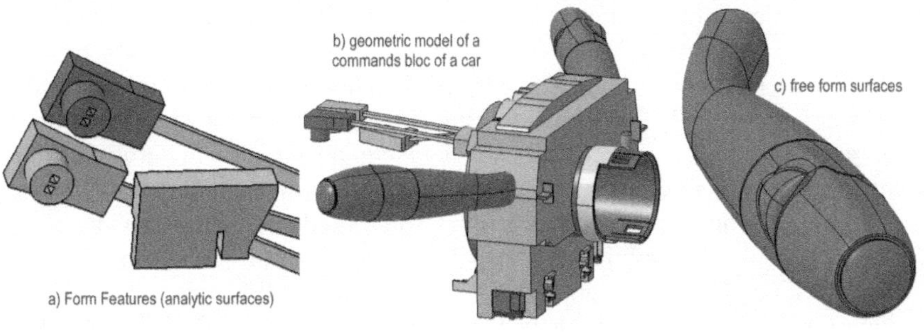

Fig. 1. Commands bloc designed with form features and free form surfaces (courtesy PSA).

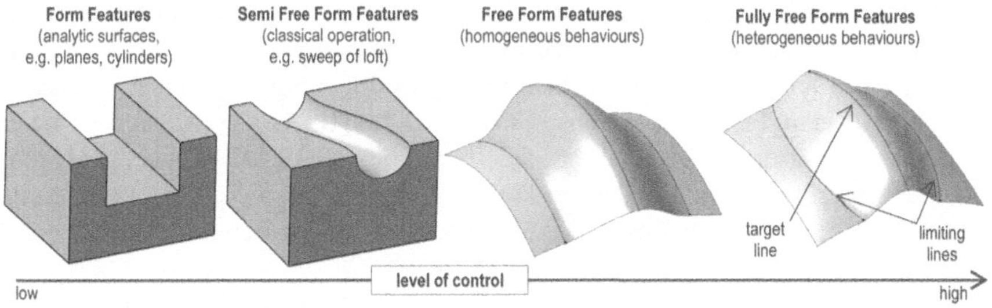

Fig. 2. Features classification based on the level of control.

[4–9]. *Free Form Features* are defined by free form surfaces, obtained through the use of adequate techniques expressing a homogeneous behaviour over the whole surfaces [10]. There is more freedom but monitoring the modified area is not always as free as possible. The insertion of the features inside the initial model often requires topological changes, e.g. trimming operations. Finally, *Fully Free Form Features* are characterised by a higher level of freedom in the definition of shapes obtained through the use of techniques prescribing heterogeneous behaviours over different areas of free form surfaces. The modified area can be bounded by any free form curve while preserving the surface initial topology (no trimming operations nor patches insertion). A complete comparison between these various techniques is proposed in [11].

Among the different Fully Free Form Features, we have been interested in those obtained by deformation. The so-called Fully Free Form Deformation Features (δ-F^4) correspond to parts of a free form surface obtained by deformation according to specific constraints and deformation behaviours [11]. Two main parameters can be used to monitor the resulting shape: the *target* and *limiting lines* give the global directions of the shape and bound the deformation area (Figure 2) whereas the *shape control parameters* enable the specification of heterogeneous deformation behaviours between these lines. Thus, a high degree of freedom is achieved through the specification of a restricted set of parameters. Here, the capabilities of the B-Spline/NURBS models are used at their highest level.

The counterpart of this high degree of freedom lies in the need of tools to master this power and to supply good quality results in terms of accuracy and smoothness. In our approach, the user-specified parameters (e.g. the curves and deformation behaviours) play a key role on the resulting shape. It is consequently essential to check their validity in order to prevent some eventual undesired consequences on the shape. For example, is it meaningful to specify a target line whose projection on the surface crosses a limiting line or a target line whose projection on the surface is outside the limiting line? Until now, such analysis processes were not available in our prototype software. When the target lines were not built very carefully, some undesired undulations appeared at the end points of the target lines, or in other cases some over-constrained configurations happened. This is due to the fact that in most of the cases, the user is unable to specify the blending areas between the various constraint lines taking part to the feature definition. In fact, these blending areas must result from a minimisation and not from a constraints specification. As a consequence, the extremities of the user-specified target lines should be considered as uncertainty areas which must not be taken into account during the deformation process. Based on this analysis, we propose an algorithm to automatically relax the extremities of the target lines defining a δ-F^4.

The paper is organised as follows. Section 2 briefly recalls the principal of the deformation technique used to implement both the δ-F^4 and the proposed automatic relaxation algorithm. The way the user can define and manipulate these entities is also recalled in Section 3. The problem of undesired undulations at the end points of the target lines is illustrated in Section 4 where possible answers are also depicted. Section 5 details the automatic relaxation algorithm whereas Section 6 illustrates its application as well as its limits. Section 7 concludes this article and open the way to new development.

2 A Generic Deformation Engine

Deforming an object or the object geometric model consists in altering or modifying its shape. The new shape results from the satisfaction of user-specified constraints according to a set of rules proper to the adopted deformation technique (see [11] for a detailed survey of existing deformation techniques). In our approach, this corresponds to the writing of an optimisation problem. First, the constraints are expressed directly on the geometric model. Since an infinite number of solutions satisfy this often under-constrained system of equations (possibly non-linear), an additional objective function has to be minimised. In order to both ease the prescription of deformation behaviours and enlarge the space of accessible shapes, we use a mechanical model of bar networks that we couple to the geometric model to be deformed. It can be coupled to any geometric model defined from a set of points (e.g. the control points of a NURBS curve/surface or the control vertices of a mesh). The adopted behaviour law requires the application of external forces to maintain the structure static equilibrium. The problem unknowns are not anymore the positions of the points but the external forces applied to these points. Various geometric and mechanical basic quantities have been imagined and can be combined inside a unique objective function φ to be minimised. Thus, the user is able to build some multi-minimisations which may produce heterogeneous deformation behaviours well suited to the aesthetic and engineering designs [12]. He/she is not restricted to the classical strain energy minimisation. Finally, the optimisation problem can be written as follows:

$$\begin{cases} \mathbf{G}(\mathbf{F}) = \mathbf{0} \\ \min \varphi(\mathbf{F}) \end{cases} \tag{1}$$

where $\mathbf{F}$ is the vector containing the unknown external forces, G is the constraint vector containing the user-specified constraints and φ the objective function describing somehow the shape behaviours. The details of the resolution algorithm can be found in [11]. Here, we would like to stress the fact that this is a generic deformation engine that works on any geometric model. It can be coupled simultaneously to curves, surfaces and meshes. Whether the user is able to specify some constraints on these geometric models, the deformation engine will be able to produce the deformation with respect to the selected deformation behaviours. As recalled in the next section; this deformation technique is at the basis of the δ-F^4 implementation. It is also used to deform auxiliary curves built during the automatic relaxation algorithm. To this aim, the constraints toolbox has been enhanced with a new set of constraints described in Section 5.

3 Fully Free Form Deformation Features Based Approach

The concept of δ-F^4 is built on top of our deformation technique. A δ-F^4 is a part of a free form surface obtained by deformation according to specific *constraint lines* and *shape deformation behaviours*. Two optional parameters can be instantiated: the *numerical parameters* used to constrain the dimension/position/orientation of the constraint lines and the *type of discontinuities* along the constraint lines [13]. All the feature parameters are gathered together inside a feature template. In order to access easily the desired δ-F^4, a feature classification has been proposed. Figure 3a shows a Basic Shape Feature (BSF) obtained by deformation according to two specific constraint lines, the *target* and *limiting lines*. The target lines give the global directions of the shape whereas the limiting lines define the shape around the target lines and bound the deformation area. Figure 3b shows a Complex Shape Feature (CSF) obtained by composition of already defined δ-F^4 according to user-specified driving lines. One can notice that the choice of constraint lines as key parameters of the δ-F^4 is not insignificant. The design/styling activity is inherently a curve-based activity

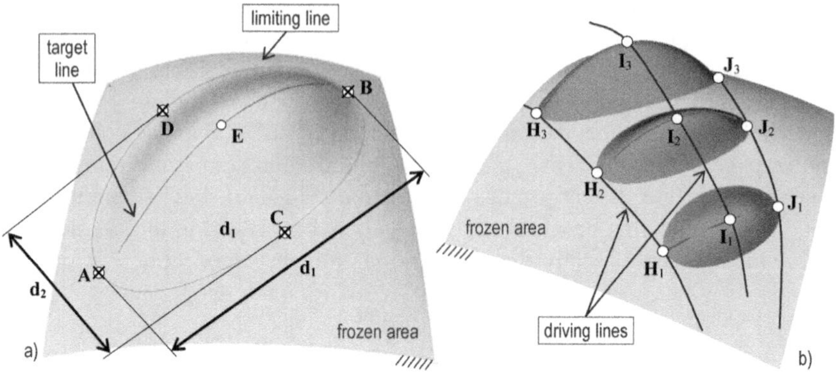

Fig. 3. Specification of Fully Free Form Deformation Features.

where the designer/stylist starts by sketching with curves. Curves are also used all along the geometric model definition (e.g. 2D sketching as well as sweep and loft operators).

Two ways for instantiating the δ-F^4 parameters have been proposed and correspond to two types of application. On one hand, the user may be interested in the use of constraint lines that he/she has built in another software or that simply come from a digitalisation. To this aim, the possibly to instantiate directly the feature parameters with user-defined constraint lines is offered. This fits well the way stylists work. On the other hand, the user may not be interested in the specification of constraint lines but in the specification of numerical parameters defining the main dimensions of the shape. In this case, the user instantiates a set of numerical parameters which are translated into a set of constraints, inputs of our curve deformation process. Here, the δ-F^4 results from a double deformation: the deformation of the curves to satisfy the user-specified numerical parameters and a surface deformation to satisfy the resulting constraint lines. This double level of deformation combined with the possibility to define local shape behaviours justifies furthermore the appellation of "Fully" Free Form Deformation Features with respect to the other categories of free form features. Such a way of proceeding fits well the way designers are used to work. It is very close from the widely adopted Form Features based approaches.

4 Problems of Smoothness and Palliative Solutions

Since a target line constrains the surface to pass through its points, its shape is very important for the smoothness of the resulting surface. It is even more critical at its end points whose positions with respect to the surface may result in over-constrained or incompatible configurations or just in simply undesired undulations. Figure 4b

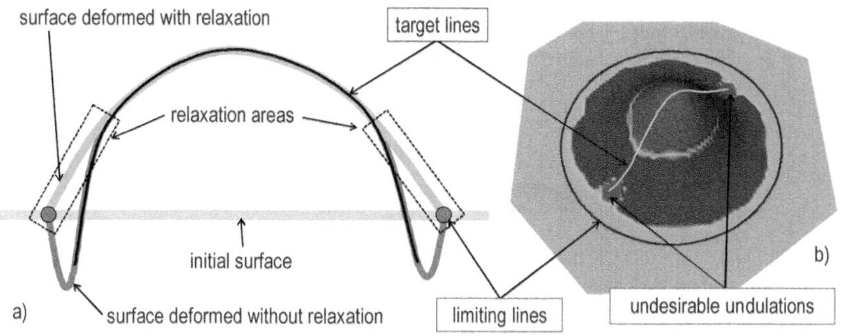

Fig. 4. Need of relaxation areas (a) to avoid undesirable undulations (b).

depicts a simple example where only one target line and one limiting line have been specified on an initially planar patch. After deformation, a Gaussian map of curvature is applied and reveals the presence of hollows around the end points of the target line. This is not acceptable for the stylists and designers.

Such a phenomenon is due to the fact that, in most of the cases, the user is unable to specify the blending areas between the constraint lines taking part to the δ-F^4 definition. Effectively, during the drawing of a target line, or during the specification of the feature template numerical parameters, the designer/stylist does not have the criterion necessary to decide the place where the target line has to end. As a consequence, the extremities of the target lines should be considered as uncertainty areas which must not be taken into account during the deformation process. It corresponds to a problem of relaxation of user-specified constraints. Figure 4a explains the effects of such a relaxation on an amplified configuration.

In order to enable the target line end points relaxation, two types of scenario can be imagined: an *a posteriori* relaxation and an *a priori* relaxation. The "prior" and "posterior" adjectives must be understood with respect to the surface deformation process. Either you do the constraints relaxation posterior or prior to the surface deformation. These two algorithms are schematically represented on Figures 5a and 5b. In the first case, i.e. the a posteriori relaxation, the final shape results from several loops of successive surface deformation, surface analysis and constraints modification. The surface is deformed several times which may slow the overall relaxation process. The surface analysis requires the computation of geometric magnitudes highlighting the surface undulations (e.g. map of curvatures or reflection lines). Their interpretation as well as the definition of a quality criterion may be difficult. One can finally notice that the a posteriori relaxation may also result from a surface correction as proposed in [14]. Unfortunately, the correction processes do not ensure that the surface still satisfies the various constraint lines. In the second case, i.e. the a priori relaxation, the user-specified constraints are analysed and relaxed according

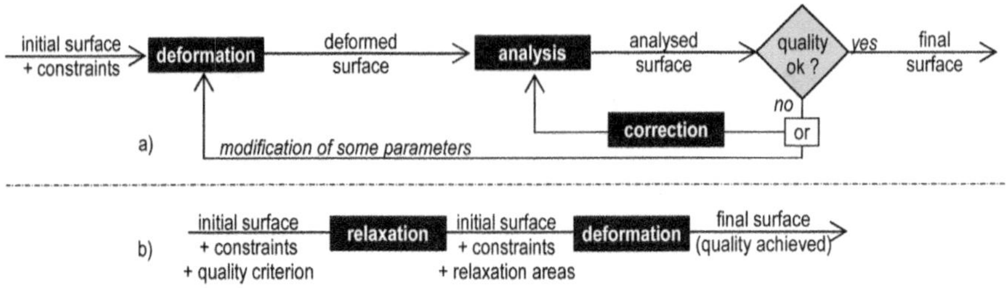

Fig. 5. A posteriori (a) and a priori (b) relaxation algorithms.

to a simpler quality criterion. The surface is deformed one time which reduces significantly the computation time. The surface can be eventually analysed but this is not mandatory since the quality is ensured by the a priori quality criterion. The a posteriori algorithm corresponds more or less to what was done before this paper in which we propose an a priori relaxation algorithm which works before the surface deformation takes place.

Before entering the principle and details of our a priori relaxation algorithm, let us come back on the relaxation strictly speaking. In fact, the constraints relaxation can either be seen as a constraints removal operation or as a minimisation process. The first type is easy to understand. After an analysis step, the constraints which may cause some undesired results are removed from the constraint vector **G** of equation (1). In the other case, a Lagrangian relaxation is performed through a transfer of the constraints from the constraint vector **G** to the functional φ to minimise [15]. The process does not anymore try to satisfy exactly these constraints but it tries to minimise a new objective function which takes into account both the initial objective function and some of the constraints. Such a way of proceeding does not fit well the philosophy of our approach where the constraints and the deformation behaviours are decoupled. Effectively, in this case, the minimisation of the mixed objective function does not anymore represent the user-specified deformation behaviours. Moreover, how can one balance the relative influence between the constraints and behaviours inside a unique functional? These are the reasons why our relaxation operation consists in a simple constraint removal. All the problem is to determine which part(s) of the target line should constrain the surface and which part(s) should not. Solely the part(s) constraining the surface will then be transformed into a set of point constraints for the surface deformation process.

5 Relaxation Areas Automatic Detection Algorithm

This section describes how the relaxation areas, i.e. the parts of the target line which must not be taken into account as constraints for the surface deformation process,

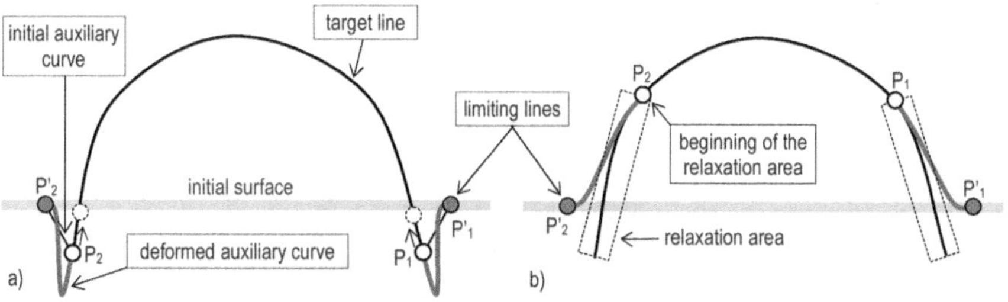

Fig. 6. Principle of the automatic relaxation process.

can be automatically determined. Section 5.1 presents the principle of the algorithm whose various steps are detailed in the next subsections.

5.1 Principle of the Algorithm

The principle of our algorithm lies in the deformation of auxiliary curves connecting both the target and limiting lines and whose deformation mimics the shape of the future deformed surface. The surface quality analysis is replaced by a curve quality analysis which speeds up the relaxation. This process is iterative and starts with two auxiliary curves attached to the end points P_1 and P_2 of the target line and connected to the limiting line at P_1' and P_2' (Figure 6a). These initially straight curves are then deformed according to specific constraints enabling a blending between them and both the target line and the initial surface. At each step, if the curves do not satisfy the quality criterion, their initial points Pi are modified and slide along the target line (dot circles of Figure 6a) until they reach a satisfactory position with respect to the prescribed criterion (Figure 6b). Thus, the relaxation areas are obtained.

This algorithm is transferred on Figure 7. The rate of displacement of the points P_i along the target line is determined before entering the iterative process. It can come from a discretisation either according to the w parameter of the target line or according to its curvi-linear abscissa. At the jth step, the position of the extremities P_i of the two auxiliary curves is computed from the equation of the target line $P(w)$. The position of the two other extremities P_i' is computed from the equation of the limiting line (see Section 5.2). Thus, the two auxiliary straight curves can be created and deformed according to specific constraints ensuring a blending between them and both the target line and the initial surface. The curvature of the various curves is then computed and enables the analysis of what could give the surface deformation with such a restricted target line. If the criterion is satisfied, the surface deformation process can start. Otherwise, if the maximum number of step is not achieved, a next step $j + 1$ starts with two new initial points. Even if relative to the jth step, the variables P_i, P_i', w_i, u_i, v_i, u_i' and v_i' will not be tagged with an additional exponent j in order to simplify the writing.

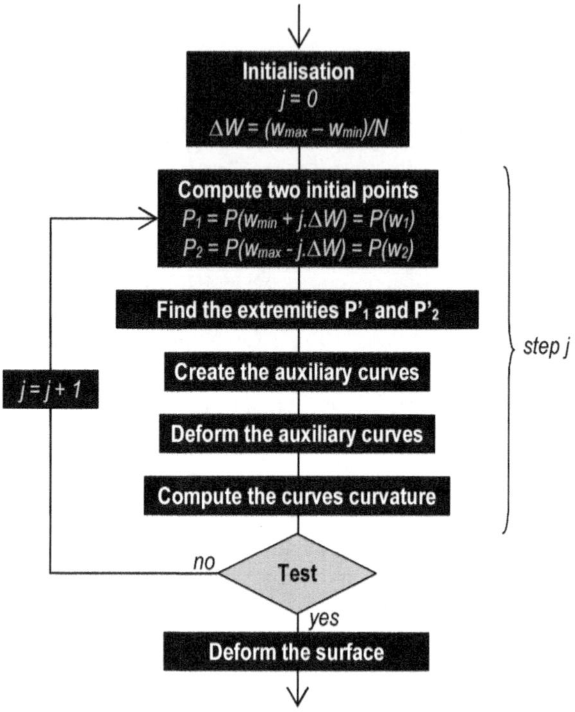

Fig. 7. The automatic relaxation algorithm.

5.2 Determination of the Auxiliary Curves End Points P_1' and P_2'

Let us suppose that we are at a given step j where two initial points P_1 and P_2 have been computed using the target line equation (Figure 8a). The goal is now to determine the position of the second extremities P_1' and P_2' of the future auxiliary curves. Since we want to simulate the blending of the future surface between the target line and the initial surface, it is mandatory that the extremities P_1' and P_2' are on the limiting line, i.e. at the boundary between the modified and the unmodified areas.

This treatment is performed in the patch parametric space. The part of the target line between P_1 and P_2 is first projected on the surface in order to find its image inside the patch parametric space. The extremities (u_i, v_i) of this projection are then prolonged according to the vectors tangent to the curve (Figure 8b). The parametric positions (u_i', v_i') of the points P_i' result from the computation of the intersections between these prolongations and the limiting line image. One can notice that this step assumes that the initial target line is not closed; otherwise it would be impossible to find an intersection between the target and limiting lines. If the target line lies in a

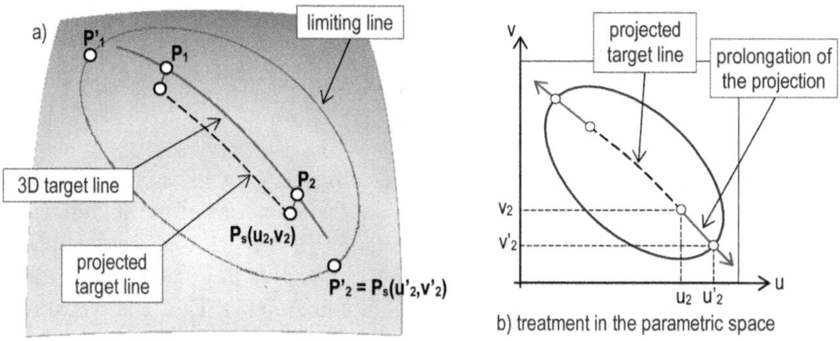

Fig. 8. Use of the parametric space to find the auxiliary curves second extremities.

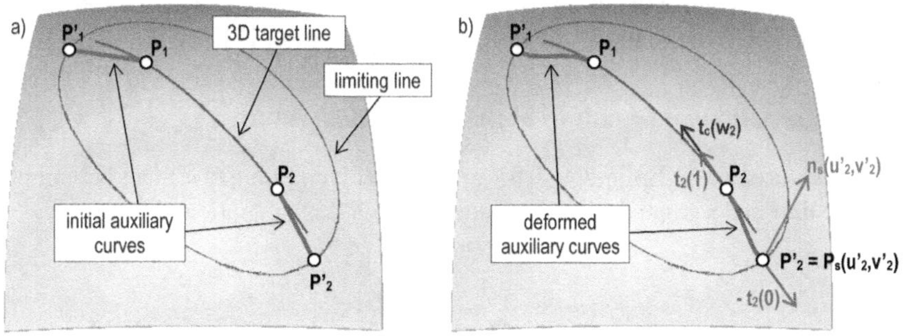

Fig. 9. Deformation of auxiliary curves (b) initially straight (a).

plane containing the projection vector, the two extremities P_1' and P_2' does not need to be recomputed at each step.

5.3 Creation and Deformation of the Auxiliary Curves

Once the points P_i and P_i' are computed, the two auxiliary curves can be created and deformed in 3D space (Figures 9a and 9b) using the mechanical model of bar networks introduced in Section 2. These B-Spline curves are of degree three and are defined by eight control points. Here, two bar networks are built from the control points of the two auxiliary curves. In order to deform them simultaneously, a single optimisation problem is written. The following constraints express the way the second auxiliary curve has to be deformed:

$$\begin{cases} \mathbf{P}_2(1) = \mathbf{P}_c(w_2) \\ \mathbf{P}_2(0) = \mathbf{P}_s(u_2', v_2') \\ \mathbf{t}_2(1) \wedge \mathbf{t}_c(w_2) = \mathbf{0} \\ \mathbf{t}_2(0) \cdot \mathbf{n}_s(u_2', v_2') = 0 \end{cases} \tag{2}$$

with $\mathbf{P}_2(w)$, $\mathbf{P}_c(w)$ and $\mathbf{P}_s(u, v)$ the equations of respectively the auxiliary curve, the target line and the surface. $\mathbf{t}_2(w)$ and $\mathbf{t}_c(w)$ are the equations of the vectors tangent to respectively the auxiliary curve and the target line. $\mathbf{n}_s(u, v)$ is the equation of the normal to the surface [16]. All these constraints plus those concerning the first auxiliary curve form the constraint vector **G** of equation (1). They can be expressed as a function of the external forces **F**, unknowns of our deformation problem. In order to simulate the behaviour of the future δ-F^4, the objective function $\varphi(\mathbf{F})$ used for the minimisation comes from the δ-F^4 shape behaviour parameters. In the rest of this paper, we will use the minimisation of the variation of the external forces which tends to preserve the shape of the initial auxiliary curves [12].

5.4 Computation and Analysis of the Curves' Curvature

The curvature of a B-Spline/NURBS curve at a given point $\mathbf{P}(w)$ can be computed from its first and second derivatives using the following equation [16]:

$$\rho(w) = \frac{1}{R(w)} = \frac{\left| \frac{d^2\mathbf{P}}{dw^2} \wedge \frac{d\mathbf{P}}{dw} \right|}{\left| \frac{d\mathbf{P}}{dw} \right|^3}. \tag{3}$$

Curvature analysis enables at once local as well as global studies of the shape behaviours. The evaluation of the curvature at a point represents the local behaviour of the curve (rather small or important shape variations) and the concatenation of this information over the entire curve highlights the imperfections in the global appearance of the curve. Thus, the undulations are characterized by an alternation of positive and negative values of the curvature. As a consequence, we have based our a priori quality criterion on the minimisation of the number of inflexion points, i.e. the minimisation of the number of times the curvature changes its sign. More precisely, while the *number of inflexion points is strictly superior to two*, the test fails and the step $j + 1$ starts (Figure 7). When the *number of inflexion points is inferior or equal to two*, the test becomes true and the relaxation areas can be defined from the position of the end points P_1 and P_2. Figure 10 shows a configuration where no relaxation has been performed (dot auxiliary lines) thus producing a curvature evolution which becomes null four times. As soon as the curvature does not present more than two inflection points, the curve undulations disappear (pink auxiliary curves).

One can note that nothing prevents the appearance of discontinuities between the three curvature evolutions that are gathered together on the single graph of Figure 10b. This is due to the fact that the constraints between the auxiliary curves

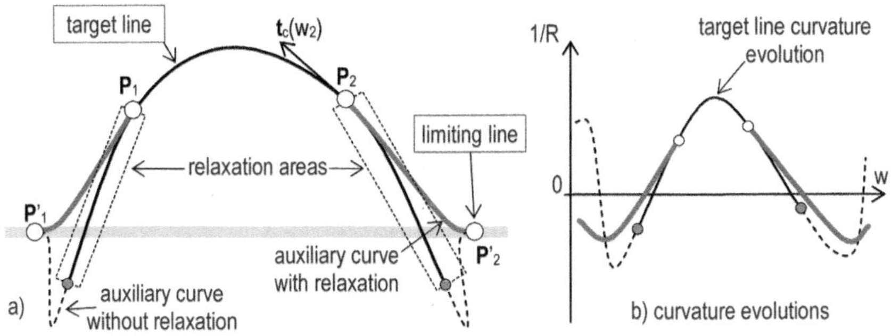

Fig. 10. Quality criterion based on the number of inflexion points.

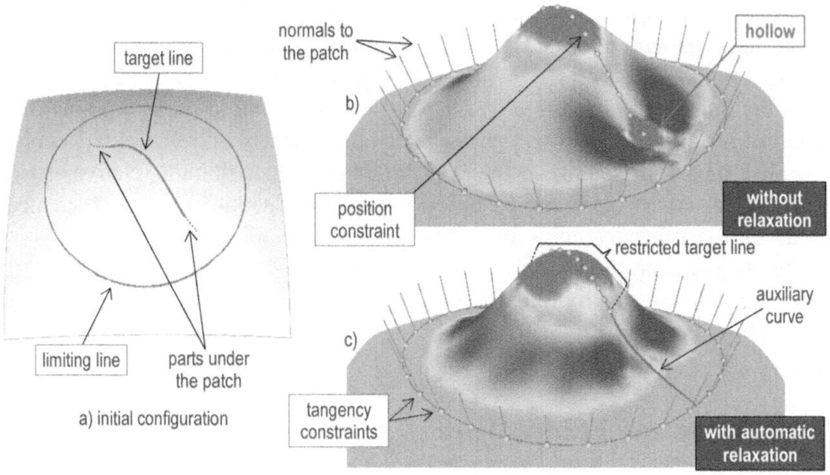

Fig. 11. Example with a target line crossing the initial surface.

and the target line ensure solely a G^1 continuity (tangency) and not a G^2 continuity (curvature). This point is further developed in the next section.

6 Results and Discussion

In the two proposed examples, we assume that the user has instantiated the δ-F^4 parameters (see Section 3) and that the constraint lines are already dimensioned, positioned and oriented on the free form surface.

The first example shows a configuration where the target line crosses the initial surface (Figure 11a). The target line is discretised in 10 position constraints whereas the limiting line is discretised in 30 points prescribing both position and tangency

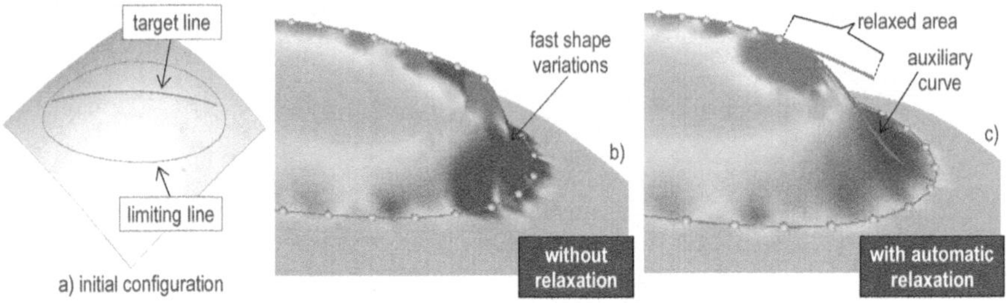

Fig. 12. Example with a target line whose projection is very close to the limiting line.

constraints. The tangency constraints impose that the direction of the normal to the surface along the limiting line is not modified during the deformation. If no relaxation is performed, the entire target line is discretised thus producing undesired hollows at the end points (Figure 11b). When the relaxation process is run, the considered part of the target line is reduced and the surface behaviour at the end points of the target line is improved. The two Gaussian curvature maps have been limited similarly to enable a meaningful comparison. One can finally notice that the relative distance between the auxiliary curve and the deformed surface is about 3×10^{-2} which means that the auxiliary curve mimics quite well the surface deformation.

The second example shows a configuration where the end points of the target line are very close to the limiting line (Figure 12). In this case the relaxation process relaxes a lot.

7 Conclusion and Future Work

In this paper, we have proposed a new algorithm for the automatic relaxation of the end points of the so-called target lines used in our δ-F^4 implementation. Such a tool is a way to manage the uncertainty the user has when defining the target line end points. This algorithm uses an a priori quality criterion based on the analysis of the smoothness of an auxiliary curve deformed in a pre-processing step. Thus, the surface deformation is performed one time which speeds up the automatic relaxation. Of course, this process is optional when the user is sure of the end points of its target lines. In the future, we would like to test both tangency and curvature constraints between the target line and the auxiliary curves. This could be a way to avoid G^2 discontinuities between the curves and to improve the robustness of the curves' curvature analysis.

More generally, we have noticed that the undulations due to the uncertainty the user has when defining the shape of the target line are difficulty differentiable from the undulations due to over-constrained configurations (Figure 12b). In the future, it could be very interesting to have a complete toolbox for both analysing the user-specified constraints and alerting him/her when something gets wrong. This is a very difficult problem on free form surfaces and more generally when the constraints are expressed on hybrid models made of free form surfaces, curves and meshes. These future investigations could take advantage of the work of Lesage [17] who has already proposed an approach to analyse a set of constraints expressed on 2D sketches.

References

1. AIM@SHAPE: *Advanced and Innovative Models and Tools for the Development of Semantic-based Systems for Handling, Acquiring, and Processing Knowledge Embedded in Multidimensional Digital Objects*, European Network of Excellence, Key Action: 2.3.1.7 Semantic-based Knowledge Systems, VI Framework, URL: http://www.aimatshape.net.
2. R. Bidarra and W.F. Bronsvoort, Semantic feature modelling, *Computer Aided Design* **32**, 2000, 201–225.
3. J.J. Shah, *Parametric and Feature-Based CAD/CAM*, Wiley Interscience Publication, John Wiley Sons, Inc., 1995.
4. G. Vosniakos, Investigation of feature-based shape modelling for mechanical parts with free form surfaces, *Int. J. Adv. Manuf. Technol.* **15**, 1999, 188–199.
5. J.S.M. Vergeest and I. Horvath, Parameterization of FreeForm Features, in *Proceedings of the IEEE Shape Modelling International Conference*, 2001, pp. 20-28.
6. Y. Song, J.S.M. Vergeest and I. Horvath, Reconstruction free form surface with parameterized features, in *ASME, DETC'02*, New York, 2002.
7. C.K. Au and M.M.F. Yuen, A semantic feature language for sculptured object modelling, *Computer Aided Design* **22**, 2000, 63–74.
8. E. van den Berg, H.A. van der Meiden and W.F. Bronsvoort, Specification of FreeForm Features, in *ACM Symposium on Solid Modelling and Applications*, 2003, pp. 56–64.
9. J.C. Cavendish, Integrating feature-based surface design with freeform deformation, *Computer Aided Design* **27**(9), 1995, 703–711.
10. M. Fontana, F. Giannini and F. Meirana, Free Form Features for aesthetic design, *Int. J. Shape Modelling* **6**(2), 2000, 273–302.
11. J-P. Pernot, Fully Free Form Deformation Features for aesthetic and engineering designs, Ph.D. Thesis, INPG (France) and University of Genoa (Italy), 2004, www.3s.hmg.inpg.fr/ci/doc/.
12. J-P. Pernot, S. Guillet, J-C. Leon, B. Falcidieno and F. Giannini, Shape tuning in Fully Free Form Deformation Features, *Journal of Computing and Information Science in Engineering (JCISE)* **5**(1), 2005, 95–103.
13. J-P. Pernot, B. Falcidieno, F. Giannini and J-C. Leon, Fully Free Form Deformation Features for aesthetic shapes design, *Journal of Engineering Design (JED)* **16**(2), 2005, 115–133.
14. P. Veron and J-C. Leon, Semi-global deformation and correction of free-form surfaces using a mechanical alternative, *The Visual Computer* **13**, 1997, 109–126.
15. A.M. Geoffrion, Lagrangean relaxation for integer programming, *Mathematical Programming Study* **2**, 1974, 82–114.

16. G. Farin, *Curves and Surfaces for Computer Aided Geometric Design*, 5th Edition, Morgan Kaufmann, 2002.
17. D. Lesage, Un modèle dynamique de spécifications d'ingénierie basé sur une approche de géométrie variationnelle, Ph.D. Thesis, INP-Grenoble, France, 2002.

DESIGN AND COMMUNICATION

Rapid and Accurate Data Integration Method for Reverse Engineering Applications

Jafar Jamshidi, Antony Roy Mileham and Geraint Wyn Owen

Innovative Manufacturing Research Centre (IMRC), Department of Mechanical Engineering, The University of Bath, Bath BA2 7AY, UK;
E-mail: {j.jamshidi, a.r.mileham, g.w.owen}@bath.ac.uk

Abstract. In many cases it is desirable to precisely extract the design information and to generate a full CAD model from an existing part or prototype to be able to analyse its functionality or to replicate it using available manufacturing techniques. The process of Reverse Engineering (RE) helps generate design information from a part, for which such information is unavailable or mislaid. Various scanning systems are available for use in RE that can be divided into contact, and non-contact systems. Contact systems like Coordinate Measurement Machines (CMM) with touch-trigger probe are highly accurate but time consuming and costly when the number of required data-points is high. Laser Scanners (LS), as non-contact systems, are fast but less accurate or expensive if in higher resolution versions. Also higher resolution equals higher level of noise in cloud-data, requiring extensive modelling operations. Moreover feature extraction from cloud-data is always an approximation of the actual feature, which might not be ideal for precision engineering applications. In this paper a data integration method for these two types of data is introduced. The accurate data captured by CMM touch-trigger probe is used to accurately create geometric features and the low-resolution data gained from the non-sensitive features by LS are used to fill the gaps between the accurate features. This method enhances the speed and flexibility of scanning while maintaining the accuracy of the results. Experimental results for method verification are presented. This method can be employed automatically, fully or partially, depending on the complexity of the component.

Key words: reverse engineering, CAD modelling, digitisation, parametric and non-parametric feature, laser scanning.

1 Introduction

In many cases it is desirable to digitise the design geometric information from the surface of an existing part. Such process of generating digital models from real parts is known as Reverse Engineering (RE) [1]. This is common practice where the design has to be an exact replica of natural objects (see [2] for a case example) such as those

S. Tichkiewitch et al. (eds.), Advances in Integrated Design and Manufacturing in Mechanical Engineering II, 163–175.

in artificial joints or in cases where the component is from an old paper-based design data-centre, or if the design information is simply lost.

1.1 Scanning Systems

Various digitisation technologies have been developed for use in measurement and RE that can be mainly divided into two groups of contact and non-contact systems. A Coordinate Measurement Machine (CMM) with touch-trigger probe as a contact system is capable of extracting accurate 3D data-points from the part surface and creating standard geometric features using normal vectors of the data-points. For instance three points with similar normal vectors within the set tolerance generate a plane while three points with various normal vectors out of the set tolerance, generate a fitting circle to those points without user interaction.

Laser Scanning systems digitise a part surface by using a Laser beam reflection either by generating points or using a sweeping line method. The nature of the result in the point method is similar to that of the CMM touch-trigger probe, but the superior sweeping line method, results in 3D cloud-data with faster digitisation speed. The cloud-data is then triangulated and used for modelling processes and geometric features' extraction.

1.2 Problems in CAD Modelling

A common problem in RE is the creation of the complete CAD model with coherent geometry and full tolerance and dimensional information. Data captured by a CMM on its own is a set of tessellated geometries in 3D space, which can provide dimensional information from the part; however it is incapable of exposing the topological relationship of those features. This matter becomes critical when such model has to be verified by non-experts. In addition the lack of solid model characteristics or surfaces between features in this type of models makes them unsuitable for computational analyses such as assembly, Finite Element Analysis (FEA), stress and heat analyses.

Cloud-data captured by Laser Scanner (LS) also cannot be used without passing through several modelling processes including noise reduction, surface fitting, feature extraction, etc. These processes lead to feature approximation and dimensional information estimation, which might not be reliable enough for precision RE applications due to their inexact nature.

1.3 Data Integration System

The advantages of the two scanning systems can replace the disadvantages of one another in an integrated system. Contact (tactile probe) and non-contact (LS head)

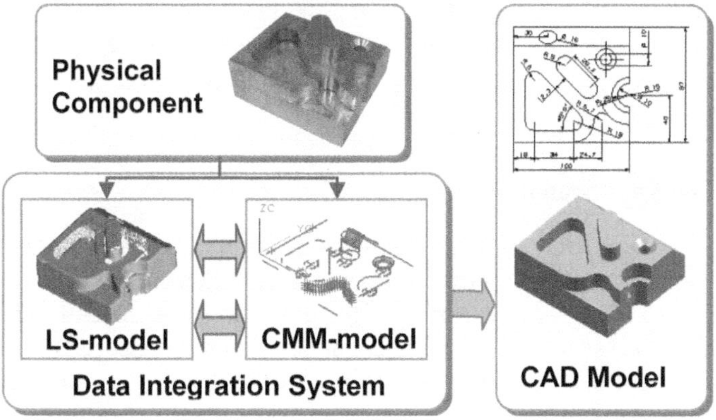

Fig. 1. CAD model creation using high and low resolution data.

are integrated in modern CMMs, but they are mostly used for individual probe usage or basic applications for instance part position finding [4] or probe path planning [5].

In this paper, a rapid and accurate CAD modelling method is introduced by which complex geometry (B-spline, complex curves, or non-parametric geometry) components with high precision but standard geometry features (Figure 1) can be fully and precisely digitised and improved, resulting in a complete CAD model with exact dimensional and tolerance information. This method is independent from the hardware set up as it is working purely on the data attained from the LS and CMM. Therefore it can be used with virtually any high and low resolution 3D data captured by contact and non-contact scanning systems.

2 Related Works

New CMMs have enhanced flexibility by their integration with different scanning probes including the LS head. Li and Liu [6] have developed a method for optimal alignment of the measured data by CMM touch-trigger probe with the design data in a CAD directed inspection system. Similarly, the Laser Scanning ability of CMM is used [6] to determine the probing points for accuracy enhancement in the scanning of a B-spline surface by the touch-trigger probe of the CMM. In this method, which is superior to random points' scanning, the cloud-data of the B-spline surface is scanned by the LS head to guide the touch-trigger probe for scanning the sample points from the surface.

In the integrated system, the complex geometry surfaces are normally scanned with the non-contact LS in a very short time, resulting in patches of low resolution cloud-data, and the accurate features are scanned by the touch-trigger probes in the form of accurate data-points, which are normally transformed into geometric features by the CMM software. However, more research is required for modelling process after the scanning of the part to achieve a practical CAD model, as the existing modelling processes and optimisation are time consuming and/or not accurate enough for RE of precise components.

2.1 Full Coverage Scanning

Complete scanning coverage of the component requires the appropriate hardware and software in order to bring the desired surfaces of the object to the view of the Laser light in the case of LS, or to the reach of the probe tips if the touch-trigger probe is used. In this process the necessary alignment between different sets of scanned data has to be either fixed or rectifiable. Tarbox and Gottschlich [7] developed a complete coverage scanning in inspection for which a CAD model is required for the initial planning. In other methods [8] full Laser Scanning coverage of a component apart from the underside features is developed again based on the CAD model, therefore unsuitable for RE applications. In complete scanning coverage, even the non sensitive features are scanned and it may be advisable not to scan those views at all for time and memory saving purposes. Such featureless faces can be added to the model efficiently later, in or at the end of the modelling process if required.

2.2 View Registration

View registration is the logical integration of the scanned views into the same coordinates system, conventionally done between two neighbouring views at a time. Iterative Closest Point (ICP) [9] is by far one of the most popular registration techniques, which works based on finding the minimum square deviation of the closest set of points on the registering views. Successful view registration by this method depends on the topology and the number of points in the common areas of the registering views. Therefore only the views located in each others neighbourhood with overlapping areas can be registered with this method. This can result in cumulative error when a series of views are registered in sequence, similar to the consequential algorithm for the registration of pairs of scanned views [10].

If the relationship of the two views is known, the registration can be achieved by rigid transformation of the merging views to the base view. Son et al. [8] have used the position of the setup fixture that can provide the appropriate location and orientation for the part in each view angle, with respect to the focal length and coordinates of the LS camera, to calculate the orientation matrix of each scanned view. The flexibility of this method is limited to the advanced planning for scanning views, which varies for each component. Accurate and flexible positioning equipments are also available but they are not cost effective. A more flexible registration technique was developed by the authors [11], in which a set of 3 coordinate balls is positioned with respect to the proposed object that can help to robustly calculate the coordinates' relationship of the paired merging views.

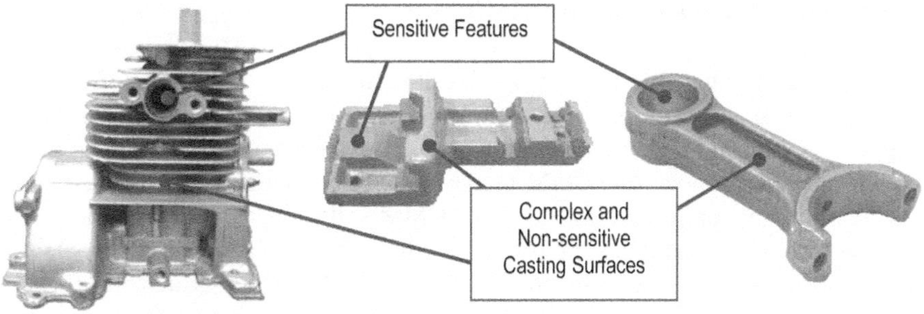

Fig. 2. Examples of casting components with sensitive and machined features.

2.3 Integrated System Calibration

In CMMs with LS head, current practice is that different probes are calibrated to the same coordinates, which is set to the centre of a single and fixed calibration ball attached to the CMM table prior to scanning [12]. Researchers have used the tip of the probe for 3D LS camera calibration in order to establish a common coordinate system [13]. Although this calibration scheme results in accuracy of more than 0.1 mm, it is inflexible if the features on the part are far away from the initial calibration position, which inevitably forces the scanning result into a lower resolution, or else would require a new camera calibration after each distance and scanning view set up.

2.4 Solid Modelling

Geometric feature creation from data-points that can be combined into a useable CAD file is a common problem in RE. The scanned model generated by the Laser scanner is viewed as scattered data-points or a triangle mesh. Many researchers in this area are focused on feature extraction [14] and surface fitting using sliced and layered cloud-data [15]. The main drawback of these methods is their limitation in feature extraction of inward geometries. Moreover these methods tend to produce abnormal surfaces in the missing information area in the form of flattened surfaces or filled holes in the cloud-data.

The scanned model generated by a conventional CMM with touch-trigger probe is in the form of islands of features with empty space in between as shown in Figure 1 (CMM-model). To make this type of model suitable for manufacturing purposes it is required to fit the appropriate surfaces into the empty spaces between the features, which could be an effortless task if the surfaces are featureless. This matter becomes more complex if the actual surfaces are of complex geometry features that are common in many automotive and aerospace parts with casting surfaces. Examples of this type of parts are shown in Figure 2.

3 Information Integration Method

The method introduced in this paper solves the mentioned problems of integrating two types of scanned models generated by LS (LS-model) and CMM with touch-trigger probe (CMM-model). After the preparation and planning, the part is scanned to generate the LS-model and CMM-model as the stage-models. Then the stage models are accurately optimised in the LS and CMM software environment. The results are then exported to an IGES file format as a base for translation into the CAD part file. In the translation process the models are imported into their designated layer for the convenience of further modelling processes. The main stages of the modelling processes are described in more details as follows.

3.1 Preparation and Planning for Scanning

Sensitive features recognition is the requirement of planning for modelling. Researchers [8] have generated scan plan by dividing the part into functional surfaces within a CAD model. However, this is a difficult task when no design information for the component is available. Having some knowledge of the assembly where the proposed part sits is beneficial, alternatively the operator can treat the machined surfaces, especially the fine finish one, as the sensitive feature.

The next problem in planning refers to datum selection, for which the manufacturing requirements of the desired part have to be known. The planning process is therefore left for the system operator to perform manually. The main scanning planning remains similar for LS and CMM model generation and since both of the models are required for the main data integration there is no preference in their order of application. Furthermore the method flexibility permits the operator to switch between different scanning systems as and when required i.e. if a particular feature needs higher resolution scanning by LS, it is possible to create a single view scanned model and import it to the LS-model layer for data integration in the CAD environment.

3.2 Preparation for Model Registration

The actual registration process takes place after the LS and CMM stage-model creation. The auxiliary registration object, which is a set of three fixed coordinate balls (3CB), needs to be in a fixed position, with minimised obstruction for scanning, with respect to the part.

The 3CB should be fixed in position with respect to the proposed part throughout the scanning operations (Figure 3). This guarantees the accuracy of the registration between the two types of models up to the accuracy level of coordinate balls and hardware i.e. the registration process does not require any overlapping area, and can be completed with minimal user interaction. The 3CB makes the method flexible

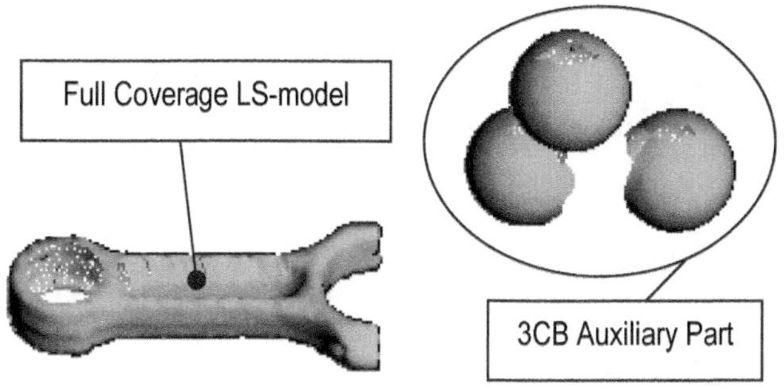

Fig. 3. Shaded view of a full coverage scanned model.

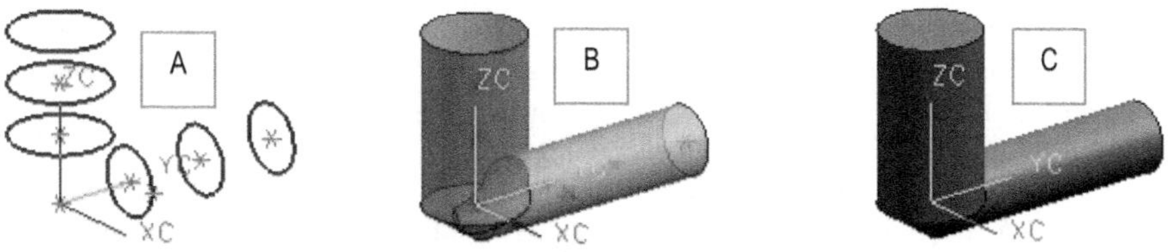

Fig. 4. Simple hole feature in CMM-model.

so that any desired arbitrary view angle and direction can be used for LS-model creation as long as a section of each of the balls is visible in the scanning views. In the unlikely event of a particular and required view, if one of the balls is hidden completely behind the others, a second set of 3CB can be used.

3.3 Measurement and 3D Digitisation

The CMM measurement of the sensitive features should also include the scanning data of the 3CB. Although the 3CB information has to be removed from the final model and is an extra digitisation operation, it is necessary and helps improve the ease and accuracy of the registration process. Also, the LS-model can be used as a quick guide for faster CMM digitisation if imported into the CMM software, provided that the CMM software supports graphic interface.

In the creation of the CMM-model the user can decide between creating standard surface model from the sensitive features (Figure 4A) or more basic geometries (Figure 4B) and leaving surface and volumetric model (Figure 4C) creation in the CAD environment. In this paper, however, the focus is more on surface models from the features.

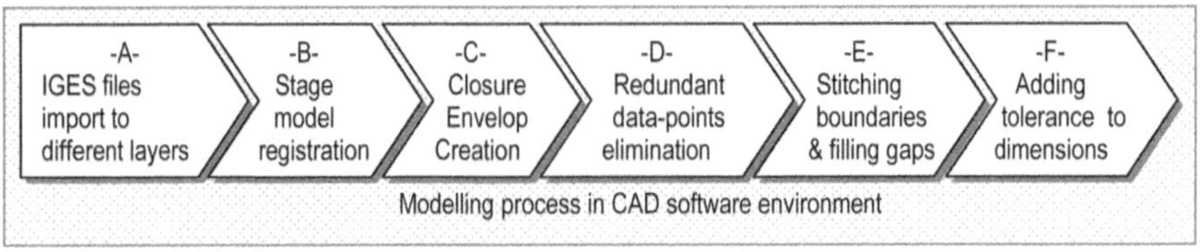

Fig. 5. Modelling processes in information integration method.

3.4 Stage Model Registration

Prior to model integration the LS and CMM scanned models need to be registered into the same coordinate similar to registering the scanned views in the creation of full coverage scanning (Figure 3). The relationship of stage models in 3D space is achieved by using the centre of balls in 3CB to replace the rotation matrix elements in equation (1) by x, y and z values of the centre of balls.

$$R = \begin{bmatrix} \cos\phi\cos\theta & \cos\phi\sin\theta\sin\psi - \sin\phi\cos\psi & \cos\phi\sin\theta\cos\psi + \sin\phi\sin\psi \\ \sin\phi\cos\theta & \sin\phi\sin\theta\sin\psi + \cos\phi\cos\psi & \sin\phi\sin\theta\cos\psi - \cos\phi\sin\psi \\ -\sin\theta & \cos\theta\sin\psi & \cos\theta\cos\psi \end{bmatrix}, \quad (1)$$

$$\begin{bmatrix} x_{i+1} \\ y_{i+1} \\ z_{i+1} \end{bmatrix} = R_i \begin{bmatrix} x_i \\ y_i \\ z_i \end{bmatrix} + T_i. \quad (2)$$

Then all the points in the 3D cloud-data of LS-model and the features from CMM-model are translated using equation (2) in which R_i is the known rotation matrix and T_i is the distance between the centres of coordinates of the merging model to that of the base model or the datum coordinates.

3.5 Information Integration and Model Merging

After the successful registration of the stage models, the redundant 3D data-points in the close neighbourhood of sensitive features need to be highlighted for deletion. For this a features' closure envelope, as shown in Figure 5 stage C, is created that is on a larger scale, with the same features, in terms of their direction and location in the model. The scale can be determined based on the feature tolerance and as an input by the operator. The closure envelop should be able to contain the real feature, therefore they are similar in shape regardless of the feature inward or outward characteristics, i.e. a round stud with 10 mm diameter and 20 mm length and a hole with the same dimensions have the same size closure envelop.

In the next stage, the closure envelopes select all their inclusive points, and in a self destructive approach, they eliminate themselves and also those redundant data-points from the triangulated LS-model, Figure 5, stage D. The resulting cavities and

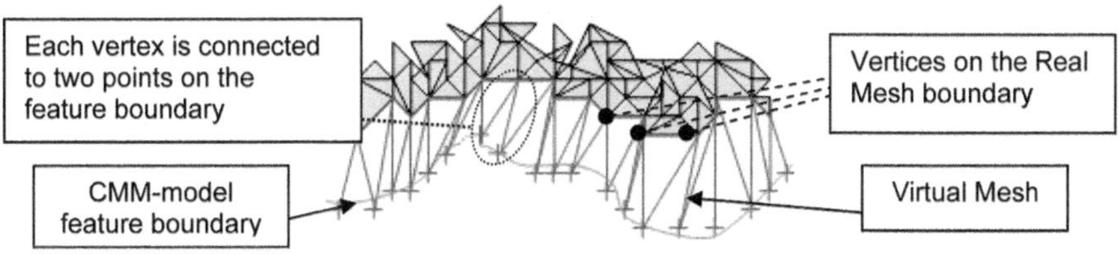

Fig. 6. Filling gap between LS-model and CMM-model.

empty spaces then are filled with the real size features from the CMM-model. The remaining model in this stage has gaps between the triangulated mesh boundaries of the LS-model and the feature edge of the CMM-model, the size of which is a function of tolerance and size of the closure envelops.

3.6 Model Ornamentation

The missing information areas in non-sensitive surfaces can be filled by stitching the boundaries of missing information area and creating an unreal mesh if such areas are within the part overall tolerance. However if the missing information creates excessive gaps, it is advisable to carry out additional scanning of that area by LS and import the scanned view into the CAD file for manual (if the 3CB had been removed) or automated registration using similar approach as described in Section 3.4.

The gaps between features in the model surfaces are stitched using a sewing algorithm (stage E, Figure 5), as shown in Figure 6. The stages of this algorithm are as follows:

1. Identify the number of vertices in the desired triangulated mesh boundary (i_b).
2. If the triangulated mesh is surrounding the feature boundary, create the same number of points on the corresponding feature edge (i_e) in equal intervals as in equation (3) then go to 4.
$$i_e = i_b. \tag{3}$$
3. If the triangulated mesh is not surrounding the feature boundary completely, create one point more than i_b on the corresponding feature edge in equal intervals as in equation (4).
$$i_e = i_b + 1. \tag{4}$$
4. Starting from the minimum distance between points on feature edge and triangulated boundary vertices; connect each two points (only) on the feature edge to one closest vertex on the triangulated mesh boundary.
5. Create surfaces between new virtual mesh and sew the surfaces together.

The accuracy of the virtual mesh stays within the accuracy level of the real mesh, which can be filtered prior to its file import to CAD environment. In the case of

Table 1. Systems used for CAD modelling experiments.

Hardware	Accuracy (mm)			Software
	X	Y	Z	
Minolta 900 Laser Scanner with Minolta tele-Lenz	±0.22	±0.16	±0.19	Minolta Polygon Editing Tool Version 1.1
Brown & Sharpe Global CMM with TP20 Renishaw Probe	±0.035	±0.035	±0.036	PC-DMIS Version 3.5 MR1 Beta
Auxiliary 3CB set	Ball Diameters = 28.63			Unigraphics NX3

filtering after file import, the virtual mesh needs to be involved in the filtering process for better uniformity.

3.7 Tolerance and Dimensional Information

The final stage of modelling is in fact similar to the conventional parametric modelling in CAD environment as all the dimensional information is directly extractable from the features. Since these features are created using the real measurement results of the CMM touch-trigger probe, without any approximation, their accuracy is assumed to be sufficient for precision RE applications. The tolerance information however should be recognised by assembly knowledge, quality requirements and design initiatives, and tacked by the user to the final CAD model.

Further model improvement such as cloud-data reduction and smoothing surfaces can be done on the final model if required, provided that the attached boundaries to the CMM-model stay intact.

4 Method Experiments

The information integration and CAD modelling method is most suitable for RE of components with insensitive complex casting surfaces that also have precise standard geometric machined features, which are typical in automotive and aerospace industries. Several components were reverse engineered to validate this method using the hardware and software described in Table 1.

The stages of data integration of a simple hole features is shown in Figure 7, which is similar in principal for any other feature. If the critical feature is adjacent to another sensitive feature, the sewing algorithm can be employed on the features after their sewing and merging. Alternatively the triangulated mesh boundary can be split into two sections opposite to the features joining point by selecting the minimum deviation approach.

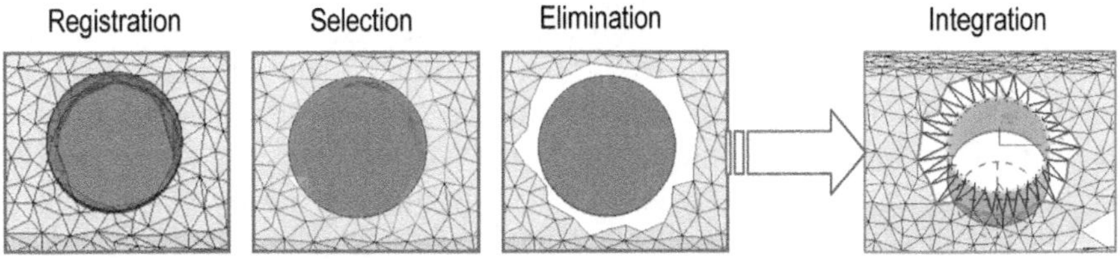

Fig. 7. Modelling stages of a simple hole with data integration method.

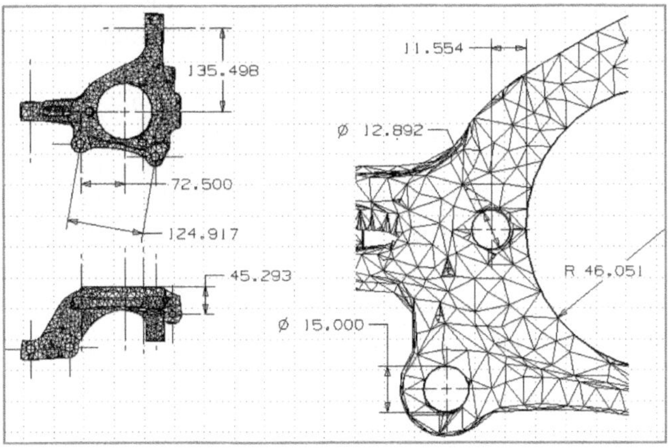

Fig. 8. CAD modelling result using the data integration method.

Figure 8 exhibits the CAD drawing of a sports car part, which is made from an aluminium casting with several simple machined features. The accuracy of the machined features were important for manufacturing and assembly of the part, however the complex geometry cast surfaces did not require to be of high precision but were necessary for casting purposes.

The information integration method has resulted in a reliable and fast RE CAD model by using fast digitisation speed of the LS to generate low-resolution data and the high accuracy of the CMM touch-trigger probe in micron level for precision data extraction. In this case the accuracy of the casing features scanned by LS was ± 0.53 mm, which can be improved up to the camera level by using a larger 3CB. The accuracy of the machined features scanned by CMM touch-trigger probe was in the micron level.

5 Conclusion and Further Work

A novel method for data integration between low-resolution 3D data captured by Laser Scanner and the high-resolution data gained from CMM with touch-trigger probe is presented. This method is most suitable for reverse engineering of parts with complex casting surfaces that also have high precision machined features. Data captured from different scanning tools are imported as stage-models into different layers in the CAD software environment. The models are then registered to the same coordinates. The high precision features are extracted from the CMM generated model and the spaces between these features are filled by the triangulated mesh of the cloud-data obtained from Laser scanning. The accuracy of this modelling method is within the limit of Laser scanning resolution for non critical features and within the accuracy of the CMM touch-trigger probe for more sensitive features. This method is superior compared to the conventional reverse engineering methods since it does not use any feature approximation and surface estimation while maintaining the speed and flexibility. The method development consists of automation of the modelling processes to minimise the modelling error caused by human interaction. Also the future research work includes reverse engineering tolerances for the model and its integration with the semi-automated data integration system in order to create a fast, accurate and complete CAD modelling system.

Acknowledgements

The work has been carried out as part of the EPSRC E-IMRC at the University of Bath under grant No GR/R67597/01. The authors wish to thank Phil Williams of Renishaw and Peter Smith from Konica Minolta Photo Imaging UK for their technical inputs.

References

1. T. Varady, R. R. Martin, J. Cox. Reverse engineering of geometric models – An introduction, *Computer-Aided Design* **29**(4), 1997, 255–268.
2. D. Driksen, C. Runte, H. Delere, C. Thomas, Z. Borocz, F. Bollmann, G. Von Vally. Computer-aided design in the modelling of facial prostheses from optical impressions of facial defects, *Biomedizinishe Technik, Germany* **47**(4), 2002, 85–90.
3. V. H. Chan, C. Bradley, G. W. Vickers. A multi-sensor approach to automating co-ordinate measuring machine-based reverse engineering, *Computers in Industry*, **44**, 2001, 105–115.
4. T. S. Shen, J. Huang, C. H. Menq. Multiple-sensor integration for rapid and high-precision coordinate metrology, *Transaction on Mechatronics* **5**(2), 2000, 110–121
5. C. H. Menq, H. T. Yau, G. Y. Lai. Automated precession measurement of surface profile in CAD-directed inspection, *IEEE Transaction on Robotics and Automation* **8**(2), 1992, 268–278.
6. Y. F. Li, Z. G. Liu. Method for determining the robing points for efficient measurement and reconstruction of freeform surfaces, *Measuring Science Technology* **14**, 2003, 1280–1288.

7. G. H. Tarbox, S. N. Gottschlich. Planning for complete sensor coverage in inspection, *Computer Vision and Image Understanding* **61**(1), 1995, 84–111.
8. S. Son, H. Park, K. H. Lee, Automated laser scanning system for reverse engineering and inspection, *Machine Tools and Manufacture* **42**, 2002, 889–897.
9. P. J. Besl, N. D. McKay. A method for registration of 3D shapes, *IEEE Transactions on Pattern Analysis and Machine Intelligence* **14**(2), 1992, 239–256.
10. D. W. Eggert, A. W. Fitzgibbon, R. B. Fisher, Simultaneous registration of multiple range views for use in reverse engineering of CAD models, in *Proceedings of ICRP*, University of Edinburgh, 1996, pp. 243–247.
11. J. Jamshidi, A. R. Mileham, G. W. Owen, A laser scanning registration technique for reverse engineering applications, in *International Conference on Manufacturing Research, IMRC*, Bath University, UK, Cranfield University, UK, 2005.
12. C. Bradley, V. Chan, A complementary sensor approach to reverse engineering, *Transactions of ASME* **123**, 2001, 74–82.
13. T. S. Shen, C. H. Menq, Automated camera calibration for a multiple-sensor integrated coordinate measurement system, *IEEE Transactions on Robotics and Automation* **17**(4), 2001, 502–507.
14. C. Oblonsek, N. Guid. A fast surface-based procedure for object reconstruction from 3D scattered points, *Computer Vision* **69**(2), 1998, 185–195.
15. Y. F. Wu, Y. S. Wong, H. T. Loh, Y. F. Zhang, Modelling cloud data using an adaptive slicing approach, *Computer Aided Design* **36**, 2004, 231–240.

Observations on Data Relating to User Browsing Behaviour and Usage of Electronic Information

Duncan Campbell, Steve Culley and Chris McMahon

Department of Mechanical Engineering, University of Bath, Bath BA2 7AY, UK; E-mail: {d.r.campbell, s.j.culley, c.a.mcmahon}@bath.ac.uk

Abstract. In this paper the implementation of a software test rig that facilitates the capture of computer-based "activity profiles" is discussed. A case study involving the analysis of profile data for the activities of a designer working in the engineering division of a medium sized enterprise is also presented. It is shown that the capture of data including the time and location of electronic information resources browsed or manipulated during the course of an activity can be achieved with a relatively high level of autonomy. It is proposed that the results will be used to provide for a better understanding of designers' interactions with electronic information but also aid in the generation of beneficial applications of profile data for the support of designers' tasks on computer.

Key words: information management, activity profiling.

1 Introduction

The overall research is investigating various aspects of electronic information usage and browsing behaviour that can be captured in the background of the user's working environment. The general objective is to identify the potential uses of the resulting activity profiles that consist of (i) data captured automatically including references to electronic information accessed, changed or created, and (ii) data provided by the user on the motivations for certain actions taken during the course of their working session.

Before introducing the general objectives of the research in more detail, the concept of an "activity profile" is defined within the context of this work to add clarity to the subject of discussion.

S. Tichkiewitch et al. (eds.), Advances in Integrated Design and Manufacturing in Mechanical Engineering II, 177–188.

2 Working Definition of the Activity Profile Concept

The term "profile" has different meanings depending on the context in which it is used. A data profile can simply be defined as a formal summary or analysis of data [1]. In some cases, the act of profiling can also refer to the categorisation or classification of the data being profiled.

In this paper the term "activity profile" is used to refer to the data collected as a result of the computer-aided observation of activities. This activity profile will represent a summarised description of the actual tasks being carried out. It is not the categorisation or classification of the observed task into a pre-defined schema based on evidence of actions observed. However, this is a related area that may be investigated in future research.

In the next section the objectives are discussed in detail, with particular emphasis on how this research complements existing research in the literature.

3 Objectives

Court [2], Lowe [3] and others have investigated the sources and types of information design engineers refer to through observational studies of various engineering activities. Examples of the data collection methods used in these studies include questionnaires, interviews, action research and protocol analysis.

The proposed research differs from previous studies in that a novel method for collecting data is used, which employs a software tool monitoring aspects of the user's working environment. In one sense this is restrictive as it only allows for the collection of data related to the use and access of electronic information resources and not paper-based resources for example. However, a strength of the method is that it enables the capture of a rich and highly detailed view of all the users' computer-based interactions, in parallel with electronic information being used or manipulated.

In particular, the empirical research method discussed in this paper should allow for a better understanding of:

- *The behavioural characteristics or traits exhibited by the computer user*, specifically in relation to the multi-tasking aspects of day-to-day computer use where the consumption and manipulation of electronic information often involves various motivations and a number of software applications.
- *Electronic information search processes*. The method requires that the computer user provide information related to their intentions at the beginning of a search, their progress during a search and their success in satisfying an information need at the end of a search. It is anticipated that this will allow for the effective reconstruction and modelling of this aspect of computer-based activities.
- *Electronic information inputs and outputs in design tasks*. The method monitors the access or editing of local computer files of interest as well as remote resources

accessed on the Internet. These records should enable the accurate depiction of inputs and outputs.

In addition to the enhanced understanding of computer use in engineering design tasks it is anticipated that a modified version of the software could be used as a tool as part of a computer user's normal working routine to capture detailed records of important tasks carried out on computer. This possibility, an aspect of future work, is discussed in Section 6.

4 Method

In this section various aspects of the monitoring method are discussed in detail. The primary focus relates to the design and function of the software used to collect data from the computer user. The section ends with some notes on the procedure used to conduct the monitoring activity with the software.

4.1 Overview

As stated previously the overall monitoring method enables the capture of both user-provided data and data captured automatically. In both circumstances, the data is recorded when pre-determined events are triggered by the computer's operating system or the software applications being used. Figure 1 shows the sorts of events and data types involved.

More detail on the auto-captured data and the user provided data is presented in the following sections.

4.2 Auto-Captured Data

The current version of the software consists of 3 components; (i) an application usage watcher that captures data on the applications open or in focus, (ii) an Internet browser watcher that captures data related to page navigation events in a proprietary web browser and (iii) a file system watcher that captures data on the access, creation or manipulation of local files.

The data elements captured automatically for each component include a time-stamp reference and a number of event descriptors that can be extracted without any need for interaction with the user. For example when a new application is initialised, component (i) captures the application title and the various window titles running under the application process.

A different set of data elements is captured depending on the nature of the event. As searching and the search process is of particular interest, component (ii) responds to page navigation events occurring in the web browser by capturing data including the web page title and text-based content. Finally, component (iii) captures the name

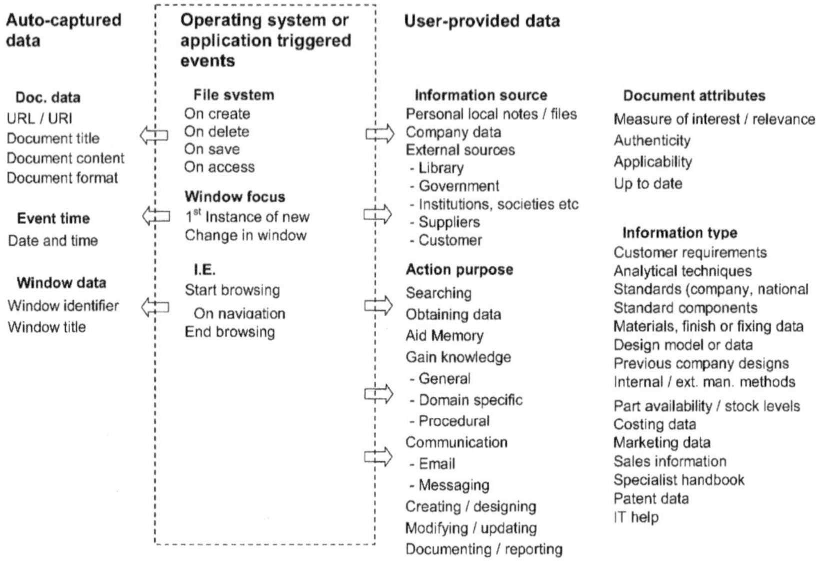

Fig. 1. Illustration of data capture triggered by operating system or application events.

and location of files read or written to on the local machine. The user is able to define the root folders to be monitored by this component at the beginning of the session.

4.3 User Provided Data

In some circumstances, in order to achieve a complete description of the activity, the user is prompted for additional information using an onscreen form. In this section, the types of data requested are described. Figure 2 shows example screenshots of the forms for each of the main components, which appear on the right hand portion of the user's screen when certain events are triggered.

The following data is requested for each of these components:

(a) Application usage watcher: The user provided data for this component includes the type and purpose of the activity being carried out at a given time during the session. When an application is initialised the user is required provide a definition of the task purpose, which can be selected from checkboxes relating to 5 categories: (i) communication, (ii) design synthesis, (iii) document editing, (iv) data management and (v) information seeking or browsing. Further detail may be provided if necessary by selecting checkbox options from associated sub categories. Once the initial definition of the application purpose has been made no further user data is required unless the user decides to update this information, as the periods when the application is active or inactive can be tracked automatically.

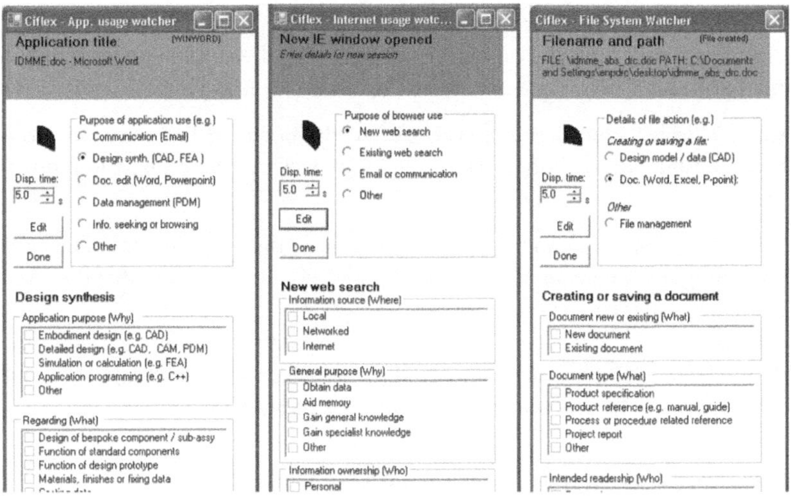

(a) (b) (c)

Fig. 2. Screenshots of the user data forms.

(b) Internet browser watcher: The user provided data for this component includes details on information searches carried out over the Internet. When a browser is initialised the user is typically required to make a definition of their information need by providing details of: (i) the general purpose of their search, (ii) a category relating to the vendor of information and (iii) a category relating to the type of information. In subsequent navigation the user can optionally rate the pages of interest and signify the page(s) that satisfy the original information need.

(c) File system watcher: The user provided data for this component includes details on motivations for the creation or manipulation of existing files and a definition of the associated information.

4.4 Software Design Considerations

The design of the method for capturing user-provided data is of particular importance to the success of this empirical research method. Unlike the automatic capture of data for the activity profile, the prompting and capture of user data needs to be restricted in order to minimise the impact of the resulting intervention on the computer user's activity. The features of the form designed to address this issue include:

- *Optional form completion*: the form was designed so that completion of the data entries is not mandatory, and that the computer user can choose to ignore the screen prompt if it is felt that data entry is unnecessary.

- A passive dialog: After a specified period of time, if no response is made, the form disappears until the next event occurs. The dialog also only appears on a small portion of the screen outside of the main viewing area.

Another design consideration is related to the potential for too much invasion of privacy on the user's actions. The following measures were taken to reduce this problem.

- *Controllable capture functions*: The operation of the software including controls for turning the data capture off and on are easily accessible so that the user can temporarily stop capture if necessary.
- *Traceable profile data logs*: All the data captured by the software is written to log files that are easily accessible to the computer user. The user is also able to view or edit the files to remove unwanted or irrelevant data.

4.5 Procedure Used in the Observational Studies

Having discussed the software function and design considerations, this section briefly introduces the procedure used to collect data. A summary of the main steps is shown below.

1. The software is installed on designers workstation
2. The designer is provided training on the main functions and usage of the software
3. The designer is asked to initiate the data capture functionality in any future "working sessions" where the activity was likely consist of predominantly computer-based work. (There was no formal restriction of the definition of a working session although it typically ranged from half a day to 1 days work and consisted of a variety of sub tasks.)
4. The designer is asked to return the text based logs files for analysis on completion of a working session.

5 Initial Observations on Activity Profile data Collected in an Industrial Setting

In this section the analysis of data collected from a detailed designer at a small to medium size engineering division of a local firm is presented. The study was carried out over a 2-day period of predominantly computer-based work. The work involved tasks relating to the end stages of the design of a harvester machine, which in summary, included the geometric modelling of detailed design features, and reference to a number of local and remote information sources. The data captured during the investigation was a combination of auto-captured data and user-provided data as described in the previous section.

Table 1. Working session summaries, constructed from captured profile data.

		Application Category						
Date Time	**Application title**	**Communication**	**Design Synthesis**	**Document editing**	**Data or file management**	**Information seek / browse**	**Other**	**Further detail**
Tues 4th October								
10:51	Solid Edge		✓					**App. purpose (Why):** Detailed design, **Regarding (What):** Function of standard components
10:53	Excel spreadsheet			✓				**Doc. Type (What):** Project report, **Readership (Who):** Personal use
10:56	Solid Edge V14 - Assembly		✓					**App. purpose (Why):** Detailed design, **Regarding (What):** Function of standard components
11:49	Microsoft outlook	✓						**Correspondent (Who):** Internal, **Regarding (What):** Other
13:56	Solid Edge V14 - Assembly		✓					
15:42	One Touch			✓				**Doc. Type (What):** Other, **Readership (Who):** Internal
16:18	Solid Edge V14 - Assembly		✓					**App. purpose (Why):** Detailed design, **Regarding (What):** Design of bespoke component
Wed 5th October								
10:31	Internet Explorer					✓		**Information source (Where):** Internet, **Purpose (Why):** Gain specialist knowledge, **Info. Ownership (Who):** Institution, **Regarding (What):** Other
11:03	Solid Edge V14		✓					**App. Purpose (Why):** Detailed design, **Regarding (What):** Design of bespoke component
11:13	Windows explorer			✓				**Document type (What):** Product specification, **Readership (Who):** Personal use
13:47	Solid Edge V14		✓					**App. Purpose (Why):** Detailed design, **Regarding (What):** Design of bespoke component

5.1 Overview

In the initial review, the primary objective was to ascertain whether it was possible to convert the captured data into the form of a detailed description of the work carried out. The component monitoring application usage was particularly helpful in this respect. Table 1 shows a summary of each day constructed from the records. Here it can be seen from the detail that the major task objectives recorded at different points in each day can be extracted to form a rich description of activities throughout the working session.

In the following subsections observations relating to the data captured by each component of the software are presented in more detail.

5.2 Data Collected Using the Application Usage Watcher Component

In addition to the summary data shown in Table 1 an alternative means for the visualisation of the application usage data has been developed to aid in the understanding of how a user switches between open applications to meet various task objectives.

Figure 3 shows a snapshot of the tool that converts the text-based log files into a computer-generated chart indicating the application in use at different times throughout the session.

An aspect of particular interest for the automated generation of activity descriptions is an understanding of whether it is possible to infer which instances of application use are common to a specific task objective (of which there may be several) that

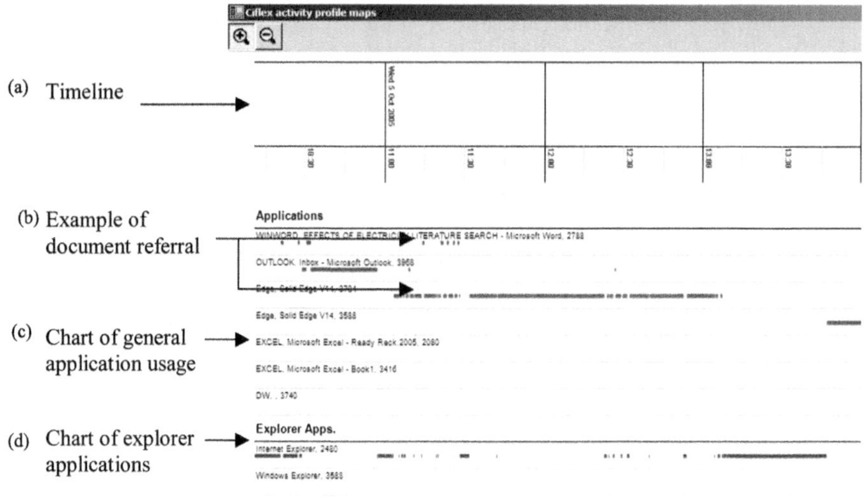

Fig. 3. Computer generated chart of application usage.

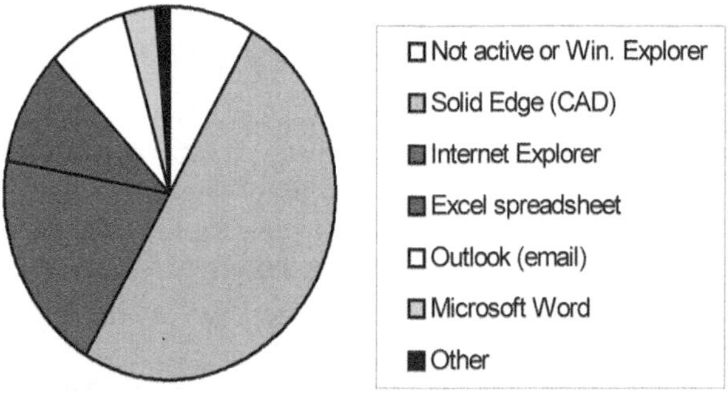

Fig. 4. Proportion of time spent using different applications.

the user is pursing over the course of the working session. A good example of where this may be inferred without any extra information from the user is illustrated in Figure 3 (Arrow (b)). It can be seen from the chart that the user is repetitively switching between an application containing a document and an application containing a CAD model. A likely cause for this behaviour is that document is being used for reference to aid with an action being carried out on the CAD model and so the inference may be made that both applications relate to the same task objective. However, it is noted that such inferences cannot be made with 100% certainty.

The pie chart in Figure 4 shows the proportion of time spent using different applications during the working session. In this case it can be seen that although the

Table 2. Tabulated summary of local files accessed changed or created (extracted from the 1st day of the case study).

	Files accessed	Files changed	Files created
CAD (Solid Edge)			
Assemblies (asm)	14	7	
Parts (par)	74	8	1
Drawings (dft)	7		
General office documents			
Microsoft word (doc)	6	3	2
Portable Document Format (pdf)	1		
Mircrosoft Excel (xls)		1	

CAD system is in use for the largest proportion of time, a number of other applications are also involved in the working session. An interesting observation from this analysis is the large diversity in application use. An aspect of future work may involve the comparison of this sort of data over a variety of work sessions from a number of designers in different stages of the design process.

5.3 Data Collected Using the File System Watcher Component

Table 2 shows a summary of the files that had been used or edited during the course of the working session. The full description of where these files reside is contained within the complete data logs along with the time they were accessed or edited. This information is of importance as it provides for an effective description of some of the inputs and outputs of the work carried out – the inputs typically being the files accessed and the outputs being files changed or created. Conventionally, no record of this sort of information is taken or used during the course of normal working routine. It should also be noted that this particular type of data can be obtained with little or no additional burden on the user's activities.

5.4 Data Collected Using the Internet Browser Watcher Component

Table 3 shows a summary of data captured during the course of an Internet-based search for information related to insect control devices utilising electric fields. This was a task extracted from the data profile on the 2nd day of the case study.

Along with the original definition of the information need, which was stated as being a search for specialist information, the content and usefulness of subsequent pages can be effectively retraced. As the user also specifies how certain web pages impact on the status of the search, the outcomes of the search activity can also be traced to some degree. In this case, the designer was unable to find the information required to satisfy the original information need.

Table 3. Tabulated summary of web browser pages visited (extracted from the 2nd day of the case study).

		User rating						Search status			
Host domain:	Pages visited:	5 (High)	4	3	2	1 (Low)	N/A	Page ends search	More info needed	Give up search	N/A
www.auto.search.msn.com	1						1				
www.imeche.org.uk	9			1					1		
www.engineeringvillage2.org	40			1	1	1			3		

It is anticipated that the search profiles and other profile data discussed above, can form the basis of a useful resource for future users revisiting the subject area. This possibility is discussed further in the next section, which relates to future research.

6 Future Work

There are two areas that will be pursued in future research. Firstly a greater number of profiles will be collated across a range of computer-based activities in order that comparative studies may be carried out. The other aspect of work will involve the investigation of how activity profile data can be represented or manipulated so that it can be readily reused. The main issues to be addressed here are covered in the following sections.

6.1 Means for the Effective Representation of Profile Data

In order for the data captured during the course of activities to be used effectively for manual reference it has to be in a form accessible and understandable to the user. Part of the research in this area involves the investigation of different ways of representing profile data including the appropriateness of the application of well-established modelling languages including the IDEF range of languages.

6.2 Minimising the Burden on the User

In order for activity profiling to be used as part of normal working routine the process should be automated to an extent that minimises the burden and disruption of a user in their task, but without detracting from the usefulness of the resulting data. An aspect of future research involves the analysis of what information is required to produce a sufficiently detailed activity profile and the level of autonomy that can be employed to achieve this.

6.3 Activity Profiles for the Referral of Related Information

An area of current and future research being pursued by the authors is to establish whether meaningful relationships between documents used in working sessions can be captured from activity profiles.

In our initial work, two types of document relationships to be extracted from activities have been defined. Firstly, a bidirectional *Common Utility Dependency* where two documents have use or are of interest in the same task carried out by the computer user. For example, where documents x and y have both been useful in satisfying an information need in a search carried out. Secondly, a unidirectional *Reference Dependency* where one document has been a reference source in the creation or editing of another.

In both circumstances, the objective will be to establish whether the relationships can be stored and indexed within information retrieval systems to improve their ability to respond to future information retrieval tasks. It is envisaged that such a system might have potential applications in two types of retrieval process to be investigated. In conventional manual retrieval it may provide a mechanism for the retrieval of a ranked list of contextually related documents, given an instance of a document that is of interest in the current search. This type of application could possibly be used in combination with conventional content-based retrieval algorithms. Alternatively, it may provide a mechanism for dynamic retrieval scenarios where a combination of the document in use and previously indexed relationships can provide an automated system with all the information needed to provide the user with additional related documents without any need for a search query.

The greatest difficulty associated with the extraction of such relationships is to make accurate assessments of document interest or utility without placing additional burden on the computer user's activities. It is anticipated that some of the data provided in the activity profiles such as the time a document is in view and other metrics associated with the applications used for browsing and editing documents maybe useful for achieving this goal. More detailed proposals on the possibilities for the automation of information-based support are discussed a related publication [4].

7 Conclusions

This paper presents a novel, low intrusion approach for carrying out empirical research in the design information field. It also shows, through an initial case study, some interesting results about the diversity of activities undertaken by an engineering designer.

Currently, a number of further activity profiles are being collated for investigation. Initial indications are that as well as promoting a greater understanding of information use in computer-based work the data can be processed in ways to assign relationships between previously disconnected items of information.

Lowe [3], Hales [5] and others note the integral role information search and retrieval plays in design tasks, particularly in the embodiment and detailed design phases. Other research indicates the value of past project data to designers and their requirement for a contextual understanding of the subject of their work. Notable references to these points include:

- "An information searching strategy based on accessing past projects is preferred, particularly by senior designers, as it provides examples of the applications of their information needs" [6].
- "In 45% of design searches in the aerospace industry the trigger for the search has been to gain an understanding of an area rather than as a response to a specific question or information need" [7].

It is anticipated that activity profile data can be used in a variety of ways to provide better support for designerŠs information needs, particularly in these sorts of searches. It is thought that the results have implications in design, but also in information usage generally.

References

1. Online reference, www.dictionary.com, Definition of the term "profile", cited 16/10/05.
2. Court, A. (1995) The modelling and classification of information for engineering designers. PhD Thesis, University of Bath, 61 pp.
3. Lowe, A. (2002) Studies of information use by engineering designers and the development of strategies to aid in its classification and retrieval. PhD Thesis, University of Bath, UK, Chapter 3.
4. Campbell, D., McMahon, C., Culley, S. (2004) Information push: Strategies for improving efficiency of design information management, in *Management of Engineering Design Complexity*, EDIProd 2004, R. Rohatynski (Ed.), Zielona Gora, Poland, ISBN 83-89712-26-1.
5. Hales, C. (1991) *Analysis of Engineering Design Process in an Industrial Context* (Second Edition), Grant Hills Publications, UK.
6. Aurisicchio, M., Langdon, P., Ahmed, S., Wallace, K. M. (2003) Investigating knowledge searches in aerospace design, in *International Conference on Engineering Design ICED 03*, Stockholm, August 19–21, pp. 293.
7. del-Ray-Chamorro, F., Wallace, K. (2003) A study of information retrieval in the aerospace domain, in *International Conference on Engineering Design ICED 03*, Stockholm, August 19–21, pp. 293.

ICT System Requirements in Collaborative Design with Suppliers

Eric Blanco and Marie-Anne Le Dain

G-SCOP, INP Grenoble, UJF, CNRS, 46 avenue Félix Viallet 38031 Grenoble, France;
E-mail: {eric.blanco, marie-anne.le-dain}@g-scop.inpg.fr

Abstract. In order to benefit from the possible advantages of an early supplier involvement (cost savings, reduction of time to market), many companies have outsourced activities and sought to build long-term collaborations with their suppliers. During the last decade more and more responsibility in designing and engineering components was transferred to supplier. The aim of this paper is to focus on the need of information management tools within co-design activities. This research is one of purpose of ISOCELE Rhône Alpes research project. In this paper, we present two models to help to define a suitable collaborative ICT (Information and Communication Technology) system. The first one is to analyse the different situations of early supplier involvement in new product development (NPD). These situations are characterised through two axes: risk of the product development and supplier autonomy. Each situation needs a specific approach of coordination between design team of supplier and customer. The second model proposes a four levels framework that allows sharing preliminary information within a collaborative design process. This framework illustrated by a tool PIQUANT is presented. In a case of a *"critical" co-design* situation with a supplier, we argue that this framework can support the design activities information and project management.

Key words: early supplier involvement, new product development, preliminary information, collaborative ICT system

1 Introduction

The paradigm shift that occurred at the beginning of the 1990s has led us to change our thinking about a company's strategy by considering the involvement of extern partners. In order to face shrinking product life cycles, increasing costs of technologies, and accelerating product complexity customers and suppliers are becoming more and more involved in the process of product development [1, 2]. The target sectors of this study are enterprises which have to face this strong competitive pressure in terms of costs but above all in terms of product variety and time-to-market

S. Tichkiewitch et al. (eds.), Advances in Integrated Design and Manufacturing in Mechanical Engineering II, 189–201.

and which took strategic decisions to outsource operations in new product development. In this context, the process of creating the offer will not be considered as a purely internal process but as a shared process involving partners (customer and supplier) who have to be coordinated. The aim of our research is to investigate the coordination mechanisms of inter-firm relations when they are carrying out collective development. In this article, we focused on information exchanged during the design phases between partners (customer and suppliers) and we are interested on the suitable management way for the coordination and communication in a collective product development approach.

This study presents first a typology of supplier involvement in new product development (NPD) project depending on two variables: the *supplier's autonomy* and the *product's development risk*. Furthermore, we present the different natures of design information in collaborative activities to understand which information is exchanged inside a design team. Then, we identify the different characteristics of a collaborative Information and Communication Technology (ICT) system that are needed to manage in an effective way the different customer-supplier relations in collaborative activities. The article concludes with a discussion about the relevance of PIQUANT [3, 4] system as ICT system to support customer-supplier distant and asynchronous interactions in collaborative design.

2 Modelling Supplier Involvement in NPD

When we contemplate customer/supplier relations in projects for developing new products, the marketing mechanisms can no longer ensure the coordination of activities between the partners. Indeed, the degree of uncertainty generated in this situation makes the regulation system strictly based on price inefficient. The customer firm has to build a "meta-organisation"[1] that includes the partners involved in the project and that is able to ensure the achievement of its goals. Based on previous works [5–7], we have proposed to model the process of integrating the suppliers similar to a diachronic activity [8, 9]:

1. A *relation configuration phase* aiming to define the "meta organisation". Five principal decisions have to be taken: (1) definition of the functional need and (2) in which moment has the supplier with its special knowledge to be involved? (3) What are the selection criteria for the choice on a supplier and (4) which kind of contractual arrangement to build up? (5) What are, in the project, the resources involved and the roles played by customer and supplier?

[1] This term implies that the project is a place of co-operation with its own rules of organisation superimposed on the rules that govern the functioning of the organisations which the partners involved in the project belong to.

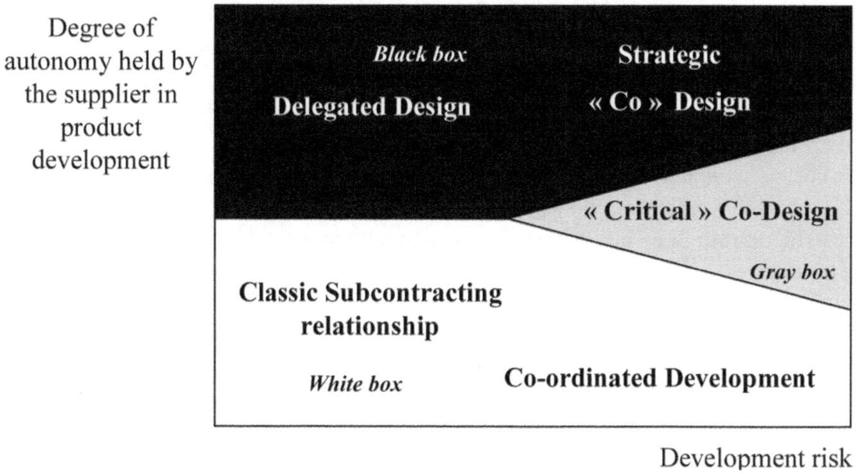

Fig. 1. Supplier involvement portfolio.

2. A *phase of defining the mechanisms of coordination* and skills to mobilise. We decided to classify them into three groups appearing during different stages of the project: coordination *ex ante*, *in cursus*, and *ex post*.

We state the hypothesis that this process of supplier involvement has to be adapted to each type of supplier/customer situations. In order to be able to distinguish the different types of supplier involvement in NPD, we adopted the theoretical model of Wynstra and Ten Pierick [10]. Already in a previous research [9], we enriched this model and proposed a matrix based on two dimensions. These dimensions are: the "degree of autonomy" that is contracted out to the supplier in the product development process (that means the level of responsibility delegated to the supplier) and the collaborative "development risk" of the outsourced component or building block. We have defined [8] a metric for each dimension that allow us to place every customer/supplier relation in our matrix (Figure 1). Finally, through an empirical and theoretic approach in a previous study [9], we defined five types of customer/supplier involvement in collaborative NPD resulting from the combination of these dimensions.

The most complex situation in coordinating supplier involvement in NPD identified in Figure 1 occurs when having a significant development risk and a medium degree of supplier autonomy. The types of dominant product risk are the development timeline, the newness of technologies of the outsourced component or building block and the systemic link with the final product. We have qualified this situation as *"critical" co-design*.[2] In this case, neither the customer nor the supplier possesses the knowledge and the ability to completely execute product design in-house. So, the concept development but also the product and process design will be carried out in an *integrated* way [11] that corresponds to a real *co-design* activity. From the beginning, the partners are thus confronted with a big uncertainty on the progress of the project as well as with a strong ambiguity in the nature of the tasks to be made. It is the reason for which we qualify this design as *"critical"*.

The definition of the information systems to support the coordination *in cursus* (phase 2), is crucial. In this paper, we focus on the requirements of ICT system to support this specific stage in the case of a *"critical" co-design situation*.

The exchange of information in the *delegated design* or *strategic "co"-design* relationship could be essentially reduced to official and validated data. The actual tools like PDM or Supplier Portals could well support this level of information exchange. In the case of *"critical" co-design*, the coordination between the two partners becomes crucial. The sharing of information has to be increase and the information change of nature. The ICT tools necessary to support this new partnership are then different. They have to take into account the designers needs, and the way they share information within design activity. The nature of information is different including both validated data and preliminary information [12]. In the section below, we will focus on the characterisation of design information exchanged within design teams.

3 The Different Natures of Design Information in Collaborative Activities

3.1 Open and Closed Intermediary Objects

Previous accounts of design work point to various ways in which design works depends on communicative activity [13]. Design work is related to the production and the use of information. We use the notion of *Intermediary Objects* (IO) [14], to describe the objects or documents that appear or are used in the process, whatever their form, their origin or their destination: schedules, minutes, functional graphs, calculation results, drafts, 2D plans or 3D models, prototypes, etc. These IO can be seen as resulting from the design work but also as supporting and highlighting it. Observers of design activity have highlighted the "mixed-use practices" [13, 14] of designers combining the electronic, paper and physical world. The term of *Intermediary Object* serves as a generic designation that is useful to include all design representations.

The IO do not all have the same characteristics in design. Those characteristics depend on the properties of the object itself (ability to be modify, transfer for

[2] The more the development risk is important, the more the customer will try to force and control collaboration between his design team and the one of supplier. This reasoning explains the triangular shape in the proposed conceptual matrix (Figure 1).

example) and on the situation of action in which it is committed (prescription or negotiation for example). Depending on the margin that is left to the user, we identify *open* or *closed* objects. A closed object transmits a strong regulation, whereas an open object is a support for negotiation. *Deliverables* produced for project miles stones are closed objects. The *drafts* and *preliminary information* are open objects that are used and exchanged to support the negotiation and the emergence of solution. The openness of the IO is not an intrinsic characteristics of the object, but depends also on the statute of the information that the actors give to it.

In collaborative design with Supplier, both open and closed objects will be used and exchanged. In the case of relationship is based on decoupled activities [14] (*delegated design, strategic "co"-design* and *co-ordinated development*), *closed objects* are preferred. In the case of *"critical" co-design*, open objects are needed to support the common decision making processes. Open objects are often published as preliminary information i.e. non-validated information, in order to support iteration, negotiation and interplay. Following Terwiesch [12] we assume that preliminary information could not be published in the same way as validated objects. Actors have to decide on coordination strategies depending on the maturity of the information. ICT systems have to take into account these different levels of maturity.

3.2 Maturity of Preliminary Information

The objects exchanged by the actors of the design are heterogeneous in term of maturity. In a CAD model, for example, some elements of design are results of decisions, whereas some are only hypotheses. Some data are frozen and stable, others are uncertain and non reliable. Many authors were interested in qualify and quantify the design information. Most of them refer to uncertainty as the main characteristic of design information [18, 21]. Other authors use different criteria to express uncertainty, completeness, reliability, ambiguity of design information [16, 17]. The limits of those formal models of uncertainty are that they are applied generally on a very low level of design parameters, generally numerical ones. We assume that the characterisation of the preliminary information is useful, but we do not believe that designers are able to characterise all the Design Parameters. Another limitation is that these approaches do not take into account the recipient point of view. The uncertainty seems to be intrinsic to the information whereas the communication process is an interplay between actors. We assume that people involved in engineering design express a judgment on the information they use and they transmit. We observed that design actors use some of the presented criteria to express the maturity of the information. We also encountered the notion of reliability, precision, confidence, etc. This characterisation takes usually the form of comments diffused by e-mail or phone or even annotations that accompany the intermediary object itself. "I'm not sure of these value, it ought to change so don't pay attention to it". But these criteria are not consensual within the company and not systematic. The misunderstandings are responsible for much iteration in the design process.

We propose to express the maturity of the design information as a set of some criteria. we claim that the choice of the criteria have to be adapted to the company's culture. The representation used to express them in an Information system have to be internal conventions within the company. Thus maturity can be defined as the judgment given by both the provider and the user on the quality of information. The provider qualify the uncertainty of the information he produces, the user qualify the usability of the information in his own context. This maturity can be valuated by set of conventional criteria fixed in the context of the company.

The knowledge of maturity of *preliminary information* is important for the users. In overlapped activities, the down-stream activity could be sensible to the quality of the output provided by the up-stream activity. Terwiesch [12] identifies three co-ordination strategies related to risks of design iteration. He explains that the co-ordination strategy should be chosen by the knowledge of precision and stability of design information.

Following Terwiesch, we assume that to improve collaboration in design, the ICT system has to support facilities to qualify information in term of maturity. But we also consider that the process of information sharing in design activity is based on different needs and social dimensions that as to be considered. In the next section we propose a model to structure the different workspaces in collaborative design.

3.3 The Four Different Workspaces in Collaborative Design

We assume that the process of sharing of information is performed through almost four levels of collaboration that we refer to as *workspace* [3, 4]. These workspaces are characterised by three aspects: the different level of maturity of shared information, the different actors involved in and who managed their inclusion and the type of activity performed on information (construction, validation, etc.).

First, the *public workspace* corresponds to the necessary coordination and plan of the design activity. It organises the sharing of public information that means deliverables of the design activities and some others intermediary objects identified in the design process. These objects are essentially closed ones. They are published in that workspace when they are validated. The actors that can access to the information can be client, other teams, suppliers, etc. The actual PDM systems offer much functionality to manage the exchange of this level of validated information. The evolution of this information could be based on specific lifecycle or Engineering Change management processes.

The fourth workspace on the opposite is the *private* workspace. In this workspace the designer arranges his drafts and personal data. These objects do not have any existence in the formal framework of the project. They are not shared and they do not have to.

Third is the workspace where the *draft* of solution is constructed as a shareable object within the design team. We refer to this state of the information as *exhibit*. This workspace is called *proximity workspace* because it includes actors of the personal network of the producer. In the proximity workspace, the designer opens himself to critics and judgment of others. It is based on loyal relationships. This workspace is built for a specific need and could evolve during the project depending on the competencies of the actors. This is the place of informal confrontation and advice. The role of this space is the construction of a robust and convincing discourse to argue the solution. It is an informal validation of the solution. The workspace is open by an actor who wants to confront his ideas with partner's point of view. Some of the actors could be inside the design team or the project some could be guest during this collaborative activity. The workspace is open at any time to an actor who needs to expose a draft as an exhibit to a group of "friendly" actors. These activities cannot be planned because they emerge from the design problems treated by the actors.

When the information is considered to be validated or when the argumentation is coherent, the information can be transmitted outside the *proximity workspace*. The designer places it in the *project workspace* (generally the common database or PDM system). Lecaille [15] uses the term of *enabled trace* to characterise the status of this information. It is non-officially validated but sufficiently convincing to be published. The actor closes the *proximity* workspace when the object could change from *exhibit* state to *enabled* state. Enabled objects are encountered in this second workspace. We call *project workspace*. This intermediary level is the level of sharing preliminary information in the constituted design team. These activities are mainly planned because of the knowledge of the design team. The *Intermediary Objects* are then identified and agreement can be built on their required maturity and dates of publication.

We want to highlight here that the publication of *preliminary information* within design teams is a process that involve social aspects that the information management systems has to take into account. We argue that the collaborative support systems should support this evolution of the information within specific workspaces. The designer is committed by the information that he delivers to other actors. He could not diffuse drafts within any design spaces. Actors take care of what *preliminary information* they diffuse in design team. This caution sometimes delays the access of information by the others. This point is increase in the case of co-design with supplier. When in *delegated design* the public workspace is sufficient to exchange validated information, in *Co-ordinated Development* the project work space is generally observable. The next section will highlight the requirements for the ICT system to support collaborative design with supplier.

4 ICT System in Collaborative Design with Supplier

The informational coordination mechanism refers to the exchange and sharing of information as well as the validation of results between customer and supplier. According to the theory suggested by Daft and Lengel [20], the choice of informational media depends on the nature of the problems to be solved. Thus, each of the relationship types identified in Figure 1 respectively induces specific needs, specifically in terms of collaborative ICT system essential to effectively support customer-supplier interactions. In the following sections, we define the required ICT system functionalities in collaborative design with suppliers and we specify the ICT system in the case of *"critical" co-design* collaboration.

4.1 A Typology of Functionalities for ICT System

Three major activities are carried out by the stakeholders (internal or external) involved in a NPD project:

- they produce intermediary objects;
- they coordinate each other; and
- they communicate information and data.

In the collaborative design with supplier, the overlapping between these three activities may be considered as interaction areas between customer and suppliers. These areas allow the identification of the four coordination modes that a customer and a supplier can be confronted with during the product development project. The ICT system must take into account the specific needs of these four different coordination modes:

- *Co-Elaboration*: This coordination mode characterises a working situation where customer and supplier need punctually an intensive coordination on a specific point to perform their respective design activity.
- *Project Communication*: The customer and supplier inquire mutually about the project management information necessary to make decisions at the different milestones of the project. They also inquire about the delay induced by the request of modification.
- *Co-Operation*: This coordination mode is characterised by open intermediary objects exchange which need, in a synchronised way, a continuous confrontation of the know-how to build shared knowledge for the collective decision-making and co-construction of solutions.
- *Result Communication*: here customer or supplier put closed IO at the stakeholders of the project disposal. For the customer, these objects are for example the contractual specifications and the request of modifications. For the supplier, these objects are for example the intermediate deliverables.

We can thus define a collaborative ICT system according to the extent of his coverage on these four functionalities (Figure 2).

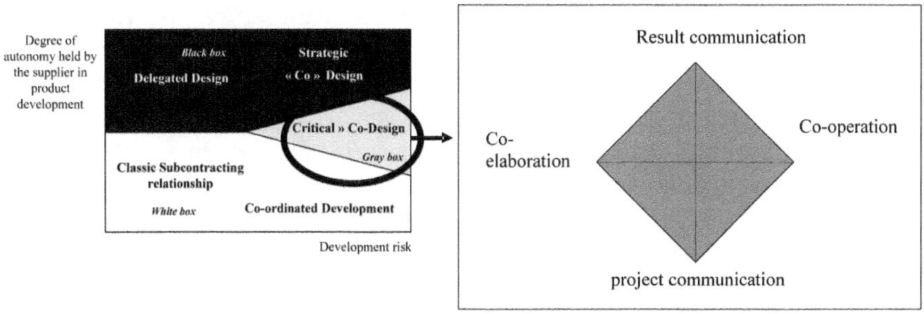

Fig. 2. Collaborative ICT system pattern in case of *"critical" co-design.*

4.2 ICT System for "Critical" Co-Design

Our basic assumption is that there is no universal communication and information system to collaborate with suppliers in NPD. The effective value-adding of the used ICT-tools in collaborative design depends of the different types of customer/supplier relationships identified in Figure 1. In previous work [19], we have proposed prescriptive suggestions about the required ICT system functionalities for each type identified in Figure 1. These suggestions have been based on former work [9] and some case studies.

We put the assumption that the ICT system support in a *"critical" co-design* requires a balanced set of tools based on four functionalities: communication, collaboration, coordination and cooperation (Figure 2). This situation requires a deep relation in terms of information, skills and know-how in order to exchange and share information. Consequently, this situation causes a strong interdependence between the customer and the supplier during the project. They need synchronous interaction to get over each milestone of the project. Then in this case, the co-operation tools are very important. In most analysed case studies, to reduce the inherent risk of this *"critical" co-design*, guest supplier engineers were integrated in the early stage of the NPD project. They were real resources of project team and this collocation is the main co-operation tool between customer and supplier. Between customer engineers and supplier engineers, information sharing relies mostly on tacit knowledge and *open objects* (draft and *preliminary information*) are used, exchanged as support on the negotiation and the emergence of solution. In this type of relation with a high level of uncertainty, the client company must create a real "meta organisation" based on not only a contract but also on organisation exchange with his supplier in order to reduce the risk. Particularly, the definition of the rules in *Preliminary Information* exchange in *project and proximity workspaces* facilitate this organisation exchange. These rules must be naturally built and shared between different actors of the project team.

The actual PDM systems offer some of the functionalities. We observed in design teams that project management and data management tools are still disconnected. One offers collaboration facilities but coordination functionalities (i.e. Workflow) are too rigid to support the variability of the design activity. Most of them offer net-conferencing facilities but very few offer co-operation facilities in an asynchronous way. This asynchronous and distant co-operation, very useful in an extended team, is still performed by e-mail. This increases the distribution of design information within the team and consequently the risk of misunderstanding. For that purpose, we have developed a prototype called *PIQUANT* (*Preliminary Information QUalification for Networked Team*). This tool implemented the proposed framework in order to improve both the asynchronous and distant co-operation and the coordination of the project activities.

5 PIQUANT: A Pertinent Co-Operation Tool

The PIQUANT application allows the definition of the maturity criteria. Based on the previous experiences, a set of maturity criteria could be defined beforehand in the system. These steps can also represent a particular agreement between the project team members under the responsibility of the project manager. These criteria should be common and comprehensible for all contributors to the project. An annotation system assists the users all along the project development. As an illustration (Figure 3), we have chosen the three following criteria to represent the maturity of the preliminary information: *stability* and *completeness*. When the *open objects* move from a workspace to another, the actor can characterise the object. This characterisation of information could be performed at different levels of the information: the object globally or only part of it. The actor that published the *preliminary information* in the shared workspace should specify the current maturity criteria for elements which he thinks are relevant.

Within the *project workspace* (Figure 3), PIQUANT offers a dashboard helping the stakeholders and the project manager to manage the project daily. On the same sheet the project manager can see the schedule of the activities, the input object or output object of each tasks. The status and the current maturity are represented for each published object.

In the *project workspace*, the planned activities are previously identified. For each IO, expected maturity levels could have been expressed by the different users. And in the same way providers can declare the target maturity, they think they can produce at the end of their tasks. In the example presented in Figure 3, the expected stability waited by the user of the object 1.2 is lower than the target stability proposed by the upstream activity for the same object. Thus, both maturity levels are not compatible to each other. This means that object 1.2 is a *critical* object. The project manager can made awarde of this and can help both the provider and the user to find an agreement

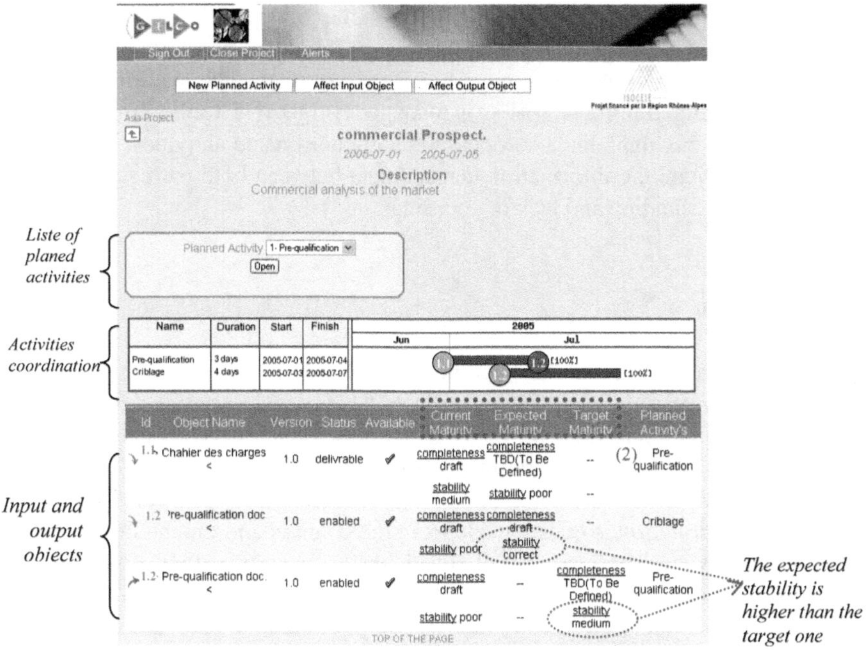

Fig. 3. View of project workspace in PIQUANT.

in order to foster coordination. The preliminary information in enabled state can be published in this workspace with a current maturity indication.

PIQUANT also supports the day to day co-operation in the *proximity workspace*. Each member of the planned activity can create a collaborative activity and access to the collaborative activities in which he is involved. The contributor to a collaboration activity can be either the host (the actor who creates the activity) or a guest (the actor who is invited by the host). There is a dashboard (similar to the project level one) that shows to all the participants of the activity the status and the different levels of maturity associated to each exchanged object in the collaboration. This level allows sharing informal and *preliminary information* (*exhibit* status) in common workspace. This information could be characterised in the same way than in project workspace, using maturity criteria.

We argue that PIQUANT can support the *"critical" co-design*. The *project workspace* offers coordination facilities to both partners. The declared target and expected maturity allow the actor to identify the critical information. They can therefore choose between different coordination strategies in order to avoid the starvation or the rework risks. The monitoring of current maturity facilitates the identification of risks of delay in the project. The knowledge of the quality of the preliminary information allows both partners to better manage their design resources. The existence of *proximity workspace* allows sharing a common workspace in order to treat the emerging problems that need closer co-operation. This workspace offers facilities to share *preliminary information* in an ad-hoc restricted team. The maturity of the information exchanged in those spaces is managed. PIQUANT tool offers continuity between day to day unplanned co-operation and coordinated activities. We assume that would facilitate the information management between both partners and avoid part of misunderstanding and lack of coordination.

6 Conclusion

In this paper, we investigate the possibility to develop ICT system for critical co-design situation of supplier involvement in NPD. First, we present a Supplier Involvement typology and then we focus on the case which needs according to our observation the closest interaction in order to avoid development risk. To manage this interaction, the point of view of ICT system is not sufficient. The sharing of *preliminary information* is questionable from the contract and the intellectual properties' point of view. These points fall outside the scope of this study at the moment and constitute a limitation of the work. The social aspect in preliminary information exchange is important. The *proximity workspaces* tend to take into account this aspect. For that, an autonomy area for the actors involved in these workspaces is necessary. But this autonomy is in disagreement with the order of transparency and traceability imposed by most of the actual ICT system.

Acknowledgement

This work is part of the project ISOCELE funded by Région Rhône-Alpes.

References

1. R.G. Cooper, S.J. Edgett, E.J. Kleinschmidt. *Portfolio Management for New Products*, Addison-Wesley, Reading, MA, 1998.
2. E.F. McDonagh. Investigation of factors contributing to the success of cross functional teams, *Journal of Product Innovation Management* **17**(3), 2000, 221–235.
3. K. Grebici, E. Blanco, D. Rieu. Framework for managing preliminary information in collaborative design processes, in *International Conference on Product Life Management PLM'05*, Lyon, France. 2005.
4. K. Grebici, E. Blanco, D. Rieu. Toward non-mature information management in collaborative design process, in *International Conference on Engineering Design ICED05*, Melbourne, Australia, 2005.

5. R. Monczka, G.L. Ragatz, R.B. Handfield, R. Trent, D. Frayer. Supplier integration into new product development: A strategy for competitive advantage, Report of the Michigan State Institute, 1997.
6. T. Johnson, D. Ford. Managing collaborative innovation in complex networks: Findings from exploratory interviews, in *The 8th International IPSERA Conference*, Manchester, 1999.
7. F. Wynstra, F. van Echtelt. Managing supplier integration into product development: A literature review and conceptual model, in *IMP Conference*, 2001.
8. R. Calvi, M.A. Le Dain, S. Harbi, M.V. Bonotto. How to manage early supplier involvement (ESI) into the new product development process (NPDP): Several lessons from a French study, in *The 10th International IPSERA Conference*, Sweden, 2001, pp. 153–163.
9. R. Calvi, M.A. Le Dain, S. Harbi. Collaborative development between client and supplier: How to choose the suitable coordination process?, in *The 12th International IPSERA Conference*, Budapest, 2003.
10. F. Wynstra, E. Ten Pierick. Managing supplier involvement in new product development: A portfolio approach, *European Journal of Purchasing & Supply Management* **6**(1), 2000, 49–57.
11. M. Peter. Early supplier involvement (ESI) in product development, Ph.D. Thesis, St Gallen, 1996.
12. Terwiech. Exchange preliminary information in concurrent engineering: Alternative coordination strategies, *Organization Science* **4**, 2002, 402–419.
13. K. Henderson.*On Line and On Paper*, MIT Press, 1999.
14. J.F. Boujut, E. Blanco Intermediary objects as a means to foster co-operation in engineering design, *Journal of Computer Supported Collaborative Work* **12**(2), 2003.
15. P. Lecaille. La trace-habilitée, une éthnographie des espaces de conception dans un bureau d'études mécanique: l'échange et l'équipement des objets grapho-numériques entre outils et acteurs de la conception, PhD dissertation in INPGrenoble, 2003 [in French].
16. L. Saint Marc, M. Callot, C. Reyterou, M. Moly, P. Girard, J.C. Deschamps, Toward a data maturity evaluation in collaborative design processes, in *International Design Conference*, Dubrovnik, 2004.
17. R. Crossland, J.H.S. Williams, C.A. McMahon. An object-oriented modelling framework for representing uncertainty in early variant design, *Research in Engineering Design* **14**, 2003, 17–183.
18. M. Stacey, C. Eckert. Against ambiguity, *Computer Supported Cooperative Work* **12**, 2003, 153–183.
19. R. Calvi, M.A. Le Dain, M. Schlipf. Customer/supplier relation in new product development: Communication and information system point of view, in *The 14th International IPSERA Conference*, Archamps, 2005, pp. 451–463.
20. R.L. Daft and R.H. Lengel. Organizational information requirements, media richness, and structural design, *Management Science* **32**(5), 1986, 554–57.
21. E. Blanco, K. Grebici, D. Rieu. A unified framework to manage information maturity in design process, *International Journal of Product Development* **4**(3/4), 2007, 255–279.

IMPROVED COMPUTATIONAL MODELS

Direct Measurement of Dynamic Structural Deformations in Order to Improve Dynamic Models of PKM

Tony Cano[1,2], Frédéric Chapelle[1], Pascal Ray[1] and Jean-Marc Lavest[2]

[1] *LaMI, IFMA/Blaise Pascal University, Campus des Cézeaux, BP 265, 63175 Aubière Cedex, France; E-mail: {tony.cano, frederic.chapelle, pascal.ray}@ifma.fr*
[2] *LASMEA – Blaise Pascal University, 24 Avenue des Landais, 63177 Aubière Cedex, France; E-mail: jean-marc.lavest@lasmea.univ-bpclermont.fr*

Abstract. In high speed machining, tool accelerations need to be increasingly high. Parallel kinematics machines present the highest ones, but combined with slimmer structures, the structural deformations cannot be still ignored. To solve this problem, some studies improve the dynamic model by taking into account the flexibility of the structure, but the identification of such model is still a difficulty. This paper proposes a method to obtain a direct measurement of the structural deformations under dynamic loading conditions due to the action of the actuators. The main difficulties are the measurement volume (1 m^3), the deformation size (a few 0.01 mm) and the movement velocity (a few 10 m/min). The principle of the method is to merge position measurement by vision and acceleration measurement by accelerometers. Indeed these two kinds of material are complementary. Accelerometers have a high frequency of acquisition, but a double integration is necessary to obtain the deformation information. This involves an accuracy problem. On the contrary, vision systems take the position information directly but the acquisition frequency is too low to complete the task. The aim is to calculate the positions with the acceleration measurements and to adjust them with the fewest direct position measurements.

Key words: PKM, machine tools, deformation, measurement, vision, accelerometer.

1 Introduction

One of the main objectives in manufacturing improvement is to obtain faster processes in order to decrease the costs. To achieve this aim, the tool accelerations need to be higher. Parallel kinematics machines (PKM) allow increasing accelerations in a significant way, partly because of lower weights in comparison with serial kinematics machines. However, such accelerations can give a large deformation of the structure during the machining [1]. Consequently, the accuracy in milling operations

S. Tichkiewitch et al. (eds.), Advances in Integrated Design and Manufacturing in Mechanical Engineering II, 205–218.

is not good. It is a reason why drilling operations constitute the main application of the machining machines [2].

To improve the efficiency of PKM, the behavior pattern of machines has been improved [3, 4]. After the introduction of inertia, flexibilities are today taken into account. The first step was to introduce flexibilities in the links [5, 6]. But given the accelerations, which are always higher, this step is no longer enough. The arms (or legs) flexibility generates non neglectable efforts on the actuators [7]. This perturbs the machine control and reduces the accuracy. Thus, patterns taking into account all flexibilities are needed [4, 8]. Moreover, the patterns must be adjusted to the behavior of the machines. The best way to adjust a pattern is to measure its predictions. Therefore this paper proposes a method able to measure the flexibilities in the working conditions of a machine tool: a measure volume in cubic meters, a deformation frequency nearing tens of Hertz and deformations values nearing a hundredth millimeter.

The paper is organized as follows. In a first part, different ways to measure, more or less directly, the deformations of a machine structure element are shown. Then, the method achieved to quantify such deformations is described. To validate this method, results obtained from an experiment using a beam in flexion are discussed before the conclusion.

2 Methods for Deformation Measurements

There are two means to obtain the deformation. The first one is a direct measurement; it can be carried out with a deformation gauge. The accuracy is very good but the measurement field is too limited. The other way consists in measuring another attribute to calculate the deformation. This can be an acceleration, velocity or position quantity.

2.1 Acceleration Measurement

It is quite simple, with an accelerometer, to measure the acceleration of one point. The displacement information is obtained using a double integration. If the measurements are taken in two points simultaneously, a subtraction between the two displacement curves gives the relative deformation between the two points.

The main advantage of such a measurement is the acquisition frequency (more than 1 kHz is common for this kind of material). Nevertheless it presents two important disadvantages. The first is a significant noise, and the second is the error of the integration. To solve the noise problem, it is enough to filter the signal. In Figure 1, an acceleration measurement filtered with a low-pass filter is shown. The second disadvantage is more annoying. A small offset error in acceleration measurement produces a skew in the velocity calculation, and the error is still more significant for

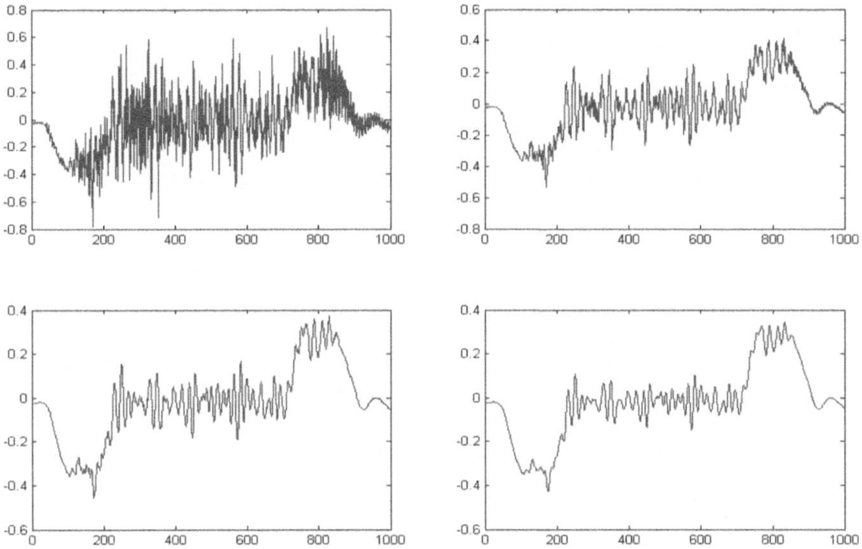

Fig. 1. Acceleration measurement filtering.

the displacement calculation. Figure 3 present the skews produced by an offset error equals to 0.01 V. This calculation error reaches 20% for a movement duration of one second. Consequently, the displacement calculation error reaches 16.5%.

Thus, it is impossible to determine the deformations only with an acceleration measurement. It is essential to have other information to readjust the results of the successive integrations.

2.2 Velocity Measurement

Velocity information allows to calculate the displacement. In this case, only one integration is needed. Consequently, an offset error is less significant. But the main problem is the obstruction. Velocimeters are very voluminous and need a retreat. In a machine tool context this is a strong limitation. Moreover, laser velocimeters impose on the target the need to follow the laser axis, which limits the movements.

2.3 Displacement Measurement

To release us from integration errors, it is only necessary to measure the displacement directly. This measurement imposes two constraints. First, the measurement field volume is nearly 1 m^3. The second constraint is a measurement without contact, because the machine tools must have free movement. Thus the only solution is to use a laser or a camera. Laser interferometers are very accurate with high frequency tasks

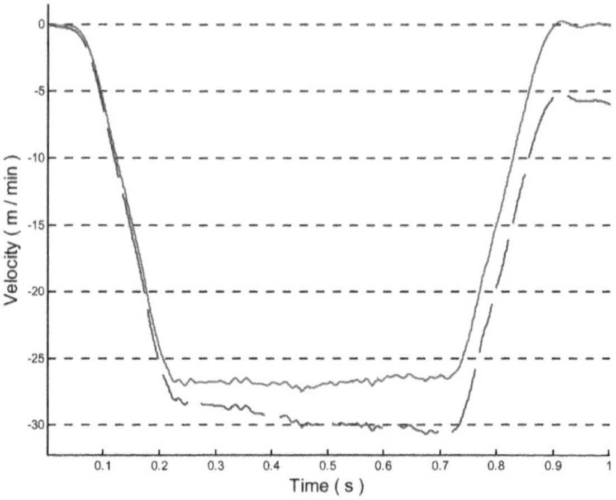

Fig. 2. Influence of an offset error in the velocity calculation.

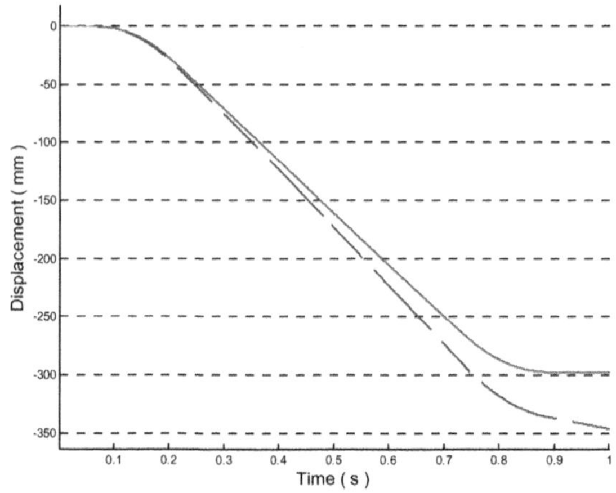

Fig. 3. Influence of an offset error in the displacement calculation.

and it is the most suitable to characterize a machine tool axis [9]. But if the movement is not on a straight line it is impossible to do any measurement. The laser tracker solves this problem, and even if the accuracy decreases it is still enough. Leleu et al. evaluate the accuracy of a SMX laser tracker equal to 5 μm in the worst direction [10]. But two problems remain: the laser must not be blocked off and the number of targets is limited to only one. It is then impossible to calculate a deformation between two points.

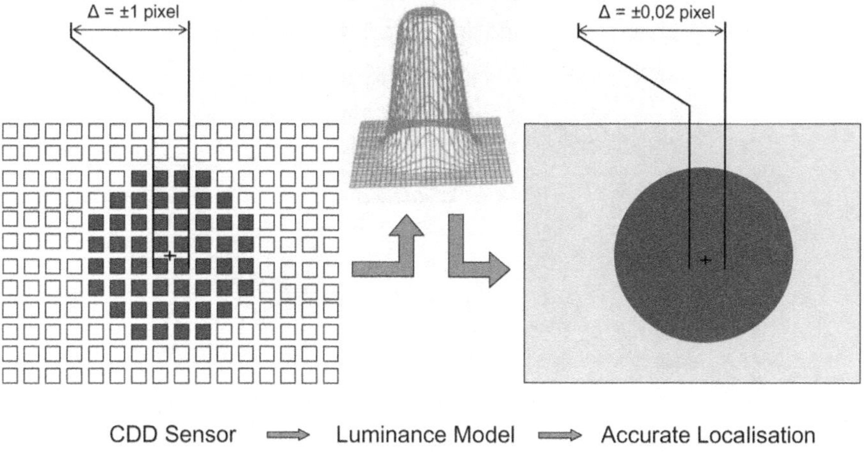

Fig. 4. Subpixelar detector.

A camera takes advantage in the obstruction problem. With a fish-eye, the retreat can be small, only a few centimeters. But the resolution is inadequate and the frequency is too low.

The first of these disadvantages can be solved using a subpixelar detector. The idea is based on a luminance model of the target (Figure 4). Using a classical target detector, the centre of the target is located with an accuracy equal to about 1 pixel. To increase the accuracy a subpixelar detector is developed. A luminance model of the target is adjusted on the luminance measured by the CDD sensor. With such a continuous model, the center of the target can be located with an accuracy equal to about 0.02 pixel [11].

In order to calculate a relative displacement between two points, two targets have to be stuck in the desired place. The movement is recorded by the camera, and after (off-line), a computer computes the successive positions of the two centers of the targets using the subpixelar detector. When the coordinates of the centers are known in the image system, it is possible to use a geometrical transformation to obtain the coordinates in the real 2D plane system. After this, a simple subtraction between the positions of the two centers allows us to know the deformation for each image.

The second disadvantage of such measurements is that the cadence of the camera depends only on the material. To increase the task frequency, there is no other possibility than buy a higher speed camera.

2.4 Conclusion

We have seen that no solution is perfect. The accelerometer is accurate with high cadence data acquisition, but the double integration introduces a significant error.

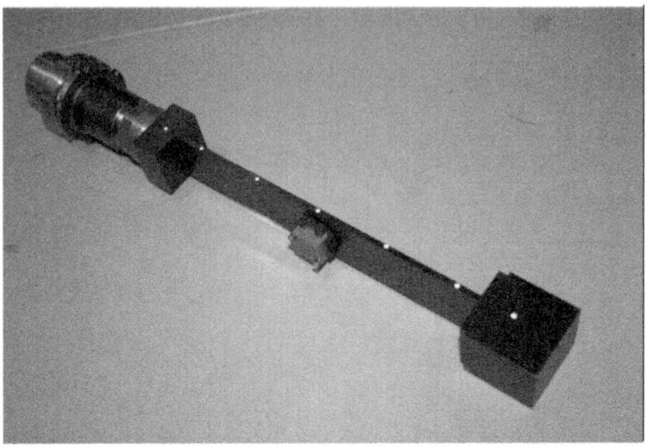

Fig. 5. Simple geometry: a beam.

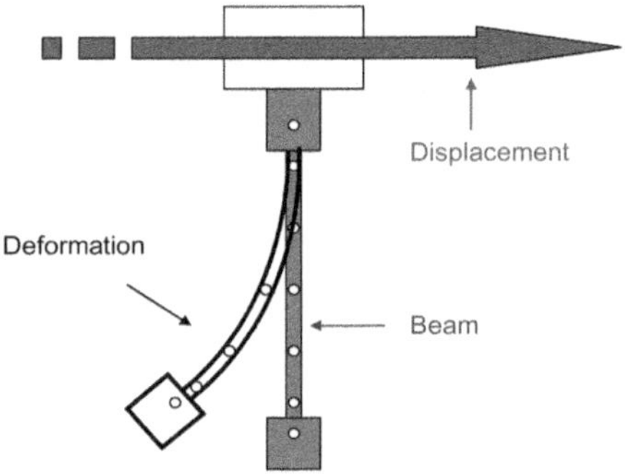

Fig. 6. Movement and deformation of the beam.

The laser velocimeter and the laser interferometer are too voluminous and require that the target moves along the laser axis. The camera is sufficiently accurate but not enough fast. Consequently, we decided to mix accelerometer and camera. The accelerometer will have a high working cadence, and the camera will allow us to adjust the result of the double integration of the acceleration.

3 Experiment

The final aim of the study is to measure the structural deformation of a PKM under dynamic conditions. In a first time, we are interested in the deformation of a beam under dynamic conditions created by a mono-dimensional movement (Figures 5 and 6). This step allows us to validate our method and seek its limits.

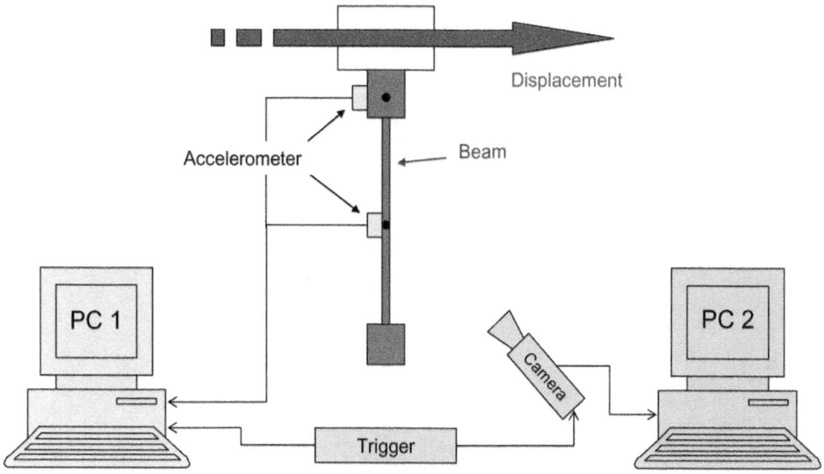

Fig. 7. Measurement device.

3.1 Protocol

The main difficulty of this experiment is to synchronize the two measurements which have different cadences. Moreover, the accelerometer measurements are recorded by an acquisition card and the camera is managed by a PC with an IEEE connection.

To synchronize the two measurements, an external trigger is used. This trigger is a square signal (0, +4V) generated by a low frequency generator. This signal is sent to the rear panel of the camera and to the acquisition card. Figure 7 shows the experimental device. At the beginning the trigger is switched off, and the two PC are waiting. The two home-made softwares are designed to start at the first pulse of the trigger. To start the record, the trigger has just to be switched on.

3.2 Calculation Algorithm

Figure 8 describes the algorithm designed to calculate the final position curve. The calculation can be done in three steps.

Firstly, trigger data are treated to detect the pulses, and then they are dated in agreement to the acquisition frequency. In the same time, the images can be treated: after a manual initialization, an automatic target detector scans all images. The targets coordinates are then transposed in the 2D plane of the movement. And, to finish the first step, the acceleration data are filtered with a low pass filter (50 Hz) to suppress the measurement noise. Actually, according to an Adams Flex Model of our beam, the deformation frequency does not exceed 15 Hz. And then, like the pulses of the trigger, the filtered acceleration data are dated.

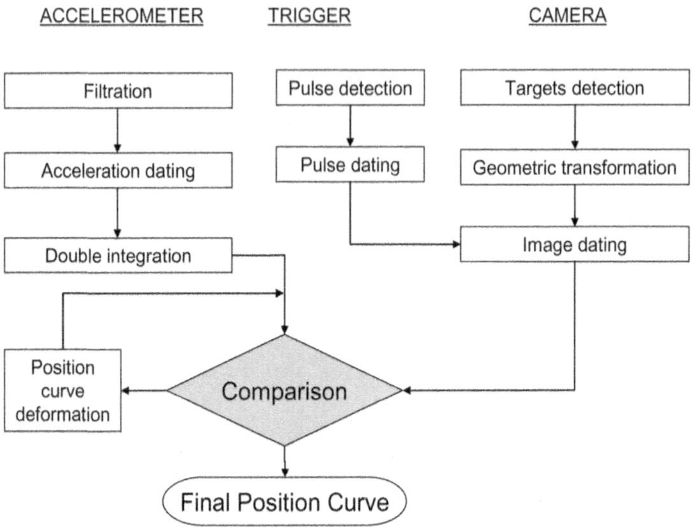

Fig. 8. Final position curve calculation.

In the second step, the acceleration data are integrated twice in order to obtain the original position curve. And the images, or the successive targets coordinates given by the mean of the camera, are dated using the pulses dates.

To finish, the coordinates of each target are compared to the yielding positions computed using the acceleration. If there is a difference, the position curve is deformed to agree with the fewest direct position measures (coordinates given by the camera).

When the position of each target is accurately known (final curve position), a subtraction is sufficient to calculate the relative position between each target and the target chosen to be the reference. The variation of this position is the deformation of the beam relative to the reference target.

4 Results

4.1 Experimental Conditions

Since the experimental conditions have to be similar to the usual ones in the machine tools field, it is interesting to carry out the experiment with a machine tool. The used machine is a KX15 from Huron corp. The main advantage of this machine is to have the Y-axis on the spindle. This allows to mount the camera on the table and the beam on the spindle. The feedrate is 30 m/min and the acceleration is 3 m/s^2. The

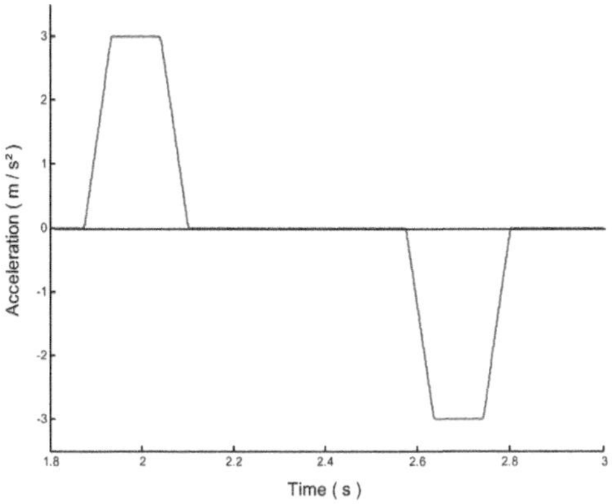

Fig. 9. Acceleration order.

Fig. 10. Experimental device

acceleration law is a trapeze generated by the CNC (Sinumerik 840D from Siemens Corp.). Figure 9 shows the acceleration and deceleration steps.

The accelerometer is a capacitive one; it is designed to measure continuous accelerations and not vibrations, on the contrary of the majority of them. The measurement field is about 3 g, it is sufficient because the Spindle acceleration does not exceed 0.3 g and the Adams prediction for a point in the middle of the beam is below 0.4 g.

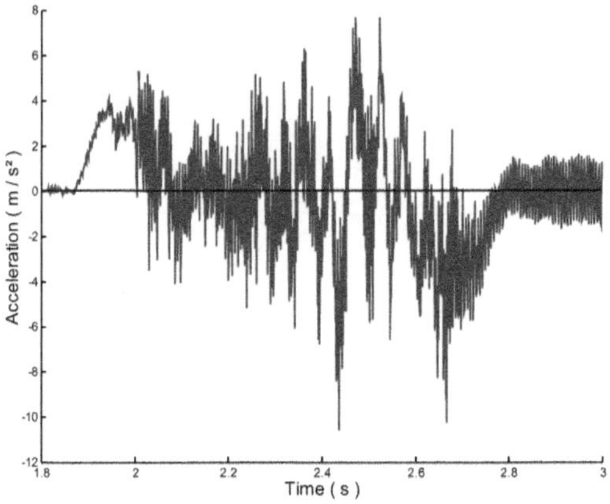

Fig. 11. Raw signal of the accelerometer.

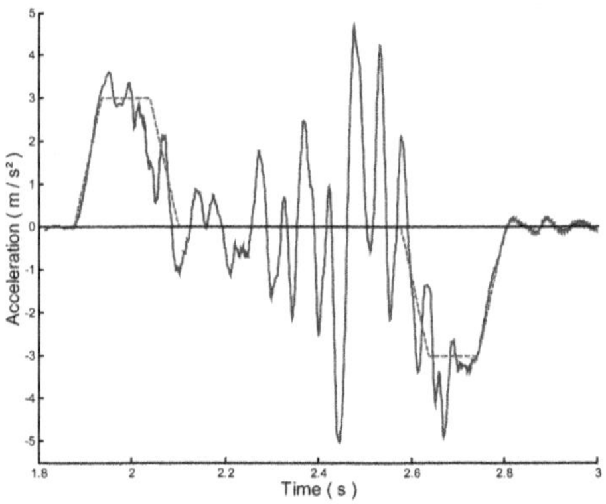

Fig. 12. Filtered signal.

The resolution of the camera is 1024 × 768 pixels, and its cadence is 30 images per second. To increase the accuracy, an annular light is used.

The experimental device is shown in Figure 10. There is only one accelerometer. The acceleration law of the Y-axis is used to emulate the second accelerometer.

Only two targets are used for the vision: the one close to the accelerometer and the other nearest to the spindle, on a part considered as rigid.

4.2 Data Obtained by the Measurements

The raw signal of the accelerometer is very noisy (Figure 11). When filtered, oscillations can be clearly seen around the acceleration law of the CNC (see Figure 12).

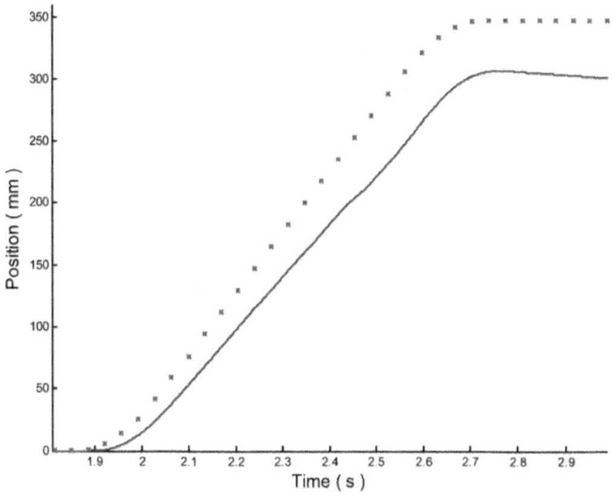

Fig. 13. Superposition of the curve and the position points.

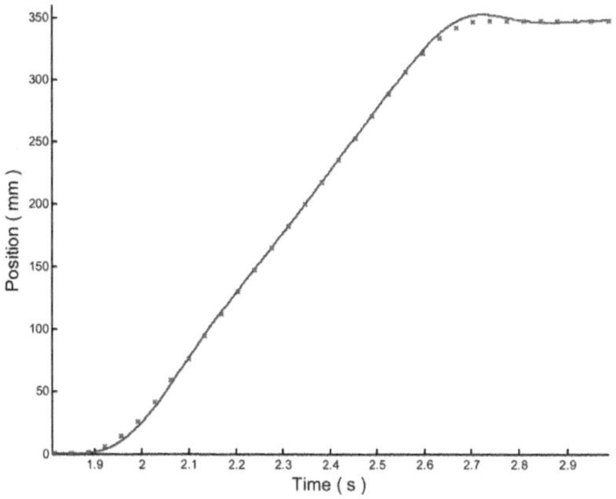

Fig. 14. Distorted curve.

The first step consists in integrating twice the signal in order to obtain the displacement curve. Then this curve is superposed with the positions calculated with the help of the camera (Figure 13). The next step is to distort the displacement curve so that it goes through the points captured by the camera (Figure 14). The distortion is achieved by successive adaptations of the acceleration offset. The offset is adapted so that the curve goes through the first captured point. The displacement curve is then distorted to match this first point using a new double integration. The offset value is determined for the second point, and only the remaining part of the curve is then distorted again to make it go through the second captured point. This operation is repeated till the last point.

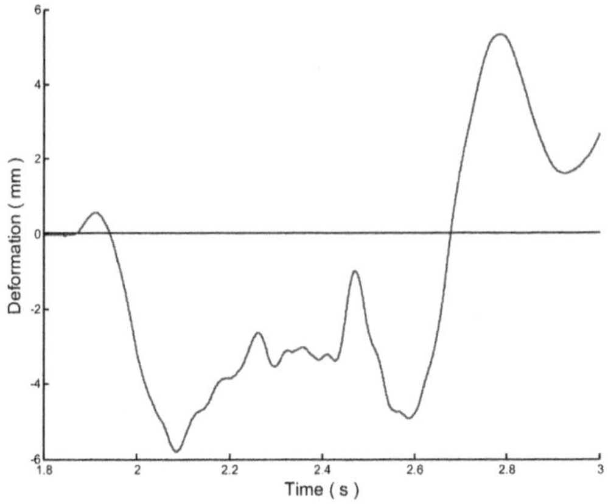

Fig. 15. Deformation between the two targets.

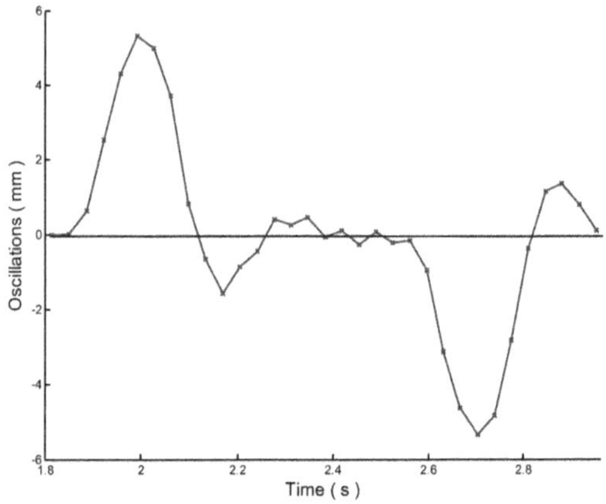

Fig. 16. Position curve oscillations.

The same treatment is applied to the second target. The deformation between the two targets is given by subtracting the two position curves (Figure 15).

Nevertheless, a precise examination of Figure 14 shows oscillations. This, and the fact that the curve does not always match exactly the camera points, is the consequence of the numerical applications of the algorithm. If the camera points and the position curve are subtracted (Figure 16), we can see that the oscillations have the same order of magnitude that the deformations determined in Figure 15. Thus, the distortion method will have to be improved (for instance, by means of B-Splines).

4.3 Conclusion

The position can be measured, but for the moment the curve distortion is not enough accurate. This problem will be solved by changing the distortion method. The reliability of the system in the case of smaller deformations remains to be characterized in the aim to be close to real working conditions of machine tools.

5 Conclusion

We have seen that it is possible to measure the deformation of a flexible element during a long displacement, using mixed measurement system. To optimize this system we will improve the mean of the displacement curve distortion and if necessary increase the camera resolution.

In a second time the results will be compared to those of a high speed camera (160,000 images/s) that could constitute a ground truth. The system presented here remains interesting since its cost is ten times cheaper.

The next step is to use this tool to adjust the Adams pattern of the beam. The final aim is to apply this method to the Adams pattern of a PKM prototype.

Acknowledgements

This paper was written within the framework of the TIMS Research Group, using grants from the Regional Council of Auvergne, the French ministry of research, the CNRS and the Cemagref.

References

1. P.J. Barre, E. Dumetz. Dynamic behaviour modelling and control of machine tool, *Mécanique & Industrie* **3**, 2002, 315–322.
2. M. Terrier, A. Dugas, J.Y. Hascoet. Qualification of parallel kinematics machines in high-speed milling on free form surfaces, *International Journal of Machine Tools & Manufacture* **44**, 2004, 865–877.
3. A. Dequidt, J.M. Castelain, E. Valdes. Mechanical pre-design of high performance motion servomechanisms, *Mechanism and Machine Theory* **35**, 2000, 1047–1063.
4. W. Chen. Dynamic modeling of multi-link flexible robotic manipulators, *Computers & Structures* **79**, 2001, 183–195.
5. Z. Yang, J.P. Sadler. On issues of elastic-rigid coupling in finite element modeling of high-speed machines, *Mechanism and Machine Theory* **35**, 2000, 71–82.
6. W. Khalil, E. Dombre. *Modeling, Identification and Control of Robots*, Butterworth-Heinemann, ISBN 1903996139, 2001.

7. N. Rat, R. Risk, G. Gogu, M. Neagoe, Comportement des robots parallèles découplés et corps déformables, in *17ème Congrès Français de Mécanique*, Troyes, France, September 2005.
8. B.C. Bouzgarrou, B. Thuilot, P. Ray, G. Gogu. Modeling of flexible manipulators applied to HSMW machine tools, *Mécanique & Industrie* **3**, 2002, 173–180.
9. M. Rahman, J. Heikkala, K. Lappalainen. A study and analysis of machine tool errors with static and dynamic measurements, in *4th International Conference on Integrated Design and Manufacturing in Mechanical Engineering*, Clermont-Ferrand, France, May 2002.
10. S. Leleu, JM. David, G. Diolez, P. Martin. Evaluation of machine geometry, Presentation of an innovative method with a Laser Tracker, in *4th International Conference on Integrated Design and Manufacturing in Mechanical Engineering*, Clermont-Ferrand, France, May 2002.
11. M.R. Shortis, T.A. Clarke, T. Short. A comparison of some techniques for the subpixel location of discrete target images, *Videometrics III* **2350**, 1994.

Mechanical Behavior Analysis of Flexible Parts in a Real-Time Virtual Environment for Aided Design

Frédéric Druesne, Jean-Luc Dulong and Pierre Villon

Roberval Laboratory, Université de Technologie Compiègne, CNRS FRE No. 2833, BP 20529, 60205 Compiegne Cedex, France;
E-mail: {frederic.druesne, jean-luc.dulong, pierre.villon}@utc.fr

Abstract. The interaction between a designer and a virtual prototype is a promising way to optimise the design of parts. Indeed mechanical industries of automotive and aeronautics already use real-time interactive simulators to evaluate virtual prototypes composed of rigid parts. Thus, there exists an industrial need to solve the problem of real-time deformation of flexible parts.

Our method is composed of two phases: a campaign is calculated during the first phase of training, and then these results are used during the second phase of real-time immersion. In this paper, we focus on the phase of training. We present an *a posteriori* method and an *a priori* method. The *a posteriori* method need to complete calculation campaign to apply the Karhunen–Loève expansion and keep only representative data. The *a priori* method is an adaptive strategy. As a linear combination of shape functions defines the displacement, these functions are enriched during the campaign. The size of the shape functions basis increases with time, so we use a model reduction approach. These two methods allow to calculate a surface response of possible displacements using a model reduction technique.

These methods are applied on an automotive hose, and should allow to simulate complex mechanical behaviour of flexible parts representative of many industrial applications.

Key words: virtual prototype, real-time deformation, nonlinear mechanical model, Karhunen–Loève expansion, enriched method.

1 Introduction

Virtual prototyping is helpful for mechanical design, maintenance and training. During the phase of design, it is possible to test assembly or maintenance manual processes. Indeed, the immersion in a virtual environment allows an operator to check in real time the feasibility to assemble two parts, or to check the access to parts engine [1–3]. Some manual operations need to handle rigid parts [4–6], some others induce more or less large deformations of the handled part [7].

S. Tichkiewitch et al. (eds.), Advances in Integrated Design and Manufacturing in Mechanical Engineering II, 219–230.

Graphic industry [8, 9] and medical industry [10, 11] have developed many models of deformation in real time, but the mechanical assumptions are restricted (elastic behaviour, small deformations) and are not adapted to the mechanical industrial context.

We focus on these flexible parts, like an automobile hose. In that case, geometric and material nonlinearities are supposed. Classic Finite Element Method (FEM) cannot solve this problem in real-time immersion since too expensive in time. For some applications, specific strategies are then employed [12, 13] but they cannot be generalised.

Two phases compose our method, the training phase then the immersion phase. Before the phase of real-time immersion, we start a representative calculation campaign of possible handlings; we focus on this training phase in this paper. After that, superposing the deformation modes, computed from the training phase, makes the real-time phase possible.

The resolution by the FEM computes displacement u. For a significant number of degrees of freedom (DOF), the nonlinear problem resolution is expensive, so we interest in methods of model reduction.

To manage the training phase, we present two approaches, an *a posteriori* approach then an *a priori* approach.

2 Methods for Training Phase

2.1 *A Posteriori* Approach: Karhunen–Loève Expansion

For the *a posteriori* approach, we realise a preliminary calculation campaign by the FEM. A grid of S load cases composes the campaign. At each time t_s, a point of the grid evaluates the quasi-static system (1). Each resolution provides a solution $u(t_s) \in \mathbb{R}^N$ stored in a matrix $C(N \times S)$, with N the number of DOF of the structure.

$$K(u, t_s) \cdot [u(t_s)] = F(t_s). \tag{1}$$

To reduce the volume of these data, we propose [14, 15] to use the Karhunen–Loève expansion [16–18].

H was a m-dimensional subspace of $\mathbb{R}^N$ ($H \subset \mathbb{R}^N$, $m \ll N$), spanned by vectors φ_i ($i = 1, \ldots, m$) and including the point P_0. Let P_H be the orthogonal projector onto this subspace H:

$$P_H(u(t_s)) = \sum_{i=1}^{m} \varphi_i (u(t_s)^T \varphi_i) + P_0. \tag{2}$$

We seek the best subspace H to minimise the distances between the solution $u(t_s)$ and its projection $P_H(u(t_s))$,

$$J(H) = \sum_{s=1}^{S} \|u(t_s) - P_H(u(t_s))\|^2. \tag{3}$$

The cost function $J(H)$ is minimal if the vectors φ_i, representing the subspace H, are the m first eigenvectors of the covariance matrix U,

$$U = \sum_{s=1}^{S} \bar{u}(t_s)\bar{u}(t_s)^T \tag{4}$$

with the centred displacements

$$\bar{u}(t_s) = u(t_s) - \bar{u} \tag{5}$$

and the average of displacements

$$\bar{u} = \frac{1}{S}\sum_{s=1}^{S} u(t_s). \tag{6}$$

The computation of approximate displacement $\tilde{u}(t_s)$ is given by

$$\tilde{u}(t_s) = P_H(\bar{u}(t_s)) + \bar{u} = \sum_{i=1}^{m} \varphi_i(\bar{u}(t_s)^T\ \varphi_i) + \bar{u} = \Phi a(t_s) + \bar{u}. \tag{7}$$

Only the m first eigenvectors of the covariance matrix U are retained in the basis $\Phi = \{\varphi_1, \varphi_2, \ldots, \varphi_m\}$. Nevertheless, building this matrix (4) is expensive. So, computing the approximate displacement (7) requires only the storage of the matrices $\Phi(N \times m)$, $A(m \times S)$ of which each column is a vector $a(t_s) = \Phi^T \bar{u}(t_s)$, and the storage of the vector $\bar{u}(N)$. Thus, this reduced storage makes the calculation of displacements possible for the load cases (campaign), instead of extracting it on the full matrix C.

2.2 *A Priori* Approach: Adaptive Strategy

The previous method (*a posteriori*) needs the complete realisation of the calculation campaign to be able to evaluate the basis Φ on all the campaign. The *a priori* approach [19] is here adaptive.

A linear combination of m shape functions φ_i, whose amplitude is a, always defined the displacement u. But now, the basis Φ is included in the resolution of the system, since the Equation (1) becomes,

$$K(u, t_s) \cdot (\Phi \cdot a(t_s) + \bar{u}) = F(t_s). \tag{8}$$

At the iteration i and according to a Newton–Raphson (NR) method, the unknown coefficients $a(t_s)$ is solved with a reduced tangent matrix $\hat{K}_T^i$ (size $m \times m$) and with a reduced residue $\hat{R}^i$ (size m),

$$\hat{K}_T^i = \Phi^T \cdot K_T^i \cdot \Phi, \tag{9}$$

$$\hat{R}^i = \Phi^T \cdot R^i. \tag{10}$$

With a number m of modes much lower than the number of degrees of freedom N, inverting the reduced tangent matrix $\hat{K}_T^i$ is not expensive. The NR method is thus based on the convergence of coefficients a. However, the basis Φ can include too poor modes to describe the new deformation of the load case s. Thus, NR method converges even if the residue R is not small. It is then necessary to enrich the basis Φ with a new mode of displacement, which is a new possible deformation for the structure.

The *a priori* approach is adaptive, that means the resolution starts with any basis which is iteratively enriched during the campaign with the new mode of displacement $\breve{u}_e$,

$$\breve{u}_e = (K_T)^{-1} \cdot R \tag{11}$$

with K_T the tangent matrix, and R the residue of the quasi-static problem $K(u) \cdot u = F$.

The new basis Φ_{e+1} is thus the concatenation of the previous basis Φ_e with the normalised vector of enrichment $\breve{u}_e^\perp$,

$$\Phi_{e+1} = \left\{ \Phi_e, \frac{\breve{u}_e^\perp}{||\breve{u}_e^\perp||} \right\}. \tag{12}$$

The vector of enrichment $\breve{u}_e^\perp$ is the projection on the subspace orthogonal with Φ_e:

$$\breve{u}_e^\perp = \breve{u}_e - \Phi_e \cdot \Phi_e^T \cdot \breve{u}_e. \tag{13}$$

During iterations i and load cases s, the basis Φ grows rich by all the modes necessary to describe the structure deformations on all the calculation campaign.

2.3 *A Priori* Approach: Adaptive Strategy with Basis Reduction

With the previous technique of enrichment, the size of the basis Φ increases with the number of shape functions, which contribute to enrich it. Its size $(N \times m)$ can become penalising with the increase of the number of degrees of freedom (Section 3.5).

We thus propose to reduce the basis with the Karhunen–Loève (KL) expansion when the number m of modes reaches a fixed limit. The KL expansion is applied to the coefficients a, on all the computation history, to decrease the number of modes m.

The history of coefficients a is represented by the matrix $\{a(t_1), \ldots, a(t_k)\}$ at the time t_k. The principle is the same one as in Section 2.1, we calculate the average of the coefficients a,

$$\bar{a}(t_k) = \frac{1}{k}\sum_{j=1}^{k} a(t_j), \tag{14}$$

the coefficients a are centred to get the coefficients $\hat{a}$

$$\hat{a}(t_j) = a(t_j) - \bar{a}(t_k) \tag{15}$$

to build the covariance matrix $U_a(t_k)$

$$U_a(t_k) = \sum_{j=1}^{k} \hat{a}(t_j) \cdot \hat{a}(t_j)^T. \tag{16}$$

We keep the m' first eigenvectors of the covariance matrix $U_a(t_k)$ to have a basis $V = \{\nu_1, \ldots, \nu_{m'}\}$ with the size $m \times m'$. Thus, after reduction, the approximate coefficients at the time t_k are:

$$\tilde{a}(t_j) = V \cdot V^T \cdot \hat{a}(t_j) + \bar{a}(t_k), \tag{17}$$

so the approximate displacement

$$\tilde{u}(t_s) = \Phi' \cdot a(t_s)' + \bar{u}' \tag{18}$$

with the new basis Φ', the new coefficients a' and the new average $\bar{u}'$

$$\Phi' = \Phi \cdot V, \tag{19}$$

$$a' = V^T \cdot \hat{a}, \tag{20}$$

$$\bar{u}' = \Phi \cdot \hat{a} + \bar{u}. \tag{21}$$

The size of the new basis Φ' is thus $N \times m'$ with $m' < m$.

2.4 From the Training Phase to the Real-Time Immersion Phase

The result of the training phase allows to approach the structure deformation by computing displacement $\tilde{u}$ for a load cases grid (times t_s). A linear combination of the deformation modes $\varphi_i$describes the displacement $\tilde{u}$, whether the approach is *a posteriori* or *a priori*.

During the real-time immersion phase, at time t, the displacement $u(t)$ is interpolated from the approximate displacements $\tilde{u}(t_s)$ known on the load cases grid. It is thus important to minimise the volume of the data resulting from the training phase to reduce the access time to these data, and the number of operations to calculate displacement $\tilde{u}(t)$ as fast as possible.

Table 1. Imposed displacements (u_x, u_y) for each load case.

Load case number (s)	1	2	3	–	99	100
Imposed displacement (u_x)	0	20	20	–	160	180
Imposed displacement (u_y)	0	0	20	–	180	180

Table 2. Size of matrices to store to do the real-time computation.

	Matrices	Size	
Complete model	C	$(N \times S)$	337392
Reduced model	A, Φ and $\bar{u}$	$(m \times S + m(N+1))$	10524

3 Results on Training Phase

We apply previous methods, *a posteriori* and *a priori*, first with a structure test, then with an automotive hose.

The structure test studied is a rubber beam of axis z, square section 20×20 mm, and length $L_0 = 500$ mm. It is embedded at one extremity and charged at the other with imposed displacements (u_x, u_y) chosen on a grid $(180 \times 80$ mm) with a step 20 mm (see Table 1). So $S = 100$ load cases compose the calculation campaign. This grid describes a range of possible handling for the studied beam.

Incompressible neo-Hookean behaviour models the material. The FEM is defined by 630 elements H8 and $N = 3408$ degrees of freedom. The computing times are obtained with a computer PC Pentium IV 2.4GHz with 512Mo RAM memory.

3.1 *A Posteriori* Approach: Karhunen–Loève Expansion

For the *a posteriori* approach, we make the complete calculation campaign. The needed CPU time to compute the totality of the grid is 811 s. Building the covariance matrix (4) and computing its eigenvalues and eigenvectors is negligible, thanks to the snapshots method [20].

To analyze the accuracy of these approached deformations obtained by the KL expansion, we evaluate the following error estimation e^{KL},

$$e^{KL} = \max_{s=1\ldots S}\left(\max_{j=1\ldots n} |u^j(t_s) - \tilde{u}^j(t_s)|\right) \Big/ L_0. \tag{22}$$

This term e^{KL} evaluates the maximal differences between the approximate displacements $\tilde{u}(t_s)$ (from KL expansion) and the original displacements $u^j(t_s)$ (from *a posteriori* campaign), on the n nodes and the S load cases, relative with the initial hose length L_0 of the hose.

Figure 1 shows that 3 modes are enough to describe the calculation campaign (all points of the grid), and thus to evaluate approximate displacements (7) with an accuracy in displacement lower than 1%. For the studied beam and the fixed grid, the matrices to be stored are listed in Table 2. The size of their storage is divided by 32.

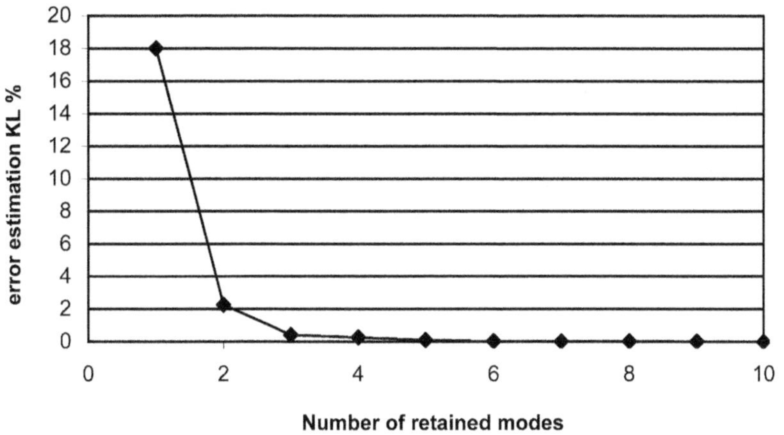

Fig. 1. Displacement accuracy between campaign results and approached displacements by KL expansion as a function of the number of retained modes.

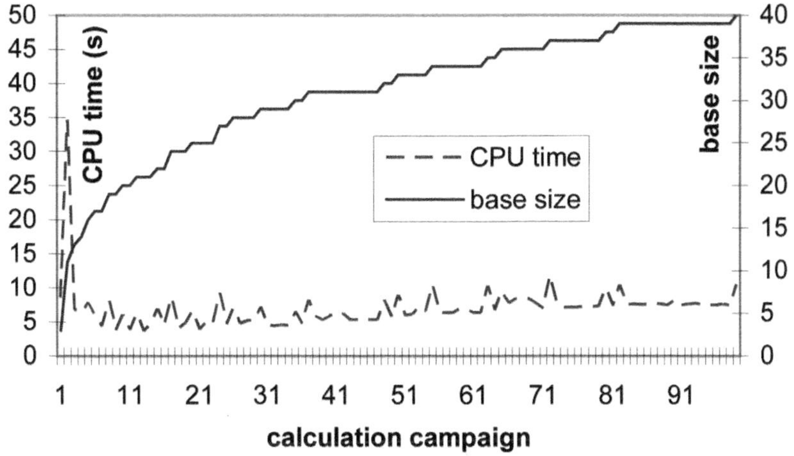

Fig. 2. CPU time of computation for each point of the grid-size of the basis during the campaign.

3.2 *A Priori* Approach: Adaptive Strategy

The same grid is evaluated for the studied beam by the enrichment technique of the basis. It needs 452 s of CPU time for the same previous precision. The gain is 2.9 compared with the *a posteriori* approach. This enrichment technique is similar to a technique of convergence acceleration [21] for a FEM.

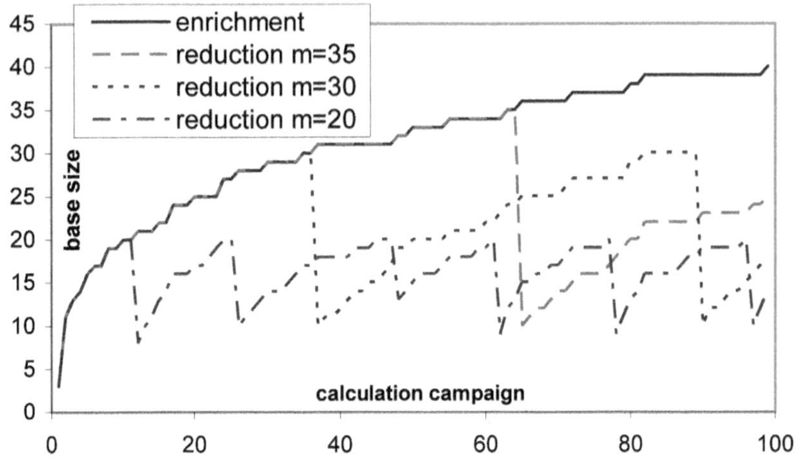

Fig. 3. Size of the basis as a function of the reduction limit m.

Figure 2 shows the first four points of the grid (that is the first four load cases) are expensive in CPU time, since they contribute to build a basis interesting for continuing the campaign.

This figure also shows the evolution of the basis size during computations. This evolution can become penalizing when the model has a large number of degrees of freedom (Section 3.5).

3.3 *A Priori* Approach: Adaptive Strategy with Basis Reduction

To manage the increase in the modal basis, we propose to reduce the basis size with the KL expansion, when the number of modes reaches a limit m. Figure 3 gives the evolution of the basis size according to the retained limit of reduction.

After each reduction, the size of the basis goes down again to 10. The most interesting reduction limit, for computing time, is $m = 30$. The CPU time is 373 s to achieve all the campaign. The gain is thus 3.5 compared with the *a posteriori* approach. We can notice that, after reduction, the number of modes increases rapidly: this enrichment is CPU time expensive, and so a low limit for reduction is not necessarily the better choice.

3.4 Influence of Number of Degrees of Freedom

We trace here the *a posteriori* (without KL computations) and *a priori* methods while varying the number of degrees of freedom of the studied beam to evaluate the cost of each method.

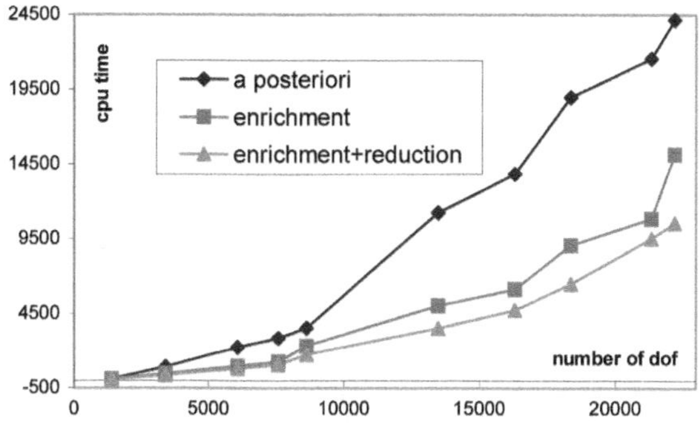

Fig. 4. Influence of number of degrees of freedom.

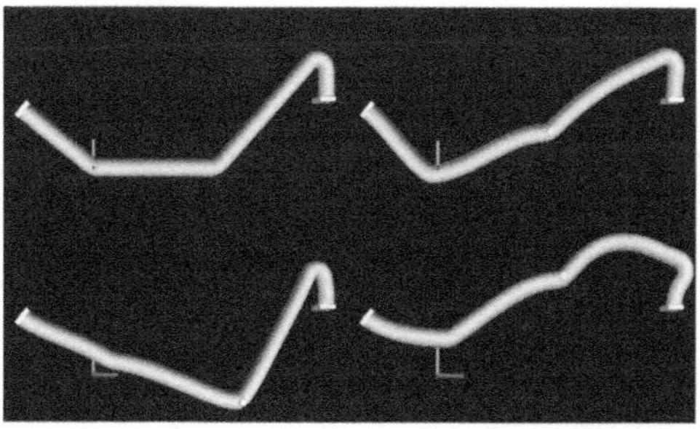

Fig. 5. Manipulation in a virtual environment: example of deformed hose for four load cases.

The gain of the *a priori* approaches compared with the a posteriori (without KL computations) approach is nearly constant with the increase in the number of DOF.

We now will interest in an industrial application (an automotive hose) to show the need to reduce the modal basis.

3.5 Industrial Application

With the aim of optimise the design of an automobile engine, it is interesting to be able to handle a hose in real time. In the phase of design, that makes it possible to check feasibility to move the hose and to check accessibility to the different parts of the engine. We have thus to start a calculation campaign representative of possible handling on a hose.

The studied hose in rubber is embedded at its two ends and it is handled in a point (Figure 5). The selected model is composed of elements H8 and has $N = 18720$ degrees of freedom.

A calculation campaign with imposed displacements (u_x, u_y) was defined on a grid (72 × 135 mm) divided into 100 observations. For the *a posteriori* and *a priori*

Table 3. Complete CPU time of the campaign for *a posteriori* and *a priori* approaches.

Approach	Complete CPU times (s)
A posteriori	21254
Enrichment	30487
Enrichment + reduction	14571

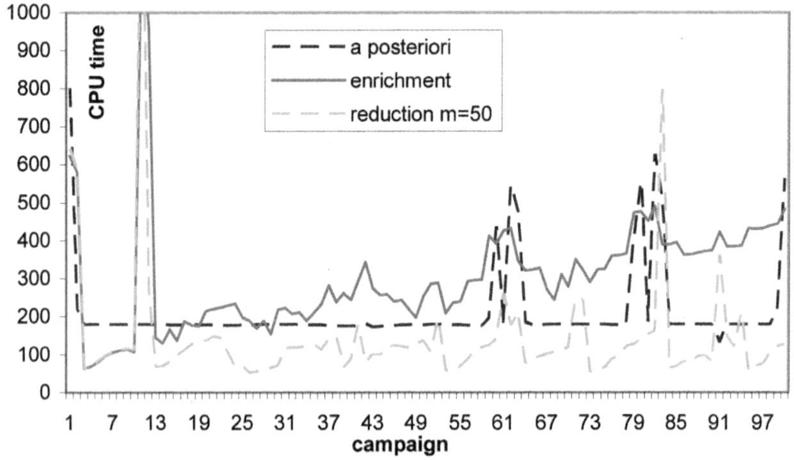

Fig. 6. CPU time for each point of the grid and each method.

approaches, CPU time of each point of the grid is evaluated. Hence, Table 3 resumes the complete CPU times.

For the *a priori* approaches, the enrichment method becomes more expensive than the enrichment method with reduction. Indeed, the modal basis becomes too big during the campaign with the number N of degrees of freedom of the model of the hose. Thus, CPU time gained by reducing the basis is interesting: it is divided by 2.

4 Conclusion

The automotive and aeronautical industries have developed real-time simulators to test virtual prototypes to optimise their design and their future maintenance. These simulators are powerful for handling of rigid parts. However many applications need to simulate the flexible parts behaviour, like a hose in an engine environment.

Solving such a complex mechanical problem is not possible in real time. Our method is based on two phases: a training phase presented in this paper and a real-time immersion phase. Two approaches were presented to realise the training phase: an *a posteriori* approach and an *a priori* approach. This paper focus on these two approaches which allow to calculate the surface response of possible displacements with a model reduction technique.

The *a posteriori* approach is interesting in term of reduction of the volume data necessary to the future real-time phase.

The *a priori* approach is an adaptive strategy, with enrichment of the modal basis during the campaign. The size of this basis increases the cost of the campaign calculation when the number of degrees of freedom is important. This was the case for the hose. Thus, during the campaign, to reduce this basis is necessary when it reaches a certain limit, with the KL expansion. The reduction is done then on the coefficients a.

This *a priori* approach provides results interesting for managing the handling of an automotive hose. It remains however to optimise this training phase by a technique of hyperreduction [19]. The resolution of the system will be done then with a reduced number of degrees of freedom.

Finally, the immersion in real-time phase will be realised by interpolation on the results of the training phase.

References

1. Ploye F. Trois ans au service de la Réalité Virtuelle, *Harvest* **85**, 2004, 31–35.
2. PerfRV Synthesis Report, http://www.perfrv.org/vitrine/resultats/commun/ColloqueFinal/accueil.en.html, 2004.
3. Amundarain A., Borro D., Garcia A., Gil J., Matey L., Savall J. Virtual reality for aircraft engines maintainability, *Mecanique and Industries* **5**(2), 2004, 121–127.
4. Son W., Kim K., Amato N.M., Trinckle J.C. A generalized framework for interactive dynamic simulation for multirigid bodies, *IEEE Transactions on Systems Man and Cybernetics; Part B Cybernetics* **34**(2), 2004, 912–924.
5. Hasegawa S., Sato M. Real-time rigid body simulation for haptic interactions based on contact volume of polygonal objects, *Computer Graphics Forum* **23**(3), 2004, 529–538.
6. Constantinescu D., Salcudean S.E., Croft E.A., Haptic rendering of rigid contacts using impulsive and penalty forces, *IEEE Transactions on Robotics* **21**(3), 2005, 309–323.
7. Krause F., Neumann J. Haptic interaction with non-rigid materials for assembly and dissassembly in product development, *CIRP Annals Manufacturing Technology* **50**(1), 2001, 81–84.
8. Terzopoulos D., Platt J., Barr A., Fleischer K. Elastically deformable models, *Proceedings of SIGGRAPH'87, Computer Graphics* **21**(4), 1987, 205–214.
9. Kry P.G., James D.L., Pai D.K. EigenSkin: Real time large deformation character skinning in hardware, in *ACM SIGGRAPH Symposium on Computer Animation*, 2002, pp. 153–160.
10. Kühnapfel U., Cakmak H.K., Maass H. Endoscopic surgery training using virtual reality and deformable tissue simulation, *Computers and Graphics* **24**, 2000, 671–682.
11. Picinbono G., Delingette H., Ayache N. Non-linear anisotropic elasticity for real-time surgery simulation, *Graphical Models* **65**(5), 2003, 305–321.

12. Mikchevitch A., Léon J.C., Gouskov A. Realistic force simulation in path planning for virtual assembly of flexible beam parts, in *Proceedings of Virtual Concept*, Biarritz, 2003.
13. Duriez C., Andriot C., Kheddar A. A multi-threaded approach for deformable/rigid contacts with haptic feedback, in *HapticSymposium*, 2004.
14. Dulong J.L., Druesne F., Villon P. La réduction de modèle pour déformer en temps réel une structure à comportement non linéaire, in *Colloque National en Calcul des Structures*, Giens, 2005.
15. Druesne F., Dulong J.L., Villon P. Real time simulation of non linear mechanical model, in *Proceedings of Virtual Concept*, Biarritz, 2005.
16. Karhunen K. Zur Spektraltheorie stochasticher Prozesse, *Annales Academiae Scientiarum Fennicae*, **37**, 1946.
17. Lumley J.L. *Stochastic Tools in Turbulence*, Academic Press, New York, 1970.
18. Ma X., Vakakis A.F., System identification by means of Karhunen–Loeve decomposition of the transient dynamics of a multi-bay truss, *AIAA Journal* **3**(2), 1999, 939–946.
19. Ryckelynck D. A priori hyperreduction method: An adaptive approach, *Journal of Computational Physics* **202**, 2005, 346–366.
20. Krysl P., Lall S., Marsden J.E., Dimensional model reduction in non-linear finite element dynamics of solids and structures, *International Journal for Numerical Methods in Engineering* **51**, 2001, 479–504.
21. Rey C., Calcul intensif et multirésolution en mécanique non linéaire, in *XVIIe Congrès Français de Mécanique*, 2005.

Knowledge Base Formulation for Aided Design Tool

Jérôme Pailhès[1], Mohammed Sallaou[2] and Jean-Pierre Nadeau[1]

[1]*TREFLE-ENSAM, Esplanade d'Arts et Métiers 33405 Talence Cédex, France;*
E-mail: {jerome.pailhes, jean-pierre.nadeau}@bordeaux.ensam.fr
[2]*ENSAM-Meknès, BP 4024, Meknès, Ismaïlia, Maroc; E-mail: mosallaou@yahoo.fr*

Abstract. In this article, a design support tool is developed and applied to a wind turbine. An analysis and structuring methodology is similarly investigated and applied. This methodology identifies structuring characteristics and provides a parsimonious overview of the design problem. The model is a set of constraints. The present approach uses functional analysis tools to get solutions at a given system level. These tools are completed by two TRIZ evolution laws and by functional flow analysis. The present analysis contributes to setting up models linked to functions and action verbs. These models can be treated as constraint-satisfaction problems. The present knowledge base is organised in terms of system and qualification levels, which leads to the generation of potentially validated solutions to continue with a detailed design process.

Key words: preliminary design, design support, functional modelling, integration of knowledge, constraints.

Nomenclature

C_e	= speed-increasing gear input torque
C_s	= speed-increasing gear output torque
ω_e	= rotational speed of speed-increasing gear input shaft
ω_s	= rotation speed of speed-increasing gear output shaft
η_m	= speed-increasing gear efficiency
q	= flow rate
k	= overall exchange coefficient
Φ	= lost power
NUT	= number of transfer units
C_p	= specific heat
T_e	= input temperature
T_s	= output temperature

S. Tichkiewitch et al. (eds.), Advances in Integrated Design and Manufacturing in Mechanical Engineering II, 231–243.

$\bar{T}$	=	average temperature
A	=	exchange surface
π_m	=	speed-increasing gear efficiency factor
ε_h	=	exchange efficiency
index a	=	air
index h	=	oil

1 Introduction

Our society dictates that companies be continually innovative, flexible and reactive as soon as the need for a product emerges. Because of the competition in the market place, and in order to respect quality standards and obtain the ISO9000 certifications [1], analysis and product management methodologies have been developed. However, design support and innovation methods have not made the same progress. Currently, in an effort to change this situation, several research projects propose procedures and tools for making the design task more efficient. These projects are all the more justified since the costs generated by the product study phase are high.

2 Design

The aim of product and system design is to satisfy users' needs as well as technical-economic constraints, while ensuring the respect for the environment, the laws and the company's profitability.

The conventional design process consists in defining successive hypotheses, and applies a trial-and-error logic to approximate a solution. It is thus based on an iterative character. Actually, before starting to search for an architecture, initial choices have to be made. Determining the characteristics of an architecture is to some extent a sequential process since it is often necessary to set the major dimensions and choose technologies, components, etc. Thus, the architecture design phase leads to an iterative process of generating product artefacts and evaluating their capacity to meet requirements [2]. These choices, which are made at first sight, lead to a "trial-and-error" mode for determining the architecture. Such irreversible decisions hide one part of the solution space. Furthermore, all the knowledge necessary to design a product is not always taken into consideration simultaneously. Finally, the experienced designer finds a solution, but it is rarely optimal [3]. This largely accounts for the increasing cost and time of the design process. An awaited improvement is the removal of the iterative character in the process.

On the other hand, some studies show that the preliminary design phase accounts for around 80% of the expenses generated by the design project itself [4]. The present research work shows that the choices made in the preliminary design phase are fundamental and should only seldom be reconsidered.

2.1 Design Process

Most of the models used in design processes propose, after a translation of the need and before the detailed design process, a preliminary design phase (from a general specification to a technical specification).

A methodology consisting in analysing and structuring a preliminary design process should be helpful in identifying the structuring characteristics of the design problem and provide a parsimonious overview of the problem that will have to be translated into constraints. To the designer, the structuring characteristics are those characteristics which allow him to start the design work and define a primary architecture. These are the criteria and functions which structure the design action [5].

The structuring characteristics of a preliminary design problem are [5]:

- real life situations which absolutely must be described;
- essential or hazardous functions;
- standard elements to be chosen or dimensioned: they generate variables and constraints;
- design variables which make it possible to define the architecture of the product;
- criteria variables which make it possible to qualify a design process;
- variation ranges for the variables;
- relevant physical behaviours which make it possible to describe how the product functions and the resulting effects; these behaviours are translated into relationships between design variables.

Systematic lists have been proposed to make the identification of the structuring characteristics of a preliminary design problem easier [7].

2.2 Approach for Analysing and Structuring the Design Problem

An approach has been developed to search for the characteristics which structure the design action, from when the concept emerges to the choice and implementation of a solution [5]. This approach consists of four phases, each phase corresponding to a different analysis level: analysis of the need, functional analysis, system analysis, and physical analysis (Figure 1). The functional analysis tools are sometimes improved, just like the Substances-Fields Diagram (SFD)[1] which completes the Functional Block Diagram (FBD). These functional analysis tools, combined with certain TRIZ evolution laws, provide indications for qualifying the design process, integrating the user's point of view, and identifying points of access to innovation. Figure 1 shows the tools used in the four phases and the structuring characteristics obtained:

- In the first step of the need analysis, the client's qualification criteria and the real life reference situations are also expressed.

[1] The Substances-Fields Diagram comes from the TRIZ theory [6].

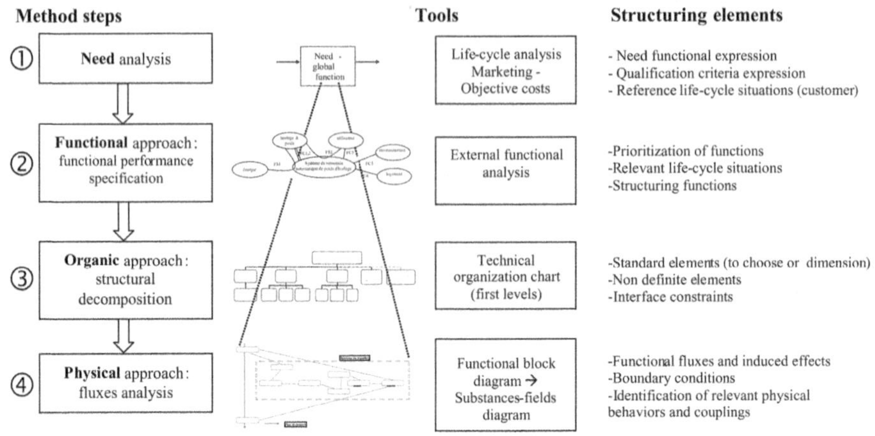

Fig. 1. Search procedure, associated tools and structuring characteristics [5].

- During the functional analysis step (Phase 2), the external Functional Analysis (FA) lists the functions linked to the product to be designed; only the structuring functions are retained.
- In Phase 3, a structural breakdown of the product is applied through the extended Technical Organisational Chart (TOC), in particular to identify the standard elements and generated constraints.
- Using the FBD and the SFD, the functional and induced flows can be analysed. Thus, the physical analysis (Phase 4) reveals the relevant physical behaviours to be taken into consideration.

To demonstrate the relevance of the design approach, it is applied to the case of producing electrical energy from wind energy, and in particular to one of its components: the speed-increasing gear.

3 Application of the Wind System Approach

3.1 Wind Energy

Wind turbines are a very old way of using wind. Yet, from 1970 to 2004, their use in the world and their design progressed considerably. The wind turbine is an extremely interesting and environmentally acceptable solution for producing renewable energy. Wind turbines can be classified into two categories (i.e. vertical-axis and horizontal-axis wind turbines). The present study focuses on three-bladed, horizontal-axis wind turbines as this construction type is the most widely used in industry.

3.2 Analysis of the Wind Turbine

3.2.1 Need Analysis

The design approach starts with an exhaustive formulation of the users' expressed and implicit need. The functional analysis translates and integrates the expression of the end user's, client's, and designer's needs [5, 8] into the design process. In the need analysis step, the client's qualification criteria and real life reference situations are also expressed. The overall function translates the reason as to why the system exists. This system is helpful both for the operator of the wind energy site as it recovers the energy of the wind and transforms it into electrical energy, and for the distributor as it supplies the network with high-quality electrical current.

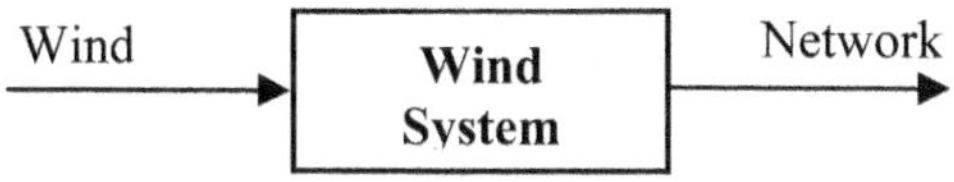

- Client qualification criteria [9]:
  - cost of the kWh produced
  - total updated cost of the project
  - quality of the electrical energy: voltage, flickers?
  - quantity of energy produced per year
  - life span.

These criteria determine the relevance of the design process and the choices made, and contribute to the final phase of sorting the solutions.

3.2.2 Functional Analysis

The Functional Analysis phase consists in applying an external functional analysis to various real life situations so as to extract the functional relationships between the products and the external environments (APTE® diagram). For the wind system and to illustrate our arguments, the relevant real life situation retained is the "normal operation" situation (Figure 2).

- Real life situation: "normal operation: production of electricity"
  - SF1: Transforms the aeraulic power of the wind into electrical power supplied to the network.
  - CF1: Resists the wind.
  - CF2: Respects the quality criteria regarding the electrical current to be supplied to the network.
  - CF3: Is connected to the ground.
  - CF4: Is environmentally non-contaminating.

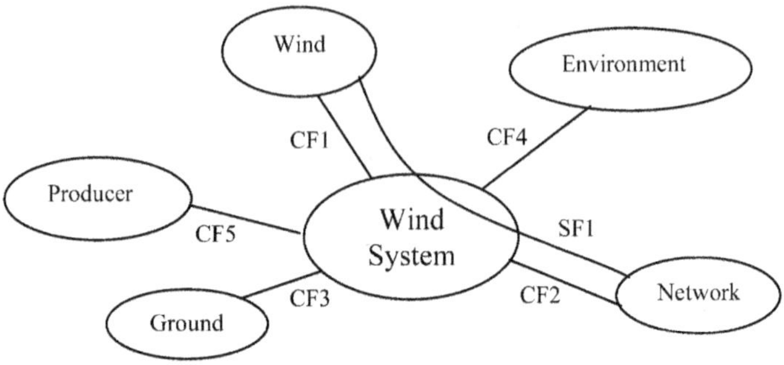

Fig. 2. APTE diagram of a wind system.

- CF5: Is profitable for the producer.

For this real life situation, the structuring functions and the appreciation criteria are as follows [5]:

- Functions with low or no flexibility:
  - CF2: Quality of the current to be supplied to the network: voltage, flickers, etc.
  - CF5: – Cost of the kWh produced and total updated cost of the project
    – Life span
- Hazardous function:
  - CF1: Resists the wind
- Mandatory constraint function:
  - CF4: Respects the environment (noise, impact on the landscap, etc.)

3.2.3 System Analysis

The third phase consists in carrying out a structural breakdown of the product through the Technical Organisational Chart extended to the external environments. The extended Technical Organisational Chart (Figure 3) describes how a wind turbine is constructed. The following structuring characteristics can thus be identified [10]:

- The standard elements (to be chosen or dimensioned).
- The non-defined elements to be dimensioned.
- The interface constraints between the functional blocks.

At the second level, one of the major elements of the wind turbine's transmission chain is the speed-increasing gear. This component connects the rotor to the generator. In the following, the present study focuses on the speed-increasing gear the FAST diagram of which is shown in Figure 4.

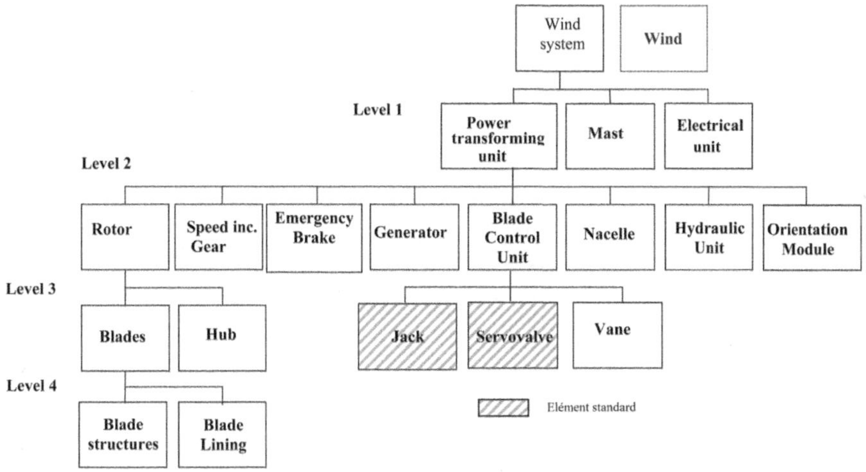

Fig. 3. Technical organisational chart of the wind system.

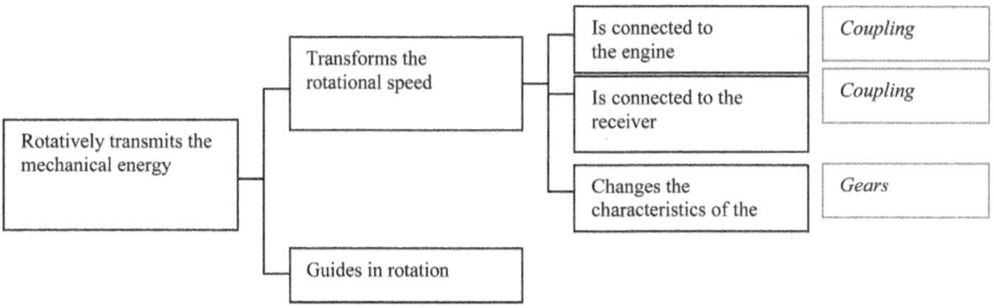

Fig. 4. FAST of the speed-increasing gear.

The FAST diagram defines the successive functions which lead to the major function of the speed-increasing gear. This diagram (Figure 4) describes a possible way of implementing the function "Rotatively transmits the mechanical energy".

The Technical Organisational Chart of the speed-increasing gear is shown in Figure 5 as extended to the external environments. As shown, in addition to the standard and non-defined components, the interface components are listed (e.g., mechanical connection, fluidics, electrics, software and hardware). This remark is important for the formalisation of the behaviour models since the distinction is made between component models and interface models, or component-interaction models [12].

3.2.4 Physical Analysis

This step is very important for the identification of the relevant physical behaviours. Using the substances-fields diagram, the functional and induced flows can be ana-

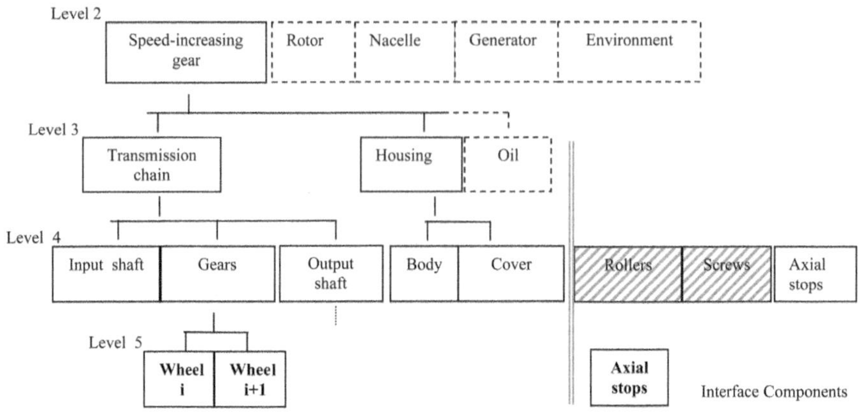

Fig. 5. Technical organisational chart of the speed-increasing gear.

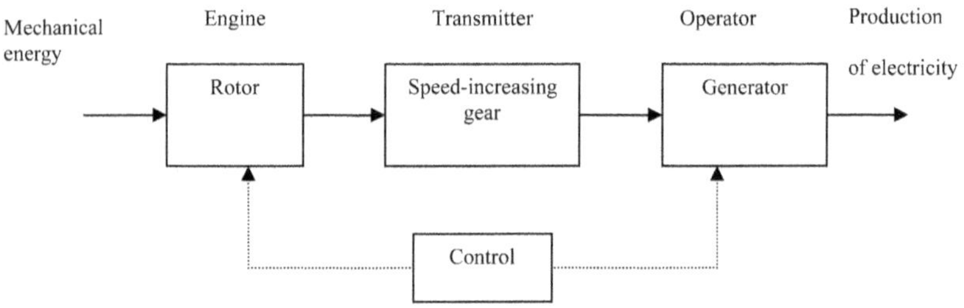

Fig. 6. Law of system completeness as applied to the wind system.

lysed [5]. The use of certain evolution laws coming from the TRIZ theory provides an exhaustive overview of the functional flow analysis [1] (Figures 6 and 7). The present study uses the first two laws which make it possible to study the rational use of energy to complete the action. These two laws are:

- Law of system completeness: it identifies a motor entity, a transmission entity, an operator entity and a control entity. The energy flow (functional flow) progresses through the entities until the required action is complete.
- Law of thermal conductivity: it specifies that the energy should flow freely through the system between all the entities when the action is being carried out.

Likewise, at level 4 in the Technical Organisational Chart, the analysis of the speed-increasing gear involves the input and output shafts and the various gears (Figure 7).

Table 1 shows the various flows linked to the implementation of the speed-increasing gear's major function: "Rotatively transmits the mechanical energy". The induced effects result in power loss in the form of heat, distortion, production of ma-

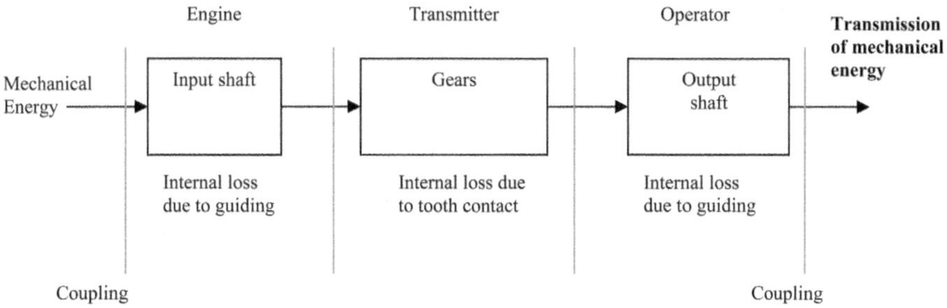

Fig. 7. Law of system completeness and law of thermal conductivity as applied to the speed-increasing gear.

Table 1. Pathway of the functional flow through the transmission chain, and induced flows.

Component	Function	Functional flows	Induced flows
Input shaft	Rotatively transmits the mechanical energy	Mechanical energy flow	- Strain energy flow (shaft torsion)
Gears	Transforms the rotational speed	Mechanical energy flow	- Thermal energy flow (contact friction, effusivity) - Strain energy flow (tooth bending, contact) - Sound energy flow - Material flow (wear)
Output shaft	Rotatively transmits the mechanical energy	Mechanical energy flow	- Strain energy flow (shaft torsion)

Table 2. Induced functions, functional flows and induced flows.

Component	Function	Functional flows	Induced flows
Housing	Rotatively guides the shafts	Mechanical energy flow (contact)	- Thermal energy flow (friction) - Sound energy flow - Mechanical energy flow (friction, resisting torque) - Material flow (wear)
Oil	Transports thermal energy	- Thermal energy flow - Material flow	- Mechanical energy flow (oil viscosity, resisting torque)

terial (wear), or noise. Therefore, other (not requested) functions are induced by the implementation of the function (Table 2).

In order to identify the relevant physical phenomena involved, the functional diagram block is drawn for the real life situation "Normal operation of the speed-

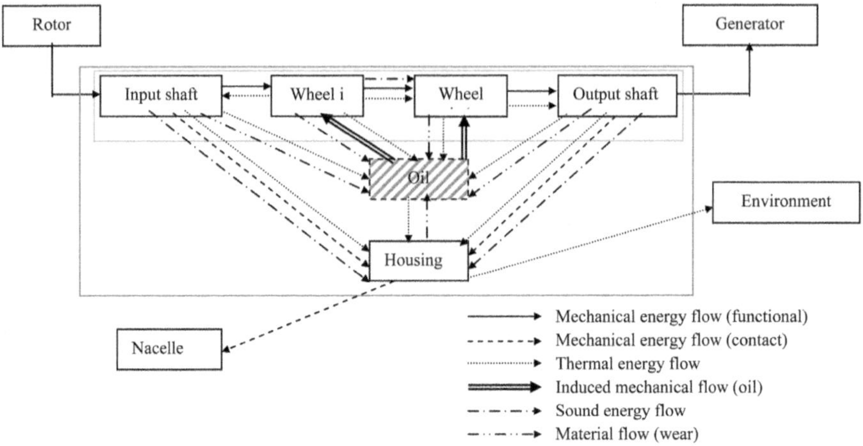

Fig. 8. Functional block diagram of the speed-increasing gear.

Table 3. Component model of the speed-increasing gear at level 2.

Function *Level 2*	**flow**	**VCr**	**VCo**	**VI**	**Relationships**
Transform mechanical power on mechanical power	Imput mechanical power Output mechanical power		r_m	ω_e ω_s η_m C_e C_s π_m P_n P	Lost power : $\Phi = (1-\eta_m).C_e\omega_e$ $\eta_m = 1 - \left[\left(1-\pi_m\right)\left(\frac{P_n}{P}+3\right) \middle/ 4 \right]^*$ $\pi_m = 0.89 P_n^{0.012}$ $r_m = \frac{\omega_{me}}{\omega_{ms}}$ *Source [11]

increasing gear" (Figure 8). The flow between the various entities appears clearly in the functional block diagram of the speed-increasing gear.

Breaking down a system into sub-systems, called components, and identifying their physical behaviours through functional block diagrams, can only be performed after a first preliminary definition of the product.

To formalise the knowledge and ensure an exhaustive design analysis, component models are proposed. These models are linked to a description level and associated with a possible solution (Tables 3 and 4).

At level 2, at the speed-increasing gear scale, the efficiency gives the power lost in the functional flow. Most of this loss exists as a thermal energy flow which must be transported towards the outside. This function appears in the immediately lower level, i.e. level 3, of the Technical Organisational Chart. The corresponding analysis

Table 4. Analysis of the two implementations of the function: "Cooling the oil in the speed-increasing gear".

Function *Level 3*		**flow**	**VCr**	**VCo**	**VI**		**Relationships**
Cooling the oil in the speed-increasing gear	Alternative 1 (use of an exchanger)	Material: oil q_h Fluid coolant Air q_a Energy: thermal $qCp_hT_{h,e}$ $qCp_aT_{a,e}$		Aera type	Φ η_m C_e ω_e	q_h q_a Cp_h Cp_a $T_{h,e}, T_{h,s}$ $T_{a,e}, T_{a,s}$ k_1 A_1 ε_h R_h NUT_h	Dissipated power $\Phi = (1-\eta_m).C_e\omega_e$ $= q_hCp_h(T_{h,e} - T_{h,s})$ $= q_aCp_a(T_{a,s} - T_{a,e})$ Exchanger efficiency $\varepsilon_h = \frac{T_{h,e} - T_{h,s}}{T_{h,e} - T_{a,e}}$ Capacity ratio $R_h = \frac{q_hCp_h}{q_aCp_a}$ Number of transfer units $NUT_h = \frac{k_1A_1}{q_hCp_h} = f(R_h, \varepsilon_h)$
	Alternative 2 (cooling via the wheel case)	Material: air $qCp_aT_{a,e}$				T_h q_{air} $T_{a,e}, T_{a,s}$ Cp_a k_2 A_2	Dissipated Power $\Phi = (1-\eta).C_e\omega_e$ $= qCp_a(T_{a,s} - T_{a,e})$ Stabilising of oil temperature $\Phi = k_2A_2(T_h - \bar{T}_a)$

is shown in Table 4 for two different solutions: "Cooling the oil outside the speed-increasing gear via an exchanger" or "Cooling the oil via the housing walls". A thermal database is required to express the exchange coefficients [13].

At level 4, through the study of the function "Rotatively transmits the mechanical energy", a detailed analysis of the contacts can be carried out, and the mechanical and thermal energy flows can be highlighted at the wheel contacts (effusive contact and local distortion). The notion of efficiency can thus be ignored (Table 3). We are currently developing more such complex models.

At level 3, the function "Guides in rotation" appears, which leads to an identical analysis of the contact flows (plain or antifriction bearings). The component and component-interaction (rollers, bearings) models make up a knowledge base. This knowledge base links the functions (expressed as: "action verb + complement") to the models which will be checked and validated at various system levels based on their qualification criteria (PEPS): Parsimony, Exactness, Precision, Specialisation (PEPS) [14].

The PEPS parameters are defined as follows:

- *Parsimony* is a measurement which is the inverse of the complexity of a model. It increases with the increasing number and level of coupling between model variables.
- *Exactness* is a measurement of the difference between the model and the reality it is supposed to represent.
- *Precision* is a measurement defined as the opposite of imprecision. Imprecision measures the vague or confused aspect linked to the distinction between several values of one variable in a model, and which is translated as a set of possible values for said variable (for example, as an interval).
- The *Specialisation* of a model is a characterisation of the hypotheses and information which restrict its application field.

This work is currently being developed in the TREFLE laboratory and will be validated by industrial applications.

4 Conclusions and Perspectives

In the scope of the development of a design support tool, an approach for analysing and structuring a design problem has been discussed. This approach, as applied to several industrial projects conducted by our team, brings out validated solution architectures and elements which allow us to progress towards a detailed design process. The tools used to carry out the physical approach and the analysis of the flows induced by the functions and their implementation at various system levels have also been discussed. In the scope of this work, the present approach has been applied to a wind system.

The multiple possible choices, the complex interaction between the various parameters of the problem and the multiple points of view to be taken into account make the definition of wind systems difficult. This example shows that models and solutions corresponding to a given system function can be formalised as a knowledge base. This analysis is applied at various system levels and makes it possible to identify the models linked to a description level and associated with possible solution alternatives.

In the continuation of our work, we will develop models at more detailed system levels. These models will be qualified based on their qualification criteria (PEPS).

References

1. J.P. Nadeau, J. Pailhès, R. Dore, D. Scaravetti, Analyser, qualifier et innover en conception par les lois d'évolution TRIZ, in *6e Congrès International de Génie Industriel*, Besançon, France, 7–10 June 2005

2. D.G. Ullman, *The Mechanical Design Process*, 3rd edition, McGraw-Hill Higher Education, New York, 2003.
3. V. Hubka, E. Eder, *Design Science*, Edited for the web by Filippo A. Salustri, 2001.
4. L. Zimmer, P. Zablit, Global aircraft predesign based on constraint propagation and interval analysis, in *Proceedings of CEAS Conference on Multidisciplinary Aircraft Design and Optimisation*, Köln, Germany, June 2001.
5. D. Scaravetti, Formalisation préalable d'un problème de conception, pour l'aide à la décision en conception préliminaire, Thèse de Doctorat, ENSAM, 2004.
6. S.D. Savransky, *Engineering of Creativity: Introduction to TRIZ Methodology of Inventive Problem Solving*, CRC Press, Boca Raton, 2000.
7. D. Scaravetti, J.-P. Nadeau, J. Pailhès, P. Sebastian, Structuring of embodiment design problem based on the product lifecycle, *International Journal of Product Development*, **2**(1/2), 2005, 47–70.
8. AFNOR. *NF X50-150/151, analyse de la valeur, analyse fonctionnelle: Vocabulaire NF*, Association Française de Normalisation, Paris, 1990-1991.
9. A. Arbaoui, J.P. Nadeau, P. Sebastian, Constraint modelling and decision support for wind energy, in *12th Seminar on Life Cycle Engineering, CIRP 2005*, Grenoble, France, 3–5 April 2005.
10. A. Arbaoui, J.P. Nadeau, P. Sebastian, L. Bchir, A. Brakez, Aide à la décision pour la définition d'un système éolien adapté à un site donné, in *Congrès International CPI 2003*, Meknès, Maroc, 22–24 October 2003.
11. R.E. Wilson, P.B.S. Lissaman, S.N. Walker, *Aerodynamic Performance of Wind Turbines*, ERDA/NSF/04014-76/1, Washington, DC, 1976.
12. Y. Vernat, Formalisation de modèles par contraintes en conception préliminaire, Thèse de doctorat, ENSAM, 2004.
13. W. Kay, A.L. London, *Compact Heat Exchangers*, 2nd edition, McGraw-Hill Book Company, 1984.
14. Y. Vernat, J.P. Nadeau, P. Sebastian, X. Fischer, Démarche de formalisation de modèles adaptés à la conception préliminaire, in *6e Congrès International de Génie Industriel*, Besançon, France, 7–10 June 2005.

Calibration Accuracy of a Parallel Structure Machine Tool with Respect to Machined Part Quality

Hélène Chanal, Emmanuel Duc and Pascal Ray

LaMI, IFMA & UBP, Campus de Clermont Ferrand/les Cezeaux, BP 265, 63175 Aubiere Cedex, France; E-mail: helene.chanal@ifma.fr

Abstract. A main limitation of the use of parallel kinematics machine tools for high speed machining tasks is their low level of accuracy compared with serial kinematics machine tools, which is largely due to geometrical transformation errors. These errors can be reduced by identifying the geometrical parameters of the inverse kinematics model integrated into the controller by exteroceptive external calibration. However, after external geometrical calibration, tool position and orientation defects may persist. Consequently, the aim of this paper is to quantify the calibration accuracy which is needed to respect the specified machined part quality. With a sensitivity analysis, we confirm that the geometrical parameters which most influence tool position defects are the leg lengths. Next, the influence of these legs on defects generated on the machined surface is computed over all the workspace for a planar profile contour. Finally, the required geometrical calibration accuracy is defined for a contoured planar surface. This work can be used to define the measurement tool and method for geometrical calibration. This study was carried out on the Verne PKM, located at IRCCyN (Nantes, France).

Key words: parallel kinematics machine tool, geometrical calibration

1 Introduction

In the automotive or aeronautical industries, few Parallel Kinematics Machine tools (PKM) are used in High Speed Machining (HSM), even though the dynamics of these structures could be of great interest for HSM as their acceleration potential is much higher than Serial Kinematics Machine tools (SKM) [1]. However, a major limitation is their low level of accuracy compared with SKM [2, 3]. This limitation is due to their large number of links and passive joints.

Several studies deal with PKM effector pose accuracy. Effector pose quality can be defined with two parameters, pose rightness and repeatability [4]. PKM exhibit much better repeatability than pose rightness, according to Merlet [5].

S. Tichkiewitch et al. (eds.), Advances in Integrated Design and Manufacturing in Mechanical Engineering II, 245–257.

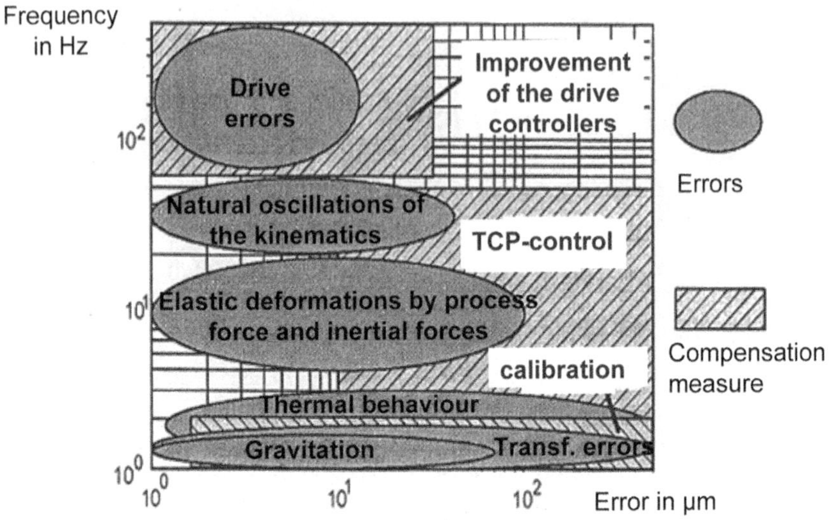

Fig. 1. Impact of error causes depending on size and frequency [6].

Tool pose defects can come from several aspects of machine tool structure behaviour (Figure 1). A major cause is geometrical transformation errors. These errors can be reduced by identifying the geometrical parameters of the inverse kinematics model integrated into the controller [6, 7]. The most accurate methods for such geometrical calibration use exteroceptive external measurements. Thus, the real tool position is measured and compared with the modelled position, in order to find the geometrical parameters which best fit the model to the measured data. However, after external geometrical calibration, a tool pose defect may persist. This depends on the accuracy of the measurement method [8, 9].

However, in the machining context, previous work has shown that the quality of the machined part is not directly linked to the tool pose defect [10]. Thus, this paper is dedicated to the quantification of the required calibration accuracy, i.e. the calibration accuracy which is needed to respect the specified machined part quality. Indeed, an appropriate measurement method should be chosen in order to ensure the machined part quality. This analysis is particularly focused on tool pose rightness. This study was carried out on the Verne PKM, which is located at IRCCyN (Nantes, France).

In this article, the Verne machine is presented first. Then a sensitivity analysis is performed by computing the sensitivity matrix. This analysis confirms that the most influential geometrical parameters affecting tool pose defects are leg lengths. Thus, the influence of one of these parameters on the defect generated on the machined surface is computed in all the workspace for a planar profile contour. If a normal

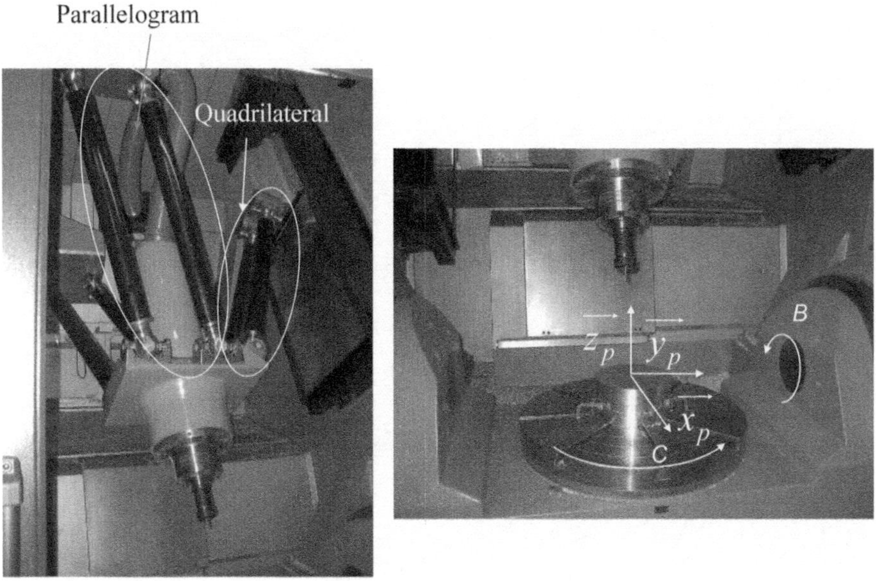

Fig. 2. Verne structure.

distribution is considered to represent the most sensitive geometrical parameter defects, the distribution of the generated defects on a contoured planar surface can be evaluated. Finally, the required geometrical calibration accuracy can be defined for a contoured planar surface.

2 Study of the Verne Machine Tool

Verne is a PKM developed by Fatronik which has equipped the IRCCyN laboratory (Nantes) since May 2004 [11].

2.1 Structure Analysis of Verne

The studied machine is a five-axis machine tool. It is composed of a parallel structure with three degrees of freedom and a tilting table with two degrees of freedom (Figure 2). The three movements of the parallel structure mobile platform are two translations parallel to the x and z axis and a translation-rotation parallel to the y axis. The serial module provides two rotations B and C.

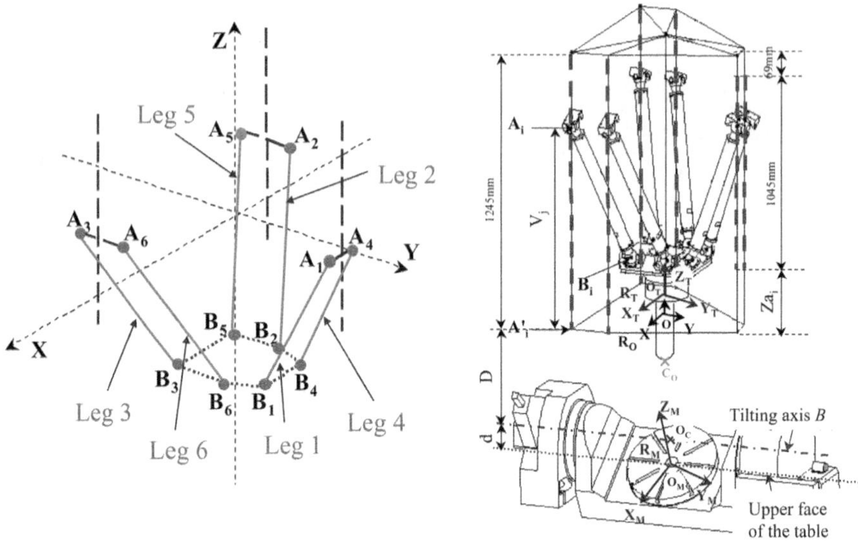

Fig. 3. Verne geometry [11].

2.2 Inverse Geometrical Model of Verne

The inverse geometrical model of Verne, presented here, is developed in Terrier's study [11]. To define the inverse geometrical model of Verne, the displacement of the motors must be expressed according to the position of the tool tip and axis. Tool position and orientation in the Cartesian (workspace) coordinates system are labelled (X, Y, Z, B, C) and the motor positions in the machine coordinates system are labelled $(V_1, V_2, V_3, B_\alpha, C_\alpha)$ (Figure 3).

Fixed leg lengths are labelled L_i; they are attached to the structure at A_i and B_i.

Thus, six equations are written:

$$L_i^2 = \|\mathbf{A}_i\mathbf{B}_i\|^2 = (X_{B_i} - X_{A_i})^2 + (Y_{B_i} - Y_{A_i})^2 + (Z_{B_i} - Z_{A_i})^2, \quad i = 1, \ldots, 6, \quad (1)$$

where $(X_{A_i}, Y_{A_i}, Z_{A_i})$ and $(X_{B_i}, Y_{B_i}, Z_{B_i})$ are the coordinates of A_i and B_i in the Cartesian coordinates system R_0.

Then,

$$\begin{vmatrix} X_{A_i} \\ Y_{A_i} \\ Z_{A_i} \end{vmatrix} = \begin{vmatrix} x_{A_i} \\ y_{A_i} \\ z_{A_i} \end{vmatrix} + \begin{vmatrix} 0 \\ 0 \\ V_j \end{vmatrix} \quad \text{and} \quad \begin{vmatrix} X_{B_i} \\ Y_{B_i} \\ Z_{B_i} \end{vmatrix} = R_1 \begin{vmatrix} X \\ Y \\ Z - d \end{vmatrix} + \begin{vmatrix} 0 \\ 0 \\ -D \end{vmatrix} + R \begin{vmatrix} x_{B_i} \\ y_{B_i} \\ z_{B_i} + L \end{vmatrix}$$

where $j = 1, \ldots, 3$ and $i = 1, \ldots 6$ with

$$R_1 = \begin{pmatrix} \cos(B_\alpha) & 0 & \sin(B_\alpha) \\ 0 & 1 & 0 \\ -\sin(B_\alpha) & 0 & \cos(B_\alpha) \end{pmatrix} \begin{pmatrix} \cos(C_\alpha) & -\sin(C_\alpha) & 0 \\ \sin(C_\alpha) & \cos(C_\alpha) & 0 \\ 0 & 0 & 1 \end{pmatrix}$$

and $R = R_1 \cdot R_2$ with

$$R_2 = \begin{pmatrix} \cos(C) & -\sin(C) & 0 \\ \sin(C) & \cos(C) & 0 \\ 0 & 0 & 1 \end{pmatrix} \begin{pmatrix} \cos(B) & 0 & \sin(B) \\ 0 & 1 & 0 \\ -\sin(B) & 0 & \cos(B) \end{pmatrix}$$

In these equations, $(x_{B_i}, y_{B_i}, z_{B_i})$ and $(x_{A_i}, y_{A_i}, z_{A_i})$ are design constants known as d and D. L is the length of the tool.

Thus the geometrical inverse transformation can be expressed if the values (X, Y, Z, B, C) are known. Then the values $(V_1, V_2, V_3, B_\alpha, C_\alpha)$ can be computed using a numerical method. This model should be the one integrated into the controller. In all there are 42 geometrical parameters to be identified, 36 joint coordinates and 6 leg lengths. This model is defined with the hypothesis that the three motor displacements are parallel.

3 Verne Sensitivity Analysis

Sensitivity analysis is often used to find out the major error sources and identification errors [9]. Moreover, it enables the choice of geometrical parameters in the inverse geometrical model to be validated [12].

To conduct this analysis, the sensitivity matrix is computed first. Then its components are analysed to determine the sensitive parameters.

3.1 Computation of the Sensitivity Matrix

In the Verne case, the inverse geometrical model can be written as:

$$V = f(X, \xi), \tag{2}$$

where V is a six-component vector $(V_1, V_2, V_3, A_\alpha, B_\alpha, C_\alpha)$ and where in the Verne case $A_\alpha = 0$. X is a six-component vector (X, Y, Z, A, B, C) which represents the six degrees of freedom to define the tool pose in the Cartesian coordinates system. ξ is composed of the 42 geometrical parameters.

To conduct a sensitivity analysis, the inverse geometrical model is first differentiated:

$$dV = \frac{\partial f(X, \xi)}{\partial X} dX + \frac{\partial f(X, \xi)}{\partial \xi} d\xi, \tag{3}$$

where

- $d\xi$ is a 42-component vector which represents the difference between the nominal value and the real value of the geometrical parameters,
- dX is a six-component vector which represents the defects in tool tip position and tool axis orientation.

However, as regards external calibration, system state is generally compared in the part coordinates system with the same motor joint coordinates defined by the vector V [12]. Thus, in our case, $dV = 0$.

Finally, this equation is obtained:

$$\frac{\partial f(X,\xi)}{\partial X}dX = -\frac{\partial f(X,\xi)}{\partial \xi}d\xi \leftrightarrow TdX = Hd\xi, \tag{4}$$

where

- T is a 6×6 matrix,
- H is a 6×42 matrix.

Thus, the first order sensitivity matrix is:

$$S = T^{-1}H. \tag{5}$$

This sensitivity matrix enables the computation of the influence of machining and assembly defects on the final tool pose with respect to the nominal tool tip position. Thus, the term S_{ij} of matrix **S** (ith line and jth column) represents the influence of the jth geometrical parameter coordinate on the ith tool pose defect coordinate.

3.2 Definition of Sensitive Parameters

Sensitive parameters have the most influence on tool tip position and tool axis orientation. For a given tool pose on the workspace, the most sensitive parameters of an ith tool pose defect coordinate are the jth geometrical parameters which have the highest absolute S_{ij} value. For example, Figure 4 represents the influence of each geometrical parameter for a tool tip position $(100, -100, 50)$ and orientation $(0, 0, 1)$. This figure presents the influence of each geometrical parameter on the tool tip position along the x axis for a defect value on each parameter equal to 1. It can be seen that the tool pose defect is proportional to the geometrical parameter defects.

The graph in Figure 4 illustrates some conclusions from the analysis. Indeed, a complete sensitivity analysis of all geometrical parameters for some tool tip positions and orientations shows that the parameters of the inverse geometrical model are well chosen. Fan's work demonstrates that the positioning accuracy of the parallel slider, the leg lengths and the tool length are the most sensitive parameters influencing the position and orientation of the tool in the case of a machine tool like Verne [9].

However, tool positioning accuracy is most influenced by the leg lengths. So in this study, we particularly focus on the influence of leg length accuracy on tool tip position and orientation defects.

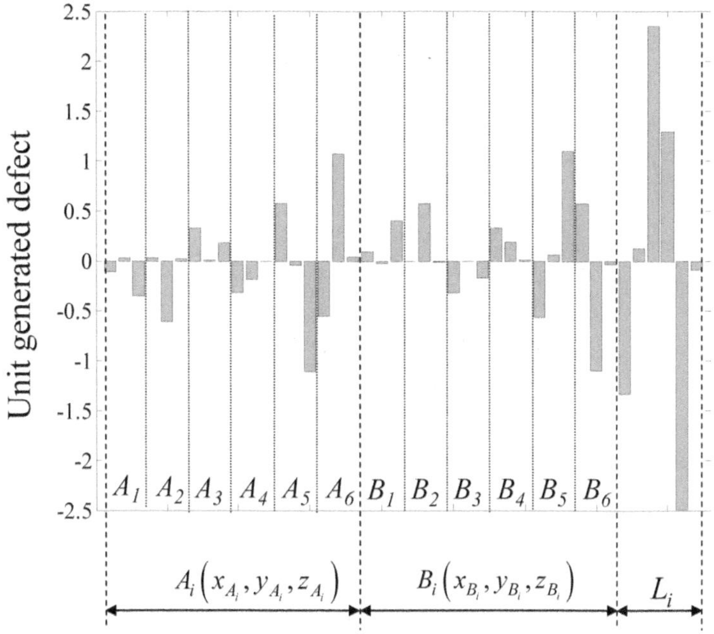

Fig. 4. Generated defect on the tool tip position projected on the x axis for a unit defect on each geometrical parameter at the (100, −100, 50) position in the workspace.

4 Leg Length Identification Accuracy with Respect to Machined Part Quality

Using the sensitivity analysis, a tool pose defect is evaluated for given leg length defects. However, previous work shows that the quality of the machined part is not directly linked to tool position defects [10]. Modelling the machined surface, using the hypotheses that the tool is a solid body, is necessary to quantify the generated machined entity defect.

In the present work, the influence of leg length defects on the generated defect is studied with particular emphasis on planar profile contouring.

First, the determination of the generated machined surface defect is presented. Second, the influence of one leg length defect on the machined part defect is computed in a given plane of the workspace. Influence maps are thus determined. Third, leg lengths are represented as having a normal distribution. Thus, the evolution of the generated defects on the machined entity can be computed over the entire workspace. Finally, the required geometrical calibration accuracy can be defined in order to ensure a given quality on a contoured planar profile.

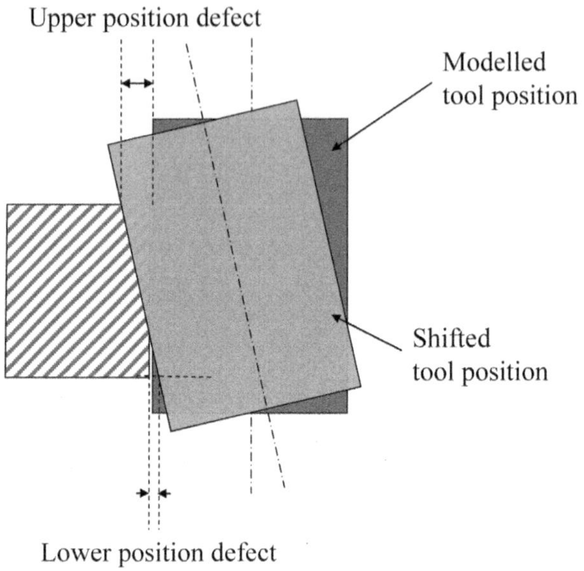

Fig. 5. Shifted tool position on the plane perpendicular to the contoured profile.

4.1 Determination of the Generated Machined Surface Defect

In this study, the machined entity is a contoured planar profile. In this case, Lartigue considers that a point M is a generative point of the profile if its speed is tangent to the tool surface at this point [13]. Then, for a given contoured tool path, tool path deformation due to geometrical parameter defects is determined using the sensitivity matrix. The contoured profile is simulated and finally the generated position defect is measured perpendicularly to the theoretical planar profile (Figure 5).

4.2 Influence of One Leg on the Generated Machined Entity Defect

To determine the influence of each leg length defect, a simulation is performed to generate the contoured profile if one leg has a unit length defect and if the modelled tool path is a straight line. The defect is computed using the sensitivity matrix. Thus, the evolution of the generated position defect is linear. Two study cases are considered with the same theoretical tool orientation (parallel to the z axis). In a first case, the tool path is considered to be parallel to the x axis and in the second case to the y axis. In Figure 6, the evolution of the generated position defect on a contoured flank profile with a height of 10 mm is presented for a given altitude z for a unit length defect on leg 1. In this figure, the upper and lower position defects are represented for each tool position on the modelled straight line tool path.

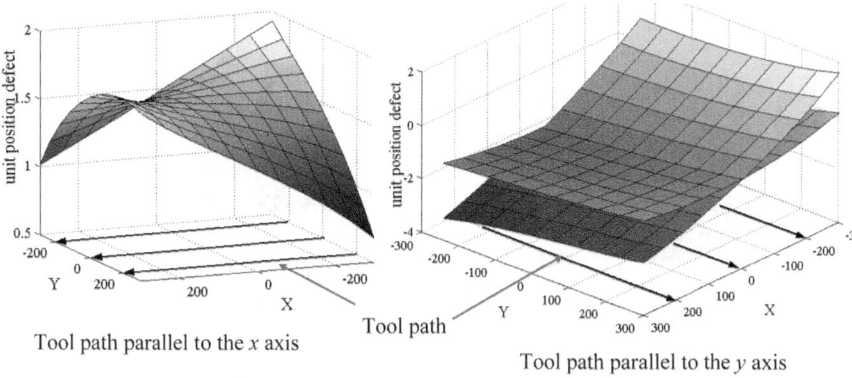

Fig. 6. Upper and lower machined surface position defects for a unit length defect of leg 1 for $z = 50$.

In the case of a contour parallel to the x axis, the surface is machined with only a position defect and the two curves are merged. This phenomenon may appear due to the topology of the structure. Indeed, leg 1 influences the orientation only around the y axis.

The same phenomenon appears if the unit defect is on leg 4.

Figure 7 presents the evolution of machined surface position defects for a unit length defect on leg 2 for a given altitude z. In this figure, we can see that the magnitude of the orientation defect on the contoured surface is higher for a tool path parallel to the y axis than to the x axis. Moreover, the unsymmetrical evolution of the two position defects for a tool path along the x axis can be explained by the particularity of Verne's structure, which produces a rotation around the y axis of the mobile platform when the tool moves along the x axis.

By computing such maps for each leg, two different behaviours can be defined according to their type of machined surface position defect evolution. The first does not influence the generated orientation defect on the machined part when the tool path is a straight line along the x axis. This behaviour appears when a defect is simulated on leg 1 or 4. The second generates an orientation defect when the tool path is a straight line along the x axis. It appears when the defect is introduced on leg 2, 3, 5 or 6. These two behaviours correspond to the two types of assembled leg configurations (Figure 2): legs 1 and 4 are assembled by forming a quadrilateral, and legs 2 and 5 (or 3 and 6) are assembled by forming a parallelogram.

However, the position defect evolution range is nearly the same for both behaviours. For a unit leg length defect the position defect is evaluated at nearly two.

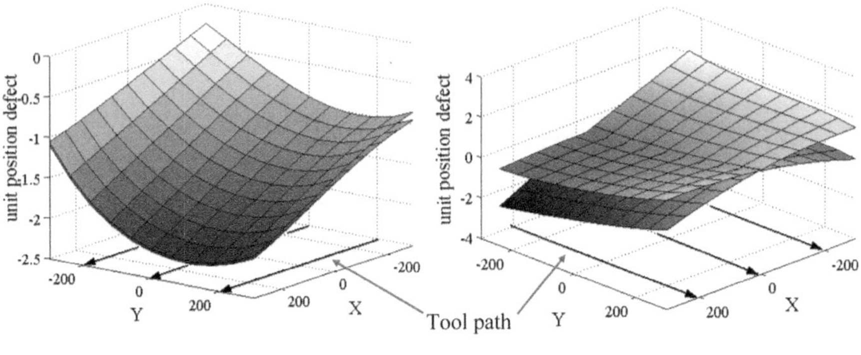

Fig. 7. Upper and lower machined surface position defects for an unit length defect of leg 2 for $z = 50$.

4.3 Evolution of the Generated Machined Entity Effect

The structure of Verne comprises six legs. In order to take into account the influence of the six legs on the accuracy of the machined profile, the same normal distribution is used to define the leg length defects. This normal distribution is chosen in order to have its standard deviation σ_{l_i} as $3\sigma_{l_i} = 1$. Then, by using the sensitivity matrix, the standard deviation of the resulting tool position and orientation defects is computed:

$$\sigma_{tj}^2(x, y, z) = \sum_{i=1}^{6} (S_{ij}(x, y, z)\sigma_{l_i})^2. \tag{6}$$

However, if six sigma capability is required for the generated defect, the tool position and orientation evolution range is $[-3\sigma_{tj}, 3\sigma_{tj}]$ for each jth coordinate of the tool. Finally, the distribution of the generated position defects for a straight line tool path along the x and y axes is computed. Figure 8 shows the upper and lower limits of the position defect evolution range.

Thus, in Figure 8, we can observe that the maximum position defect is arrived at when the tool path is along the y axis. In this case, the defect is attained if the tool path is on the boundary. However, in the case of a tool path along the x axis, the position defect is close to the maximum in the middle of the workspace. Moreover, in the case of a tool path along the y axis, there is a curve where the orientation defect is null. This curve is situated near $x = 40$.

In addition, tool path position is very influential on generated position defects. However, position defect variation is not very significant when the tool describes a tool path along the x and y axis.

The fact that each leg length defect is considered as a normal distribution gives us an optimistic value for maximum position defect values. To determine the highest

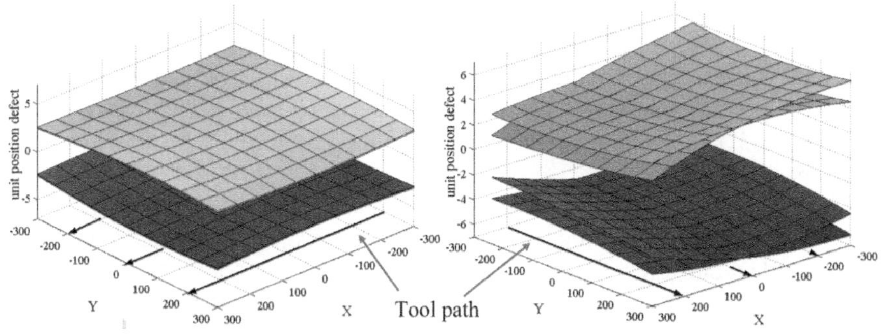

Tool path parallel to the x axis   Tool path parallel to the y axis

Fig. 8. Upper and lower limits of machined surface position defects for length defects represented with normal distribution for $z = 50$.

value of the pose defect for a contour along the x axis or y axis, each leg length defect is represented with a uniform distribution. In this case, the pessimistic values are near 12 if the tool path is along x axis or along y axis.

4.4 Definition of Required Geometrical Calibration Accuracy

The previous maps show that if a position defect of 0.1 mm on the machined profile has to be respected across the whole workspace defined in Figure 8, then the absolute value of each leg length defect has to be under 0.028 mm if the tool path is along the x axis. In the case of a tool path along the y axis, these values have to be under 0.016 mm. Indeed, generated defect evolution is linear for tool tip position and orientation defects.

Thus, if measured leg defects are close to 0.01 mm (minimum leg length is 850 mm), the absolute value of the resulting contoured surface position defect may be near to 0.036 mm along the x axis.

However, the assembly of the structure and the consideration of all 42 parameters decrease calibration accuracy. Let us consider a temperature rise of around 5°C; if the legs are steel they will expand by more than 0.05 mm, which may produce a position defect on the contoured surface along the x axis of only around 0.040 mm, but of around 1.5 mm along the y axis on the boundary of the workspace.

5 Conclusion

The aim of this paper is to define the calibration accuracy which is needed to respect specified machined part quality. A sensitivity analysis is performed to determine the sensitivity matrix and the significant parameters, which are the leg lengths. Then the influence of these parameters on the defect generated on the contoured profile is studied with the consideration that leg length defect variations can be approximated with normal distributions. Moreover, this approach enables the optimisation of the contoured part position and the workspace in order to ensure a minimisation of profile position defect values.

This work is applied to the Verne structure and is illustrated by a flank contouring operation. Then calibration accuracy is defined in order to respect a maximum position defect of 0.1 mm.

This study is a first analysis of a new calibration method which is currently being developed. This calibration method relies on the measurement of a machined part with a defined profile.

Acknowledgements

This paper was written within the framework of the TIMS Research Group, using grants from the Regional Council of Auvergne, the French Ministry of Research, the CNRS and the Cemagref.

This work was carried out within the context of the working group Manufacturing 21 which gathers 11 French research laboratories. The topics approached are:

- the modelling of the manufacturing process,
- the virtual machining,
- the emerging of new manufacturing methods.

The authors would like to warmly thank J.Y. Hascoët from IRCCyN for his help during the analysis of the Verne PKM.

References

1. M. Terrier, A. Dugas, J.Y. Hascoet. Qualification of parallel kinematics machine in high speed milling on free form, *International Journal of Machine Tool & Manufacture* **44**, 2004, 865–877.
2. W. Khalil, E. Dombre. *Modélisation, identification et commande des robots*, Hermès, ISBN 2-7462-0003-1, 1999.
3. P. Renaud, N. Andreff, G. Gogu, M. Dhome. Optimal pose selection for vision-based kinematic calibration of parallel mechanism, in *IEEE/RSJ International Conference on Intelligent Robots and System*, Las Vegas, USA, No. 3, October 2003, pp. 2223–2228.
4. M. Priel. *Les robots industriels: Caractéristique, performances et choix*, Collection AFNOR Technique, ISBN 2-1230-6211-3, 1990.
5. J.P. Merlet. *Parallel Robots*, Kluwer Academic Publishers, ISBN 0-7923-6308-6, 2000.

6. G. Pritschow, C. Eppler, T. Garber. Influence of the dynamic stiffness on the accuracy of PKM, in *3rd Chemnitz Parallel Kinematic Seminar*, Chemnitz, Deutschland, May 2002, pp. 313–333.
7. T. Huang, D.G. Chetwynd, D.J. Whitehouse, J. Wang. A general and novel approach for parameter identification of a 6-DOF parallel kinematic machines, *Mechanism and Machine Theory* **40**, 2005, 219–239.
8. B. Jockiel, J.C. Ziegert, L. Bieg. Uncertainty propagation in calibration of parallel kinematic machines, *Journal of International Societies for Precision Engineering and Nanotechnology* **25**, 2001, 48–55.
9. K.C. Fan, H. Wang, J.W. Zhao, T.H. Chang. Sensitivity analysis of the 3-PRS parallel kinematic spindle platform of a serial-parallel machine tool, *International Journal of Machine Tool & Manufacture* **43**, 2003, 1561–1569.
10. H. Chanal, E. Duc, P. Ray. A study of the impact of machine tool structure on machining processes, *International Journal of Machine Tool & Manufacture*, to be published.
11. M. Terrier, M. Gimenez, J.Y. Hascoet. VERNE – A 5 axis parallel kinematics milling machine, *IMECH International Journal of Engineering Manufacture* **219**, 2005, 327–336.
12. P. Renaud. Apport de la vision pour l'identification géométrique de mécanisme parallèle, PhD Thesis, Université Blaise Pascal, Clermont-Ferrand, France, 2003.
13. C. Lartigue, E. Duc, A. Affouard. Tool path deformation in 5-axis flank milling using envelope surface, *Computer-Aided Design* **35**, 2003, 375–382.

DESIGN AND MANUFACTURING KNOWLEDGE MODELLING AND HANDLING

An Ontology Architecture for Standards Integration and Conformance in Manufacturing

Laurent Deshayes[1], Sebti Foufou[2,*] and Michael Gruninger[3]

[1]*Lami, IFMA, Les Cézeaux, BP 265, 63175 Clermont Ferrand, France;*
E-mail: laurent.deshayes@ifma.fr
[2]*LE2i, UMR CNRS 5158, University of Burgundy, B.P. 47870, Dijon, France;*
E-mail: sfoufou@u-bourgogne.fr
[3]*DPG Group, MSID Division, National Institute of Standards and Technology*[**]*, Mail Stop 8263, 100 Bureau Drive, Gaithersburg, MD 20899, U.S.A.; E-mail: gruning@nist.gov*

Abstract. Standards reflect consensus on the semantics of terms. When used to communicate, whether between people or software systems, standards ensure the communication is correct. Different standards have different semantics for the same terms and express common concepts using different terms and in different ways. Communication between software systems based on different standards is sometimes difficult to achieve. Standards integration concerns the explicit representation of the overlapping sets of concepts in standards and the differences in their semantics to ensure that these standards are used consistently together. This in turn enables software that is based on integrated standards to interoperate, reducing the cost of software integration. Standards conformance determines whether the interpretation of the standardized terms used by software applications is consistent with semantics given by the standards. This paper proposes a general architecture to design ontologies for standards integration and conformance in manufacturing engineering. The ontology architecture is divided into four levels: vendor, standards, domain, and core. Manufacturing turning tools are used as a case study to illustrate the approach. Finally this paper offers some short examples of first order logic propositions.

Key words: product manufacturing, turning tools, standards integration, ontologies, first order logic, ontology architecture.

* Currently guest researcher at MSID Division, NIST, Gaithersburg, MD 20899, U.S.A.; E-mail: sfoufou@cme.nist.gov

** Official contribution of the National Institute of Standards and Technology; not subject to copyright in the United States.

S. Tichkiewitch et al. (eds.), Advances in Integrated Design and Manufacturing in Mechanical Engineering II, 261–276.

1 Introduction

The manufacturing phase is a very important stage of any product lifecycle. To reduce production costs, companies need standardized information exchange that allows one to represent manufacturing processes capabilities as precisely as possible. In addition, knowledge for manufacturing engineering is a very complex set of information resources that generally crosses several engineering domains of competency. Standards are, in some way, consensual models created to reduce as much as possible semantic and inferential mismatches. In manufacturing engineering, most of these standards are decades old and are mainly based on mathematical or empirical models that presume sophisticated interpretation by highly skilled people and may confuse the non expert. To avoid semantic ambiguities across organization(s), most of these standards include natural language definitions and drawings illustrating the relevant information. A formal machine and human interpretable representation of implicit information in standards is intended to avoid defining and creating redundant standards and to add more expressiveness to facilitate their use and interpretation by automated information systems.

Various groups within industry, academia, and government [1] have been developing sharable and reusable knowledge bases known as ontologies [2–4]. The purpose of ontologies within engineering is to make explicit, for a given domain, the knowledge contained in engineering software and in business procedures [5]. In manufacturing, standards are used extensively in all phases of production, design and maintenance. The data, information and knowledge represented with standards need to be sharable across the enterprise. But, despite being written by people with a common background, the definitions and relations within standards have subtle inconsistencies, which may be exacerbated by the use of data from multiple standards or the implementation of the standard within manufacturing information systems. Furthermore, manufacturers frequently define their own extensions to the standard to satisfy specific needs not covered by the standards.

In this paper, we discuss the problems of standards integration and conformance checking for manufacturing applications. A four-level ontology architecture is proposed to deal with this problem, and is illustrated with axioms of standards for turning tools.

2 Problem of Standards Integration and Conformance

According to the American Office of Management and Budget (OMB) Circular A-119 [6], the term "standard", or "technical standard" as cited in the National Technology Transfer and Advancement Act [6], includes all of the following:

- Common and repeated use of rules, conditions, guidelines, or characteristics for products or related processes and production methods, and related management systems practices.

Tool style symbol	Drawing	Process type	Cutting edge angle	Shank type
D	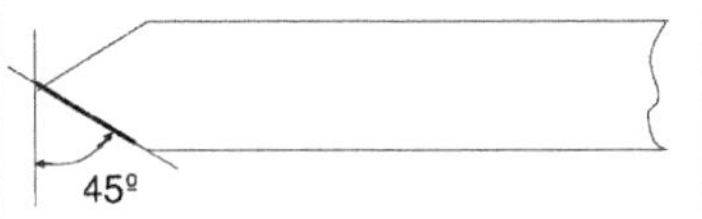	side cutting	45°	Straight shank

(a)

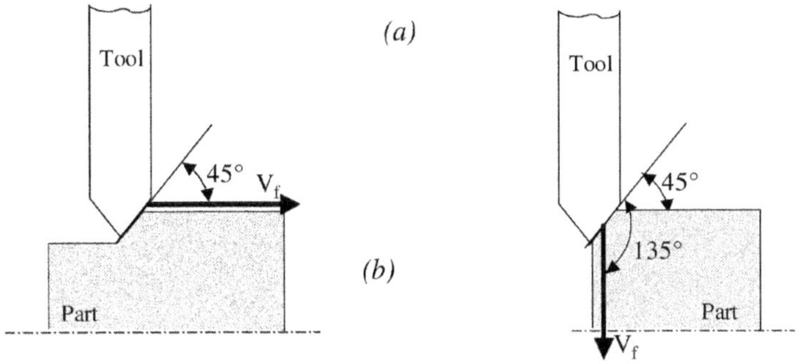

Fig. 1. The semantic conveyed by drawings for tool style defined in the ISO 5608.

- The definition of terms; classification of components; delineation of procedures; specification of dimensions, materials, performance, designs, or operations; measurement of quality and quantity in describing materials, processes, products, systems, services, or practices; test methods and sampling procedures; or descriptions of fit and measurements of size or strength.

Standards are often locally extended, but they still form an important aspect of manufacturing knowledge representation as they reflect consensus on the semantics of terms for a wide variety of industries. But standards do not always present enough specific engineering knowledge, which drives vendors to develop their own extensions of the standards or sometimes their own standards. These user-defined extensions are generally required to fully capture the specific engineering knowledge of a company's intellectual capital. In addition, due to the globalization of markets, companies need new ways of translating their own data to their clients or subcontractors to avoid costly errors at the time of interpreting the encoded information. Ontologies are one way to cover such challenges by providing sharable domain concepts and precise relations between concepts.

From the information technology point of view, standards may be seen as a list of welldefined concepts and relationships used to describe: (i) a particular manufacturing domain (resource, process, etc.) or (ii) information about the product during its manufacturing. Nevertheless implicit information only known to the manufacturing specialists is needed for manufacturing applications to achieve full interoperability and to predict feasibility of design solutions [7].

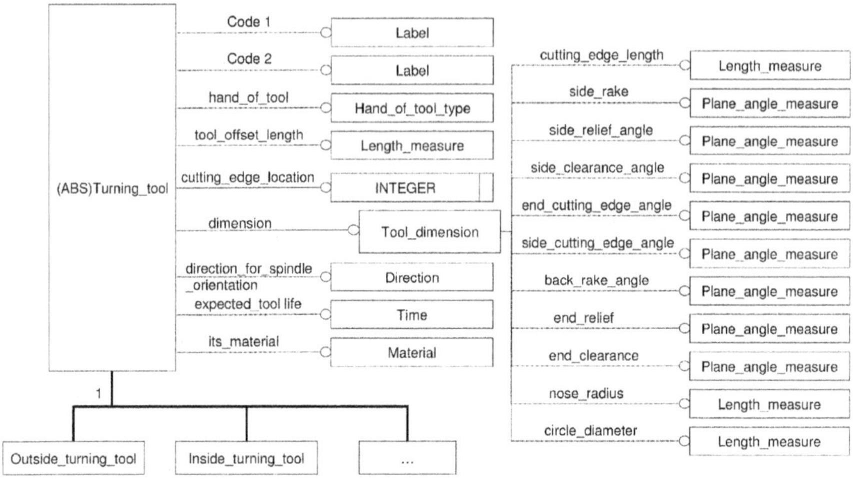

Fig. 2. A cutting tool model in *Express-G* from ISO 14649-121.

Figure 1a shows an example where adding illustrative drawings help to clarify the standard semantics. According to the ISO 5608 standard [8], the tool style is represented by a symbol and defines three characteristics of the tool: the type of cutting process that a tool can perform (Process Type in Figure 1a), the value of the cutting edge angle, and the type of shank. From the figure in column two, one can understand much more easily the meaning of the information. One can even deduce more inferences (using personal interpretation) about the context, or process, in which each tool style may be used, or the kind of machining features that each tool style is able to produce, e.g., the semantics conveyed by a tool style of symbol D is that it can perform side turning kind of operations whose features require an angle less than or equal to 45°, such as illustrated in Figure 1b. It can also be used to perform chamfers which generally require an angle of 45°, see Figure 1c. But no indication is given in the standard about this knowledge. Also, the reference one can use to determine the cutting edge angle value is not given, which may lead to some confusion, e.g., if the feed speed is taken as a reference the cutting edge angle is 45°, in the case of an extern turning operation (Figure 1b), and 135° in the case of a chamfering operation (Figure 1c).

Due to such implicit knowledge contained in standards, recent works done by standardization committees used information technology tools, such as the *Express* language or the *Extensible Markup Language* (XML) [9, 10], to define their concepts. Figure 2, extracted from the ISO/FDIS 14649–21 [11], shows a data structure modelled in *Express-G* [12] for turning tools. Notice that most of the concepts are optional (dashed lines in Figure 2 within the domain of computerized numerical controllers; the only mandatory concept is *cutting_edge_location*. More interesting, *code 1* and *code 2* in Figure 2 are specified (in natural language) as being defined according to ISO 5608 [8] and ISO 1832 [13] respectively, each code is related to the cutting tool dimensions information. However, there is no clearly defined correspondence between the codes, the corresponding standards [8, 13], and the *Tool_dimension* entities of the model. Within other domains, such as cutting process modelling [14], other geometrical aspects of the cutting tool are required. Moreover, the concept of cutting edge angle is represented by two entities: the *end_cutting_edge_angle* and the *side_cutting_edge_angle*. The ISO 3200 [15] defines them respectively as the working cutting edge angle and the working minor cutting edge angle. Here again, slightly different implications, difficult to represent even for the expert, were added to the new standard.

Since standard customizations (or extensions) and duplication of information in standards are unavoidable in practice, it is required to provide a modular and flexible approach to explicitly restrict the relations and definitions of domain-specific concepts. This needs to address two industrial needs for representing implicit knowledge: standards integration and standards conformance. *Standards integration* concerns the explicit representation of the overlapping sets of concepts in standards and the clear characterization of the differences in their semantics to ensure that these standards are used consistently together and with other systems. *Standards conformance* should determine whether the interpretation of the terminology used by software applications is consistent with semantics of the terminology given in standards. In practice both standards integration and conformance need engineering knowledge that over-crosses several manufacturing domains. Finally, mechanisms to incorporate domain knowledge are required to answer queries in a broader manufacturing level. These mechanisms include the representation of: (i) implicit knowledge that may be derived from standards; and (ii) the extensions defined by users of these standards. To answer these queries, a four-level ontology architecture is presented in the next section.

3 Proposed Architecture

Recently, ontologies have been extensively used to represent semantics and knowledges. In particular an ontology makes explicit the knowledge that people often take for granted or as implicit knowledge in a domain. Although there is a wide variety of ontology approaches, all approaches agree that there are two essential components of any ontology: (i) a vocabulary of terms that refer to the things (or concepts) of interest in a given domain; (ii) some specifications of meaning for the terms, grounded in some form of logic. What distinguishes one ontology approach from another is the way the relationships among terms are specified. An ontology supports representation of a very rich variety of structural and non-structural relations such as generalization, inheritance, aggregation and instantiation. It can supply a precise model for software applications. Ontologies can be represented using a

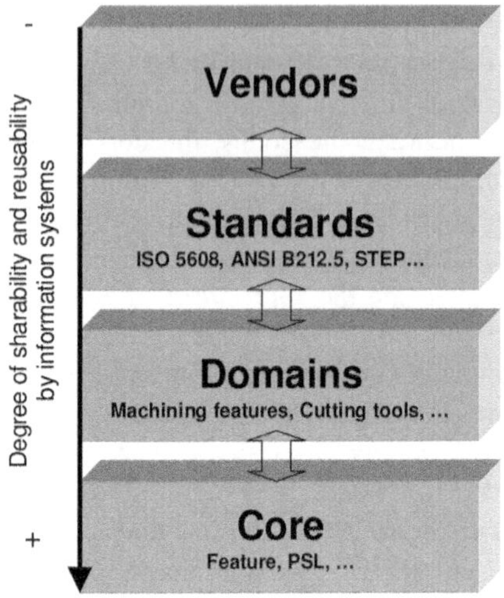

Fig. 3. The proposed fourlevel ontology.

wide variety of logical languages which are understandable both by human beings and machines [16], such as propositional logic, first order logic and semantic web languages. Propositional logic lacks the expressive power to model concisely an environment with many objects and facts. First Order Logic (FOL) has much more expressivity and can represent much more complex relations between objects [16]. The Ontology Web Language (OWL) is the language widely used by the semantic web community [16–18]. In comparison to FOL, OWL is less expressive and based on a taxonomic model, which remains the core for many ontology tools. In this paper, due to the subtleties of concepts' semantic of the proposed ontology architecture, we specify the ontologies using the knowledge interchange format [19, 20], which is a first order language designed to support the interchange of knowledge among heterogeneous computer systems.

Nelson and Schneider showed that product modelling architectures require ontology and interoperability standards to provide better functionalities to PLM systems within current commercial applications [21]. In [1] it was shown that enhanced ontology architectures will soon be required. Rather than developing large monolithic ontologies, complex systems will be supported by ontologies that can be decomposed into modules. This is particularly relevant for the problem of standards integration and conformance for manufacturing resources and processes. Modularization makes it easier to update the ontologies, since additional extensions are added without altering the existing ontology architecture. Moreover, by allowing different modules to provide alternative definitions, a modular organization makes differences in the semantics of overlapping concepts more explicit.

In Figure 3, we propose a simple architecture to overcome all these specifications for standards integration and conformance. It is divided into four levels; each level consists of different sets of ontologies. From bottom to top, ontologies of a given level are used to integrate those of the immediate upper level. This structure imposes rather strong requirements on the representational adequacy of the ontologies – they must not only be strong enough to capture the intended semantics of the terminology, but they must also be expressive enough to allow the generation of semantics-preserved mappings between the less generic ontologies.

Vendor ontologies are developed and used by providers of business applications or manufacturing resources. They are very specific to the provider's business activities. It is often the case that new terminologies are created by private companies and included with their products, such as the component technology called COM developed by Microsoft and largely used in the visual basic programming community.

Standards ontologies are created by consortia to standardize applications and provide uniform interfaces. Since a number of companies are involved in the development of these standards, terms and concepts are more generic than those defined at the vendor level. Nevertheless, they are the first level that presents some "common" agreement on the usage, definition or interpretation of terms or manufacturing systems.

Domain ontologies are used to check consistency between standards ontologies. In particular, standards are often developed for overlapping domains, which can lead to conflicting definitions for the shared terms; even if the terms used in different standards are closely related, subtle consequences can make it difficult for the terminology to be reusable and consistently interpretable. In addition, domain ontologies are used to provide rigorous definitions for the implicit assumptions made within the standards' ontologies. For example, many standards include visual drawings as part of their specification, but do not provide an explicit axiomatization of the intuitions behind these drawings; domain ontologies are used to provide such axiomatizations, which can then be used for automated reasoning such as conformance checking.

Core ontologies are the most generic concepts that cross multiple domains. They cover concepts such as process, product, resource, and geometry. In addition to providing formal specifications of the semantics of these generic concepts, core ontologies are also designed to maximize sharability and reusability, and hence do not make any ontological commitments that are not shared by all related domain ontologies.

4 Case Study

Cutting tools are important resources of manufacturing processes since they affect the quality and the cost of manufactured parts. Due to the proliferation of cutting tools for mechanical machining, standardization has been required for several decades. Today, even if the use and the codification of cutting tools is well understood

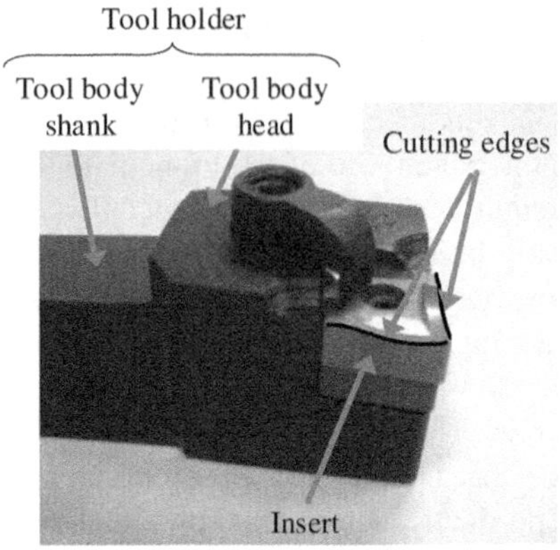

Fig. 4. An insertable turning tool.

and standardized, there are still difficulties to really fully encapsulate in a consensual way all explicit and implicit concepts used to describe these resources. These difficulties are particularly important when terminologies are exchanged between worldwide applications. In this section, the four-level architecture introduced in Section 3 is illustrated with standardized concepts used for turning tools encoding. Turning tools are a subclass of cutting tools used to machine rotational features on a class of machine tools known as lathe centers [11]. This study focuses only on insertable turning tools, although the approach can be generalized to other classes of manufacturing resources and processes. As showed in Figure 4, the cutting edge of an insertable turning tool is held by an interchangeable insert. An insertable turning tool is generally composed of two assembled parts: the tool body head, which holds the insert, and the tool body shank, which mounts the turning tool to the tool block.

The two standards considered in this study, ISO 5608 [8] and ISO 1832 [13], define codes and associated semantics for the turning tool holders and inserts respectively. The role of these codes is (i) to classify those resources; and (ii) to represent concisely their properties. Figure 5 shows an example of three codes, one for an insert and two for a tool holder. The insert code and the first tool holder code are based on their respective international standards, while the second tool holder code is obtained from a tool provider catalogue; the two tool holders have a similar geometry. Standardized as well as vendor specific codes are composed of compulsory and optional symbols. Compulsory symbols are the first height symbols for inserts, and the first nine symbols for tool holders. For the entire code to be syntactically valid and conform to the standard, its symbols have: (i) to be placed following a specific order; (ii) to respect a precise syntax; and (iii) to correspond to a specific semantics. Each symbol represents particular properties (dimensional and non dimensional) of the resource and has a semantic meaning associated with it. More descriptions of tool

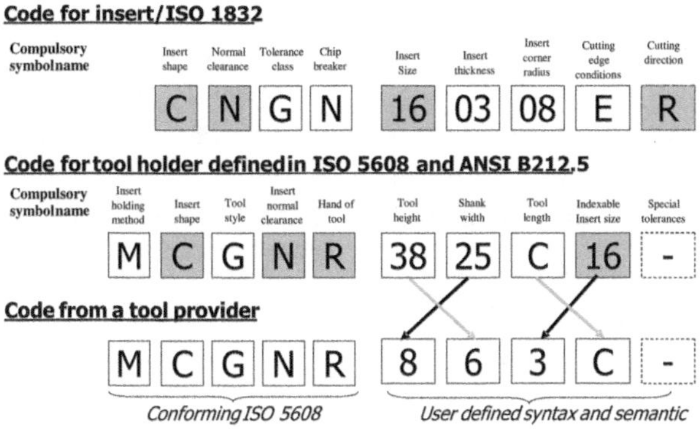

Fig. 5. Standardized insert and tool holder codes as well as differences with an industrial tool holder's code.

holders and insert properties are given in standards [8, 13], specialized handbooks [21], or tool manufacturer catalogues.

Subtle differences exist between the standardized code and the one provided by the tool provider. First, there is a difference on the dimension system, the ISO and ANSI standards use the metric, while the tool provider use the British system (inches). Therefore for a tool provider, the sixth and seventh symbols are described as: "*The sixth and seventh symbol shall be a significant two-digit number that indicates the holder cross section. For shanks 5/8" square and over, the number will represent the number of sixteenth of width and height. For shanks less than 5/8" square, the number of sixteenths of cross section will be preceded by a zero. For rectangular holder ...*", while for the metric standard we will find that these symbols are respectively the tool height and the shank cross section, see [8] for more details. In this example, not only the dimensional system is confusing, but the meaning of the symbols too. In some cases, this meaning seems to indicate that these two symbols were inverted. Concerning the eighth and ninth symbols, they are effectively inverted by the tool provider. Another aspect of these codes consists in determining whether a given insert can be mounted on a particular tool holder. In order to verify this relation, correspondences between the two codes are used. For the example of Figure 5, the encoded tool holders are mountable with the encoded inserts because symbols C, N, 16 and R (optional) in the tool holder's code correspond respectively to the same symbols in the code identifying the insert (dark symbols). These relations can be modelled easily using today's database systems. However, information such as the cutting processes that can be performed or the exact tool geometry is not explicitly represented. In fact, to be more accurate, such information as well as the symbols meaning is generally encoded in the application which does not facilitate systems interoperability.

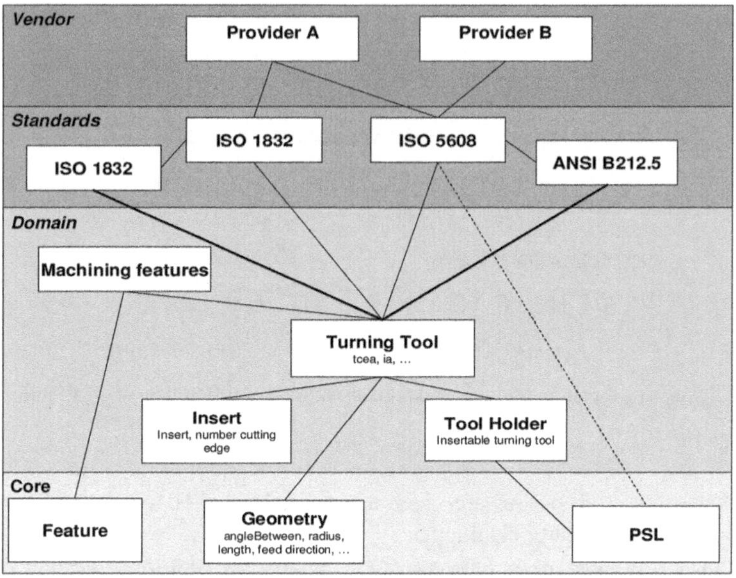

Fig. 6. The four-level architecture for cutting tool ontology.

The four-level ontology approach for the turning tool domain is illustrated in Figure 6. The top level concerns tool provider catalogues and concepts. The specific concepts used by tool vendors (Providers A and B) are defined in relation to the standardized concepts described at the standards ontology level. ISO 1832 [13] for inserts and ISO 5608 [8] for tool holders are internationally standardized concepts that may be encoded at this level. Notice that regional standards are also part of this level, such as the ANSI B212.4 and ANSI B212.5 that are the equivalent, in the United States, to the two previous ISO standards [22]. For standards integration purposes, i.e., for the standardized concepts to be more interpretable by other information systems, the domain level ontology is required to define the concepts and relationships that are generic to a wider manufacturing domain, such as turning tools, tool holders, inserts and machining features (features produced by a turning tool). Finally core ontologies include the most generic and interchangeable concepts. The design of well defined core concepts is a very intensive iterative process that may reach to standardize them, such as the PSL ontology [23].

The transition from vendor to standards ontology can be considered as a requirement for integrating specific knowledge with software applications. As a fact, it cannot be imagined that each company would build its own ontological theory without using existing standards. Therefore the mapping from vendor ontologies to the standards ontology may strongly reduce the amount of concepts to consensually define. Vendors may avoid the standards ontology level, but they risk introducing a vendor-

The third symbol of the tool holder code defined in the standard ISO 5608 is the tool style.

```
(1)        (forall (?x ?y) (implies (ISO_5608_3 ?x ?y)
                  (and   (toolStyle ?x ?y) (toolHolder ?x))))
```

Tool style is a property of a tool holder and includes the information about the tool cutting edge angle, shank type and cutting processes which, can be performed with the tool style.

```
(2)     (forall (?w) (implies (ISO_5608_3 ?x ?y)
             (exists ( ?u ?v ?w)
             (and  (toolHolder_tcea ?x ?u) (shankType ?x ?v) (cuttingProcess ?x ?w)))))
```

For all tool holder code ?x the third symbol ?y for tool holder implies that the tool holder requires performing a side cutting activity or a end cutting activity or both.

```
(3)     (forall (?x ?y) (implies (ISO_5608_3 ?x ?y)
                  (or (requires sideCutting ?x) (requires endCutting ?x)
                       (and (requires sideCutting ?x) (requires endCutting ?x)))))
```

Fig. 7. Axioms for ISO 5608 ontology.

specific ontology into the domain level. The transition with the standards ontology level guarantees that vendor specific knowledge does not cross the levels without being discussed by a wider community. In the remainder of the paper the next three levels (from standards to core ontologies) and their transitions are analyzed.

In the ISO 5608 standard, the tool style includes information about tool cutting edge angle, shank type, and the associated cutting processes (see Figure 1). Some relevant axioms for this concept are given in Figure 7. Notice that a full study case is out of the scope of this paper. The predicate ISO_5608_3 represents the third symbol of the standardized code described in ISO 5608 and is defined in axiom (1) using the knowledge interchange format. Axiom (2) specifies the properties of *toolStyle*. The *toolHolder_tcea* predicate represents the tool cutting edge angle. Axiom (3) specifies the values that instances of *shankType* and *cuttingProcess* can have. This is also particularly useful for standard conformance purposes. The predicate *requires* is part of the PSL ontology [23]. Therefore the *cuttingProcess* concept is assumed to be a PSL activity. The use of PSL in describing cutting tool knowledge is particularly interesting to provide additional inferences on processes and improve standard integration. The use of such generic concepts with the domain ontology is detailed below. Axioms such as axiom (3) are particularly useful for standards conformance checking, to determine whether the tool style given by a tool provider conforms to ISO 5608. The tool provider can also define new axioms to describe its own specific knowledge and make it more interpretable by expert agents.

More generic ontology modules for the domain ontology level may be defined. The turning tool geometric concepts used in this example are represented in Figure 8. These concepts, selected from the turning tool standards [8, 13, 15], are: the turning tool holder; the indexable insert; the tool cutting edge; the tool minor cutting edge;

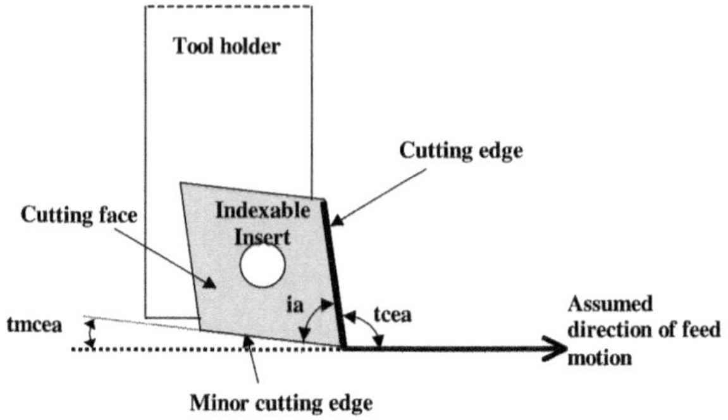

Fig. 8. Characteristics of turning tool geometry.

An insertable turning tool is the assembly (mounting) of an indexable insert and of a tool holder

(4) (forall (?x ?y) (implies (turningTool ?x ?y)
(and (toolHolder ?x) (insert ?y) (mounting ?x ?y))))

The tool cutting edge angle, tcea, is the angle included between the cutting tool edge and the feed rate direction.

(5) (forall (?x ?y ?w) (iff (tcea ?x ?y ?w)
(exists (?v ?z)
(and (turningTool ?x ?y) (cuttingEdge ?x ?y ?z)
(feedDirection ?x ?y ?v) (angleBetween ?z ?v ?w)))))

The included angle is the angle included between the cutting tool edge and the secondary tool edge.

(6) (forall (?x ?y ?w) (iff (ia ?x ?y ?w)
(exists (?z ?v)
(and (turningTool ?x ?y) (cuttingEdge ?x ?y ?z)
(minorCuttingEdge ?x ?y ?v) (angleBetween ?v ?z ?w)))))

The tool minor cutting edge angle, tmca, is the angle included between the secondary cutting tool edge and the feed rate direction.

(7) (forall (?x ?v ?y) (iff (tmcea ?x ?y ?v)
(exists (?w ?z)
(and (turningTool ?x ?y) (minorCuttingEdge ?x ?y ?w)
(feedDirection ?x ?y ?z) (angleBetween ?z ?w ?v)))))

Fig. 9. Axioms for integrating turning tools with process information using PSL.

the tool cutting edge angle, denoted by *tcea*; the tool included angle, denoted by ia; and the tool minor cutting edge angle, denoted by *tmcea*. The direction of feed motion (or the working plan) is a concept used in defining the angles and cutting edges.

Figure 9 shows part of an ontology used to describe turning tool components and geometric concepts. Axiom (4) specifies that a turning tool is composed of an insert and a tool holder. The predicates *insert* and *toolHolder* are primitives within

A tool holder is either a EndToolHolder, a SideToolHolder, or both

```
(8)        (forall (?x) (iff  (toolHolder ?x)
                              (or (EndToolHolder ?x ) (SideToolHolder ?x )
                                 (and (EndToolHolder ?x ) (SideToolHolder ?x )))))
```

An external turning activity occurs if and only if there exists a side tool holder or an end and side tool holder within the activity.

```
(9)        (forall (?x) (iff  (and (externTurning ?x ) (activity_occurence ?x))
                              (exists (?y) (9)
                                  (or (SideToolHolder ?y)
                                      (and (EndToolHolder ?x ) (SideToolHolder ?x )))))
```

A tool holder is a reusable resource within a turning process activity

```
(10)       (forall (?x ?y) (implies  (and (toolHolder ?x) (turningProcess ?y))
                                     (reusable ?x ?y)))
```

An insert is a wearable resource within a turning process activity

```
(11)       (forall (?x ?y) (implies  (and (Insert ?x) (11) (turningProcess ?y))
                                     (wearable ?x ?y)))
```

Fig. 10. Defining tool holders concepts using a core ontology defined in PSL.

the turning tool ontology. They will be defined in their respective domain ontology. Axioms (5) to (7) respectively define *tcea, ia*, and *tmcea*, which are the three major geometric angles considered for describing turning tools (see figure). Notice that in these axioms predicates are primitives provided by domain and core ontologies. For example, predicates *angleBetween* and *feedDirection*, which represent the value of an angle included between two lines and the assumed direction of feed motion, have their definition provided in the geometry ontology.

The predicate *cuttingEdge*, which represents the tool cutting edge, may be a primitive of the turning tool ontology or defined in the insert ontology, depending on the context.

Figures 7 and 9 are representative of standards and domain ontologies. The transitions and the mapping between these two levels and the core ontology serve different purposes. Domain ontologies provide better reusability and consistency between the overlapping concepts and implicit assumptions made between standards. It reduces the number of translations from standards to core ontologies. For example both ANSI B212.5 and ISO 5608 concepts may be mapped to the turning tool domain ontology (axioms (4) to (7)), and the use of the tool holder domain ontology ensures stronger integrability of these two standards with the core ontology. In fact, the use of domain ontologies to make such transition provides a rigorous structuring and guarantees better completeness of the overall ontology. Some core concepts may be used at the different levels to facilitate the integration of specific concepts but at the risk of making the integration of standards more complicated. As an example, the turning tools domain ontology may use the *require* concept from PSL to indicate that a turning process requires a particular turning tool [24, 25]. The dashed line in Figure 6 represents this relationship. Indirect definition of this relationship is also given through the domain level ontologies with the links from standards to turning tools to tool holders to PSL. In this case, more properties (e.g. the machining feature, the turning tool, the tool holder and the insert) of the process maybe inferred. Oppositely, a direct relation (and modelling) between the ISO 5608 standards ontology level and PSL core ontology would impede such knowledge reasoning and discovery.

The inferences that can be done in the indirect representation are illustrated using the axioms of Figure 10, which are defined for the tool holder ontology using a core ontology defined in PSL. From Axiom (9), it may be inferred that if a process planner defines an external turning operation (i.e., an activity called external turning) then either a side tool holder or an end and side tool holder must be used to accomplish this operation. In addition, if an insertable turning tool is used, then it is automatically inferred that a tool holder and an insert are required (axiom (4)). Thanks to PSL theory it is also inferred that a tool holder does not need to be changed between activity occurrences, since axiom (10) defines a tool holder as a reusable resource for a turning process activity. Finally, using axiom (11) it can be inferred that the insert will need to be changed after some activity occurred. These reasoning can be mapped with standards terminology when required using the mapping between standards and the domain ontology level. In the first scenario it would be only inferred that *endCutting* is an activity; the tool holder needs to be used in an activity called *endCutting*, but nothing is deduced from the activity itself and the relations between the concepts. In addition, because of important number of standards, the use of core concepts at this level needs to be reduced in order to avoid replications of inferences.

5 Conclusion

In manufacturing, standards represent consensual definitions of the concepts within a particular domain; consequently organizations' specific knowledge is not represented, which leads manufacturing actors (such as tool providers) to define their own extensions of the standards. Semantic conflicts may appear during the mapping of terms between these extensions and the standards. The next generation of manufacturing systems needs more rigorous foundations of semantics than what is currently provided by data models and architectures. The four-level ontology architecture proposed in this paper for manufacturing resources standards integration and conformance checking may provide such a foundation. The generic core ontologies of the lower level enable full integration between the proposed architecture and other existing applications. In addition, due to the modularity of the proposed approach concepts from the different levels may be used, which facilitates the integration of new concepts. The domain ontology level is certainly the most important level since it can be used as a base to build more effective core ontology by categorizing and classifying manufacturing concepts and relationship in a coherent modular architecture.

Of course, more efforts have to be done by manufacturers and application developers to provide such architecture and the consensus required by each level. The next step of this work is to develop a full application of the proposed architecture and validate it on a complete manufacturing scenario.

Disclaimer

No approval or endorsement of any commercial product by the National Institute of Standards and Technology is intended or implied. Certain commercial equipments, instruments, or materials are identified in this report in order to facilitate better understanding. Such identification does not imply recommendations or endorsement by the National Institute of Standards and Technology, nor does it imply the materials or equipment identified are necessarily the best available for the purpose.

References

1. Ray, S., Interoperability standards in the semantic web, *Journal of Computing and Information Science in Engineering* **2**, 2002, 65–69.
2. Gruber, T.R., Towards principles for the design of ontologies used for knowledge sharing, Padova, Italy, 1993.
3. Ciocoiu, M., Gruninger, M., and Nau, D., Ontologies for integrating engineering applications, *Journal of Computing and Information Science in Engineering* **1**, 2001, 45–60.
4. Antoniou, G. and van Harmelen, F., *A Semantic Web Primer*, The MIT Press, Cambridge, Massachusetts, 2004, pp. 109–146.
5. Uschold, M. and Gruninger, M., Ontologies: Principles, methods, and applications, *Knowledge Engineering Review*, **11**(96), 1996, 137.
6. American Office of Management and Budget. http://standards.gov/standards_gov/, 2005.
7. Zhao, Y., Ridgway, K., and Al-Ahmari, A.M.A., Integration of CAD and a cutting tool selection system, *Computers & Industrial Engineering* **42**, 2002, 17–34.
8. ISO 5608:1995, Turning and copying tool holders and cartridges for indexable inserts – Designation, 1995.
9. ISO 10303-11, Industrial automation systems and integration – Product data representation and exchange – Part 11. Description methods: The EXPRESS language reference manual, 1994.
10. Extensible Markup Language (XML). http://www.w3.org/XML/, 2005.
11. ISO/FDIS 14649-12, Industrial automation systems and integration – physical device control – Data model for computerized Numerical Controllers – Part 12: Process data for turning, 2005.
12. Schenck, D. and Wilson, P.R., *Information Modeling: The EXPRESS Way*, Oxford University Press, New York, 1994.
13. ISO 1832:1991, Indexable Inserts for cutting tools – Designation, 1991.
14. Tlusty, G., *Manufacturing Processes and Equipment*, Prentice Hall, Englewood Cliffs, NJ, 1999, pp. 415–610.

15. ISO 3002-1:1982, Basic quantities in cutting and grinding – Part 1: Geometry of the active part of cutting tools – General terms, reference systems, tool and working angles, chip breakers, 1982.
16. Russell, S. and Norvig, P., *Artificial Intelligence, A Modern Approach*, 2nd ed., Prentice Hall, Englewood Cliffs, NJ, 2003, pp. 193–319.
17. Daconta, M.C., Obrst, L.J., and Smith, K.T., *The Semantic Web*, Wiley Publishing, 2003.
18. Web Ontology Language (OWL). http://www.w3.org/2004/OWL/, 2005.
19. Menzel, C. and Hayes, P., *SCL: A Logic Standard for Semantic Integration*, Sanibel Island, 2003.
20. Genesereth, M.R. and Fikes, R.E., Knowledge Interchange Format, Version 3.0, Reference Manual, Computer Science Department, Stanford University, Logic Group Report Logic-92-1, Stanford, California, 1992.
21. Nelson, D.H. and Schneider, G.J., *Applied Manufacturing Process Planning: With Emphasis on Metal Forming and Machining*, Prentice Hall, Englewood Cliffs, NJ, 2001.
22. ANSI B212.4 B2212.5. http://www.ccpa.org/pubs.html. 2005.
23. Gruninger, M. and Menzel, C., Process Specification Language: Principles and Applications, *AI Magazine* **24**, 2003, 63–74.
24. Process Specification language (PSL). http://www.nist.gov/psl, 2005.
25. Deshayes, L., El Beqqali, O., and Bouras, A., The use of Process Specification Language for cutting processes, *International Journal of Product Development* **2**(3), 2005, 236–253.

Knowledge Loss in Design Reviews

Gregory Huet[1], Christopher A. McMahon[2], Florence Sellini[3], Stephen J. Culley[2] and Clément Fortin[1]

[1]*École Polytechnique de Montréal, 2500 Chemin de Polytechnique, Montréal (QC), Canada H3T 1J4; E-mail: {gregory.huet, clement,fortin}@polymtl.ca*
[2]*University of Bath, Claverton Down, Bath BA2 7AY, UK; E-mail: {enscam, enssjc}@bath.ac.uk*
[3]*Airbus UK, Filton, Bristol BS99 7AR, UK; E-mail: florence.sellini@airbus.com*

Abstract. In a design project, a great deal of information about the aspects of the product being developed and the progress of the project is exchanged during meetings. However, reports which constitute the formal records of meetings are often limited in the extent to which they capture the information exchange. New recording and facilitation technologies for meetings are improving but further work is needed to enable the efficient capture of knowledge and experience from the discourse. The work reported in this paper is part of a project, carried out jointly with Airbus UK, exploring how design experience and rationale can be captured from the discourse of design reviews. In particular, this paper will present a methodology to evaluate the knowledge lost from the formal record of design reviews based on a comparison between transcripts and official minutes. The interpretation of the results is based on a unique information mapping technique. It provides the research team with an essential visual tool reflecting the loss of useful design knowledge for each design issue identified during the meeting. All findings are based on an industrial case study and a survey on the requirements of practicing engineers in leading aerospace companies.

Key words: collaborative design, knowledge loss, aerospace design reviews, meeting minutes.

1 Introduction

Unstructured information is present in a wide variety of design situations, for example emails, face-to-face meetings, design reviews, and discussions with suppliers over the telephone. It is during these transactions that information flows, decisions are made and design rationale or intent is established. The storage and archiving of these transactions is increasingly important and has to be considered as a major issue in the improvement of information services for engineers. From these observations,

S. Tichkiewitch et al. (eds.), Advances in Integrated Design and Manufacturing in Mechanical Engineering II, 277–291.

the development of a transaction monitoring methodology, which would assist in extracting and storing the essential knowledge from these transactions, appears as an important research issue in design knowledge management. The "Design Transaction Monitoring" (DTM) research project has focused on a specific and critical situation in aerospace design activities: reviews. This work has led industrialists from Airbus UK, and researchers from the University of Bath (UK) and the École Polytechnique de Montréal (Canada) to conduct a series of case studies based on both academic and industrial projects. These studies have shown that minutes, the central document of formal records of meetings, are often limited in the extent to which they capture the information exchange that has taken place. New means are needed to capture the essential design knowledge, experience and expertise shared.

This paper reports on a systematic methodology for the evaluation of knowledge lost from the formal record in design reviews. Within the *context of aerospace design reviews*, the approach is supported by a survey of engineers from aerospace companies to assess the role of minutes in the design process. *Evaluating information loss* has previously been applied to different fields of interest, and inspired the *information mapping* technique reported in this paper; it involves mapping out decisions, rationale, and lessons learnt[1] elements, based on the comparison of two critical documents: a transcript of the meeting discourse recorded on tape and the minutes of the meeting. The interpretation of the results has helped to understand and quantify the actual *knowledge loss in minutes* of meetings. The *concluding remarks* outline fundamental issues and requirements for the effective capture of design knowledge during meetings.

2 The Context of Aerospace Design Reviews

In the specific case of this research, the monitoring of aerospace designers almost invariably gives insights on adaptive design activities held by widely distributed and multinational teams. This section will review the key collaborative aspects which aerospace companies need to master: efficient communication and a rigorous control of the engineering activities. Finally, the results of a survey on minute taking in the aerospace industry will highlight current practices for recording engineering meetings.

2.1 Communication Processes in Design

An important aspect of design communication processes is the relationship between verbal communication and sketching. It is during a specific stage in the design process, the conceptual phase, where verbal communication between design engineers

[1] For the purpose of this research, lessons learnt elements cover the problem elicitation, the story, and the recommendations related to a past company experience.

linked to sketches or diagrams is more prominent. In the design context, discourse has therefore, overall, a broader meaning than one might expect. According to Yen [1], the term implies a "verbal interchange of ideas"; this expands the notion beyond a verbal and textual exchange to include other modes of communication like sketching. Nevertheless, in the work presented here, it is the verbal discourse between participants which is at the centre of the monitoring activities.

2.2 Design Reviews

A great deal of information about the design process and the product is exchanged during meetings. Background work on the topic [2] has led to the characterization of the concept of meeting and the definition of a Transcript Coding Scheme for marking up and encoding the meeting transcripts. To give a more formal view of what can be seen as a design meeting, the following generic definition is used for the purposes of the DTM research:

Design meeting: *a set of communication processes which take place in a synchronous or asynchronous[2] manner over issues linked to spontaneous or predetermined topics. A design meeting aims at achieving general agreement over design issues by spreading information between participants with the support of specific artefacts.*

New technologies, especially the evolution of IT networks, have expanded the traditional definition of a meeting as a face-to-face situation to a temporally and physically distributed situation where participants do not need to be all at the same place at the same time. Nowadays, the word "meeting" represents a fascinating variety of situations in terms of number of participants, level of structure in the information exchanged and technologies used to communicate.

However, this research focuses on a particular type of meeting, design reviews, which are key milestones in the design process of complex products. Design reviews, from a business point of view, can be seen as Stage-Gates. This recalls the Stage-Gate process defined by Cooper [3], where a gate is a go/no go decision point which divides the product development process in discrete stages. A generic Stage-Gate process is illustrated in Figure 1.

Since the implementation of such management strategies, companies have further expanded the Stage-Gate process by inserting milestones or gates within their product development phases to increase project monitoring (i.e. Gate 4 in Figure 1). The breakdown of the process into smaller phases of activity is used to bring the risk associated with the product development down to a minimum. Of course, the number of phases will vary from a company to another.

[2] Synchronous communication is communication in which messages are exchanged during the same time interval (e.g. instant messaging). Asynchronous communication is communication in which messages are exchanged during different time intervals (e.g. e-mail).

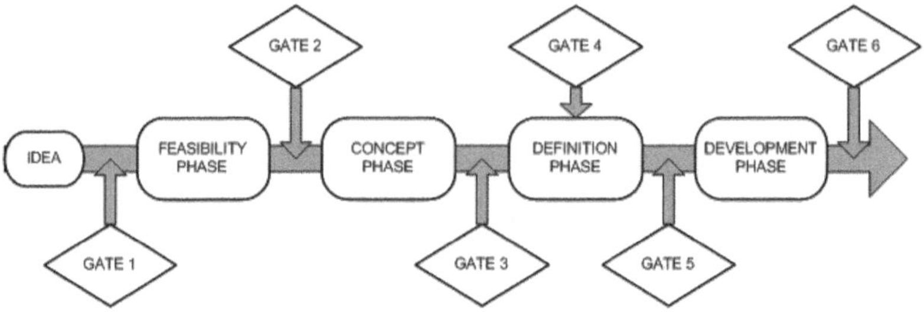

Fig. 1. A generic representation of the Stage-Gate process [3] similar to the aircraft development phases in the aerospace industry.

Cooper [3] defines a gate as a go/no go decision point:

> Gates are predefined and specify a set of deliverables, a list of criteria, both must meet (which are mandatory) and should meet (which capture desirable characteristics), and an output (for example, a decision and approved path forward).

Design reviews or gates held during the design phases of the product development are formal meetings attended by a majority of the stakeholders involved in the project. These meetings are highly structured and follow precise company guidelines imposed by international standards [4]. The definition of a design review taken from the Canadian Standards Association [4] is as follows:

> (Formal) design review: a formal and independent examination of an existing or proposed design for the purpose of detection and remedy of deficiencies in the requirements and design which could affect such things as reliability performance, maintainability performance, maintenance support performance requirements, fitness for the purpose and the identification of potential improvements.

In the aerospace product development phases where most design effort is deployed i.e. the concept phase and the definition phase (Figure 1), four typical design reviews divide the design process: the requirements review (Figure 1, Gate 2), the concept review (Figure 1, Gate 3), the preliminary design review (Figure 1, Gate 4) and the critical design review (Figure 1, Gate 5).

Although design reviews are primarily a control process, most companies also acknowledge the event as an opportunity for all involved parties to share information about the product and process. This implies that the creative input of a design review is not as secondary as it may seem. It is a place where key design decisions and their rationale are made explicit.

2.3 The Minutes Survey

A questionnaire on the role of minutes in the design process was circulated in various aerospace companies and suppliers in 2005. So far, over 50 engineers replied to a set of 17 questions. It is not the aim of this paper to outline in detail the entire results of the survey and the focus will therefore be made on certain relevant issues for the discussion of the knowledge loss data.

Of particular interest to the work presented here are the engineers' views on the purpose of minute taking. Answering the question: "which section of the minutes do you refer to most for the purpose of your work?", a clear result showed that over 96% of the respondents' first choice was the "list of actions" followed by the "list of decisions" for 65% and in third place came the "list of attendees" with 35%. When asked to choose from a list of 6 possible choices to define the purpose of keeping minutes: 96% see minutes as a "formal reminder of actions to take", 59% as a "written proof of the project's progress" and 40% as an "input/output of the design process". Strangely, only 31% of the engineers think minutes are kept for "legal purposes" when aerospace standards [5] clearly outline the legal implications of minute taking for companies working in the trade.

Another interesting aspect tracked by the survey is the training followed by engineers for minute taking. Seventy-eight percent of the respondents were not formally taught to take minutes during their engineering studies. This might explain why further down the line, they do not seem to fully acknowledge the importance of the secretary involved in recording an engineering meeting; only 51% valued the statement "it is important that minutes are taken by an engineer working on the project" as true or mostly true. From a communication process point of view, it is essential that meeting participants use common references to exchange and understand the information efficiently. It therefore also applicable to the person taking the minutes as it will undoubtedly improve the quality of the record. A secretary with a good understanding of the design issues discussed during the review coupled with training in the production of minutes should most probably be a priority in the list of requirements for the efficient capture of design rationale from reviews.

The results of the survey are unequivocal: engineers learn to take minutes by experience and only truly value the actions list, the practical side of traditional minute taking.

3 Evaluating Information Loss

This paper is about measuring the information and subsequent company knowledge lost from the records of design reviews. The study compares textual information elements from two documents (the minutes and the transcript) relating to an event of a formal but discursive nature. The minutes are the historical record of a meeting. They typically summarise the actions and decisions taken during the meeting. The transcript of a meeting is a verbatim record of all the interventions made during the event. A transcript is a more accurate record but it follows the unstructured and unpredictable nature of the verbal discourse.

The approach presented here is based on the occurrence of specific design knowledge elements in each document. Decisions, rationale, actions and lessons learnt are the four concepts tracked in the information mapping technique presented in Section 4. In general, the authors noticed that actions and decisions in meetings were usually well recorded but often the rationale and experience leading to these were not captured in the minutes. Identification of the extent of this failure is an important stepping stone in the improvement of meeting capture technologies for knowledge re-use.

3.1 Design Rationale and Experience

The need for knowledge capture and re-use in aerospace design has led to a strong emphasis on knowledge management [6]. Design rationale and experience are two of the most important knowledge elements to be captured by knowledge management systems in the engineering domain. Design rationale in its most general sense is an explanation of why an artifact is designed the way it is [7]. Moran and Carroll edited an excellent reference collection [8] where many of the issues in the capture, representation and use of rationale have been outlined. As defined by Marsh [9], the experience tracked in this study is of a corporate nature as opposed to personal experience and will be referred to throughout as lessons learnt to avoid any confusion with the notion of expertise.

The tracking of rationale and decisions in the transcript was facilitated by the Transcript Coding Scheme [2] developed for the purpose of the research; this methodology helps to monitor the types of interventions made, the role of the exchanges between participants and the topics discussed.

3.2 Related Work

The concepts (actions, decisions, rationale and lessons learnt) that this study is seeking to track through the minutes and transcripts of the design reviews are made explicit in the documents through textual information elements. The knowledge loss evaluation is then based on a measure of the occurrence and connection between predefined information entities.

The approach that has been taken is founded on Derr's [10] concept of information in ordinary discourse:

> An abstract, meaningful representation of determinations made of objects. Furthermore, it has been concluded that information has derivative properties which enable it to communicate, inform, empower and to exist in some quantity.

Information is therefore quantifiable and different approaches to the measure of information have been encountered in the literature. In [11], quantitative measures of textual information were studied based on the linguistic structure of sentences in medical documents. The results of the analyses were communicated using a form of tree diagram, based on an account of informational operators linking together the various concepts in a sentence. Although this work aimed at defining and measuring information properties of textual material, the graphical approach which was used seemed of great clarity and value.

The idea of an information mapping technique to measure the knowledge loss was effectively inspired from the work carried out by Hoffmann [12]. His quest for a suitable definition of the term information led to a comparative study of textual documents and their abstracts. Analysis of the material was based on a graphical representation of the information content. Statements were highlighted in the text and numbered and then mapped out using conceptual interrelationships found in the original text. Measures produced were based on the value of the facts or statements depicted as nodes in the diagrams. The value of a node was given by the number of conceptual connections it possessed (i.e. the higher the number of links to a node the more important the node was). Therefore nodes with an important value should be found in the abstract of the main document.

This last point is very interesting for the analysis of the information maps presented in Section 4, but it is not the intention of this research to evaluate the whole information content of the documents; the loss of certain predefined knowledge elements is the focus point of this study.

4 Information Mapping

This section will introduce a new approach to evaluate and compare the information content and structure between documents based on key knowledge elements. As an example, this study uses the meeting minutes and the transcript of a requirements review recorded at Airbus UK in November 2003.

4.1 The Information Mapping Methodology

The concept of topics has been chosen to help cluster the four design knowledge items according to the different focuses of the meeting. Each element has its own symbol and is attached to one of the main meeting topics to form a map of the document. The result therefore presents itself as a sequence of network graphs centered and sequenced following the main discussion topics.

The coding of these knowledge items and the meeting topics is based on the shape and colour of the symbol to distinguish the different elements. The size of the symbol reflects the volume in number of words of each information item and is relative to the overall size of the document. Figure 2 illustrates the coding scheme.

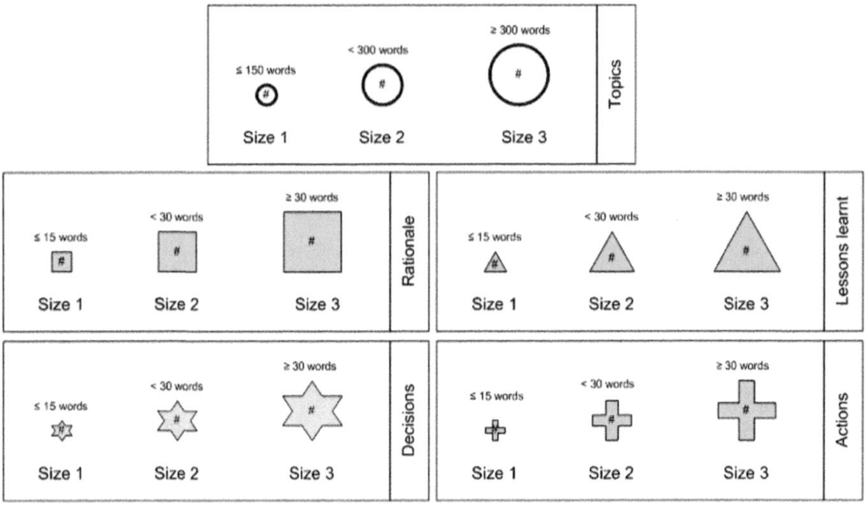

Fig. 2. Example of the information mapping coding scheme.

Table 1. Example of the structure of the table for action elements.

Actions			
#	Summary	N° of words	Coding size
1	Include quantity in the Equipment Specification	15	1
2	Clarify and confirm the top-level requirement for the deletion	35	2

Each node in each graph is associated through a reference number with a row in a table, which summarises the meeting content relative to the type of knowledge represented by the node. Other details such as the number of words and the subsequent coding size are also recorded in the tables. A separate table is maintained for each node type. Table 1 is an example of the structure of the table for action elements found in the minutes of a design review. Note that the table contents have been modified for reasons of confidentiality.

To produce the maps, four steps were followed:

1. The documents were marked-up using colour codes. This was effectively done using a Tablet PC and respecting the same colours as used in the coding scheme.
2. The topics were inferred from the structure of the minutes (section titles) and the meeting agenda.
3. The information elements were listed in 5 different tables (topics, rationale, decisions, lessons learnt and actions) of the type shown in Table 1.

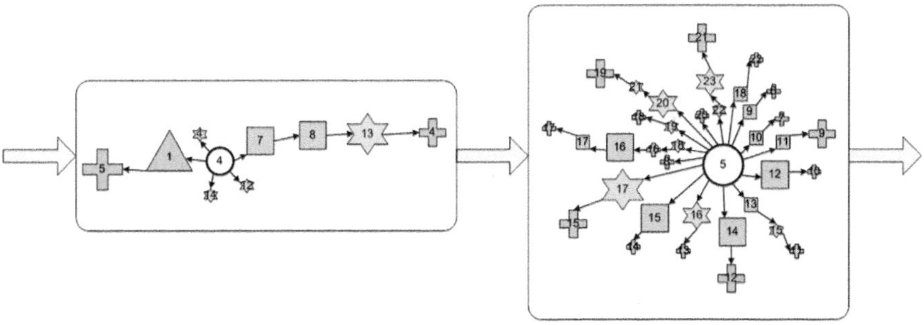

Fig. 3. An extract of the information map for the minutes of a design review.

4. An information map was built using the topics as focal points and connecting the items using the conceptual relations between them.

In order to effectively compare both the minutes and the transcript, the choice of topics was based on the minutes of the meeting. This was easily achieved as design reviews stay close to the proposed agenda and the transcript was therefore sectioned accordingly.

4.2 Example: Airbus Requirements Review

Figure 3 illustrates a part of the information map that was built based on the minutes of a requirements review meeting recorded at Airbus UK in 2003. The minutes were written by one of the participants of the design review, a junior engineer.

Figure 3 only shows two of the eight topics encountered during the meeting. The information map follows the sequence of topics. All the different knowledge elements are centred on the topics and ordered according to the thread given by the original document, in this case the minutes. The size of the shapes gives a feel of the importance in terms of number of words. This criterion alone however is far from being a good measure of the importance of an information entity. As suggested in [12] a more pertinent correlation can be found by looking at the number of links for each topic, this gives a more accurate idea of its importance. Finally, looking at the threads or branches surrounding the topics is of a particular interest: the sequence can be analysed in terms of length, volume, variety, and order.

5 Knowledge Loss in Minutes

Section 4 detailed and illustrated the information mapping approach developed for the assessment of the loss or modification of specific information objects in the minutes of design reviews. It is now important to interpret the findings from the

Table 2. Word counts for two Airbus UK design reviews.

		Requirements Review	Preliminary Design Review
Criteria	(1) Raw N° words	17101	7000
	(2) Useful N° words	9330	3850
	(3) Minuted N° words	2224	1300
Ratios	Ratio 1: (3)/(2)	0.23	0.43
	Ratio 2: (3)/(1)	0.13	0.24
	Ratio 3: (2)/(1)	0.55	0.57

monitored industrial meetings to highlight the implications in terms of knowledge loss. A measurement based on a simple count of the number of words will outline the general feel for the information lost in the documents. The analysis of some of the information maps will then seek further explanations in corroboration with the expectations of the users on minute taking, which have been assessed through a survey on the role of minutes in the aerospace design process and detailed in Section 2.3.

5.1 General Trends

Two design reviews were monitored at Airbus UK. The monitoring process involved the audio recording, the verbatim transcript of the interventions made by the participants, and collecting the meeting input/output artefacts (PowerPoint slides, reports, minutes, etc.).

From the transcript and the minutes of the design reviews, a brief comparative study has been made based on the number of words encountered in each one of the documents. The results have been summarised in Table 2. Three criteria were used according to the following definitions:

- *Raw number of words* (ref (1) in Table 2) is the number of words found in the transcript.
- *Useful number of words* (ref (2) in Table 2) is the number of words that should typically be found in the minutes. This evaluation is based on the Transcript Coding Scheme [2] where the interventions made by the participants were grouped according to predefined exchange role categories. Some of these categories will typically carry information which has no real *raison d'être* in the minutes. The authors judged that the number of words contained in "clarifying", "digressing",

"debating" and "managing" could be subtracted to the total tally in order to approximate the "useful number of words" in the transcript.
- *Minuted number of words* (ref (3) in Table 2) is the number of words referring to the content of the meeting found in the minutes.

Before analysing the results shown in Table 2, it is important to specify that the apparent difference in size of the two monitored reviews might be confusing. Effectively, both meetings lasted for over two hours, but in the case of the Preliminary Design Review (PDR) the meeting was dominated by a presentation involving the supplier contracted to produce the design. It was part of the research rationale that transcribing the presentation supported by a PowerPoint slideshow was of no great value since this portion of the meeting could be described as a monologue. Nevertheless the various interventions (questions, clarification points etc.) made by the participants during the presentation were transcribed and the PDR minutes summarised the event by inserting the slides in the document. Therefore, the word count for the minutes did not include the words which were part of the PowerPoint document.

In Table 2, ratio 2 suggests that the secretary of the meeting, in both cases engineers working on the project, tends to reduce the discourse down to around 15–25% of the original number of words spoken. Now, if the "useful number of words" figures are considered, around 50–60% of the conversations would be worth capturing (ratio 3). Finally, ratio 1 shows that the ability to capture the "useful number of words" in the minutes of these reviews varies significantly according to the secretary; the secretary for the PDR was an experienced engineer, while a junior engineer was assigned to take the minutes for the RR.

Of course, it is difficult to conclude from this intuitive evaluation based on a word count. The limitations are obvious: for one, the figures are only available for two meetings and more importantly, the count of the number of words is a purely quantitative approach which does not say much about the content and ultimately the value of the information being compared. The use of the information maps will propose a more robust analysis of the actual knowledge lost when using traditional minute taking techniques. The following section will propose an example to highlight the comparative possibilities offered by information mapping.

5.2 Results from an Information Mapping Case Study

The minutes of the Airbus UK Requirements Review were fully mapped according to the methodology detailed in Section 4. The transcript and the minutes accounted for eight main topics which were at the forefront of the conversations. For the purpose of this comparative study, only the critical topics were analysed. In the minutes, the most important topic in terms of number of words involved and highlighted knowledge items was *topic 5*; its map was therefore completely detailed both from the minutes document and from the transcript. In the transcript however, the two most important topics based on the same criteria were *topic 5* and *topic 4*. This was quite

Table 3. Comparative table of the information maps for "topics 4 and 5" of the Airbus UK Requirements Review.

Maps from the minutes	**Maps from the transcript**
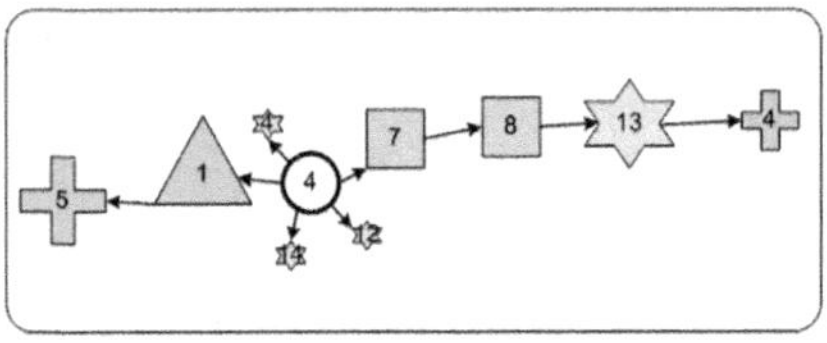	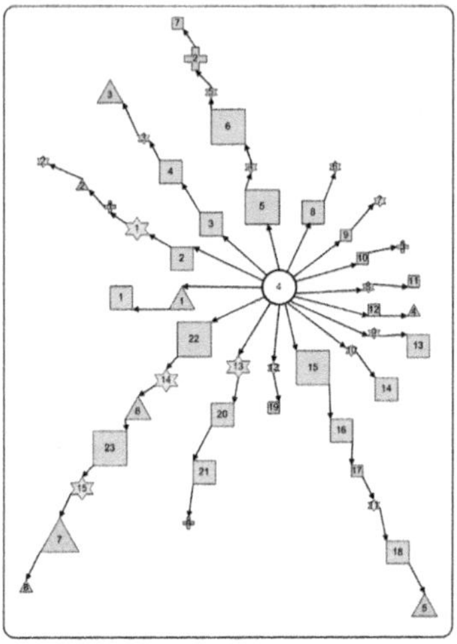
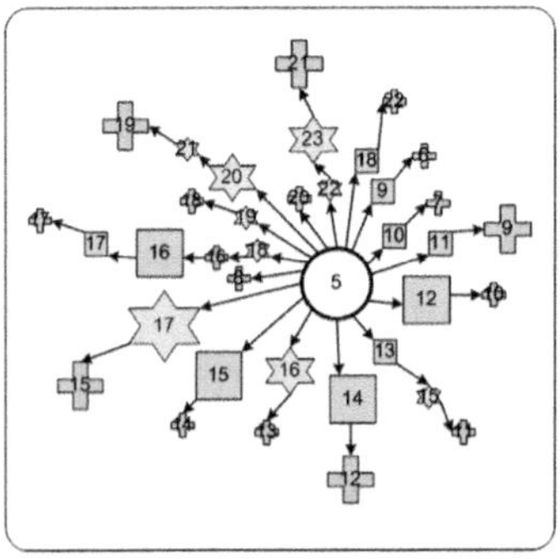	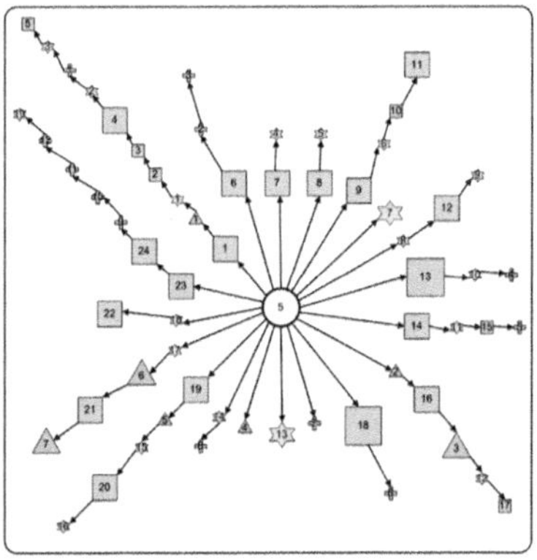

a surprise as the minutes' map suggests that *topic 4* was not of great importance. Table 3 is a comparative table where both maps (transcript map and minutes map) are shown for the two topics of interest for the knowledge loss study.

Topic 4, based on the transcript map, appears to have been badly minuted by the secretary: a high number of threads, many rationale, decision and lessons learnt elements but very few actions associated to these. On the other hand, when comparing the maps for *topic 5* it seems that the minutes give an accurate account of the discussions which took place. Looking at *topic 5*, a different observation can be made immediately: it seems that when writing up the minutes, the secretary "transformed" some of the decisions into actions. This is perfectly understandable as decisions can be interpreted as actions.

The single most important difference between *topic 4* and *topic 5* lays in the actions: in the discourse, most of the threads linked to *topic 4* do not contain actions, whereas it is quite the opposite in the case of *topic 5*. The resulting difference in the minutes' maps suggests that it is easier for the minute taker to record the meeting when actions are set out following the discussions. Experienced project managers [13] and meeting management guides [14] suggest that meetings need to be action orientated to become effective. This empirical advice seems to reach not only the quality of the meeting's progress as proposed in [13] and [14] but also the value of its record.

To complete the analysis of this information mapping case study, a list of four comparative criteria to evaluate knowledge loss have been established:

Volume and *Length*: *volume* refers to the specific number of words used to express any of the knowledge entities or topics and is visually acknowledged by the coding size of an element. The *length* is determined by the number of elements in a thread. These two criteria help to express the importance of a topic, knowledge element or thread relative to the rest of the document.

Variety: this criterion reveals the level of richness of a thread. A rich thread will therefore have a variety of explanations and outcomes. From the example in Table 3, one can immediately observe that the richness of the text based on the discourse is lost; actions and decisions on one side and rationale and lessons learnt on the other are very often merged or transformed.

Order/Sequence: analysing the *order* in which the four knowledge elements appear in a thread could give an insight on how threads are typically formulated. With more data, typical patterns would appear. One of the conclusions based on this criterion is that in the document relating the discourse, rationale is given before or after the decision or action, whereas in the minutes the sequence is invariably rationale then decision. The unstructured and unpredictable nature of speech is well reflected by these information maps.

6 Concluding Remarks

Based on the results detailed in the previous section, the following suggestions can be made to enhance the management of aerospace design reviews for effective company knowledge re-use:

- Turning decision points into actions will significantly enhance the review record

- Lessons learnt are made explicit during design reviews. They should be captured separately from the minutes to avoid merging them with rationale items.
- The secretary of the meeting has a key role to play. It would be preferable to choose an experienced engineer with little involvement in the review.

As a general rule, company guidelines and policies should seek to implement or strengthen the role of actions in design reviews. The three suggestions presented above are only simple guidelines but their implications are multiple in the fields of meeting management and meeting facilitation technologies.

This paper has developed and explored a graphical approach, which effectively measures the knowledge lost between a crucial design management activity, the design review, and its official record. The methodology has unveiled a range of properties which can be used as comparative criteria, and real life design reviews recorded at Airbus UK have illustrated these proposals. The work presented here has focussed essentially on organisational knowledge loss, but information mapping is thought to have much more to offer in the field of design research. A new form of design rationale representation could be developed and a further study of this technique could give practical insights into alternative minutes archiving strategies.

Acknowledgements

The authors are grateful for the funding provided by the EPSRC (Engineering and Physical Sciences Research Council – UK) under the CASE award No. 02300702 and Airbus UK. The authors would also like to thank the engineers at Airbus UK who readily participated in the design reviews monitored for the purpose of this research.

References

1. S.J. Yen. Capturing multimodal design activities in support of information retrieval and process analysis, Ph.D. Thesis, Stanford University (USA), 2000.
2. G. Huet, S.J. Culley, C.A. McMahon. A classification scheme for structure and content of design meetings, in *Proceedings of Design 2004*, D. Marjanovic (Ed.), The Design Society, 2004, pp. 1363–1369.
3. R.G. Cooper. *Winning at New Products. Accelerating the Process from Idea to Launch*, 2nd edition, Addison-Wesley, ISBN 0201563819, 1993.
4. CEI/IEC 1160: 1992. Formal design review, National Standards of Canada, 1996.
5. SAE AS9100 revision A: 2001. Quality systems – Aerospace – Model for quality assurance in design, development, production, installation and servicing, Society of Automotive Engineers, 2001.
6. R.H. Bracewell, S. Ahmed, K.M. Wallace. DREAD and design folders, a way of capturing, storing and passing on, knowledge generated during design projects, in *Proceedings of DETC'04*, A.R. Parkinson (Ed.), American Society of Mechanical Engineers, 2004.

7. J. Lee, K-Y. Lai. What's in design rationale?, in *Design Rationale: Concepts, Techniques, and Use*, T.P. Moran, J.M. Carroll (Eds.), Lawrence Erlbaum Associates Publishers, 1996.
8. T.P. Moran, J.M. Carroll. *Design Rationale: Concepts, Techniques, and Use*, Lawrence Erlbaum Associates Publishers, ISBN 080581566X, 1996.
9. J.R. Marsh. The capture and utilisation of experience in engineering design, Ph.D. Thesis, University of Cambridge (UK), 1997.
10. R.L. Derr. The concept of information in ordinary discourse, *Information Processing & Management* **21**(6), 1985, 489–499.
11. D.B. Gordon, N. Sager. A method of measuring information in language, applied to medical texts, *Information Processing & Management* **21**(4), 1985, 269–289.
12. E. Hoffmann. Defining information: An analysis of the information content of documents, *Information Processing & Management* **16**, 1980, 291–304.
13. T. Little. I hate reviews!, *Academy Sharing Knowledge Magazine*, **10**, 2004, available online only: http://appl.nasa.gov/ask/
14. B.J. Streibel. *The Manager's Guide to Effective Meetings*, McGraw-Hill, ISBN 0071391347, 2003.

Internet/Intranet/Extranet-Based Systems in the CESICED Platform for Virtual Product Development Environment

George Dragoi[1], Costel Emil Cotet[1], Sebastian Rosu[2] and Luminita Rosu[3]

[1] *University Politehnica of Bucharest, Splaiul Independentei nr. 313, Sector 6, Bucharest, Romania; E-mail: {gdragoi, costel}@mix.mmi.pub.ro*
[2] *Serto Invest, Bucharest, Romania*
[3] *Grup Romet S.A., Buzau, Romania; E-mail: luminita@romet.ro*

Abstract. Development teams involved in the product development are often geographically and temporally distributed. There is a high level of outsourcing in the actual product development efforts. In order to achieve a market objective that otherwise could be out of reach, enterprises, which do not have the complete knowledge for developing all the components needed for their products, need to outsource the development of specific parts; in this way a cooperation with other enterprises working in the same industrial sector starts. It is then possible to say that enterprises give rise to a special type of virtual enterprise (VE), in which each company maintains the greatest flexibility and business independence. This paper presents a vision of next generation engineering working environments and describes a core information technology which future systems can be built on. This paper gives an overview of the new Web Services technology based of the IBM Rational Software solutions, discusses the critical issues of the virtual product development, presents the roles of the virtual teams, and build a general architecture of a CESICED experimental platform for training, research and consulting in the new digital economy, located in the PREMINV center at the University Politehnica of Bucharest.

Key words: integrated and distributed product design, virtual enterprises, virtual teams, Internet/Intranet/Extranet-based systems, outsourcing.

1 Introduction

In the past, product development was often carried out within a single company by co-located design teams. In more recent years, there has been a shift in product development paradigms. Development teams involved in the product development are often geographically and temporally distributed. There is a high level of outsourcing, not only in the domain of manufacturing but also in the actual product development

S. Tichkiewitch et al. (eds.), Advances in Integrated Design and Manufacturing in Mechanical Engineering II, 293–307.

efforts. In order to achieve a market objective that otherwise could be out of reach, SMEs, which do not have the complete knowledge for developing all the components needed for their products, need to outsource the development of specific parts; in this way a cooperation with other enterprises working in the same industrial sector starts. The cooperation often implies not just a simple made-to-order development but also a real collaboration among the companies involved in the definition of a new product. It is then possible to say that enterprises give rise to a special type of virtual enterprise, in which each company maintains the greatest flexibility and business independence. Each member contributes its specific core know-how to the virtual enterprise. In the virtual enterprising context, it is possible to define the term virtual product development as a temporary cooperative activity of independent and distributed partners concerned with the translation of customer requirements into system functionalities. As such partners deal with large products, complex design processes and engineering knowledge, the management and maintenance of the relevant information is diverse. Phases in the process of product development, particularly the early ones, are characterized by dynamic, evolutionary, vague, incomplete and occasionally conflicting product specifications as well as imprecise product defining data within distributed and heterogeneous IT-system environments. As a result of this new product development paradigm, there is a greater need for software tools to effectively support the formal representation, collect and exchange of product information during the development stage. As the complexity of products increases and the product development becomes more distributed, new software tools will begin to cover a broader spectrum of product development activities than the traditional mechanical CAD systems. Accordingly, the ability to collect, in an effective and formal manner, additional types of information will become a critical issue. This paper presents a vision of next generation engineering working environments and describes a core information technology which future systems can be built on. The concepts of a virtual product development environment and virtual teams, enabled by a new generation of Internet/Intranet/Extranet-based services [1, 15] are discussed here, as means to stay competitive and to thrive in a turbulent market. This paper gives an overview of the new Web Services technology based of the IBM Rational Software solutions, discusses the critical issues of the virtual product development, presents the roles of the virtual teams, and build a general architecture of a CESICED experimental platform for training, research and consulting in the new digital economy, located in the PREMINV [2, 3] center at the University Politehnica of Bucharest.

2 E-Economy Paradigm

Also, today, what do you want in an e-team with multidisciplinary optimization [1, 4]?

First, for your enterprise: low risk, low cost, single point of support; reap benefits as soon as possible, ready for unpredictable demand and growth, future capabilities.

Second, for your customers: excellent response time, ability to build competitive advantages from a wide variety of applications (use specialized products for unique features, integrate them for seamless customer support, etc.), ensure optimal customer experience to build loyalty, customer trust (security features build trust: validate who is doing business with whom; secure financial transactions, protect internal assets, halt spread of viruses, protect against hackers), high availability, etc.

A virtual product development in a virtual enterprise is a temporary alliance of teams that come together to share skills, abilities and resources in order to attend a project opportunity and whose cooperation is supported by computer network and adequate tools, competencies and special application software [3]. Virtual Enterprise operates as nodes in a network. A different architecture, engineer and construction organization, a fresh virtual team [5], is needed every time for every new project. Innovative techniques to co-ordinate and manage information, resources and documents need to be developed to integrate successfully and reduce lead times, increase quality and keep within budget constraints. Consequently, the partners in the virtual enterprise need to exchange legacy data and migrate with other systems outside their own secure corporate boundary. In order to achieve collaboration between different actors in the Virtual Enterprise, there needs to be common processes supporting the distributed product development process [4].

The global market today increase the need for common and collaborative processes and sharing information seamlessly between companies involved in extended enterprise. All forecasts predict very rapid growth in e-commerce as part of e-business and e-services that attract increasing attention because of the impact of new information and communication technologies on firms, markets, employment, and development. There are fundamental effects on the organization of business flows and processes, transaction costs, the creation of new business models, and changes in the boundaries of firms across sectors [6]. All this effects are defining the new digital economy perspectives on business, products, individuals and technology. Today, digital economy encompasses e-business, e-commerce, and e-services as services or resources that can be accessed through people or businesses, using network technologies [2, 3, 5] (see Figure 1).

For business, e-service is going to be a new way to save money, to revenue growth, and faster development model. For end-users, e-services increase productivity and simplify life, take advantage of more sophisticated and specialized services on as needed basis. At the level of production dedicated enterprises, e-services are (see Figure 1): Business-to-Business (B2B), Intra-Business (IB), and Business-to-Customer (B2C). In a real meaning, an e-business is any business that uses Internet or Internet technologies to attract, retain and cultivate relationships with customers, streamline supply-chain, manufacturing, and procurement systems and automate corporate processes to deliver the right products and services to customers quickly and cost-effectively, also to capture, explore, analyze, and automate corporate processes information on customers and company operations in order to provide better busi-

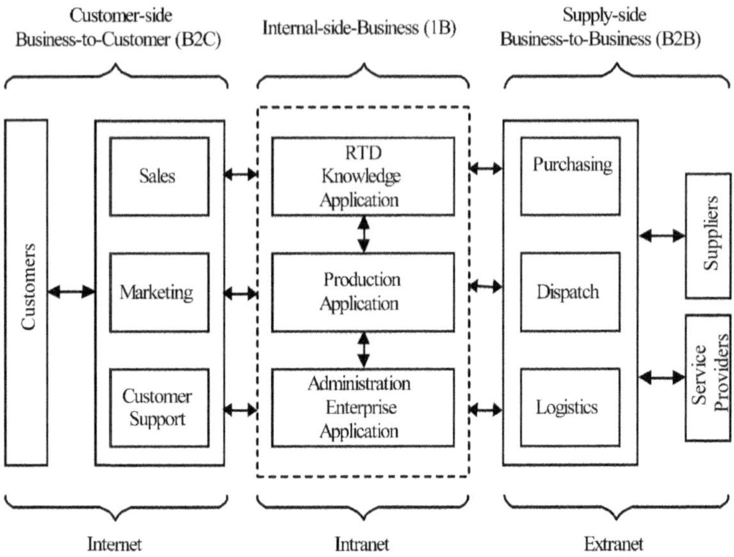

Fig. 1. The exchange of information services and goods through networks.

ness decisions. For the future, e-services and e-business, as were defined, require the enterprise re-thinking and re-modeling, with the system and applications design for an efficient use of new network technologies. The perspectives of this kind of manufacturing and economy, named shortly new digital economy, we can see the product perspective (holistic product view, product life-cycle, value-network integration, etc.), business organizational perspective (new organizational form, customers and suppliers integration, collaborating organization, etc.), the technology perspective (technological building blocks, infrastructures, interoperability, etc.) and the individual perspective (skills, workspaces, collaborating individual, different roles: worker, consumer, citizen). As a general requirement for an infrastructure to support virtual enterprise it can be underlined that the companies must be able to interoperate and exchange information in real time so that they can work as a single integrated unit, although keeping their independence/autonomy.

3 Virtual Networks Support for VE in the CESICED Platform

A hierarchical network design model (see Figure 2) breaks the complex problem of network design into smaller, more manageable problems. Each level, or tier, in the hierarchy addresses a different set of problems so that network hardware and software can be optimized to perform specific roles. Devices at the lowest tier of the

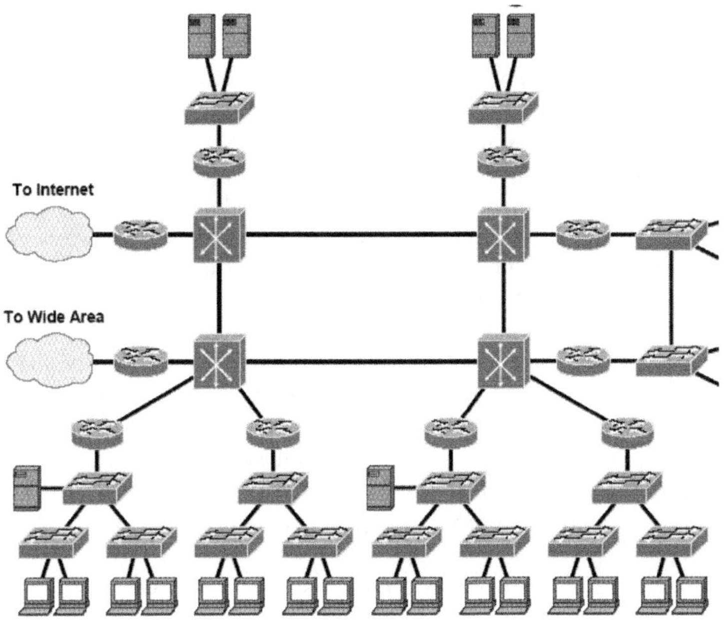

Fig. 2. A hierarchical network design model in the CESICED platform.

hierarchy are designed to accept traffic into a network and then pass traffic up to the higher layers [6].

The core of the network has one purpose: to provide an optimized and reliable transport structure by forwarding traffic at very high speeds. In other words, the core layer should switch packets as fast as possible. Devices at this layer should not be burdened with access- list checking, data encryption, address translation, or any other process that stands in the way of switching packets at top speed [6]. The distribution layer sits between the access and core layers and helps differentiate the core from the rest of the network. The purpose of this layer is to provide boundary definition by using access lists and other filters to limit what gets into the core. Therefore, this layer defines policy for the network. A policy is an approach to handling certain kinds of traffic, including routing updates, route summaries, Virtual Local Area Network (VLAN) traffic, and address aggregation. You can use policies to secure networks. The access layer feeds traffic into the network and performs network entry control. End users access the network via the access layer. As a network's "front door," the access layer employs access lists designed to prevent unauthorized users from gaining entry. The access layer can also give remote sites access to the network via a wide-area technology, such as Frame Relay, ISDN, or leased lines. A reliable and available network provides users with 24-hours-a-day access.

In a highly reliable and available network (see Figure 2), fault tolerance and redundancy make outages and failures invisible to the end user. The high-end devices and telecommunication links that ensure this kind of performance come with a steep price tag. Network designers constantly have to balance the needs of users with the resources at hand. Multicast traffic can also consume a large amount of bandwidth. Multicast traffic is propagated to a specific group of users. Depending on the number of users in a multicast group or the type of application data contained in the multicast packet, this type of broadcast can consume most, if not all, of the network resources. As networks grow, so does the amount of broadcast traffic on the network. Excessive broadcasts reduce the bandwidth available to the end users and force end-user nodes to waste CPU cycles on unnecessary processes. In a worst-case scenario, broadcast storms can effectively shut down the network by monopolizing the available bandwidth. Two methods can address the broadcast issue for large switched LAN sites: the first option is to use routers to create many subnets and logically segment the traffic (This scenario can create a bottleneck in the network); a second option would be to implement virtual local area networks (VLANs) within the switched network. A VLAN is a group of end devices that populate multiple physical LAN segments and switch ports; they communicate as if they were on a single LAN segment. One of the primary benefits of VLANs is that LAN switches (by creating VLANs) can be used to effectively contain broadcast traffic and manage traffic flows.

The Virtual Local Area Network (VLAN) is the best support for the virtual design offices or the virtual teams [6, 7]. The interconnection of VLANs is realized on the level two (ELAN – Emulation Local Area Network), and on the level three (MultiProtocol Over AT – MPOA). MPOA Client (MPC) and MPOA Server (MPS) are obtained the configuration by LECS (LAN Emulation Configuration Server). The virtual networks MPOA is named IASG (Internet Address Summarization Groups). In this case a design team is formed with members located at different geographic locations [7]. A virtual local area network is created for the project (see Figure 3).

In addition to the team's full-time members, the team also includes contributing members who are recruited for specific components of the project. As such, a core group is responsible for leading the project and a sub-group is involved in specific components of the project [7]. While the full time employees form the central core of the team, experts in the different problems of the project (control systems, mechanic systems, electronic systems, programmers, etc.) are also team members (see Figure 3).

Virtual teams for engineering design are becoming more commonly used in industry and the engineering education community must prepare graduates to be employed in such work environments. It is inevitable that multidisciplinary teams for product design, with members located in different geographic locations, will become more common place in the future (see Figure 3). It is widely understood that successful design is often a highly collaborative team based activity [8].

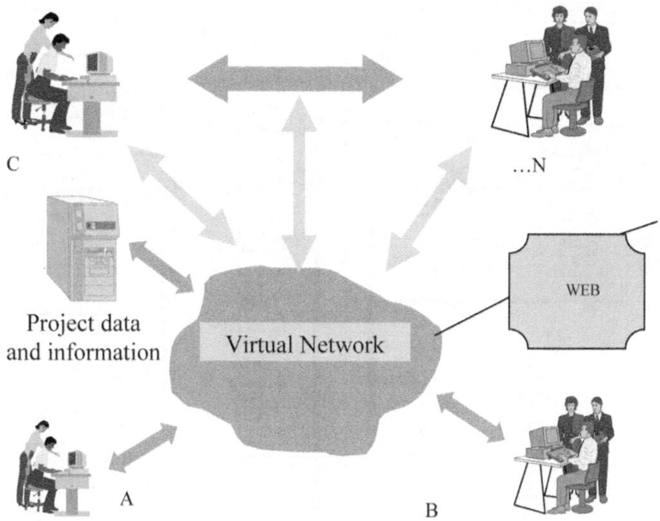

Fig. 3. Virtual teams in collaborative processes within virtual organization.

3.1 Virtual Teams

To be effective, a virtual team (see Figure 3) must be able to communicate, collaborate and coordinate, all at distance. Though some corporation are practicing a form of distributed design, a documented procedure for conducting distributed design and product development has yet to be created, tested and distributed. However, the same set of skills that guide design teamwork for a team where all members are in one location is different from that set of skills needed to lead virtual team.

The term "virtual team" is a misnomer as although it makes reference to virtual reality and the concept of creating a Virtual space that can be experienced it also suggests that the virtual team is not actually a team and as such can lead to a loss of performance. In the initial stages it became clear that whilst a distributed computer based-platform could support distributed teams it could not completely replace face-to-face contact [8], The reasons behind this are complex but lie largely in the operational cultures of the organizations and individual apprehension to the process of decision making and conflict resolution in a distributed environment. In particular it is recognized that face-to-face contact at the beginning of a project leads to significant advantages in the areas of building trust, establishing the team working methods and communications protocols, and defining a common vision of the product. This presented an opportunity to explore how collaborative tools and technologies can be used to support the gathering of distributed teams within a co-located environment for key decision making sessions [8]. To illustrate the infancy of distributed virtual design, the method of distributed team members working on any project cur-

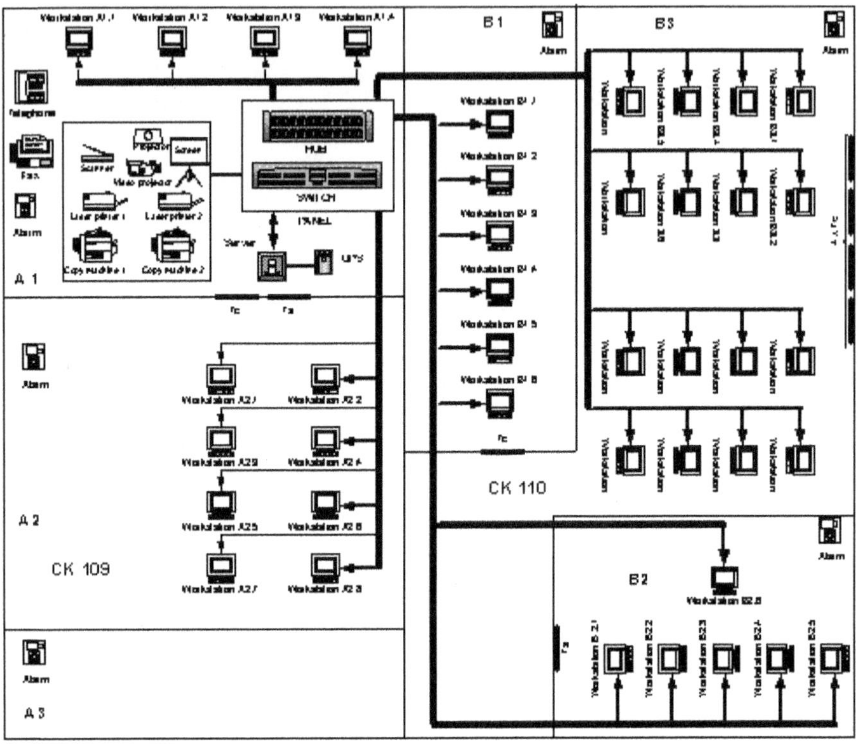

Fig. 4. General infrastructure of the CESICED platform.

rently lacks a universal name [4]. Terms such as virtual teams, collaborative learning groups, geographically and temporally dispersed teams, globally distributed teams, distributed design, e-design and e-teams, are all used to describe various Internet based design activities [9, 10].

4 The CESICED Platform

The general infrastructure of the CESICED platform is presented in Figure 4. As a result of a new product development paradigm, there is a greater need for software tools to effectively support the formal representation, collect and exchange of product information during the product development stage. The concept of virtual product development environment that is enabled by a new generation of Internet based services is discussed [1]. The Internet is a world-wide conglomerate of different networks that communicate among each other via a common protocol, independently of the hardware type used. Various network services can be used by everyone, either supplying or demanding them.

A large range of distribution, the platform independence, an big number of user friendly services that are easily accessible through the World Wide Web as well as the open standards used and free or budget-priced products (such as browsers, html editors, software updates) have lead to a high and continuously growing proliferation of the Internet [1]. The advantages offered by the Internet for covering the information needs are held to be the following [2]: reduction of local barriers by means of world-wide information offers; reduction of time barriers by means of permanently available information; reduction of (transaction) costs by way of automation of information processing on the supply and/or the demand side; improved coordination and cooperation with external partners using an integrated information and communication platform (e.g. platform independence, information exchange without media ruptures). However, the application-to-application communication problem still exists. Businesses have needed a standardized way for applications to communicate with one another over networks, no matter how those applications were originally implemented [3]. Web Services, the latest evolutionary step in distributed computing, represent exactly this solution by providing a standardized method of communication by means of which different applications can be integrated together in ways not possible before. Different applications can be made to call on each other's resources easily and reliably, and the different resources that applications already provide can be linked together to provide new sorts of resources and functionality. Moreover, the application integration becomes much more flexible because Web Services provide a form of communication that is not tied to any particular platform or programming language. At the core, the Web Services represent a unit of business, application, or system functionality that can be accessed over the Web. Web Services are applicable to any type of Web environment [5], Internet, Intranet, or Extranet, and be focused on business-to consumer, business-to-business, department-to-department, or peer-to-peer communication (see Figure 5).

A Web Service consumer could be a human user accessing the service through a desktop or a wireless browser; it could also be an application program or even another Web Service [6]. The following issues have been identified as crucial [2]: integration of different applications within the same organization; integration of organizations, applications and content between different organizations in the same virtual enterprise; integration of partners.

5 Collaboration and Communication in the CESICED Platform

A central point of future product development is therefore collaboration and communication. This is based on consistent, integrated data sets and on tools that support the collaboration (see Figure 6).

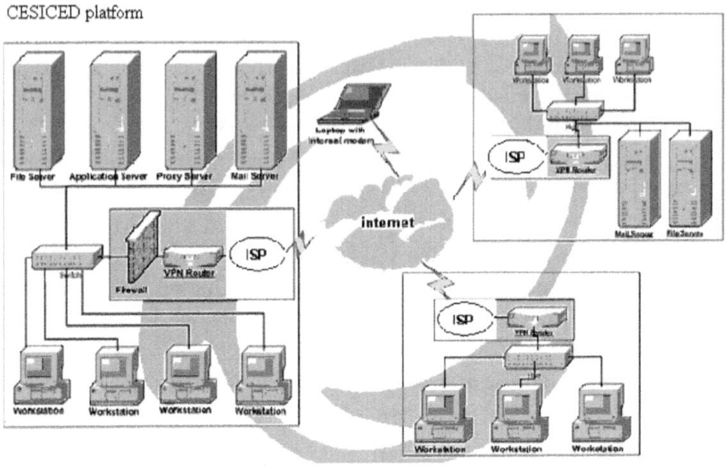

Fig. 5. Internet/Intranet/Extranet based tools in the CESICED platform.

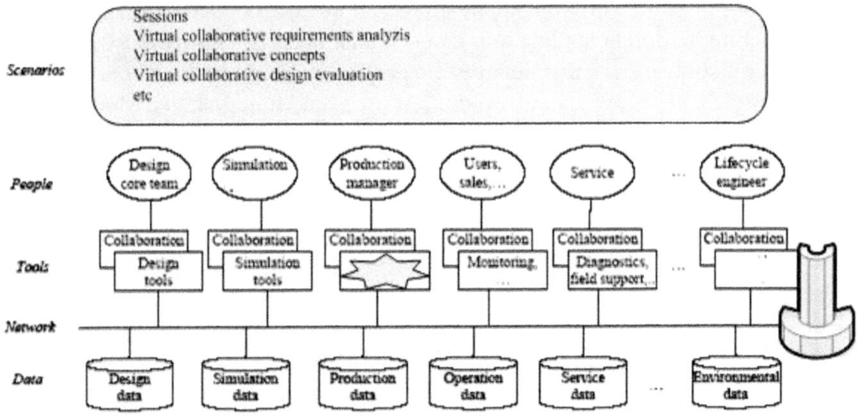

Fig. 6. The key elements of collaboration and communication in the CESICED platform.

Traditionally, only data about the designed product are stored, made available for the product development team and archived for later product development projects [11]. With product lifecycle management, information coming from all product lifecycle phases is to be integrated, including sales, operation and service data about sold/installed products. Product data management technologies are extended to manage more complex and more dimensional data. Upon this data management, tools used by different people throughout the product lifecycle are integrated, including requirement analysis, reverse engineering, production planning, resource planning, logistics and traditional design and simulation systems [12]. An environment supporting collaborative design would comprise the following components:

- integrated data sets, including description of the product and all related processes, and including data collected from the lifecycle of this and other related products;
- interface to detailed planning, analysis and simulation systems, as some of the product and process features need to be derived from the collected data;
- simulation and visualization engines, to obtain an virtual environment containing easy understandable representations of product features and scenarios;
- multi-modal, intuitive user interfaces, which allow people to easily view and manipulate the product features.

Collaborative design is based on the simulation of scenarios [9]. The key elements of a scenario are the desired product, the targeted environment, the lifecycle processes and the people involved in the processes [12].

5.1 Role of the Web Services in the CESICED Project

The benefits of using Web Service technology as a core IT platform for the CESICED, have been identified [11]. They comprise minimized processing times and costs and the improvement of the following features: functionality (system supporting and/or fully automating product development); process; integration (system to (internal or external) system communication); usability (effortless communication between the human user and the system); security (protection of the enterprise knowledge); flexibility (easily adjustable to a fast-changing business environment). The IBM Rational software solutions [13] implemented in the CESICED platform (see Figures 7 and 8) [13] aligns projects and resources with business priorities. It is about planning and managing projects individually and as a comprehensive portfolio of assets which must meet enterprise objectives [13].

There are some inherent advantages that make Web Services the ideal platform for realizing CESICED infrastructure. They include the ease of integration. Web Services can be used in product development process for [11]: organizational mode; vocabularies definition; integration of processes in product development; actions in the product development process; communication of the product development process status.

5.2 Knowledge Management in the CESICED Platform

Besides the four conversions of the knowledge (i.e. sharing experience, conceptual knowledge, systemic knowledge, and operational knowledge), many other steps were added in order to increase the knowledge of the company and, consequently, to improve the product development process (see Figure 9) [12–15]. Also, the additional steps are: obtain and use, learn and contribute, evaluate and sustain, discharge and

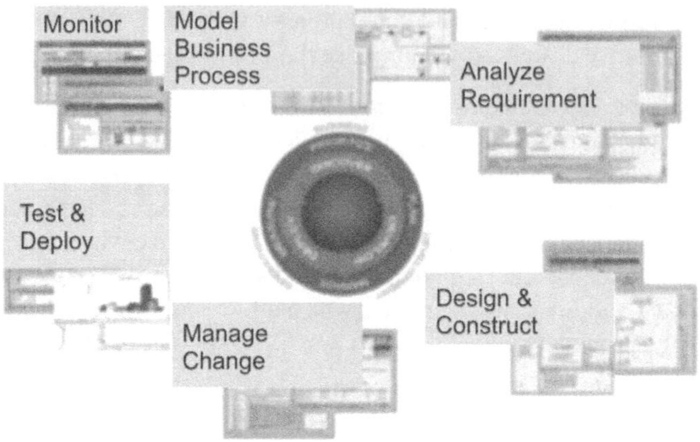

Fig. 7. Build, integrate, extend, modernize and deploy software (IBM Rational Software).

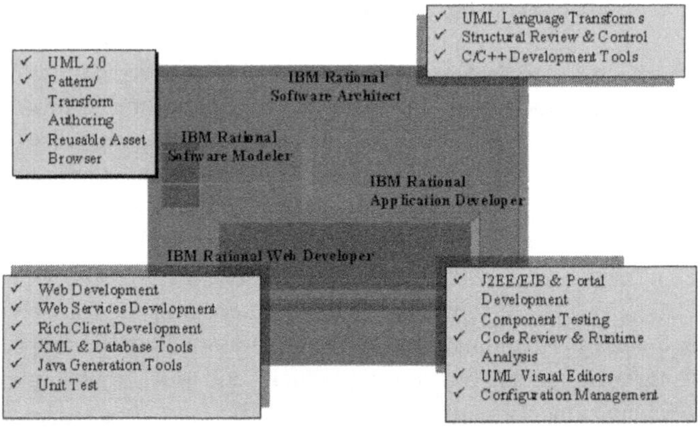

Fig. 8. Design and construction tools with IBM Rational Software [13].

support. The knowledge of the organization is composed by the sharing knowledge of each individual. Based on the steps described in Figure 9 an efficient and effective management of the intellectual capital of the company is obtained. The steps obtain and uses are well known within organizations. People always seek information and use them later to solve their problems, to take decisions or to create new products. Therefore, new technologies (e.g. intranet/internet/extranet) allow that the large amount of information that flows within organizations can be correctly managed. The steps learn and contribute are relatively new for organizations. For example, it has been difficult to convince employees to contribute to the organization's knowledge base. New technologies have helped companies easily organize, send and transfer certain types of information. However, this facility has been seen by the em-

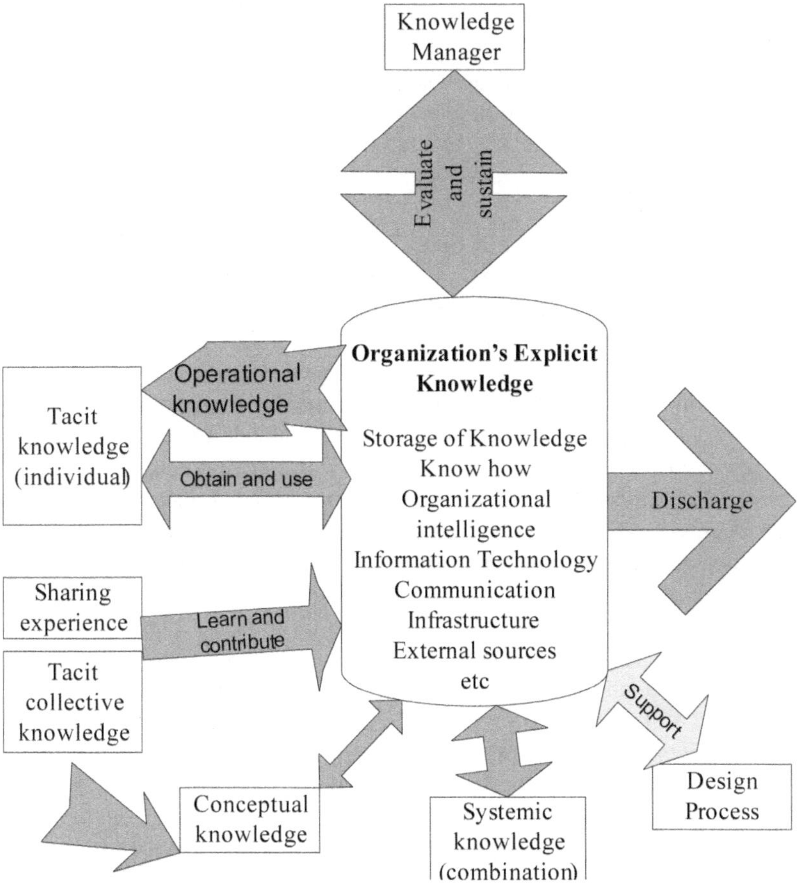

Fig. 9. Organization knowledge management model in the CESICED platform.

ployee as a threat for his/hers own job security. The most difficult task is to convince individuals that their contribution will give return to their organization as well as to themselves.

The step evaluate indicates that the organization should define its own necessary knowledge for its mission and classify its own currently intellectual capital. In other words, the knowledge manager does more than organize the content in system online, he/she should understand and foresee the community's needs. The step sustain or maintain should assure that the future intellectual capital will maintain the organization viable and competitive. Organizations tend to build their own intellectual capital through their relationships with customers, employees, suppliers, etc.

The knowledge manager should also be responsible for the maintenance of the organizations knowledge base. The step discharge excludes any useless knowledge from the organizations knowledge base. However, some knowledge can be more valuable if it can be transferred to outside of the organization. The step support can be used for the continuous improvement of the product design process

6 Conclusions

The presented solution is based on CESICED platform that defines a standard for developing multi-tier, portable and platform neutral enterprise applications based on the technology of Web Services. Striving to discover the main problems of collaborative product data management, the theoretical backgrounds in the particular research area were analyzed, and interviews with the members of the design departments of SMS companies were carried out. On the basis of the collected findings, the main elements of the product informational metamodel, that presents a foundation for building a Web Service were determined [5]: information binding product data carriers (physical components of the product and structured documents defined by hierarchy, revisions, variants and status); information binding subjects that create, own and manipulate product data (users, projects, organization, design tasks); information binding activities that use the product data and that are coordinated by subjects (hierarchical structure management of products and documents, engineering change management, approval and authorization procedures, design process documentation and history, product configuration processes). Specific interfaces have been designed for different types of users and other applications to activate functions that are provided for them by the integrated platform, i.e. engineering change management or search for needed information. The developed service and methodology have been tested in pilot projects carried out in the design process. A fundamental topic of the testing has been the validation of the concept of the product data management among co-design partners using the technology of the Web Services. The results of the testing process showed that the traditional approach to the integration of engineering data between different systems can be improved by using the new technology.

References

1. G. Drăgoi, *Întreprinderea integrată: metode, modele, tehnici şi instrumente moderne de dezvoltare şi realizare a produselor*, Editura Politehnicapress, ISBN 973-8449-25-1, 2003.
2. G. Drăgoi, M. Guran, C. Cotet, The PREMINV platform for training in management and engineering for virtual enterprise, in *Collaborative Business Ecosystems and Virtual Enterprises*, Luis M. Camarinha-Matos (Ed.), Kluwer Academic Publisher, Dordrecht, ISBN 1-4020-7020-9, 2002, pp. 571–579.

3. G. Drăgoi, C. Cotet, Virtual laboratories for the collaborative design to improve product development, *International Journal of Simulation Modelling-IISJMM* **3**(1), 2004, 29–40.
4. G. Dragoi, B. Radulescu, S. Tichkiewitch, A cooperative system for the design of the mechatronic products with the multidisciplinary optimization, in *Methods and Tools for Co-operative and Integrated Design*, Kluwer Academic Publisher, Dordrecht, ISBN 1-4020-1889-4, 2004, pp. 281–290.
5. G. Drăgoi, C. Cotet, L. Roşu, S. Roşu, Virtual teams in collaborative systems in networks and corporations, in *Proceedings of the 7th Conference on Management of Innovative Technologies (MIT2004)*, 4–5 October , ISBN 973-700-028-5, Constanta, Romania, 2004, pp. 119–124.
6. G. Drăgoi, *Infrastructura informaţională şi de comunicaţii a îmtreprinderii moderne*, Editura Politehnicapress, Bucuresti, ISBN 973-8449-73-1, 2005.
7. G. Drăgoi, C. Cotet, L. Roşu, S.M. Roşu, Role of the virtual networks in the virtual enterprise, in *Proceedings of the 8th Conference on Management of Innovative Technologies (MIT2005)*, 24–26 September 2005, Fiesa, Piran, Slovenia (www.fs.uni-lj.si/lat/MIT2005), 2005.
8. S. Szykman, S.J. Fenves, W. Keirouz, S.B. Shooter, A foundation for interoperability in next-generation product development systems, *Computer-Aided Design* **33**, 2001, 545–559.
9. K. Furst, T. Schmidt, G. Wippel, e-Service based business information systems, in *Proceedings of the 8th International Conference on Concurrent Enterprising – ICE2002*, Centre for Concurrent Enterprising, University of Nottingham, Nottingham, UK, 2002, pp. 381–38.
10. S. Khoshafian, *Web Services and Virtual Enterprises*, Tect, Chicago, USA, 2002.
11. G. Drăgoi, C. Cotet, G. Patrascu, L. Roşu, S.M. Roşu, Internet/Intranet/Extranet-based systems for virtual product development environment in the CESICED platform, in *Proceedings of the 8th Conference on Management of Innovative Technologies (MIT2005)*, 24–26 September 2005, Fiesa, Piran, Slovenia (www.fs.uni-lj.si/lat/MIT2005), 2005
12. G. Drăgoi, C. Cotet, L. Rosu, S. Rosu, The CESICED platform for training, research and consulting in the new digital economy, in *Proceedings of the Second International Conference of Integrated Engineering (C2I-2005)*, 16–18 October 2005, Timisoara, Romania, URL: http://www.eng.utt.ro/ccii/c2i2005, 2005.
13. IBM Software Development Platform, *IBM Rational Software Solutions*, IBM Corporation, Bucharest, Romania, 2005.
14. S. Tichkiewitch, B. Radulescu, G. Dragoi, K. Pimapunsri, K., Knowledge management for a cooperative design system, in *Proceedings of the International CIRP 2004 Design Seminar – Design in the Global Village*, May 16–18, with CD-ROM support, Cairo, Egypt, 2004.
15. F. Noel, D. Brissaud, S. Tichkiewitch, Integrative design environment to improve collaboration between various experts, *CIRP Annals, STC Dn* **52**(1), 2003, 109.

Which Representation for Sets of Product and Associated Services during the Design Process?

Nicolas Maussang, Daniel Brissaud and Peggy Zwolinski

G-SCOP Laboratory, University of Grenoble, BP 53, 38041 Grenoble Cedex 9, France; E-mail: {nicolas.maussang, daniel.brissaud, peggy.zwolinski}@g-scop.inpg.fr

Abstract. Since a couple of years, services are taking more and more importance beside the physical products. The market is shifting from selling product to providing utility. The Product-Service System (PSS) is an emergent concept that promotes the embedding of services with product. But the inclusion of services within the product will add some modifications to the design process of the product. Consequently, to encourage enterprises to create PSS, it is important to have during the design process of these sets a common representation of product and services. In this article, the PSS principle will be explained. Then, a PSS case study will be represented with two different representations. One is Agent-Based Model which has been developed in Service/Product Engineering and Functional Analysis. It will be discussed which one could be the more adapted for PSS in order to help and guide designers.

Key words: design, PSS, representation model.

1 Introduction

Since a couple of years, there is a shift on the market economy because services take more and more importance beside the physical products (service contract in the automotive industry, taking back of electronic goods, etc.). Moreover customer's needs cannot be only fulfilled by the physical product alone and services have to be included to products. In some selling cases, customer only pays for the service provided by the product rather than the product itself. The most famous example is the photocopier: the customer buys a number of photocopies instead of the copier. In this case, the customer buys a functionality integrating services. These services beside the product are linked to breakdowns, maintenance, and upgrades, etc. These services, which are managed in the background, are as important as the physical product itself. For the provider, he will look for the effectiveness of services at the best cost and services will guarantee the satisfaction of customer.

S. Tichkiewitch et al. (eds.), Advances in Integrated Design and Manufacturing in Mechanical Engineering II, 309–323.

During the purchase, customers buy a set of product and services and services are as important as the product. But, services are created separately from the product. Thus, it is important to take them into consideration during the design process of the product in order to sell a complete set of product and services. These new services will have to be integrated as early as possible during the design of the product included in the Product-Service System (PSS) [1].

In this article, the PSS principles used to describe the sets product-services will be presented. Among the existing service representations and more particularly the Agent-Based Model, it will be discussed the usefulness to model PSS during the design process. In the last section, the Functional Analysis, which is a current approach used by designers to create either products or services, will be used to identify a representation to help designers of PSS.

2 PSS Principle

The emergence of services added to the product leads to the emergence of different concepts that embed product and services. Two principles are mainly used which are Extended Product and PSS.

These two concepts emerged from supplementary customer's needs and from a new selling approach. According to Goedkoop [2], "PSS can be seen as a marketable set of product and services capable of jointly fulfilling the customer's needs". It means that both product and services are used to fulfill the customer's needs, but the ratio between the "product value" and the "service value" can vary. On the one hand, the product value can be more important than added service, but on the other hand services can satisfy customer's needs rather than the physical product.

But other considerations can justify this market shifting. In a market more and more competitive, enterprises have to developed new solutions or improved products in order to keep in touch with competitors. Service selling appears as a good alternative and a new business for both customers and enterprises. According to Thoben [3], service selling can have two goals. For the customer, it permits to propose an attractive package and enables providers to stand out from competition.

From environmental considerations, the inclusion and sale of services instead of the physical product could lead to turn away the customer from the physical product. It can be a way to promote dematerialization that could lead to decrease the environmental burden [4]. In addition, the integration of services with the physical product could contribute to the creation of sustainable products [5]. If we imagine that a customer buys a PSS with a specific service, which has to upgrade the product, the customer will be sure to have at anytime an efficient product and moreover that respects regulations.

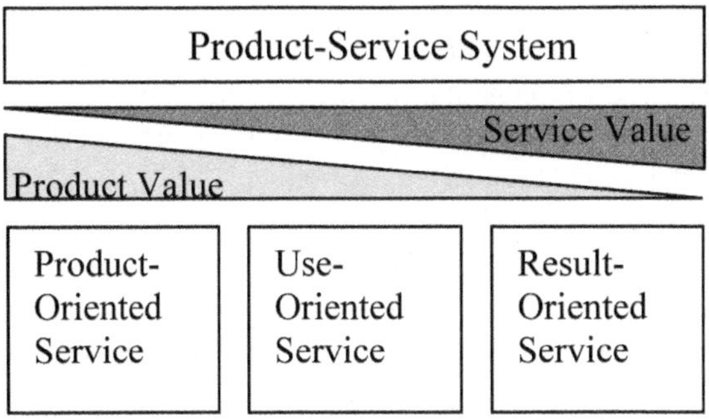

Fig. 1. PSS characterization according SusProNet.

2.1 Typology of PSS

As said previously, product value or service value can vary within a PSS. This variation comes from different customer's needs and it influences the type of services that is added to the physical product. SusProNet, which is a European network of industries and institutes focusing on the design of product and services for sustainable competitive growth, tries to promote PSS and made a characterization of PSS from a customer point of view [6]:

- Product-Oriented Services (POS)
- Use-Oriented Services (UOS)
- Result-Oriented Services (ROS)

For this categorization, it can be seen that the ratio between product value and service value is different (see Figure 1).

Each category can be illustrated with the photocopier example. For the first category called Product-Oriented Services (POS), a customer buys mainly the physical copier and can add some services if he wants. But the product is predominant on the associated services and belongs to the customer. In the second category Use-Oriented Services (UOS), a customer buys only the use of a set of product and services. Services will ensure to the customer the working of the product. With the photocopier example, the customer buys a number of photocopies and services will ensure him to use product without breakdown, or as less as possible. In this case, product does not belong anymore to the customer, but it is the provider who owns the product and the PSS in general. In the last category Result-Oriented Service (ROS), the customer only pays for the result provided by the PSS and services are predominant. In this case, a customer brings a document to a photocopying service, which will provide him the photocopies of his documents.

From the designer's point of view, another categorization can be made with only two categories. In one, product is sold alone to a customer and some services can be added. In the other one, it is a homogeneous set of product and services that is proposed to the customer. Indeed, the ROS category is just a use of a PSS by another person instead of the final customer. According to expected services, it will be necessary to reformulate the design problem of the product since the definition of services will lead to modifications on the product.

2.2 The Concept of "Service" in a PSS

The PSS acronym means Product-Service System, but generally the term service is not very obvious. Some people talk about service selling in the tertiary sector for example, others focus on specific services like maintenance or upgrading, or else one can discuss about the service provided by the product during the use. Thus it is necessary to identify which services will influence the design of the product.

With services modifying the product (e.g. maintenance) and services available but not linked to the product, it is often difficult to conclude on the real modifications on the design problem. For example, a service that deals with maintenance could be separated in two "sub services". On the one hand, one would operate on the physical product which would be the "front office" for the customer. On the other hand the other one would be the "back office" and would deal with the organizational side of the service that splits the different technical agents according to customer requests. In this case, two services are gathered in one and the border for the whole service is not very clear for designers. The organizational side could permit to identify new people included in the PSS and lead to detail new characteristics. The technical side of service will bring some additional specifications and constraints for the design of the product. By integrating services, the design problem will be changed and it is necessary to find the boundary of this new design problem. Thus, a representation of the PSS is needed in order to integrate services the earliest possible during the design process [7].

In order to clarify the design problem, we will describe one of the existing models that permits to describe a PSS.

3 A Representation for a PSS: The Agent-Based Model

The Agent-Based Model (ABM) has been developed in Service Engineering. This model is more focused on designing "pure" services rather than physical product but its representation could permits to represent PSS. This model permits to represent agents involved in the service and then the links between each others are detailed in order to increase the customer satisfaction. We will describe this model and will try to apply it on the PSS example "Call a Bike".

3.1 Model Description

Among existing service representations, we have considered the Agent-Based Model developed with Service/Product Engineering [8]. Service engineering is a design methodology used to increase the value of artifacts and decrease the load to the environment by focusing on services. Services can be seen as "an activity that changes the state of the customer". In the PSS concept, services are embedded with the product and it is the overall PSS that will satisfy the customer. In this PSS, several agents are involved to provide satisfaction to the customer.

ABM is made up of three level of representation:

- The flow model represents all agents involved in the PSS. Existing relationships can be identified and described in the scope model;
- The scope model is focused on a particular relation between two agents. These relations between agents represent the supply of value between a provider and a receiver whom is not necessarily the final customer. From this relation, parameters relating to receiver's satisfaction can be highlighted. They are called RSP (Receiver State Parameter);
- The view model details each RSP and leads to solutions used to fulfill the RSP. Then, technical solutions can be improved and optimized to increase customer's satisfaction.

3.2 Application of ABM on "Call a Bike"

Call a Bike is a German company that provides mobility to customer by the way of bikes. Actually, Call a Bike shares bicycles in Cologne, Berlin, Frankfurt or Munich. Bicycles are located at the most important crossroads of these cities. Everybody can use a bike, and pays only for the time of use. To use it, the customer calls a number (Call a Bike) marked on bikes and asks for using this bike (each bike has a particular serial number). Then the call centre gives him a code and the customer can unlock the electronic bolt bike and use the bike. As soon he has finished using the bike, he calls back Call a Bike to terminate the contract. Consequently the bike is locked again and can be used by another people. The customer only pays for use and he has not to bring back the bike at the initial location where he took it.

By applying ABM on the "Call a Bike" case, the flow model can be depicted as in Figure 2. The actors involved in the PSS are the city, Call a Bike, a maintenance service and the final customer.

Some relations exist between two agents and come from the influence between both. The city will be the provider of value in two different relations. In the first one between the city and the customer, the customer will be satisfied if sharing points will be indicated in the city. In the second one, the city will provide some space on the main crossroads to Call a Bike in order to leave bikes. The maintenance service will have to satisfy Call a Bike by fixing broken bikes owning to Call a Bike. Finally,

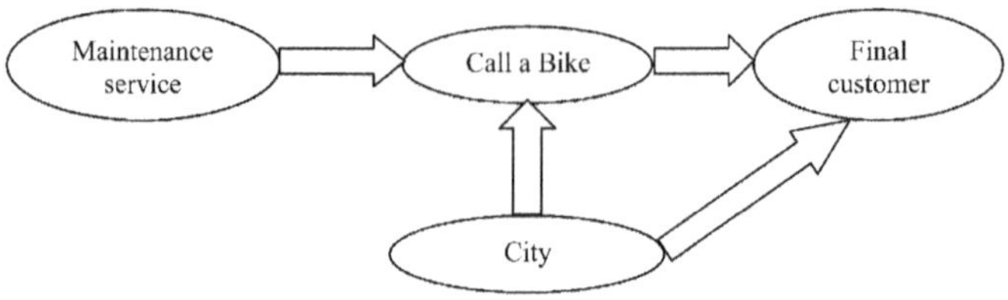

Fig. 2. Flow model of Call a Bike example.

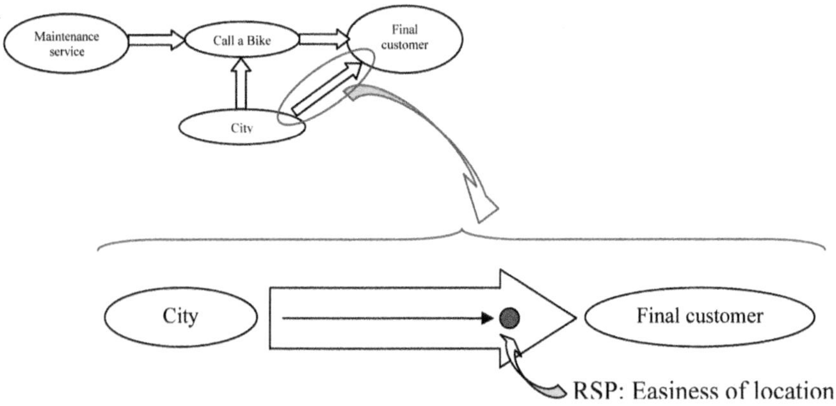

Fig. 3. Scope model between the city and the final customer.

Call a Bike will satisfy the final customer by providing working bikes, having an easy renting phone system, etc.

Then, each relation can be detailed in the scope model. In Figure 3, the relation between the city and the final customer is detailed in order to describe RSP linked to this relation.

In each scope model, an agent influences the other one by providing value. For example, the city will influence the customer and will increase his satisfaction. This increased satisfaction can be characterized by RSP, which are represented by points. For example, one RSP for the final customer in this model could be "easiness of location of sharing points" in the city.

Then, each RSP is detailed in a view model. As seen in Figure 4, the beginning of the tree in the view model is the RSP "easiness of location". Thus, we can describe this RSP by several functions and then these functions will lead to components. For example, in this case the RSP can be made by "indicating the sharing point in the city" to the customer. But in order to indicate these points, we can imagine several functions. One is to indicate these points on a map, another function is to have a special sign for sharing point, and the last can be to signpost the sharing

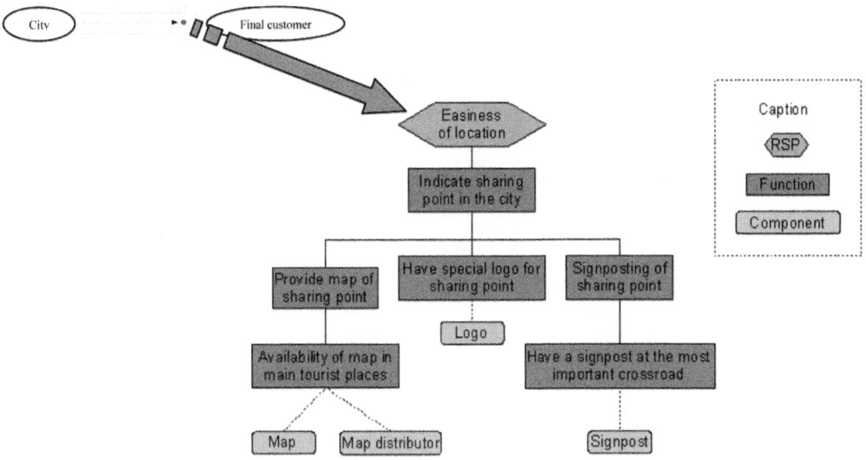

Fig. 4. View model of the RSP "easiness of location of sharing points".

points. Moreover for each solution we can detail further as for example to have maps available in the most principal tourist places, or to place the signposting at the most important crossroads of the city. By this model, we can reach components which are used to fulfill a specific RSP.

3.3 Conclusion

The Agent-Based Model can be used to represent all the agents involved in a PSS such as the providers of the PSS, agents supplying technical services and customer. This representation takes into account all agents and enables the designer to highlight characteristics between each others with RSP. These RSP are then detail in the view model and lead to technical solutions. But in this representation and specifically in the view model, technical solutions appear but the engineering designers will not necessarily take them into consideration. For example, the designer of the bike (the product designer) will not mind where signposting will be placed in the city in order to indicate the sharing point to the customer. One the other hand, as a wide picture of agents is drawn, it is possible to integrate new characteristics that the product will have to fulfill in this whole system coming from agents involved in the PSS.

4 Modeling of PSS during the Design of the Product

Designers who have to represent a product during the whole design process, can use a functional approach and in particular a Functional Analysis [9–11]. The Functional Analysis method considers the expression of the needs [12] of the various actors of the product life cycle as essential. This method proposes concepts (functions, constraints, assessment criteria) and tools (Graph of Interactors, FAST, etc.) to make them work and construct relations between needs and constraints, service functions, technical functions and solutions, stage by stage. Functional Analysis can thus be considered as a guideline for the whole design operation, as a dynamic way of looking at a problem to which all designers involved attempt to respond in a collaborative manner.

4.1 Description of Functional Analysis

The objective with Functional Analysis is to characterize the needs that have to be satisfied for the end-user or the buyer, by contributing to a complete description of the functions and their relations. Those are systematically characterized, classified and assessed. A function is described as the action of a product or one of its constituent parts. It is expressed, only in terms of finality, by a verb followed by a complement (e.g., "transform energy"). The method uses the concepts of External Functional Analysis and Internal Functional Analysis:

- External Functional Analysis lists the services that the product has to provide irrespective of the means available to provide them. It can be carried out on the basis of the consideration of the different situations in the product life cycle (use, manufacture, maintenance, recycling, etc.). The product[1] is, at this point, considered as a black box, which is able to satisfy customer needs and those of the professionals involved in making it [13]. It is described in terms of service functions and constraints. The service functions define (Figure 5):
  - The interaction functions (IF) which correspond to the services provided by the product during the product life cycle.
  - The adaptation functions (AF) that reflect reactions, resistance or adaptations to elements found in the outside environment (E1, E2, E3).

  These functions are obtained from systematic analysis of the possible relations between:
  - The different elements of the outside environment via the product for IF.
  - The elements of the outside environment and the product being designed for AF. Different tools are put forward (e.g., graph of interactors, Figure 5) to allow these needs to be expressed.

  A constraint is defined as a design characteristic, effect or provision for design, which is made compulsory or forbidden for whatever reason. A note specifies that two types of constraints have to be taken into account: those concerning solutions and those concerning the aim (in this case, the functions).

[1] The "product" designed with Functional Analysis can be a physical product, a service or an organization.

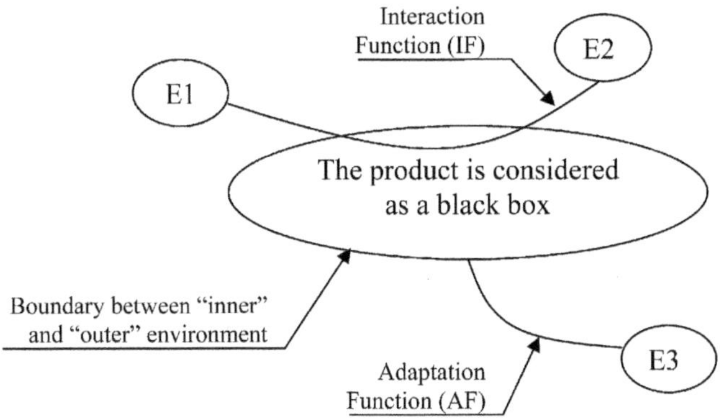

Fig. 5. External functional analysis and graph of interactors.

- Internal functional analysis, or technical functional analysis, enables us to analyze the resources necessary to provide the service required, identified in the external analysis. They are several tools to make an internal product analysis. The FAST is an example of a tool made available to identify the technical functions that could potentially fulfill the service functions, and then to match up those technical functions with potential solutions. The Functional Bloc Diagram (FBD) gives a schematic formalism of the product with the different components, the links between them, the functional flow and the flow concerning design choices. It is a mean to consider not only functional and technical components characteristics but also the interaction between components and especially the main functional flow of the product. With the FBD (see Figure 6) we can have information on:
  - The different components of the product and particularly, those necessary to the function realization.
  - The contacts between the components (simple lines)
  - The functional interaction and physical connection between components (black lines)
  - The design choices (buckles) that are for example technical solutions used to assemble components or to put them in position.

4.2 Application of the Functional Analysis

In order to determine what are the main differences and obstacles between a simple product and a PSS using functional analysis description, we have applied it on two cases: the bike, and the "Call a Bike" PSS. As said previously, the Functional Analysis method can be used to design a product or a service, but separately. Thus, it

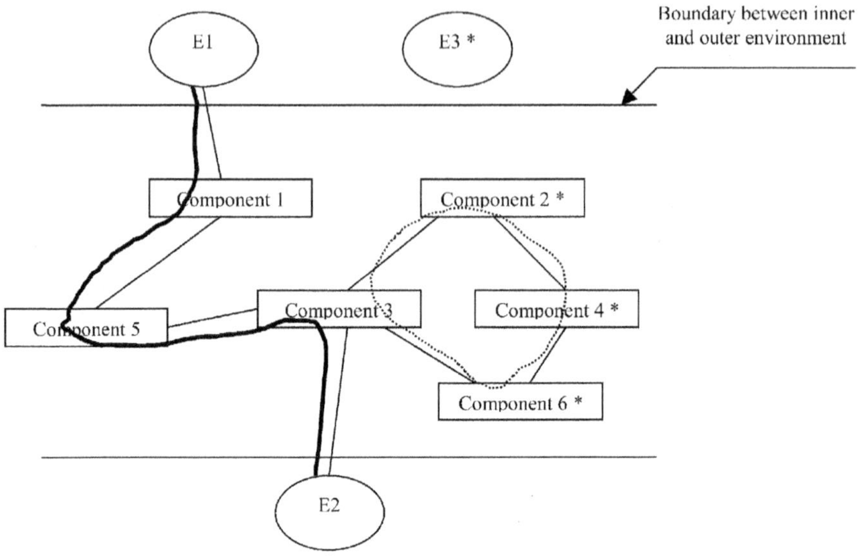

Fig. 6. Internal functional analysis and the Functional Bloc Diagram (FBD).

will be seen if it would be possible to represent a homogeneous set of product and services by this way.

4.2.1 Functional Analysis for a "Bike"

In this first application (see Figure 7), the graph of interactors is applied on a bike, which can be bought by a customer in a classical store. The bike allows an interaction between a customer and the city because the product permits the customer to move in the city. Then, we can have a maintenance service, which is proposed to the customer with the bike. Thus, we have an adaptation function that is "the bike must be adapted to the maintenance service". It means that if maintenance gets only a hammer and a screwdriver, the bike will have to be repairable with only those two repairing tools. The adaptation function provides constraints from outer environments.

A FBD (see Figure 8) can be realized after the graph of interactors. In this representation, links between outer environment and physical part can be made. The customer is physically sit on the saddle and drive with the handlebars and the rear and front wheel are in contact with the city. Links are described between customer and saddle (and handlebars) and between wheels and the road of the city. Moreover, the maintenance service is in relation with all components of the product because repairing is made if a breakdown occurs. As soon as a physical contact between parts or if information flow goes through a component, link have to be model in the FBD.

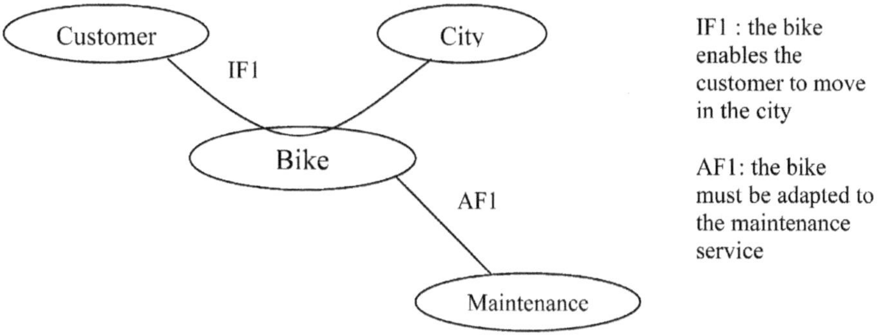

Fig. 7. External Functional Analysis on the product bike.

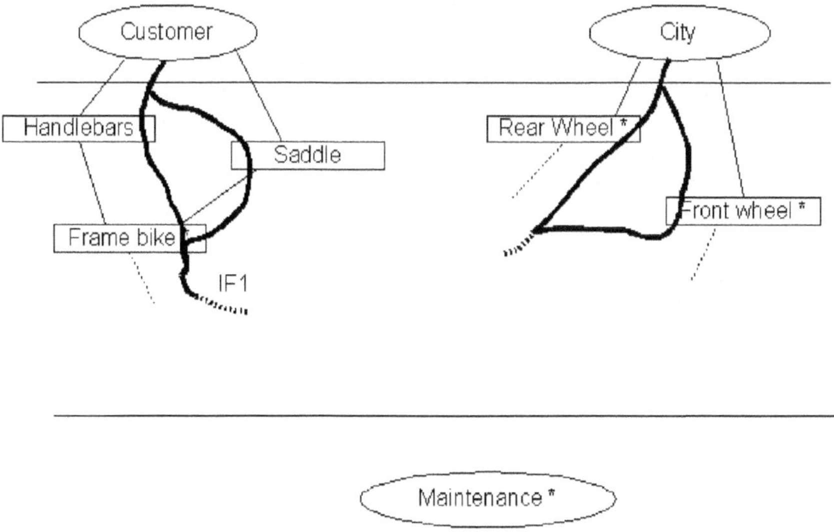

Fig. 8. FBD of the bike.

4.2.2 Application on PSS

In the previous paragraph, it was possible to model the interaction between the physical product alone and the outer environment with two types of functions: the interaction function and the adaptation function. But, with the addition of services, some others specifications will appear in the design problem. Here, the set of product and services will be represented. In the Call a Bike example, the customer calls for a bike and can use it as soon as Call a Bike gives an access to him. A new function has to be developed: "the bike enables the customer to establish a relationship with Call a Bike to use the bike". But for this function provided by the service (which

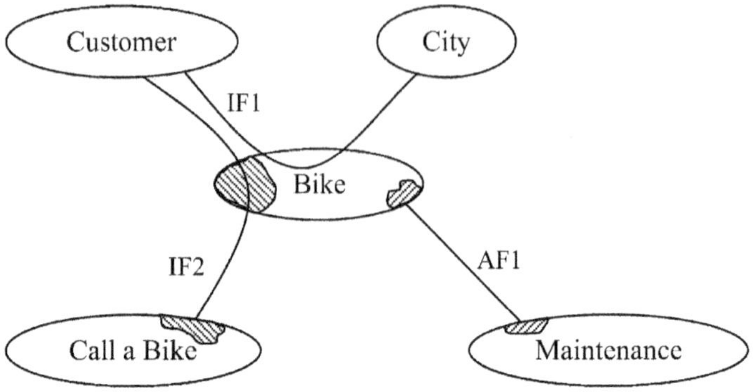

Fig. 9. External Functional Analysis on Call a Bike.

is the outer environment), some elements have to be added both on the product and the service to enable the relationship. Two hatched areas show that a new component must certainly be added on the product and on the outer environment "Call a Bike". For the adaptation function between maintenance service and the bike (AF1), we can imagine that new components have to be added to the product to optimize the maintenance operation. For example, if a breakdown occurs, the maintenance service can plug a tester on a recorder data system. Consequently, parts must be added onto the bike and the maintenance service so hatched areas have to be added on the both.

The functions of the interactors graph can be detailed as follows:

- IF1: the bike enables the customer to move in the city;
- IF2: the bike enables the customer to have a relationship with Call a Bike and to use the bike;
- AF1: the bike must be fixed by the maintenance service.

Now, if the FBD is considered (see Figure 10), new items appear. All of the bike components have been gathered in a black box called "Bike" and it is possible to think about the new components which have to be added on the product and the service in order to fulfill the new functions provided by the added services. On the one hand, if we look at the locking mechanism, an intelligent bold can be added on the "original bike" and consequently a new component part between the customer and the bike. On the other hand, Call a Bike needs a system that creates a code to unlock the bike. In this case, designers have to create for Call a Bike a system that generates the code to open the bike bolt. But the bolt system is linked to the bike, and the component that creates the code to open the bolt is linked to Call a Bike because it will be used by Call a Bike. With this diagram, it is obvious that the design problem will be bigger because of the new services proposed to the customer. These new services provide more satisfaction to the customer, but interfaces between the product and services are required, that is why the designer has to take into account more than the

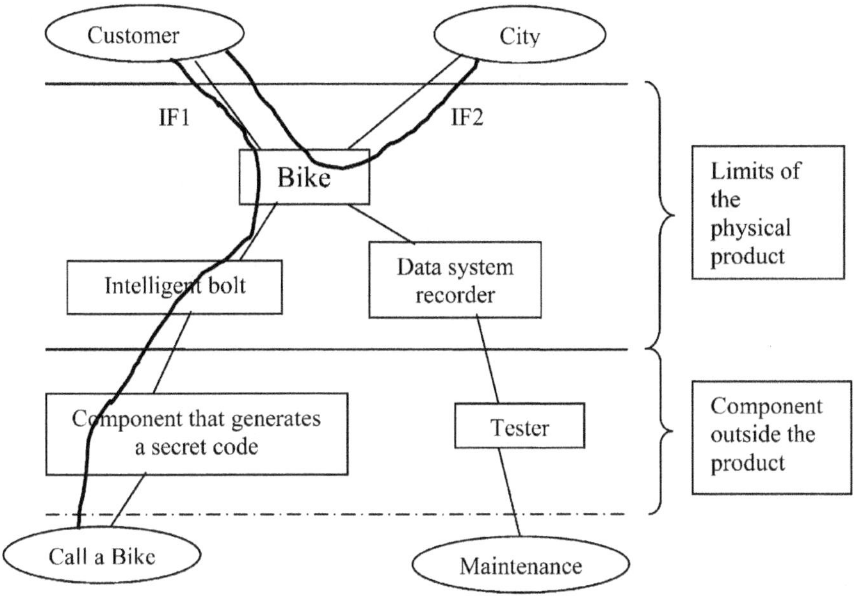

Fig. 10. FBD on Call a Bike.

physical product alone. New items are added and must be designed, but the border of the problem is larger and no more restricted to the physical product. For example, the component that creates the secret code is designed and added to Call a Bike, so it is outside of the product. In the FBD representation this item is not included in the problem, thus it is outside the border represented by the line. Indeed, new borders have to be created to limit the problem, which is not restricted to the physical component linked to the product. Now, if we consider the adaptation function between the maintenance and the bike, we can imagine that Call a Bike wants to improve and increase this service delivered to the customer. So, a recording data system could be integrated to the bike, and a tester could be used by the maintenance in order to detect breakdown. In this case, a component is added on the physical product, and another one in the maintenance service.

To represent a PSS is not easy because there is a mix between physical product and services. New services enlarge the design problem and new components have to be added on both the physical product and services. That is why it is necessary to represent the overall set of product and services during the design process. With the functional analysis, a representation of the function required between the product and services is possible in order to determine where supplementary parts must take place to fulfill the functions of the PSS.

5 Conclusion

The current market economy is shifting more to services added to physical products in order to stand out from competitors. PSS concept comes from both enterprises and new customer's needs. Nevertheless, to optimize "Quality-Cost-Time to market", it is necessary to include services during the design process of the product because if the services are developed separately from the product, modifications will have to be made on the product to include the added services.

In a current design problem, the first step is to analyze customer's needs and then to find the border of the problem. But in PSS cases, the design problem seems to be bigger and difficult to mark a boundary. During the design process of PSS, the product is the core problem for designers, but services will enlarge this problem. Designers will carry on the product involved in the PSS, but other physical components will be added on the service. Indeed, a PSS representation during the design process is necessary in order to help designers.

For the moment, we used the Agent-Based Model to represent existing PSS. This representation ensures to represent existing PSS solutions in order to optimize weak points to increase customer's satisfaction. It is more a design-oriented representation for PSS, which starts with an existing solution and that tries to optimize it from a particular point of view.

In order to represent PSS, we have considered the functional analysis tools and bring some modifications for PSS cases:

- Services included in the design phase can be described within the graph of interactors. It is possible to highlight services link and interaction with the product that have to be taken into account by designers.
- The FBD will list, in a more detailed way, components onto the product and the services necessary to increase customer's satisfaction.

Perspectives are to use Agent-Based Model and Functional Analysis representation in order to highlight PSS specifications. The Agent-Based Model enables to detail all people involved in the PSS and then highlights characteristics between each others. The Functional Analysis permits to detail external and internal functions linked to the product. Then, functions are described with criteria. But it should be more efficient to detail a functional analysis of the whole system rather than separate elements included in the PSS. Finally, the ABMs characteristics could be integrated in the criteria detailed in the Functional Analysis in order to integrated services during the design process of PSS's product.

References

1. Maussang N., Zwolinski P., Brissaud D., Design of product-service systems, in *10th European Roundtable on Sustainable Consumption and Production*, Antwerp, 2005.

2. Goedkoop M.J., Van Halen C.J.G., Te Riele H.R.M., Rommens P.J.M., Product service systems, ecological and economic basis, Technical Report, Pre Consultants, 1999.
3. Thoben K.D., Eschenbächer J., Jagdev H., Extended products: Evolving traditional product concepts, in *ICE 2001*, Bremen, 2001.
4. Bhamra T., Evans S., Van der Zwan F., Cook M., Moving from eco-products to eco-services, *Journal of Design Research* **1**(1), 2001.
5. Manzini E., Vezzoli C., Clark G., Product-service systems: Using an existing concept as a new approach to sustainability, *Journal of Design* **1**(1), 2001.
6. Tischner U., Verkuijl M., Tukker A., SusProNet report – First draft report of PSS review, Technical Report, SusProNet, October 2002.
7. Zwolinski P., Prudhomme G., Brissaud D., 2003, Environment and design. Methods and tools for integration and co-operation, in *Methods and Tools for Cooperative and Integrated Design*, S. Tichkiewitch and D. Brissaud (Eds.), Kluwer Academic Publishers, 2003, 11 pp.
8. Sakao T., Shimomura Y., Lindahl M., Sundin E., Applications of service engineering methods and tools to industries, in *12th CIRP Seminar on Life Cycle Engineering*, Grenoble, 2005.
9. Normes NF X 50–100, Analyse fonctionnelle, caractéristiques fondamentales, AFNOR, Paris, 1996.
10. Normes NF X 50–150, Analyse de la valeur, analyse fonctionnelle, vocabulaire, AFNOR, Paris, 1990.
11. Normes NF X 50–150-1, Vocabulaire du management, de la valeur, de l'analyse de la valeur et de l'analyse fonctionnelle, AFNOR, Paris, 1996.
12. Normes NF X 50–151, Expression fonctionnelle du besoin et cahier des charges fonctionnel, AFNOR, Paris, 1991.
13. Prudhomme G., Zwolinski P., Brissaud D., Integrating into the design process the needs of those involved in the product life cycle, *Journal of Engineering Design* **114**(3), 2003, 333–353.

Knowledge Management Aspects for Business Process Management: An Approach through the Information Management within and between Processes – Case Study at STMicroelectronics

Hendrik Busch[1], Mickael Gardoni[2] and Michel Tollenaere[2]

[1] *STMicroelectronics, Group ICT-KM, 850, rue Jean Monnet, F-38926 Crolles Cedex, France; E-mail: hendrik.busch@web.de*
[2] *INPG-GILCO, 46 avenue Félix-Viallet, F-38031 Grenoble, France; E-mail {gardoni, tollenaere}@gilco.inpg.fr*

Abstract. In the last few years, Business Process Management and its support by workflow tools became more and more important. It can be considered as the coordinated work and orders of tasks execution in order to produce a product or a service. The knowledge produced during a business process is also used during its execution in order to produce a final product. However, this use of knowledge occurs often locally within a process and not within the whole process, nor is it used among different production processes. As it is important to improve the information flow within and between these processes, we propose that process actors could take profit from approaches of introducing knowledge management activities in a business process management environment.

Key words: knowledge management, knowledge capitalization, knowledge retrieval, business process management, reuse of existing, return of experience.

1 Introduction

Knowledge Management gained popularity in the recent years and became almost a buzzword after 1995, where a lot of works were based on the research results from Nonaka [1]. Many enterprises introduced knowledge management in a variety of different forms whilst more and more scientific work was concentrated on knowledge management.

It is still however difficult to measure the return on investment and the gain of knowledge management activities, as many different definitions exist. Additionally, the application of knowledge management is often related to information technology (IT). As a lot of IT projects fail – because of different factors as lacks in identification

S. Tichkiewitch et al. (eds.), Advances in Integrated Design and Manufacturing in Mechanical Engineering II, 325–339.

of user needs, resource problems, acceptance problems, etc. – enterprises stopped a lot of knowledge management activities; but currently knowledge management concepts come back as support for innovation processes.

Furthermore, some scientific work models knowledge as objects. Concepts, method or other proposition to use Knowledge Management are often presented as stand-alone discipline. As a result, enterprises tried often to build up a new knowledge management system or knowledge base as a stand-alone tool, or as a separate activity, and many times at the end of the project or process. This KM activity as end phase was often considered as an additional workload by the user with no "added value" as this value is not recognized by the same user, and lead to the future of many industrial implementations.

In fact, knowledge management is not a stand-alone activity and has to be integrated in daily work, otherwise it will be considered as overload, and will encounter high barriers to acceptance. Empirical studies have shown that users of such KM systems do not enter information even if they have a personal gain in the future [2]. This is why knowledge management tools have to be integrated in daily work in order to first of all respond to a user need for the daily work. This requirement will guarantee the user acceptance and secondly provide knowledge management functionalities to introduce and promote a knowledge sharing culture in the enterprise.

Our focus is on knowledge sharing aspects in an asynchronous, delocalized environment. As a direct human contact does not therefore exist, knowledge is transferred in the form of information. The asynchronous, delocalized knowledge sharing is often supported by Information Technology (IT). Therefore, we think that – in this context – knowledge management cannot be separated from the information technology and by introducing knowledge management supported by information technology (IT), aspects from related domains as ergonomic, Data-Base, Information Management, etc. must be respected.

In this article, we will explain the context and characterize our research methodology. Then we will explain the observed actual practices in the scientific and industrial field as well as to present our *PIFA* method for analyzing knowledge intensive business processes that was developed and applied in a case study at STMicroelectronics to develop an IT-tool to support experiment processes management and integrate knowledge management activities in daily work.

2 Context

This research started in the R&D environment of STMicroelectronics. The microelectronic domain is a very dynamic and agile field where new technologies or manufacturing standards become quickly obsolete and have to be replaced or redesigned: Products could be obsolete after six months, new technologies appears every two years, conception phases for similar products are executed in parallel. Therefore new technologies and production methods (called a technology platform) will be developed. Even if this platform concerns only one specific technology, new fabrication methods could also be reused for current technologies and replace old production methods as some phases of the manufacturing method could be the same or similar.

In the microelectronic field, new ideas will immediately be tested by producing a prototype to verify if theoretical ideas are producible – to detect production limits that have not been taken into account in the theoretical model as well as detect measurements that are different from the theoretic calculations. Therefore, theoretical concepts will immediately be tested by producing a prototype (executed by an experiment process).

This research work is focused on this experiment process that is a monodirectional process where some tasks could be executed in parallel and it is characterized by:

(a) the technology that has to be qualified (process context),
(b) the machine that is used for the fabrication (process context),
(c) the employees who execute the experiments and analyze the results (actors),
(d) the number and type of wafers[1] that are used for the experiments (process context),
(e) the tasks to do (dynamic process structure),
(f) intermediate and final results (process data and information).

As it is a dynamic field, context conditions could change and cause the need to handle the dynamics aspects (tasks have to be executed again, tasks will be added, and actors change or wafers could be scraped[2] during the experiment and cannot be used any more). Therefore, it is necessary to mention that often not only the goal of the experiment (experiment description) change during the experiment, but the context information of the experiment.

3 The Research Methodology

This research started with a field study in order to understand the practices and problems encountered in the execution of the experiment processes and to identify the aspects and needs of handling the information and the dynamic of a process. This work was complemented by the definition of a proposal to satisfy the needs and has four mains phases:

[1] A wafer of virgin silicon sliced from a 4, 5, 6, 8 or 12 inch diameter silicon bar (2.54 cm is equal to 1 inch), which is used as the foundation to build semiconductor products on.

[2] A decision statement that a product that does not comply with the legal, statutory, contractual, technical requirement etc., cannot be used or recovered after reworking and must be destroyed.

Fieldwork: The aim of this phase is to understand the practices and problems by executing the experiment process as well as to understand which role knowledge management can play for handling such a knowledge intensive process where information changes rapidly and influence the execution of the process. The process was studied in two different ways:

- Firstly, the working methods of employees were analyzed as well as the teamwork. This allowed to understand the different work and different processes employees are involved in as well as to analyze where the information is coming from and going to. Additionally, it helps to formalize which information is mandatory and which one could additionally be interesting.
- Secondly, a process was followed from the beginning to the end to analyze the execution and the information flow aspects. This field was studied for six months, and we performed interviews with 31 different process actors. This analyzing phase also helped to produce the PIFA-method to analyze knowledge intensive business processes.

Understanding and abstraction of the gathered information: here, the aim was to understand and analyze the gathered information to propose some knowledge management elements to take into account and to integrate knowledge management aspects in business process management.

Comparison of the needs and existing workflow tools: the aim was to understand where functionalities of classic workflow tools do not respond to the needs as well as do not handle the information aspects.

Instrumentation of the proposal: this phase consists of the design of a tool to support the information management combined with business process management and integrates knowledge management aspects. As the tool proposition is often designed for a specific application field, others aspects have to take into account as the immediate surplus value, ergonomic rules, and specific functionalities besides the functionalities that are responding to the scientific concepts. A good way to built up an IT is to propose a sampler (html code only with simple functionalities) to get a feedback to the propositions from potential users. But only using the tool can detect missing functionalities, divergence from user need or prove an acceptance by the users.

4 Some Features of Knowledge Intensive Dynamic Business Processes

The fieldwork, together with the study of the literature about the functioning and handling of business processes, has allowed highlighting some important characteristics of such processes: Business Processes are often supported and managed by workflow management systems.

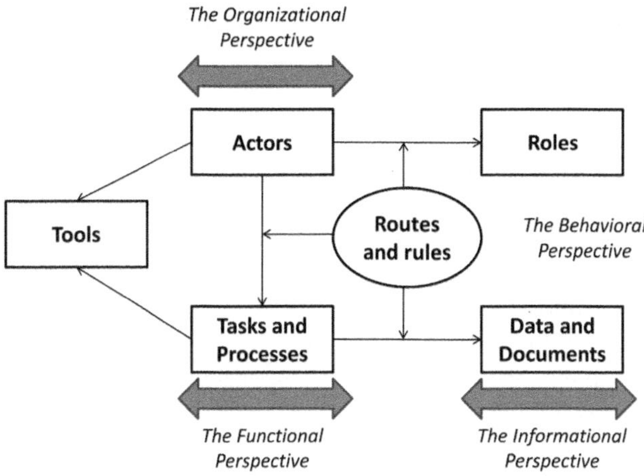

Fig. 1. Four perspectives od a workflow model framework [9].

The commercial activities in this area have increased dramatically in the last few years since the start of the Workflow Management Coalition in 1993 [3]. Most workflow management systems are based on the four perspectives of business process representation, i.e., functional, behavioral, organizational, and informational [4–8].

The diagram in Figure 1 illustrates the relationship between the four perspectives. This diagram is essentially a meta-model for workflow management systems.

- The *functional perspective* indicates that workflow management systems need to specify the tasks and the underlying rationale of a workflow by decomposing high level functions into tasks that can be allocated to human or software agents.
- The *behavioral perspective* refers to the need for specifying when and how the tasks are performed; these can be specified using process logic in Petri nets,[3] or other process models [10] like IDEF or UML.
- The *organizational perspective* seeks to answer the question of who performs what tasks and with what tools. In workflow management systems, the organizational perspective involves actors, roles, resources, and resource management rules that can be modeled with organization charts and object hierarchies.
- The informational perspective relates to the business data and documents that are the subjects of workflow activities. In workflow management systems, informa-

[3] Petri nets were invented by Carl Adam Petri to model concurrent systems and the network protocols used with these systems. The Petri nets are directed bipartite graphs with nodes representing either "places" (graphically circles) or "transitions" (graphically rectangles). When all the places with arcs to a transition (its input places) have a token, the transition "fires", removing a token from each input place and adding a token to each place pointed to by the transition (its output places).

tion is usually organized in object hierarchies or networks and stored in databases or file systems.

Media-Mora [11] defines Business Processes as market-centered description of organization's activities, implemented as information processes (create, process, manage, and provide information) and/or material processes (assemble physical components and deliver physical product). A business process is triggered to fulfill a business contract or satisfy a specific customer need. Georgakopoulos [12] defines a workflow as a collection of tasks organized to accomplish some business process. A task can be performed by one or more software systems, one or a team of humans, or a combination of these. Human tasks include interacting with computers closely. It is composed as a predefined order of tasks. Each task is assigned to a role. A role can be assigned to a group of person or to only one person.

The dynamic aspects of business processes add another aspect to take into account: frequent changes in all of the four perspectives. Bachimont [13] already emphasized that models do not model the reality, but propose instruments to explore the sources that humans put in relation with its situation of use. A contextual, unpredictable interpretation of the sources canŠt be modeled. Van der Aalst [14] distinguishes two different types of changes that have to be supported:

- context Changes for a process: "handling modifications of specifications because of changing conditions", i.e. duration, actors, data;
- unpredictable changes for the process execution: "dealing with unanticipated events resulting from an incomplete process flow model.

An efficient process structure is important for a fast and optimized process execution. The produced knowledge at every step as well as the final product present the added value of the process and the Intellectual Property (IP) producing this product. As the context or process may change, it is difficult to merge the right information to the right task.

5 The Current Knowledge Management Practices Implemented in Business Process Management

The traditional workflow systems provide an "action inbox" with open tasks to each user. The process structure or the state of other processes is always a "black box" for the user. Nowadays, it becomes more and more important to introduce knowledge retrieval in current processes (i.e. actual state, problems, implicated actors, etc.).

Zhao [9] has another approach to the previous defined four perspectives. He defines a knowledge management model in business process management systems while concentrating in particular on three different types of knowledge:

- *Process knowledge* that contains the description of tasks, roles, rules, and routes.

- *Institutional knowledge* that describes the roles, the actors, and business procedures and regulations.
- *Environmental knowledge* that describes the business environmental factors such as governmental regulations, industrial associations, competitors, and customers.

We retain that these categories also match to the defined *functional → process knowledge, organizational → institutional knowledge and behavioral → environmental knowledge.*

We propose to enlarge these perspectives with another one as our field study suggested that the added value of a process is also the produced result:

- *Information perspective → Content knowledge* that represents all data and documents produced during the execution of a business process.

Plesums [23] emphasizes that the "rendez-vous" aspect (automatic matching of incoming information to the work that is suspended) of workflow management system becomes more and more important as today's information is produced with a large number of different tools. Actual knowledge retrieval is only concentrated on the three described types of knowledge (named as process, institutional and environmental), often to look for the context (process state, due date, actors, ...). During the execution of a process, huge amount of knowledge is produced and used in later tasks. It is difficult for an employee to match the incoming information to the right process as well as to check each time if all needed information already exist for a task to start its execution.

Today, more and more processes are executed in parallel and more and more information are produced in different tools. For each task type in a process, a specific information type will be needed and produced. The dynamic aspect of these business processes is a difficulty to take into account. Context is essential to understand and to internalize the knowledge correctly. It is very knowledge intensive and important for the understanding of the knowledge. As the business process is dynamic – meaning that often changes occur – it is first of all difficult to manage the process correctly. On the other hand, it is difficult to provide the right context information to describe the produced knowledge correctly, as the knowledge depends on the context used for its internalization [15, 17]. It is therefore important, not only to merge information to the right task, but also to give enough context to this information for an immediate use for a task as well as for a later reuse.

Assuming the acceptation of workflow management systems in the industrial field, knowledge management activities can also be implemented in Workflow Management Systems. The information flow is part of the process, but it could be different from workflow. The focus of the knowledge management activities is on the process execution and process improvement. Presently, there is no implementation of "cross-over" knowledge management sharing activities between processes, meaning that the produced knowledge during a process can be retrieved by different actors, especially from actors involved in parallel, similar processes.

As the exchange of "final" and "intermediate" knowledge objects become more and more important for a concurrent engineering environment where processes are executed in parallel, we assume that capturing this knowledge at its source by integrating knowledge management aspects in workflow management systems could be one of the best practices to capture a maximum of "useful knowledge" as well as to introduce a knowledge sharing be tween processes. Some propositions will be explained in the next sections.

6 Approach: Implementation of Knowledge Management in Business Process Management

We are interested in finding ways to benefit from the knowledge [1] produced throughout the execution of processes to improve the knowledge flow within a process and between processes. However, Wunram [18] indicates that "the approaches that start with the goal of capturing all the knowledge of the employees are predetermined to fail". Knowledge Capitalization has to be introduced in order to respond to a knowledge need in order to manage the knowledge [19]:

Therefore, two guiding ideas have been followed:

- Capitalized knowledge must respond to an identified need.
- Capitalization activities must be combined with an immediate added value for the daily work to break down the barriers of knowledge capitalization.

In order to develop a knowledge management method that could be integrated in a business process management, a method to analyze the process and its related knowledge flow has been developed; it is described in the following section.

7 Process Information and Functionality Analysis (PIFA) – A Process Analyzing Methodology for Knowledge Intensive Business Processes

In usual daily work, more and more information tools are used and implemented in order to make tasks easier and improve the working conditions. These changes also impact the business processes as information needed for the process are not delivered with the task description, but have to be retrieved from different sources as humans and IT tools. Introducing knowledge management aspects in Business Process Management can provide the necessary information to the right people as well as reuse information from existing and already executed processes. Therefore, the information flow must also be established between different processes and forwards and backwards within the processes.

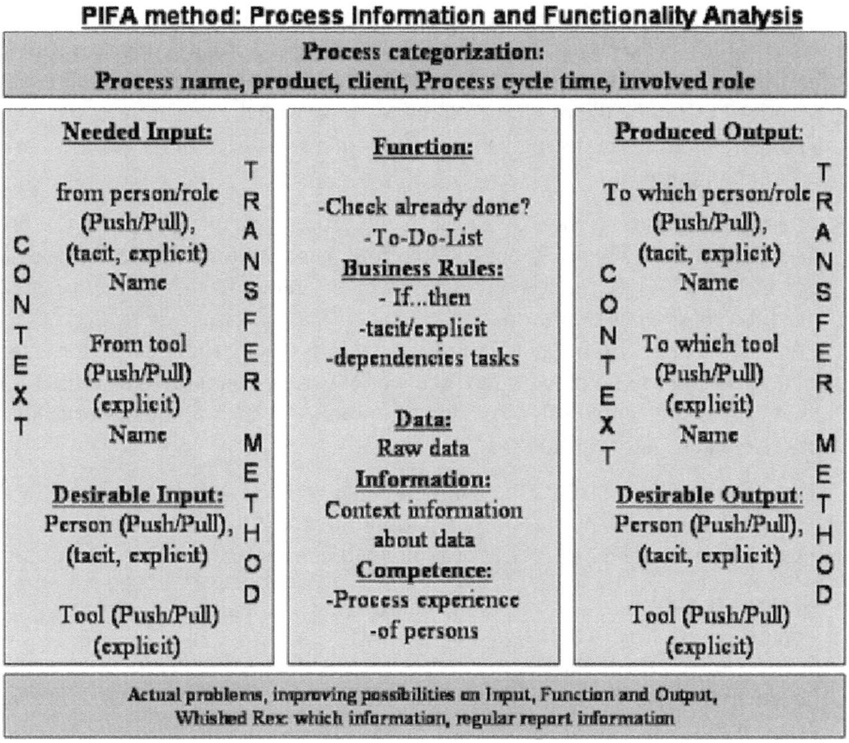

Fig. 2. PIFA method – Process information and functionality analysis.

Most of the time, information merging to the right tasks exists, but a knowledge sharing backwards within a process or cross-over between processes does not exist. Therefore, we distinguish two different kinds of knowledge: *Needed and desired knowledge*. Theses terms are discussed below.

Figure 2 proposes our method Process Information and Functionality Analysis (PIFA) that is inspired by the existing methods from the project management institute [20] and method "H" [21]. Processes are a structured execution of tasks. Each task is composed in three parts: the *Input*, the *Functional* and the *Output* part. Each task is related to a process and has a specific context. A process is unidirectional to produce a good or a service, but it might be convenient to introduce an information flow backwards the process to give a *return of experience (REX)* to all involved actors as well as to introduce a cross-over knowledge sharing between processes. This analysis is based on the tasks and its execution based on conditions (business rules/events). Therefore, this analysis could especially be used for a dynamic environment where process structure changes.

These five aspects are described in the following:

Input, Functional, Output: These three parts are the core of each task and their structured order of execution build up the process flow. For each task, we can describe which functionality must be done and on which data the functionalities are based on. Additionally, the produced output to go on with the workflow is the result of the functionalities. We distinguish as already said two different types of information as input or output:

- Information that is *needed* to go on in the process chain. Information is produced to keep on with the process. Therefore, our proposal includes an analysis of which kind of information is used and where it comes from in order to analyze the different sources of information.
- Information that is *desirable* to improve the results or save time. This information is not necessary to go on with the process flow, but is an additional value if it is used. The source could often be a similar process to have first ideas or analyze positive and negative results.

The source of this information can be human or IT and it is transferred in an explicit or implicit way by pushing or pulling methods. In the following list, we summarize the identified categories for Input/Output information:

- Name (free text)
- Context description (free text and categories)
- Source (Person (Role)/Tool)
- Information Type (Necessary/Desirable)
- Transfer Type (Explicit/Implicit)
- Transfer Method (Push/Pull)
- Transfer Support (Mail, phone, tool)
- Conditions for launching the activities

The conditions for opening the actions are very important for the process flow to determine when an action can be open (start to execute). Additionally, information and context can change; therefore some tasks probably have to be done again ($\rightarrow$ reopen a process activity). A categorization by each of the categories in the list could help to identify the right information in a process.

The heart of a task is the work to do, described as the functions a task owner must execute. These functions are based on the input and on different business rules that define how the functions have to be executed to transform the input to an output based on business rules. These functions require a certain competence of the person who is in charge of the task.

Context: The process flow can be categorized in order to distinguish different processes. This process description gives the context of a process and helps to identify different groups of processes. We have used the process family name as main category to characterize and distinguish process instances. In order to describe the process more explicitly, other categorization data could be "client's name", "process cycle time", "involved employees", "responsible roles", "description", etc. Used categories should be determined and adapted to each applied case context. In order to allow a better information sharing and retrieval, it is considered to categorize the process in the most detailed way. As best practice we can retain to characterize the process context of each task. Therefore, we distinguish three types of categories:

- Process family name – a value that describe the activities in the process and helps to distinguish the different processes.
- Process characterization – process annotation by prefixed categories.
- Process characterization – process annotation by free text annotation.

REX: Return of EXperience: Furthermore, the capitalized information of a process or part of them could be interesting for all involved actors as well as for different actors or the management. Therefore, it is interesting to define the return of experience, they want to have about their tasks: needed information about the following process execution and results in order to see if their work were satisfied or not. This REX could help to improve the personal competence of an employee by putting done work in the context to the process and compared to the result of the process. Therefore, the process view must be analyzed, the employees want to have about process-follow-up, synthesis, etc.

The process flow is often mono-directional, but the information flow can be bi-directional as well as it is necessary to provide a summary-view on the process progress at any time. As the actors have different needs, it will be necessary to define filters for this information access that implies in which way and with which kind of view they want to access to the information as well as which information should be displayed.

8 Solution's Principle to Improve the Knowledge Sharing and the Feedback Loops

Our PIFA-analysis has shown that the transfer type of information is in most cases already explicit, but some information is not enough formalized and people have problems to internalize them. Therefore, they call people directly to catch the missing knowledge that is often a more precise context to understand the transferred document. This context should not remain implicit as it could also be used for a better information retrieval in an asynchronous, delocalized environment by annotation of the process to give more context.

Our goal is to improve the knowledge sharing (push (passive) and pull (active)) for process information. To optimize the Information Retrieval for a formalized environment, we rely on the annotations of documents by free text annotations [22] and by values from predefined categories structured in a kind of ontologies, taxonomies or similar [23].

Based on the facts, that the experiment processes could be categorized by a finite word-list, experiment process structures are repetitive and context is similar, the context of an experiment can be described in a standardized way.

We built up an ontology for each of the involved domains in the experiment process (manufacturing, R&D, engineering) to capture the different categories and their values. In a second step we overlapped the different "ontologies" to determine a common vocabulary (shared ontology) that will be understandable to each employee of the three domains and that will be used for a common annotation of the processes. We proposed as well to keep the specialized vocabulary of each domain to annotate the experiment more precisely for each of them.

For a problem to be solved, the source of the problem will be analyzed more precisely: (machine problem, recipe problem, volume effect, etc.). After the problem is solved and a new solution was found, the result will automatically be spread to other people already worked or who are currently working with these or similar parameters. This could avoid to do the same errors again as well as to improve the personal competence of the employees. This diffusion is not only useful for a feedback-loop between the engineering and the R&D part, but also between the different generations of technology.

9 Knowledge Sharing and Process Management for Experiment Processes

As already mentioned in this paper, in an asynchronous and delocalized environment, IT is often used with the aim to support knowledge management by manage the information flow with the goal to transfer and share the existing knowledge. In the present case study at STMicroelectronics, we designed a software tool to support experiment processes that build microelectronic chip prototypes for validating theoretical ideas.

To guarantee a user acceptance, the tool must provide the classic workflow functionalities to manage a process (support daily work activities) as well as the "rendezvous" functionality to merge information to the right task in the process flow direction. A user will explain all needed information to start an experiment in the designed tool and the information will be spitted and merge to the right tasks in the process flow.

To provide "knowledge" retrieval methods, the tool must provide push functionalities to inform actors about new tasks, available information for tasks or information about results (as a return of experience). Otherwise, it must provide "pull" (on demand) functionalities to allow at any time, actors, clients, managers to search for information like build up global, general reports as to search concrete knowledge items. As in our context, technology platform development and working methods are similar; we suppose that the originator and the recipient interpret the information

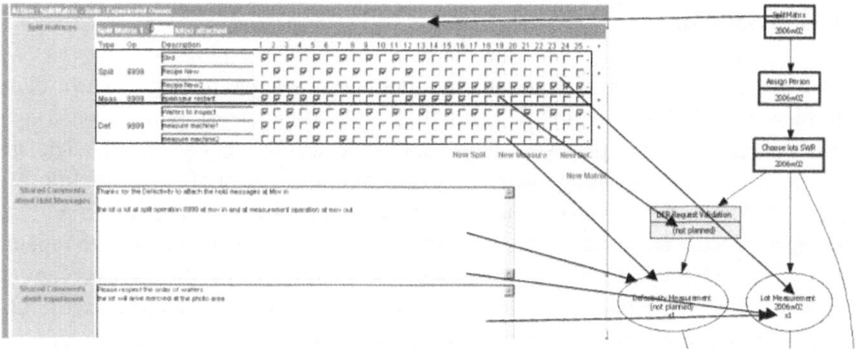

Fig. 3. Split the information and merge them to the right task.

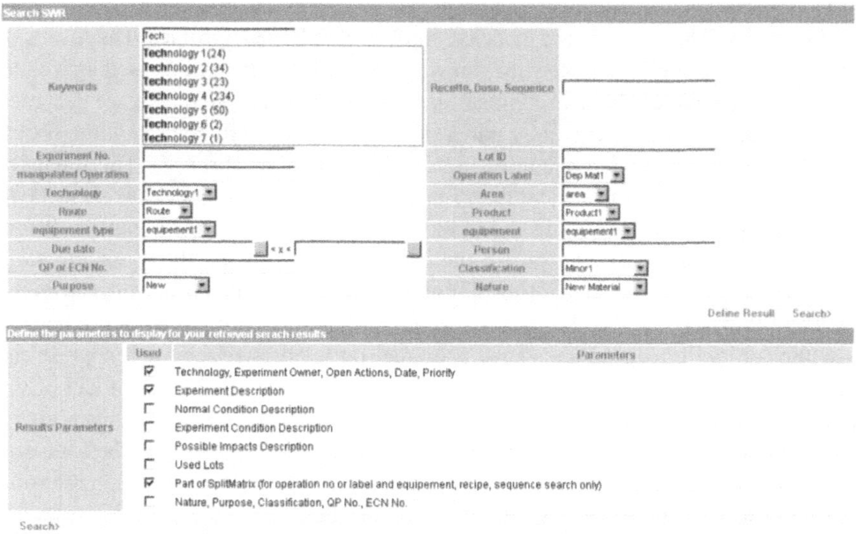

Fig. 4. Knowledge sharing between processes (ex. for an knowledge retrieval interface).

in the same way as the context is similar. The knowledge interfaces are necessary to introduce a cross-over knowledge sharing between processes.

This tool is deployed since January 2006 in a pilot phase. The return of the users shows that we successfully solved and combined the problems of dynamic process management as well as the knowledge capitalization. The knowledge retrieval efficiency is actually be analyzed by observing the use of the knowledge retrieval interfaces.

10 Conclusion

This paper shows the importance of introducing knowledge management in daily activities instead of introducing KM activities as stand-alone discipline. Since a context of a large number of projects, organizational project barriers prevent the delocalized, asynchronous knowledge sharing by direct contact, we are focused on improving and reducing the organizational barriers supported by information technology. Therefore, we developed first of all the PIFA-method that helps to analyze repetitive knowledge intensive business processes. The analysis results give an overview of which knowledge is produced and used for the process and which knowledge could be interesting and be used for similar processes. Since we are focusing on supporting knowledge management functionalities by information technology – based on this analysis – in a delocalized and asynchronous environment, we decided to design an IT tool that includes the described aspects.

Our method has been tested in a case study at STMicroelectronics. The results of an analysis permitted to identify the knowledge that is produced and used as well as the knowledge that could be used in similar process context. Therefore, we designed up a software tool that is currently under development and will be used at the Front-End Technology and Manufacturing R&D Site in Crolles (France). The intensive work with the potential users and their reaction on the IT-tool model has shown two different aspects: the interest in the workflow functionalities shows that the PIFA-method has given a good analysis of the daily work and the produced information. The interest in the knowledge retrieval functionalities shows that users agree with the need and they are interested to test the tool. On the other hand, it is not predictable how the tool will be used and if the provided retrieval interfaces will support the knowledge need sufficiently. The PIFA-method has given a sufficient model of the domain to propose an IT-tool to support daily work as well as knowledge activities. As the model is always abstracted from the reality, the tool cannot cover the complete needs. It is therefore important to keep in mind that the implicit knowledge sharing between users always exists and could sometimes be more efficient than a software tool.

References

1. Nonaka, I. and Takeuchi, H., *The Knowledge-Creating Company*, Oxford University Press, 1995.
2. Creß, U., Von der Schwierigkeit, Wissen zu teilen – eine psychologische Sichtweise, *Revue Wissensmanagement*, Büro für Medien Verlag, Augsburg, 03/2004.
3. WfMC, http://www.aiai.ed.ac.uk/WfMC/index.html, http://www.wfmc.org/information/introduction_to_workflow02.pdf: Workflow Management Coalition, 1996.

4. Bussler, C. and Jablonski, S., An approach to integrate workflow modeling and organization modeling in an enterprise, in *Proceedings of the 3rd Workshop on Enabling Technologies*, 1994, pp. 81–95.
5. Curtis, B., Kellner, M.I. and Over, J., Process modeling, *CACM* **35**(9), 1992, 75–90.
6. Gruhn, V., Business process modeling and workflow management, *Int. J. Cooperative Information Systems* **4**(2–3), 1995, 145–64.
7. Kwan, M.M. and Balasubramanian, P.R., Dynamic workflow management: A framework for modeling workflows, in *HICSS97*, 1997.
8. Zhao, J.L., Knowledge management and organizational learning in workflow systems, in *Proceedings of the Fourth Americas Conference on Information Systems*, Association for Information Systems, 1998, pp. 647–649.
9. Kumar, A. and Zhao, J.L., A framework for dynamic routing and operational integrity controls in a WFMS, in *HICSS96*, 1996.
10. Medina-Mora, R., Wong, H. and Flores, P., The action workflow approach to workflow management, in *Proceedings of the 4th Conference on Computer-Supported Cooperative Work*, June, 1992.
11. Georgakopoulos, D. and Hornick, M., An overview of Workflow Management: From process modeling to workflow automation infrastructure, *Distributed and Parallel Databases* **3**, 1995, 119–153.
12. Bachimont, B., Pourquoi n'y a-t-il pas d'expériences en ingénierie des connaissances?, in *15ème Journées Francophones d'Ingénierie des Connaissances*, Lyon, 2004.
13. Van der Aalst, W.M.P. and Jablonski, S., Dealing with workflow change: Identification of issues and solutions, in *Computer System Science and Engineering* CRL Publishing Ltd, 2000, pp. 267–276.
14. Busch, H., Gardoni, M. and Frank, C., Conception of an information retrieval system based on meta-data with an access according multiple points of view, in *International Conference on Integrated Design and Manufacturing in Mechanical Engineering*, 2004.
15. Herrmann, T., Kontextberücksichtigung als Kernaufgabe der Wissenskommunikation, in *Wissenskommunikationen in Organisationen*, Springer Verlag, 2004.
16. Reinhardt, R., *Wissenskommunikation in Organisationen – Methoden, Instrumente, Theorien*, Sammelband, Springer Verlag, 2004.
17. Wunram, M., Weber, F., Pawar, K.S. and Gupta, A., Proposition of a human-centered solution framework for KM in the concurrent enterprise, in Pawar, K., Weber, F., Thoben, K-D. (Eds.), *Proceedings of the 8th International Conference on Concurrent Enterprising – Ubiquitous Engineering in the Collaborative Economy*, Italy, 2002, pp. 151–158.
18. Prax, J.-Y., *Le guide du Knowledge Management*, Dunod, Paris, 2000, 226 pp.
19. Project Management Institute, *A Guide to the Project Management Bbody of Knowledge (PMBOK guide)*, Newtown Square, Pennsylvania, USA, 2000 Edition.
20. Turbit, N., Business analysis using "Method H", White Paper, http://www.projectperfect.com.au/downloads/info_method_h.pdf, 2005.
21. Denoue, L. and Laurence Vignollet, L., L'importance des annotations – Application à la classification des documents du Web", *Indexation Conceptuelle et Structurelle* (Special Issue) **1**(1), 1998.
22. Sure, Y., Factsheet: Ontologies, in *KT Web*, 2003.
23. Plesums, C., Introduction to workflow, Computer Sciences Corporation, Financial Services Group, Workflow Management Coalition – WfMC.

PROCESS PLANNING

Machining Strategy Choice: Performance Viewer

Laurent Tapie[1], Kwamivi Bernardin Mawussi[1,2] and Bernard Anselmetti[1,3]

[1] *LURPA – ENS Cachan, 61 avenue du Président Wilson, 94235 Cachan Cedex, France; E-mail: {laurent.tapie, kwamivi.mawussi, bernard.anselmetti}@lurpa.ens-cachan.fr*
[2] *IUT de Saint Denis, Université Paris 13, Place du 8 Mai 1945, 93206 Saint-Denis Cedex, France*
[3] *IUT de Cachan, Université Paris 11, 9 avenue de la Division Lecler, 94234 Cachan Cedex, France*

Abstract. Nowadays high speed machining (HSM) machine tool combines productivity and part quality. So mould and die maker invested in HSM. Die and mould features are more and more complex shaped. Thus, it is difficult to choose the best machining strategy according to part shape. Geometrical analysis of machining features is not sufficient to make an optimal choice. Some research show that security, technical, functional and economical constrains must be taken into account to elaborate a machining strategy. During complex shape machining, production system limits induce feed rate decreases, thus loss of productivity, in some part areas. In this paper we propose to analyse these areas by estimating tool path quality. First we perform experiments on HSM machine tool to determine trajectory impact on machine tool behaviour. Then, we extract critical criteria and establish models of performance loss. Our work is focused on machine tool kinematical performance and numerical controller unit calculation capacity. We implement these models on Esprit® CAM Software. During machining trajectory creation, critical part areas can be visualized and analyzed. Parameters, such as, segment or arc lengths, nature of discontinuities encountered are used to analyse critical part areas. According to this visualization, process development engineer should validate or modify the trajectory.

Key words: high speed machining, technical knowledge, performance evaluation.

1 Introduction

The high quality and high productivity in machining of complex parts has been of primary interest in manufacturing industries these last years. To improve the efficiency of NC machining, high speed machining (HSM) is now widely used in moulds and dies manufacturing industries [1]. The complex shapes of these moulds and dies are rather defined as complex features which are mainly represented by their geometric parameters [2]. The geometric representation of features has been widely ex-

S. Tichkiewitch et al. (eds.), Advances in Integrated Design and Manufacturing in Mechanical Engineering II, 343–356.

plored in several works [3, 4]. In conventional machining, the geometric features are associated to machining features through a recognition process [5, 6]. In an HSM context, the machining strategies applied, such as Z-level or parallel planes [7], lead to the manufacturing of the whole part in a single sequence [8]. At this stage, the geometric parameters are not sufficient for the development of an efficient machining process. From the results of the machining simulation, A. Dugas [9] showed that it is necessary to take into account security, technical, functional and economical constrains in the development of machining processes. During the machining of the complex features, it is difficult to keep the programmed feed rate in some areas. The feed rate reduction which is observed in these areas is closely linked with the geometric variation of complex features shapes and technical constraints of the machine tool. Several works were devoted to these technical constraints [10] which can be classified as follow [9]: machine tool, numerical controlled unit (NCU) and cutting tool limits.

When analyzing the geometric variation of shapes, the technical constraints of the machine tool must be integrated in the development of the machining strategy. This integration is generally difficult because machine tool and NCU limits are not well known. The specialist does not have tools such as monitoring to associate the technical constraints to the geometric variation of the shapes. In this paper, we propose an original procedure to make easier the choice of machining strategies. This procedure is based on experimental observations of our machine tool, tool paths analysis and machining performance visualization. The experimental observations are conducted according to kinematical constrains and NCU calculation capacity. They lead to the modelling of the machine tool behaviour. The tool paths are analysed according to the instantaneous feed rate which is computed. The visualization of the performance is done directly in the work space of the Esprit® CAM software in order to assist the specialist.

2 HSM Technical Constrains

When machining complex features, important variations of the shapes geometrical characteristics lead to a feed rate reduction according to machine tool and NCU limits. This feed rate reduction decreases the part accuracy as the programmed tool path is not respected. The geometrical characteristics are mainly tangential and curvature discontinuities. When these discontinuities appear in a program, the NCU adjusts in an anticipated way the feed rate in order to compensate for its brutal drop: this is the static look ahead phenomena.

To underline the static look ahead phenomena and to elaborate the indicators useful for the analysis of tool paths, we carried out experiments in a milling center Mikron UCP 710 with Siemens controller 840D. Records of the feed rate were carried out thanks to its integrated oscilloscope reading output encoder data. The experimental results are presented afterwards.

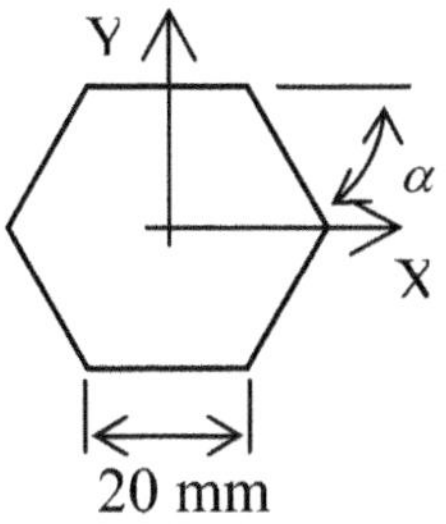

$$\alpha \in [5-10-20-30-40-60-90-120°]$$

$$Feed\_speed \in [2-4-8-12m/\min]$$

Fig. 1. Programmed trajectories.

2.1 Machine Tool Cinematic

2.1.1 Tangential Discontinuities

To highlight the machine tool behaviour during tangential discontinuity crossing between two consecutives segments, we carried out the experiments proposed in [11]. The tool path is a polygonal shape and the factors considered are the corner angle α (given by the number of segments in the polygon) and the programmed feed rate (Figure 1). The segments have the same length.

Theoretically to cross a tangential discontinuity with a given feed rate, the machine tool must produce an infinite acceleration [9]. To avoid this impossible increase in acceleration, a solution consists in forcing a null feed rate at the end of each program block. In this case, vibrations can affect the machining accuracy. Another solution is to modify the tool path in order to keep the programmed feed rate. But the modification of the tool path introduces variations which can also affect the machining accuracy. Most of HSM controllers integrate a hybrid solution whereby the programmed feed rate is reduced in order to control the variations induced by the modification of the tool path.

Our experimental records show a significant reduction of the feed rate (Figure 2a) at the tangential discontinuities. The 840D controller of the Mikron UCP 710 milling center integrates the hybrid solution presented above. The tool path is modified by the insertion of a curve which we represented by an arc of a circle and the reduction of the feed rate is not dependent on that programmed. The analysis of the kinematic behaviour shows that at the discontinuity the feed rate is constant and the tangential acceleration is null (Figure 2b). Moreover, it highlights the influence of the angle between two consecutive segments of the tool path (see Figure 1, angle α) on the feed rate. Indeed, the feed rate decreases especially as the angle is small. This angle thus represents an indicator for performances evaluation.

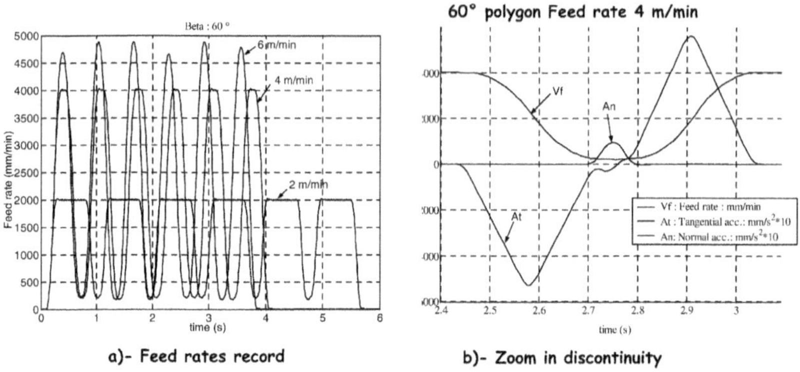

Fig. 2. Records for a 60° polygon.

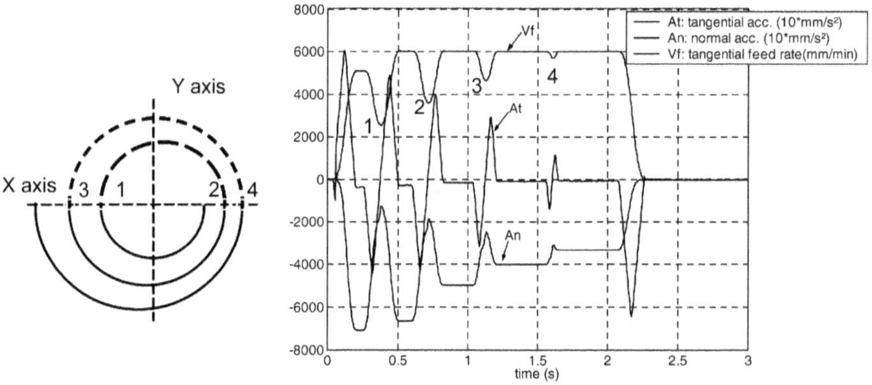

Fig. 3. Analysis of curvature discontinuities.

2.1.2 Curvature Discontinuities

Experiments are based on spiral tool paths programmed with the circular interpolation mode (G02 type). Different radii of curvature and feed rates are tested. Spirals are constructed with semi circles or quarter circles (example of semi circle spiral is shown in Figure 3). At each radius change 1, 2, 3, and 4 (Figure 3), a curvature discontinuity is created. The acceleration and feed rate records highlight the influence of curvature change. The decrease of the tangential feed rate is accompanied by variations of the normal acceleration.

As in the case of tangential discontinuities, the feed rate at the curvature discontinuity is independent of that programmed. Nevertheless its variation is linked to the curvature radius change. So, the curvature radius variation can also be retained as indicator.

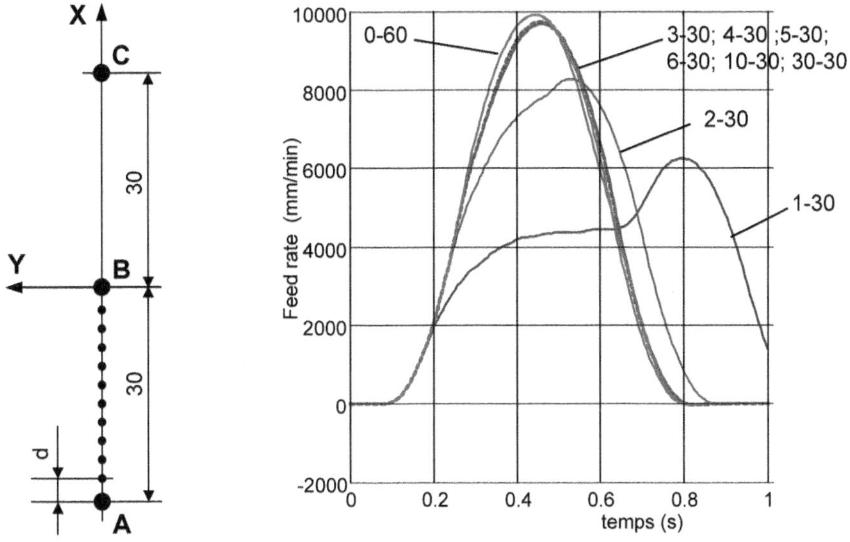

a)- Tool path b)- Feed rate records

Fig. 4. Instantaneous feed rate surveys.

2.2 NCU Calculation Capacity

Experiments are based on straight line tool path defined along each axis of the machine tool. Figure 4a shows an example of X axis. Between points A and B, the tool path is defined by successive points separated by a distance d. Between points B and C, only one segment of 30 mm is defined. The distance d takes different values as it is shown in Figure 4b. Tests are carried out with 10 m.min^{-1} feed rate.

Over 3 mm for the distance d, the programmed feed rate is reached. Under this value, the instantaneous feed rate decreases. For 1 mm distant points, the reduction of the instantaneous feed rate at point B is about 55%. The comparison between the two parts of the tool path (AB and BC) shows the insufficiency of the interpolation time to process blocks of 3 mm with a feed rate of 10 m.min^{-1}. Indeed the controller reduces, if necessary, the feed rate to allow the axis to cover the distance d with a time equal or more than the interpolation time. On the basis of these experiments, we choose the calculation capacity of the controller as indicator for the performance evaluation.

3 Performance Viewer

The main objective of the performance viewer is to assist process development engineers during machining strategy elaboration. It is integrated after tool path computa-

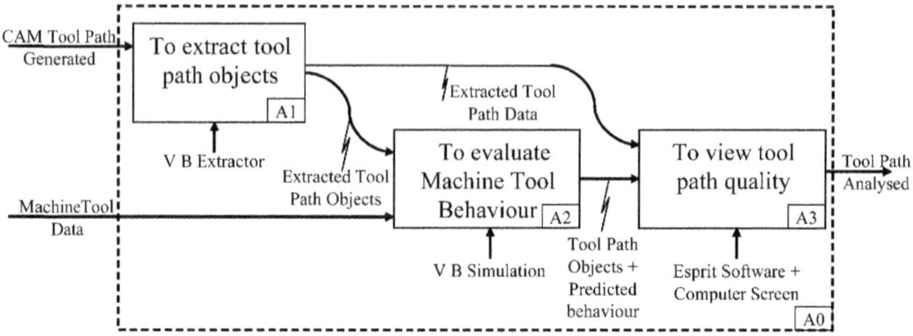

Fig. 5. Performance viewer: general structure.

tion with a CAM software. Indeed, to analyze the machining performance according to the chosen indicators, we use the machine tool technical constraints and the data resulting from the computation of tool paths. In the scope of our work the performance viewer is implemented in the Esprit® CAM software. Esprit® is programmed with objects defined in Visual Basic language. We implemented three main tasks (Figure 5). First, tool path objects are extracted (task A1). Then machine tool behaviour is evaluated according to the indicators (task A2). From the results of tasks A1 and A2, the machining performance is displayed in the Esprit® workspace (task A3).

3.1 Tool Path Objects Extraction

Tool path extractor is divided in two sub-tasks which explore the tool path (object in VB language) selected in the workspace of Esprit®. Indeed, the tool path object is composed of tool path type, feed segments and rapid segments. The algorithm developed analyses the tool path object in order to extract geometrical parameters (segment and arc) and machining data such as feed rates.

3.2 Machine Tool Behaviour Evaluation

Machine tool evaluation behaviour integrates feed rate modelling. The models used are based on two main works [11, 12] and our experiments. As presented below, the data processed by the models are derived from the tool path object.

3.2.1 Tangential Discontinuity Crossing Model (Model 1)

In this work, we consider that a circle arc is inserted at each tangential discontinuity, in the basis of the results of our experiments. So, the first step of machine tool behaviour modelling is to determine the arc radius. The value of the radius given by

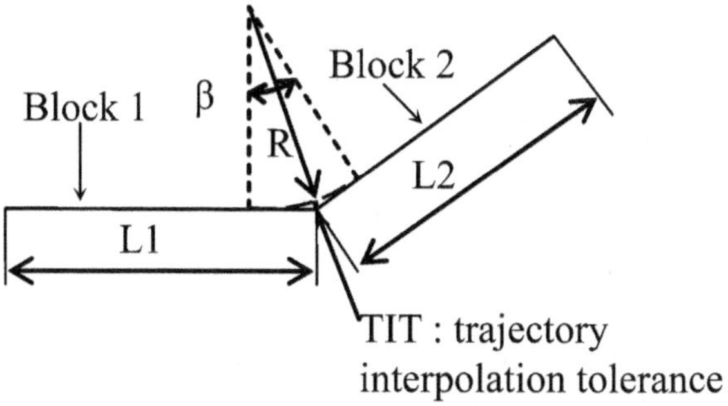

Fig. 6. Model 1: radius model calculation.

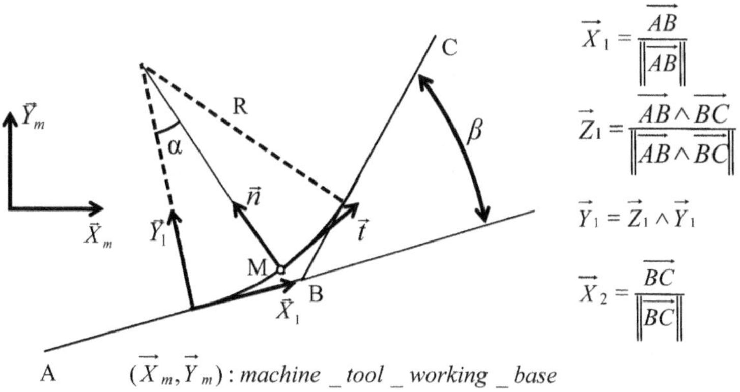

Fig. 7. Crossing arc radius inserted in XY plane.

equation (1) depends on the trajectory interpolation tolerance (TIT) (Figure 6). TIT identification on our machine tool is based on the test protocol proposed in [9] and our experiments.

$$R = \min\left(\text{TIT} \times \frac{\cos(\beta/2)}{1-\cos(\beta/2)};\ \frac{l}{2 \times \sin(\beta/2)} - \text{TIT}\right) \quad \text{with } l = \min(l_2; l_2). \tag{1}$$

The minimal feed rate is reached when the tangential acceleration is null. This phenomenon has been explained in literature and highlighted in our experiments (see Figure 2).

In Figure 7, we consider that the minimal feed rate (V_f) is reached at the point M on the arc of a circle. So, we can deduce that at the M point the feed rate is constant and the normal acceleration can be calculated as follows:

$$\mathbf{A} = a \cdot \mathbf{n} = \frac{V_f^2}{R} \cdot \mathbf{n}. \tag{2}$$

The jerk value can also be calculated as follows:

$$\mathbf{J} = \frac{d\mathbf{A}}{dt} = \frac{da}{dt}\mathbf{n} + a\frac{d\mathbf{n}}{dt} = \frac{V_f^2}{R} \cdot \frac{V_f}{R} \cdot \mathbf{t} = \frac{V_f^3}{R^2} \cdot \mathbf{t}. \tag{3}$$

According to equations (2) and (3), the feed rate is limited by the normal acceleration or the tangential jerk. The feed rate limited by the normal acceleration is deduced from equation (2). Its value is given by:

$$V_{f\_a} = \sqrt{a \times R}. \tag{4}$$

The feed rate limited by the tangential jerk is deduced from equation (3):

$$V_{f\_\mathrm{Jerk}} = \sqrt[1/3]{\mathrm{Jerk} \times R^2}. \tag{5}$$

The feed rate can be also limited by the maximal tangential feed rate ($V_{\max}$) according to the machine tool feed rate capacity. By taking into account the machine tool kinematic characteristics, we can deduce the feed rate limitation:

$$V_f = \min(V_{\max};\, V_{f\_a};\, V_{f\_\mathrm{Jerk}}). \tag{6}$$

The kinematic characteristics (maximum feed rate, maximum acceleration and maximum jerk) given by machine tool manufacturers or evaluated are different for each axis. So, the tangential feed rate, the normal acceleration and the tangential jerk defined in the model 1 must be computed according to each axis capacity. We developed an algorithm to compute the values of these kinematic characteristics.

For instance in the case described in Figure 7, the minimum feed rate is reached at the point M. The feed rate can be limited by the X axis or the Y axis capacity. To present the algorithm developed, we consider that the limitation is introduced by the jerk of the X axis. This limitation is reached if:

$$\mathbf{t}(\alpha) \cdot \mathbf{Y}_m = 0 \Leftrightarrow \cos(\alpha) \times \mathbf{X}_1 \cdot \mathbf{Y}_m + \sin(\alpha) \times \mathbf{Y}_1 \cdot \mathbf{Y}_m = 0. \tag{7}$$

From equation (7) the restrictive angle (α_{J_X}) associated to the jerk limitation for the X axis can be deduced. Then the tangential jerk at point M is computed as follows:

If $\alpha_{J\_X} \in [0, \beta]$ then Jerk $= J_X$ where J_x is the maximal jerk of the X axix. (8)

$$\text{Else if } \mathbf{t}(0) \cdot \mathbf{Y}_m > \mathbf{t}(\alpha) \cdot \mathbf{Y}_m \text{ then Jerk} = \frac{J_X}{|\mathbf{X}_1 \cdot \mathbf{Y}_m|}. \tag{9}$$

$$\text{Else Jerk} = \frac{J_X}{|\mathbf{X}_2 \cdot \mathbf{Y}_m|}. \tag{10}$$

The same computation process can be applied for the maximal feed rate and acceleration for the Y axis in our example. Finally, the jerk values given by equations (8), (9) or (10) for the two axes X and Y are compared. The value of the tangential jerk used in model 1 equation (5) is the minimal one. The same reasoning is made to deduce the tangential feed rate and the normal acceleration defined in model 1.

3.2.2 Curvature Discontinuity Crossing Model (Model 2)

To calculate the feed rate when crossing a discontinuity between a segment and an arc, we use the corner radius model developed in [12]. As is shown in Figure 3, the tangential acceleration is null at the curvature discontinuity. The normal acceleration also changes at the curvature discontinuity and its value can be written in a Frenet frame ($\mathbf{t}$, $\mathbf{n}$) as follows:

$$|\mathbf{A}(t + \delta t/2) \cdot \mathbf{n} - \mathbf{A}(t - \delta t/2) \cdot \mathbf{n}| = \text{Jerk} \times \delta t. \tag{11}$$

δt is an elementary time variation necessary to pass the discontinuity. This elementary time has been evaluated for our machine tool during the works carried out by Pateloup [13]. It has also been identified during our experiments.

When crossing the discontinuity the jerk is tangential and the feed rate V_p is constant. So, only the normal acceleration can be calculated with the programmed arc radius R as follows:

$$|\mathbf{A}(t - \delta t/2) \cdot \mathbf{n})| = 0 \quad \text{and} \quad |\mathbf{A}(t + \delta t/2) \cdot \mathbf{n})| = \frac{V_P^2}{R}. \tag{12}$$

The feed rate of the model 2 given by equation (13) can be deduced from equations (11) and (12):

$$V_P = \sqrt{\text{Jerk} \times \delta t \times R}. \tag{13}$$

A curvature discontinuity model for two consecutive arcs was developed by Pateloup [13]. Its observations about the tangential acceleration and the normal acceleration correspond to those we did for an arc and a segment. Thus, by applying the reasoning of Pateloup, the feed rate at the discontinuity can be calculated as follows:

$$V_{fr} = \sqrt{\frac{R_{r1} \times R_{r2} \times \text{Jerk} \times \delta t}{|R_{r1} - R_{r2}|}}, \tag{14}$$

where jerk is the tangential one, R_{r1} and R_{r2} are respectively the radius of curvature before and after the discontinuity, δt is the elementary time variation necessary to pass the discontinuity.

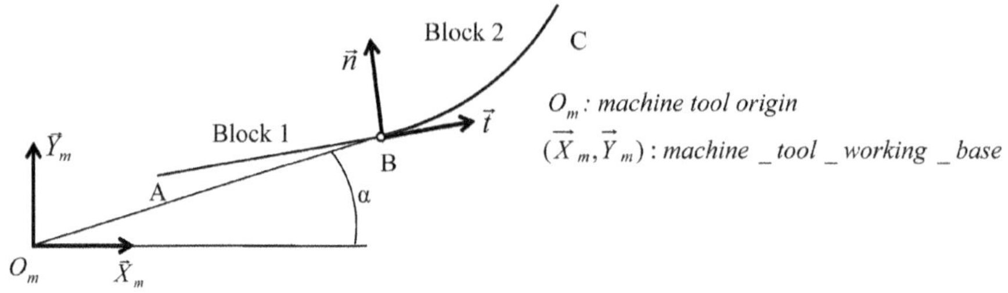

Fig. 8. Crossing curvature discontinuity in XY plane.

For each model, the tangential jerk is computed according to its capacity. In the example shown in Figure 8, the discontinuity appears at the point B. The tangential jerk can be calculated as follows:

$$\text{Jerk} = \min\left(\frac{J_X}{|\cos(\alpha)|}; \frac{J_Y}{|\sin(\alpha)|}\right). \tag{15}$$

3.2.3 NCU Capacity Model (Model 3)

As we observed in our experiments (Figure 4) the controller calculation capacity is linked to the minimal distance L between to consecutive points on the tool path. For a given programmed feed rate, the distance L can be evaluated from the interpolation time t_{int} of the NCU as follows:

$$L = V_P \times t_{\text{int}}. \tag{16}$$

The interpolation time is difficult to evaluate with experiments [9]. To calculate the distance L, we use the value of tint given by Siemens for the 840D controller ($t_{\text{int}} = 12 \times 10^{-3}$ s).

3.3 Tool Path Quality Display

The tool path quality evaluation is based on the comparison of the programmed feed rate and the predicted feed rate. According to a colour code, this comparison is displayed on the Esprit® workspace. The three indicators retained are used for the evaluation of the tool path: NCU calculation capacity and discontinuity crossing (tangential and curvature discontinuities). Figure 9 shows the general structure of the quality display module.

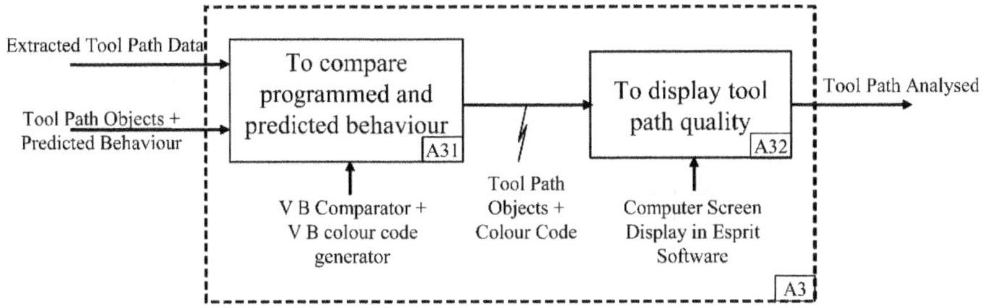

Fig. 9. Tool path quality display structure.

3.3.1 NCU Calculation Display

Each segments or arcs length is read and according to equation (16) the length is compared with the minimal distance L which corresponds to the programmed feed rate. Then each segments and arcs are classified according to its length. Figure 10 shows an example of the evaluation carried out during a Z-level spiral finishing tool path.

The machining strategy performance analysis is made according to the capacity of the Mikron UCP 710 milling center. Two main observations can be made from this analysis. First, several small segments or arcs have been generated with the chosen machining strategy. This observation is illustrated by the high number of portions of tool paths beneath the minimal distance *L*. Second, some areas generated more small segments and arcs than others. These areas correspond either to features having small dimensions, or with the inadequacy of the selected machining strategy. From these observations, process development engineer can locally modify the tool path or globally change the machining strategy.

3.3.2 Discontinuity Crossing Display

Figure 11 shows the machine tool behaviour visualization at discontinuities. The predicted feed rates computed according to models 1 and 2 are compared with the programmed feed rates. Each discontinuity crossing point is classified according to the programmed feed rate reduction percentage.

The visualization underlines the areas already identified during controller calculation capacity (encircled areas in Figure 10). Indeed, the areas in which the length of the segments and the arcs are lower than the minimal length *L* correspond in general to a reduction of more than 90% of the programmed feed rate. This observation highlights the complementarity between the two types of visualization and indicators relevance. It also makes it possible to retain the inadequacy of the machining strategy for these areas.

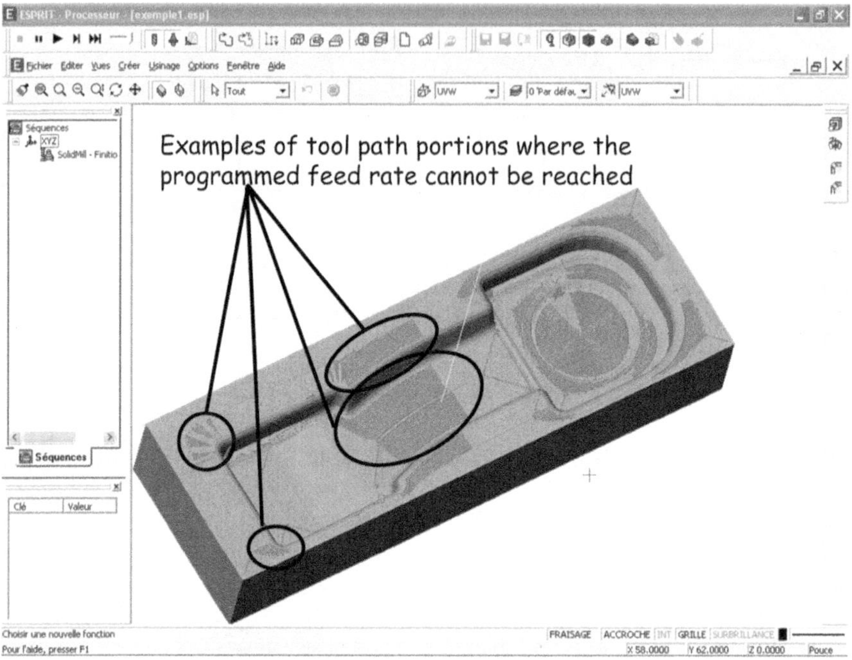

Fig. 10. NCU capacity display.

4 Conclusion

In this paper we propose a performance viewer developed in Esprit® CAM Software. Three main modules have been developed: tool path object extraction, machine tool behaviour evaluation and tool path quality visualization. As the performance viewer is integrated in the Esprit® workspace, the development of the modules is carried out using the Visual Basic language. The machine tool behaviour evaluation is based on the prediction of the feed rate according to geometrical elements (segments and arcs) found in the tool path. The indicators identified for this evaluation are controller calculation capacity and two machine tool kinematical behaviours: tangential and curvature discontinuities. These indicators have been modelled and experiments have been performed to determine some parameters of the models.

Our work is focused on three axis tool path often used in mould and die machining. In this context, linear and circular interpolated trajectories are analysed with our performance viewer. Thanks to this viewer, die and mould makers are able to analyse machining strategies and to evaluate the tool path quality according to the machine tool capacities.

The future applications of the work presented in this paper are directed towards the decomposition of topology of moulds and dies in order to improve the quality of

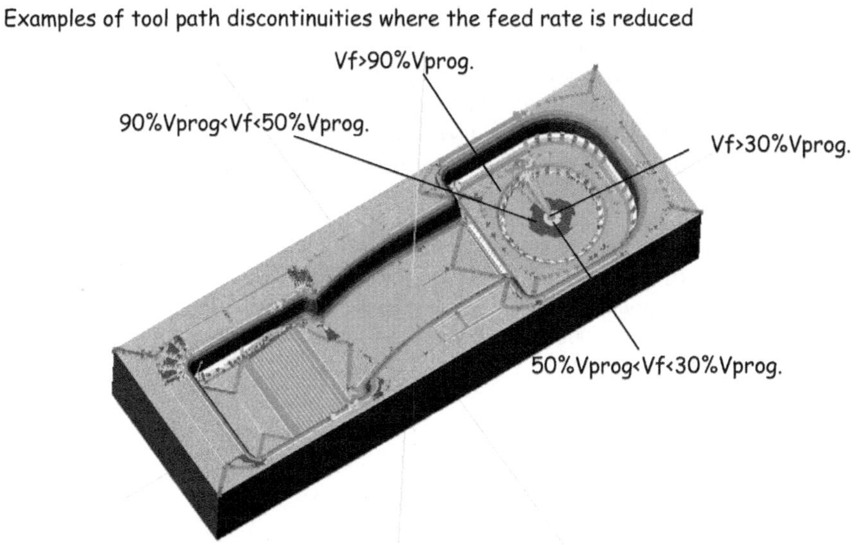

Fig. 11. Crossing discontinuity display.

machining. This decomposition is an alternative to the machining features recognition which is very difficult to be implemented in the case of complex parts machining.

References

1. P. Fallböhmer, C.A. Rodriguez, T. Özel, T. Altan, High-speed machining of cast iron and alloy steels for die and mold manufacturing, *Journal of Materials Processing Technology* **98**, 2000, 104–115.
2. K. Mawussi, A. Bernard, Complex dies representation and definition with complex feature-based model, in *Proceedings of the 15th International Computers in Engineering Conference (ASME)*, September 17–21, Boston, USA.
3. E. van den Berg, W.F. Bronsvoort, J.S.M. Vergeest, Freeform feature modelling: concepts and prospects, *Computers in Industry* **49**, 2002, 217–233.
4. V. Sundararajan, P.K. Wright, Volumetric feature recognition for machining components with freeform surfaces, *Computer-Aided Design* **36**, 2004, 11–25.
5. X. Yan, K. Yamazaki, J. Liu, Recognition of machining features and feature topologies from NC programs, *Computer-Aided Design* **32**, 2000, 605–616.
6. X. Zhang, J. Wang, K. Yamazaki, M. Mori, A surface based approach to recognition of geometric freeform surface machining, *Computer-Aided Design* **36**, 2004, 735–744.
7. C.-M. Chuang, H.-T. Yau, A new approach to Z-level contour machining of triangulated surface models using filet endmills, *Computer-Aided Design* **37**, 2005, 1039–1051.
8. A. Flutter, J. Todd, A machining strategy for toolmaking, *Computer-Aided Design* **33**, 2001, 1009–1022.

9. Dugas A., Simulation d'usinage de formes complexes, Thesis graduate dissertation, IRCCyN, Ecole Centrale Nantes, 2002 [in French].
10. K. Mawussi, A. Larue, P. Ryan, E. Mougel, W. Belluco, Experimental characterisation of a high speed machine tool, in *7th CIRP International Workshop on Modeling of Machining Operations*, CDROM paper, Cluny (France), May 4–5, 2004.
11. M. Monreal, C.A. Rodriguez, Influence of tool path strategy on the cycle time of high-speed milling, *Computer Aided Design* **35**, 2003, 395–401.
12. V. Pateloup, E. Duc, P. Ray, Corner optimization for pocket machining, *International Journal of Machine Tools & Manufacture* **44**, 2004, 1343–1353.
13. V. Pateloup, Usinage à grande vitesse des poches : prise en compte du comportement dynamique de la machine outil dans le calcul de la trajectoire, DEA graduate dissertation, LURPA ENS, Cachan, 2002 [in French].

Towards an Approach for Rapid Copying of Free-Form Surfaces in 5-Axis Machining

Pierre Breteau[1], François Thiébaut[2], Pierre Bourdet[1] and Claire Lartigue[2]

[1]*Ecole Normale Supérieure de Cachan, 61 av. du Pt Wilson, 94235 Cachan Cedex, France; E-mail: {pierre.breteau, pierre.bourdet}@lurpa.ens-cachan.fr*
[2]*Laboratoire Universitaire de Recherche en Production Automatisée, IUT de Cachan, Université Paris-Sud 11, 9 avenue de la Division Lecler, 94234 Cachan Cedex, France; E-mail: {francois.thiebaut, claire.lartigue}@lurpa.ens-cachan.fr*

Abstract. The paper deals with the various steps of free-form copying from surface acquisition to machining using five "positioned" axes. The originality of the proposed approach is to be free from surface reconstruction.

The object surface acquisition is performed using an optical digitizing system in order to obtain a representative image of the real surface. Relative positions and orientations sensor/surface are chosen to guarantee the completeness of the digitized points as regards the whole object surface. The next step concerns the pre-processing of the points. Indeed, the data are discrete, non-homogeneous, dense and noisy. Cleaning, filtering and densification are essential steps. A voxel representation of the data is carried out with the objective to recreate "pseudo-continuity". Therefore, the calculation of local characteristics such as local normal to the discrete data is thus possible. The last step is the choice of optimal part set-ups for the machining using five positioned axes. To initialize the automatic research of the best set-up, each relative orientation sensor/surface may define one set-up. For each set-up, the trajectory of the ball-end cutter tool is calculated in three axes, using the offset inverse method. The feasibility of the whole approach is illustrated through an example.

Key words: surface machining, digitized data, 5-positioned axis machining, voxel representation.

1 Introduction

Rapid copying of complex-shaped parts is the process that consists in directly machine data points collected from the measurement of a physical model. More generally, existing methods to reproduce objects with free-form surfaces are made up of the two independent steps [1]: (1) surface reconstruction from the acquired data points; and (2) tool path generation using the constructed surface model. The first

S. Tichkiewitch et al. (eds.), Advances in Integrated Design and Manufacturing in Mechanical Engineering II, 357–371.

step, also well-known as Reverse Engineering, includes preliminary steps such as data pre-processing, data segmentation and data fitting. It is generally time consuming and strongly linked to the possibilities offered by the CAD system used. The resulting surface model is thus an approximation, more or less correct, of the digitized point cloud. The second step, which is performed independently, relies on a CAM system. Henceforth, tool path generation using a CAM system is an activity well-controlled, offering numerous possibilities in 3-axis machining as well as in 5-axis machining. However, the calculated tool path is also an approximation of the CAD model [2]. The combined procedure suffers from error accumulation and inefficiency [1].

In this context, direct copying, that means directly calculating the tool path from the discrete data points, appears to be an interesting way to explore. Indeed, the step of surface reconstruction is removed, limiting approximation errors and time consuming. Nevertheless, few works propose direct tool path generation from data points. The main reasons are linked to the discrete nature of the data, the lack of data structure and the data quality. This point will be discussed in the next section. Furthermore, few approaches address the problem of 5-positioned axis machining. For each part set-up, the portion of surface accessible by the tool (most generally a ball-end cutter tool) is machined according to a 3-axis strategy. The issue here is to find the minimal number of set-ups allowing the whole surface to be machined.

This paper proposes a method for rapid copying of complex shape part in 5-positioned axis machining. The approach deals with the whole process of free-form copying, from the surface measurement to the actual machining of the surface (Figure 1). The surface measurement is performed using an optical digitizing system: a CMM equipped with a laser plane sensor mounted on a PH10 head from Renishaw giving the sensor a large surface accessibility. In this first approach, the scan planning is not automated. Nevertheless, relative positions and orientations sensor/surface are chosen to guarantee the completeness of the digitized points as regards the whole object surface. As the data obtained are discrete, non-homogeneous, dense and noisy, operations such as cleaning, filtering and densification are essential steps. For this purpose, a voxel representation of the data is carried out with also the objective to recreate "pseudo-continuity". Therefore, the calculation of local characteristics such as voxel normal and barycentre is thus possible. The last step is the choice of optimal part set-ups for the machining using five positioned axes. This is performed by examining for each voxel its accessibility by the tool for a given set-up. An approach is proposed to optimize the number of set-ups allowing the part to be completely machined. Note that, for each set-up, the trajectory of the ball-end cutter tool is calculated in three axes, using the offset inverse method.

The paper is organised as follows. Section 2 exposes the different stages of tool path generation and describes existing methods to generate tool path from discrete data. Section 3 deals with the digitizing and the data pre-processing based on a voxel representation. Then, Section 4 details the method leading to optimised tool set-ups

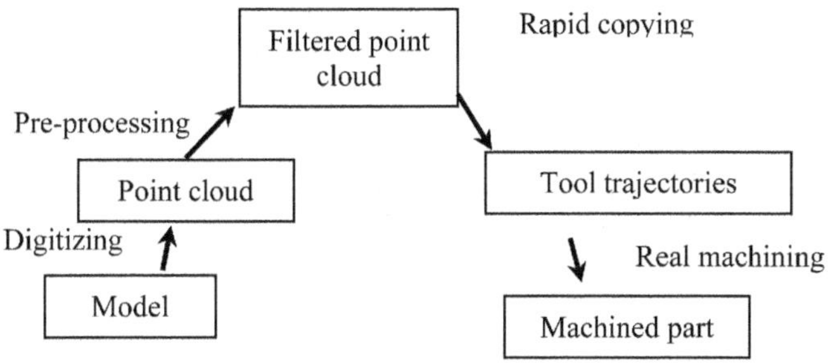

Fig. 1. Free-form copying process.

necessary to mill the part. Finally, Section 5 is dedicated to an illustration of the whole process through an example. The paper is ended by some conclusions. It is important to note that steps of data processing and tool path generation are home-developed and rely on the voxel representation of the points.

2 Tool Path Generation from Discrete Data

The tool-path generation consists in the calculation of a set of successive points (Figure 2). Basically, for 3-axis machining using a ball-end cutter tool, the calculation relies on the surface geometry, the tool geometry and the machining strategy. The machining strategy defines the driving direction of the tool and the CAM parameters. Common calculation methods consist of three main steps [2]:

- Calculation of the tool location on the surface
- Calculation of a single path in the driving direction according to the machining tolerance
- Calculation of adjacent single paths in the perpendicular direction according to the transversal step.

Although various approaches exist and are efficient when the surface geometry is defined through a CAD model, few methods are proposed in literature when the surface geometry results from a measurement. Obviously, the main problem is the lack of continuity and structure of the data making difficult the calculation of differential geometry properties. The normal for instance is necessary to position the tool onto the surface [2]. Besides, classical parameters, the machining tolerance and the scallop height, defining the level of approximation between the CAD model and the tool path have less sense in such a case. More generally, the machining strategy retained is the usual parallel plane method, the size of the grid giving the level of precision. Nevertheless, different approaches for tool path generation from discrete data have been developed.

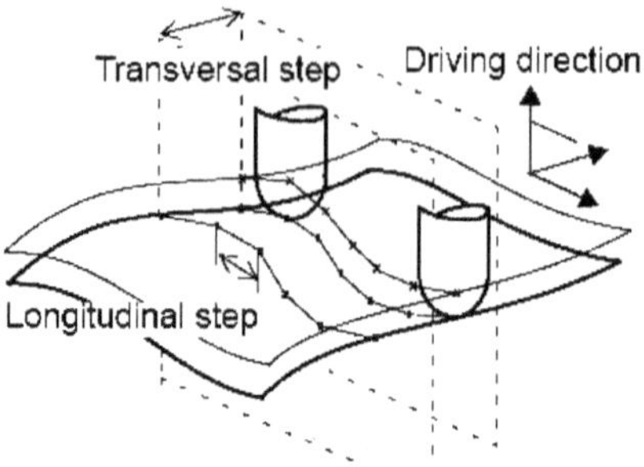

Fig. 2. Tool path generation.

Lin and Liu [3] create a rectangular mesh from the digitizing lines obtained with contact or non-contact measuring devices. The rectangular mesh is oriented according to the driving measuring direction, and is the support for tool path generation. Hence, classical methods based on surface meshing can be used. This approach has the advantage to be close to classical methods. On the opposite, a mesh of the data involves a new approximation of the points which is not satisfactory. Indeed, it is difficult to create a mesh from massive point clouds, generally noisy and inhomogeneous. Park and Chung [4] based their approach on the digitising lines obtained with a laser plane sensor. The tool trajectories are directly built on the digitising lines. When different views are superposed, the difficulty is to define what a digitising line is. So this method fails when the object surface requires more than one scan path. Feng and Teng [1] present a method to generate tool path for which machining error and surface finish are explicitly considered. They base their approach on the definition of a CL-net (Cutter Location net) which consists in 2D cells sequentially organised. The size of the cell net is defined relatively to the machining tolerance (for its length) and the maximum scallop height allowed (for its side size). Each cell node corresponds to a CL point. Due to the absence of continuity, authors evaluate machining errors based on two adjacent tool swept surfaces of the ball-end cutter tool. An optimization method of the position of the cell node is thus proposed to minimize the machining error. This method seems efficient but requires solving an optimization problem. Steps are numerous and difficult to implement.

The method developed by Osty et al. [5] can work whatever the nature of the point cloud (Figure 3). If z is the tool axis, a grid in the xy plane is built so that each grid nod represents a cutter location. The organisation of the grid defines the machining strategy. When using a parallel plane strategy, the y direction may define the driving direction and thus the x axis corresponds to the perpendicular direction. The sampling in both directions gives the level of precision in both directions.

Authors determine the z cutter locations using a method similar to the offset inverse method. They search points in the neighbourhood of the tool axis the location of which is defined by the grid nod. Then, for each point belonging to the neighbour-

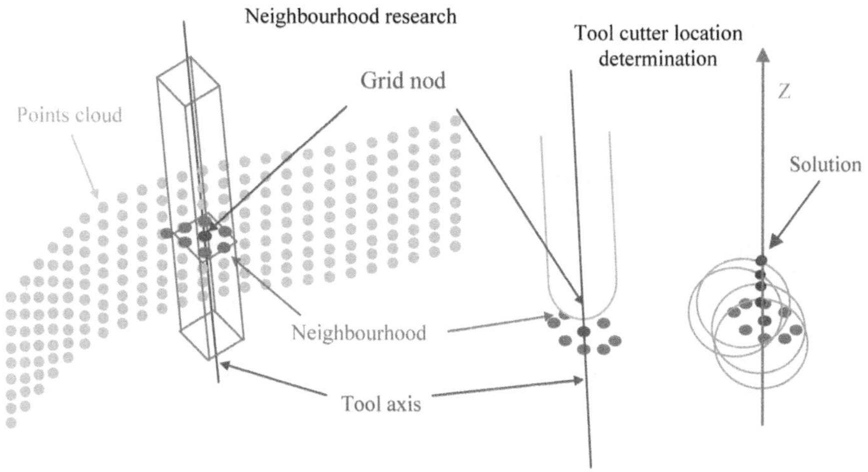

Fig. 3. Tool cutter location on discrete data.

hood, authors calculate the z intersection between the tool axis and the spheres, the radius of which is equal to the tool radius, centred on each point. The CL point is defined by the maximal z value (Figure 3). As the grid parameter is fixed, the final result is not optimized as regards the part quality. Indeed, to avoid facets onto the machined surface, the grid parameter is determined in function of the most discriminating part of the point cloud (in the point of view of the chordal deviation). This may involve over-quality for other portions of the surface. The use of an adaptive grid should optimise the method. On the other hand, this method is simple to implement.

We base our approach of direct copying in 5-positioned axis machining on the method proposed by Osty et al. for 3-axis tool path generation from discrete data points using a ball-end cutter tool. As for most methods, the step of data pre-processing is required. This point is discussed in the next section.

3 Digitizing and Data Pre-Processing

The proposed approach relies on digitized points resulting from 3D measurement of object surfaces. The measuring system used is a Coordinate Measuring Machine (CMM) equipped with a laser-plane sensor mounted on a motorized indexed head. The sensor is moved in translation according the CMM axes, and the indexed head gives the sensor two additional possibilities of rotations. With such a configuration, each fixed orientation (given by the two rotation angles) defines a sensor setting relatively to the object surfaces. For each setting, the sensor is moved according to a certain direction and with several paths. To ensure the complete sweeping of the object surfaces, different settings are required. As a result, data acquired are dense, non-homogeneous and noisy (the noise varying for each digitized point in function of the sensor setting) [6].

To work efficiently, direct copying requires a structure of the data, in order to reduce the amount of data to be treated, to clean the data (filter) and to give data a pseudo-continuity allowing various calculations. For this purpose, our approach is based on a voxel representation.

3.1 Voxel Spaces

The voxel representation consists in cutting the volume including the point cloud into small cubes. Each voxel is thus classified in two classes: empty voxels and non-empty voxels. Each non-empty voxel contains a minimum number of digitized points from which voxel attributes are calculated. The most usual voxel attributes are the barycentre, the normal and the quality indicators. The latter quantifies the point cloud quality which is essential for direct copying [7].

For this specific application the following three quality indicators are relevant: the completeness, the noise and the density. Indeed, the point cloud must be complete free from digitizing gaps in particular for CL point calculation. The noise is also an important factor as it directly influences the normal calculation. To ensure that the normal calculation is efficient, the noise must be as small as possible. Digitizing with laser-plane sensors implies that the density of acquired points is not homogeneous: density is more important for digitizing lines than in other directions. Besides, overlapping zones exist resulting from the various paths for a given setting. Previous works highlighted that the density must be homogeneous for a good surface finish.

In the next, we consider that a preliminary analysis of the point cloud has been performed, leading to data, complete as regards the actual surface, homogeneous, and with a digitizing noise admissible as regards predefined thresholds.

3.2 Normal Calculation

The calculated normal can be interpreted as a local normal, representative of a small portion of the surface (a small sub-set of points). Therefore, the normal is calculated as a voxel attribute. The size of an elementary cube is fixed in function of the free form surface geometry and the dimension of the tool radius used for the milling. The main constraint is linked to the tool radius which gives a superior limit to the voxel size: the size is smaller than the tool radius. If the majority of the surfaces are tensed surfaces, the normal direction does not change abruptly. As a result, the voxel size can be large. At the opposite, when the normal direction bluntly changes the size of the voxel must be small in order to account for abrupt changes of normal directions.

The normal is calculated as the normal to the least square plane that best fits the points including in the voxel. This method works well as long as the point number is sufficient to calculate the normal. When locating close to the point cloud frontier,

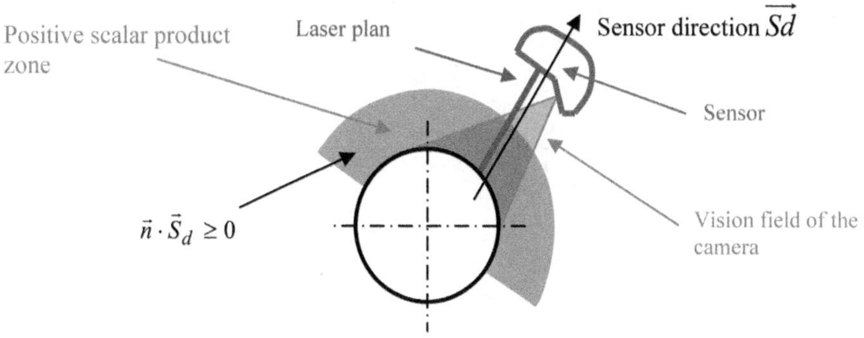

Fig. 4. Sensor vision field.

a voxel may only contain one or two points which are not enough for plane association. In this case, the normal is calculated using the voxel and its 27 neighbours. Indeed, the voxel representation allows a rapid identification of a voxel neighbours. If a voxel is identified using the three indices (i, j, k), voxels neighbours are found by combining $i, i-1, i+1$ with $k, k-1, k+1$, and $j, j-1, j+1$. Once the voxel neighbours are identified, each voxel barycentre can be calculated. The normal thus corresponds to the normal to the plane that best fits the set of 27 barycentres.

To minimise the calculation time of the least-square plane, the method used is that of Shakarji [8] which leads to the exact solution of the plane. If *pts* is the matrix defining the collection of the n_{pts} points on which the plane is fitted. *Pts* is a matrix n_{pts} lines and three rows:

$$pts = \begin{bmatrix} x_1 & y_1 & z1 \\ x_2 & y_2 & z_2 \\ \vdots & \vdots & \vdots \\ x_{n_{pts}} & y_{n_{pts}} & z_{n_{pts}} \end{bmatrix}.$$

As the sum of the variations is null, the barycentre of the n_{pts} belongs to the least square plane. Then the origin of the n_{pts} coordinate referential is moved to the barycentre. In this configuration, the normal vector of the least square plane is the eigenvector associated to the smallest eigenvalues of the 3×3 matrix defined by $pts^T \times pts$.

For tool path calculation, the normal must be directed outward material. To correctly orient the normal we take advantages of the knowledge of the sensor orientation $\mathbf{S}_d$. If a point is seen by the sensor, the scalar product between the normal at this point and the sensor orientation must be positive (Figure 4). Therefore, if $\mathbf{n} \cdot \mathbf{S}_d \geq 0$, where $\mathbf{n}$ is the normal at the considered point, then normal orientation is correct, if not the normal orientation is inverted.

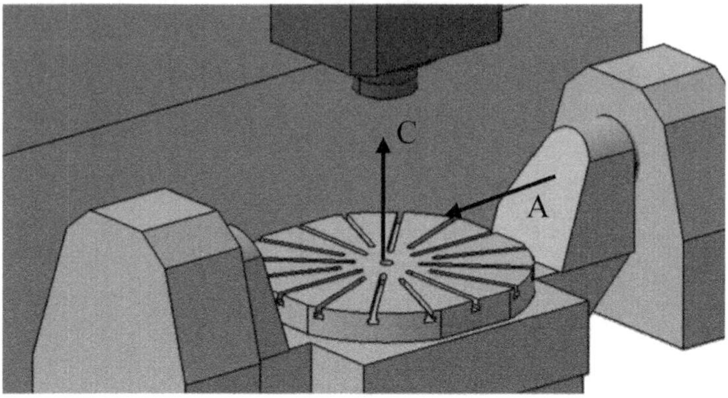

Fig. 5. The machine tool axes.

At this stage, a couple (normal, barycentre) is affected to each voxel. Therefore, the whole point cloud can be substituted by its voxel representation making simple further steps. It gives data structure, pseudo-continuity and attributes allowing calculations such as accessibility.

4 Point Cloud Partitioning

Our application lies on free-form copying from surface acquisition to machining using five "positioned" axes. The machine tool we use is a five axis milling center "Mikron UCP710". In 5-positioned axes, the two rotation axes (A and C) of the milling machine are fixed for a given set-up (Figure 5).

The two objectives of the partitioning are to obtain the collection of set-ups that permits to machine the whole part surface and to minimize the number of set-ups. The efficiency of the partitioning lies on the use of the voxel representation and its associated attributes. The proposed two step method consists in searching (1) to find admissible set-ups that allow milling the points which are included in a given voxel and (2) to group voxels according to the admissible set-ups. The first step of the method is called the voxel visibility determination.

According to the tool radius value, some details of the part surfaces cannot be machined whatever the set-up (Figure 6). Consequently, a second voxel representation is used for which the cube size is equal to the tool radius. The whole study developed in Section 4 take advantage of barycentres and normals of the voxels of this second voxel representation.

Collision between the ball-end tool and the points: points cannot be milled with this tool axis direction.

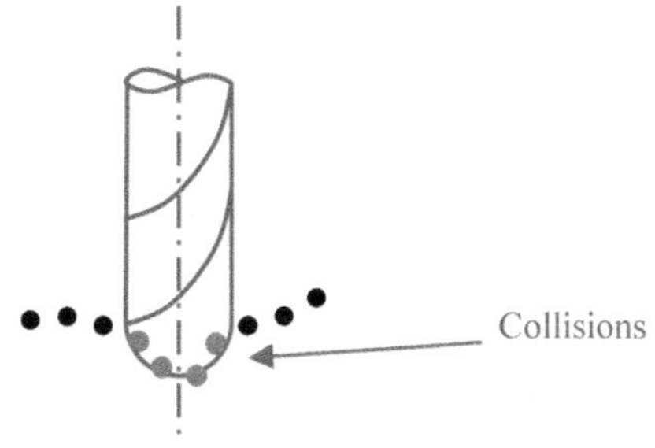

Fig. 6. Collisions between the ball-end tool and the point cloud.

For this second voxel representation, the normal is calculated as the average of the normals assigned to the voxels of the first representation that are included within the new sized voxel.

4.1 Admissible Set-ups

For a given voxel, a set-up is admissible if the points included in the voxel can be machined. As an approximation, we consider that a set-up is admissible if the least-square plane fitted to the points can be machined. Therefore, a set-up is admissible for a given voxel if the following two conditions are satisfied:

1. The tool axis belongs to the visibility cone built around the normal of the voxel and which vertex is the barycentre of the voxel.
2. There is no collision between the tool and the point cloud.

The goal of the visibility is to determine which set-ups may be admissible for each voxel according to machining criteria. That is called voxel visibility since the tool radius is not taken into account. Further, a voxel could be visible but non-accessible. The machining criteria are linked to the tool geometry and the quality of the machining. For a ball end tool, the contact between the tool and the part may be theoretically allowed all over the extremity of the tool. Practically, vibrations occur when machining close to the tool body. The resulting admissible orientations of the tool relative to the part must be included into the cone presented Figure 7 which is called the visibility cone.

For each voxel, the visibility cone is defined by the voxel barycentre which is the cone vertex and the voxel normal which its axis. The angle α defines the cone dimension (Figure 7). On the other hand, a domain including all the tool/part orientations attainable by the machine is defined. This domain is called the tool axis domain. Theoretically to determine the tool/part orientations which make it possible to see a voxel, the intersection between the tool axis domain and the visibility cone must be done. In practice, that requires heavy calculations owing to the fact that the

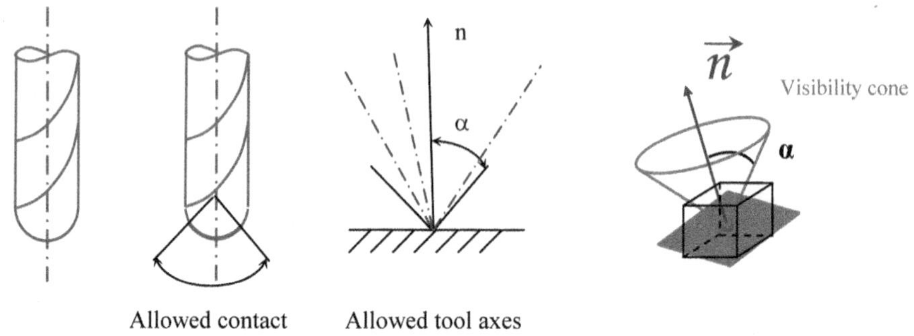

Fig. 7. Determination of the visibility cone.

tool axis domain includes infinity of tool/part orientations. To simplify calculations, the tool axis domain is sampled.

To sample the tool axis domain, the two angles associated with the two rotation axis of the machine are used. An increment must be correctly chosen for both angles. The forms of the model which is copied influence the choice of the increment value. Indeed, for a model presenting many details accessible with particular tool/part orientations (a statuette of elephant for instance), a small value for the increment must be chosen. For the application developed in Section 5 this increment is chosen equal to 15°. Each set-up is referenced thanks to a set-up number.

To determine the voxel visibility, a table is built. The row of this table contains one voxel number followed with the number of each admissible set-up. The table is filled out in the following way. For each set-up, if the tool axis is included within the visibility cone, the set-up reference is stored in the table on the voxel row.

The second stage of the determination of admissible set-ups consists in searching for each set-up if there is a collision between the tool body and the voxel barycentres (Figure 8). A collision is detected when some voxel barycentres are included in the volume occupied by the tool. All the set-ups for which a collision is detected are removed from the table.

4.2 Voxel Grouping

The result of the previous section consists in a table of the admissible set-up for each voxel. The last step of the partitioning consists in determining the collection of set-ups that permit to machine the entire part. The best partitioning is obtained while respecting the following constraints:

- the number of set-ups is minimal,
- the machined area for a given set-up is larger as possible,
- the number of small machined zones for a given set-up is smaller as possible.

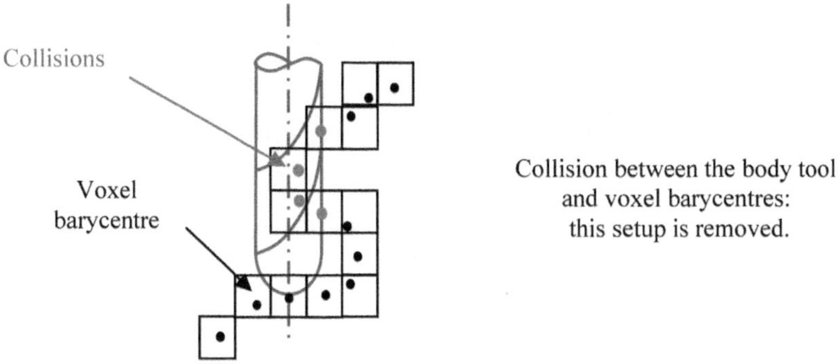

Fig. 8. Collisions between the tool and the barycentres.

The partitioning is performed through voxel grouping. Three methods are proposed to achieve the voxel grouping.

The first one consists in searching which set-up permits the machining of the maximum number of voxels. This method implemented by Germain [9] gives results when the part surface is tensed. If the surface is not tensed, the set-up that corresponds to the grouping of the maximum of admissible voxels may correspond to a large number of small machined zones which is not acceptable.

The second method consists in using a growing method. This method is initialised by choosing an initial group composed of a unique voxel and its associated admissible set-ups. Each non-empty voxel of the neighbourhood of the group is added to the group while a common admissible set-up exists. This method that depends of the initial voxel has not yet been implemented.

The third method consists in manually imposing set-ups. This method permits to group voxels for set-ups that seem obvious to the operator according to the shape of the part. The advantage of this third method is to drastically decrease the amount of calculation. This third method should advantageously be used as an initialisation method of the two other methods.

For each one of the three methods, voxels that are machined for a retained set-up are removed and the method is applied to the remaining voxels.

5 Application

In order to validate the efficiency of the proposed approach, its application is performed to the rapid copying of a little statue representing a cartoon character.

The first step is the object measurement. The scanning of the object is carried out thanks to a laser plane sensor mounted on a CMM equipped with a laser plane sensor mounted on a PH10 head from Renishaw. The relative positions and orientations of

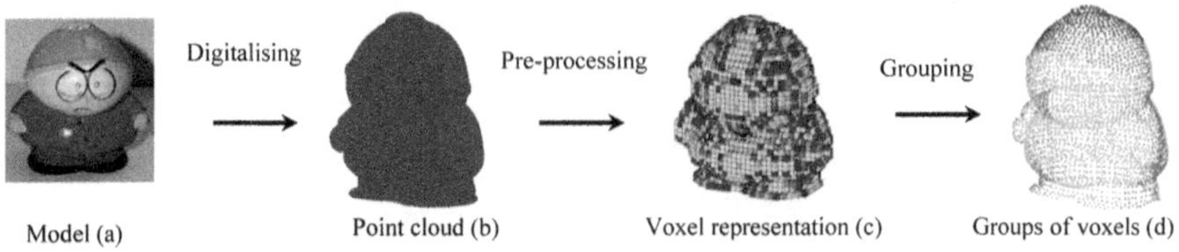

Fig. 9. Pre-processing and voxel grouping.

the sensor relative to the surface are chosen to guarantee the completeness of the digitized points as regards the whole object surface. To obtain a complete point cloud, five sensor directions have been used; $z, x, -x, y, -y$. The point cloud resulting from the measurement is dense (221746 points), noisy and inhomogeneous (Figure 9b).

The second step is data pre-processing to substitute the point cloud to its voxel representation with attributes. The size of each cube is chosen equal to 2 mm, which leads to 1437 voxels. The normal and barycentre are thus calculated for each voxel (Figure 9c).

The third step consists in the set-up determination, we take advantage of the knowledge of the five sensor/surface settings used to digitize the object surfaces. As presented in Section 4.2, the five sensor/surface settings are imposed as tool/surface settings. Therefore, for each voxel, the accessibility by one of the five measuring directions is tested. The tests prove that the five imposed set-ups allow the surface to be entirely machined. The machine angle values that correspond to the five orientations are obtained: $(A = 0, C = 0)$, $(A = 90, C = -90)$, $(A = 90, C = 90)$, $(A = 90, C = 180)$, $(A = 90, C = 0)$.

The groups of voxels are now defined for each set-up and the tool paths can be generated. The base of the algorithm developed for the tool-path calculation is the one developed by Osty and presented in Section 2. In this method, the coordinates (x, y) of the successive CL points correspond to the nod coordinates on a rectangular grid.

Concerning the method we propose and for each set-up, the tool-path is calculated as follows:

- a plane, normal to the tool axis, is created (Figure 10). Let us call (x_p, y_p) the coordinates of a point that belongs to this plane,
- the projection of all the voxel vertex that belong to the selected set-up is performed and a envelope rectangular grid is created (Figure 10),
- the order the nods are selectioned from the grid is defined through the choice of the machining strategy,
- for each nod, if the tool axis intersects the group of voxels, the CL point is calculated,
- the successive CL points are collected that gives the tool-path (Figure 10).

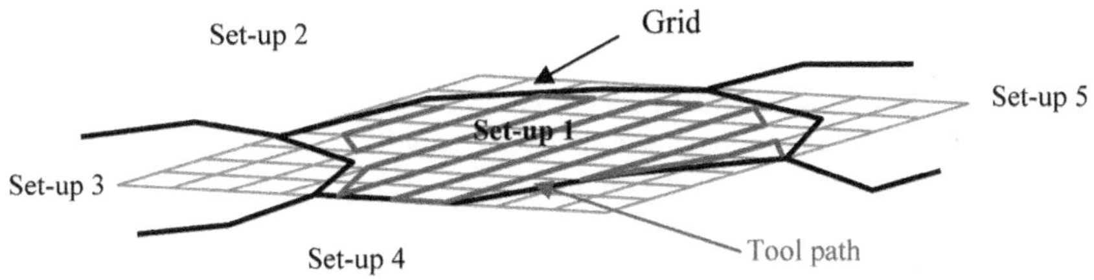

Fig. 10. Machining zone.

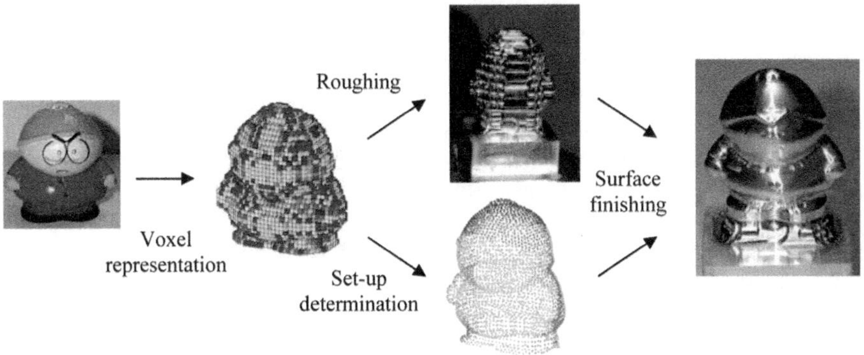

Fig. 11. Copy of the part.

Once tool paths are generated in 3-axis machining for each set-up, virtual machining is performed in order check the calculated tool-paths using the module of simulation of CATIA V5$^{©}$. The purpose of this virtual machining is to check that the tool-paths are free from collision with external systems. It also allows validating the machining covering of the whole surface using the different tool set-ups.

Finally, the surface machining is carried out on the machine Mikron UCP 710. A phase of roughing must be added in order to make operant the finishing phase. The method proposed by Lin [3] has been developed by Blat [10] and adapted by Breteau [11] to take into account the specificities of high speed machining.

Figure 11 shows the result of the milling, which includes both phases, roughing and finishing using the five set-ups. Due to the limit of the tool radius, some surfaces cannot be milled such as the neck of the little character. Nevertheless, some details like the eyes or the wind-breaker buttons are visible which brings out the efficiency of this approach.

6 Conclusion

In this paper, various steps of free-form copying from surface acquisition to machining using five " positioned" axes have been presented.

The first step concerns the pre-processing of the points. A voxel representation of the data is carried out. The cloud of point is certified as acceptable for the copying since the completeness, the noise and the density indicators are within acceptable ranges. Therefore, the calculation of local characteristics such as local normal to the discrete data is performed. The last step is the choice of optimal part set-ups for the machining using five positioned axes. A visibility cone and a collision research method are presented that allows defining the admissible set-ups for each voxel. The research of the retained set-ups for machining is performed while grouping voxels. Three methods are proposed to group the voxels.

The whole method has been developed and tested using the model Cartman©. The relative orientations sensor/surface define initial set-ups. For all voxels, at least one of the initial set-ups belongs to the admissible set-ups. So, voxel grouping is performed using these initial set-ups. For each retained set-up, the tool-path of the ball-end cutter tool is calculated in five "positioned" axes machining, using the offset inverse method. The copying of the part has been performed on a five axis milling centre that proves the feasibility of the whole method.

Improvements, in particular those concerning automated scan planning and tool path generation algorithms are in progress.

References

1. H-Y. Feng, Z. Teng, Iso-planar piecewise linear NC tool path generation from discrete measured data points, *Computer-Aided Design* **37**, 2005, 55–64.
2. E. Duc, C. Lartigue, Algorithms for the evaluation of errors in the calculation of the NC tool paths, in *2nd International Seminar on Improving Machine Tool Performance*, La Baule (France), CD-Rom Paper, 20 pp, July 2000.
3. A-C. Lin, H-T Liu, Automatic generation of NC cutter path from massive data points, *Computer-Aided Design* **30**, 1998, 77–90.
4. S-C. Park, Y-C. Chung, Tool-path generation from measured data, *Computer-Aided Design* **35**, 2003, 467–475.
5. G. Osty, A. Contri, E. Duc, Free-form copying from point clouds; A tool-path calculation method using B-Spline interpolation, RCFAO, Rapid Production, 2000, pp. 367–382.
6. G. Osty, Extraction de particularités sur données discrètes issues de numérisation 3D: Partitionnement de grand nuage de points, PhD. Thesis, ENS de Cachan, January 2001.
7. A. Contri, P. Bourdet, C. Lartigue, Quality of 3D digitised points obtained with non-contact optical sensors, *Annals of the CIRP* **51**(1), 2002.
8. C-M. Shakarji, Least-squares fitting algorithms of the NIST algorithm testing system, *Journal of Research of the National Institute of Standards and Technology* **103**, 1998, 633–641.

9. F. Germain, Contribution à l'usinage direct d'un nuage de points: Partitionnement par visibilité et accessibilité, DEA de Production Automatisée, ENS de Cachan, July 2004.
10. B. Blat, Numérisation et copiage de formes complexes, DEA de Production Automatisée, ENS de Cachan, July 1999.
11. P. Breteau, Copie de forme par usinage cinq axes positionnés; usinage direct d'un nuage de points, Mémoire de Master, ENS de Cachan, 2005.

Manufacturing Plant Layout Supported with Data Mining Techniques

Bruno Agard[1] and Catherine Da Cunha[2]

[1]*Département de Mathématiques et de Génie Industriel, École Polytechnique de Montréal, C.P. 6079, Succ. Centre-Ville, Montréal (Québec), Canada H3C 3A7; E-mail: bruno.agard@polymtl.ca*
[2]*Laboratoire GILCO, 46, avenue Félix Viallet, 38031 Grenoble, France; E-mail: da-cunha@gilco.inpg.fr*

Abstract. The question of plant layout is central in a manufacturing process. This question becomes even more important in a mass customization context, when a large product diversity has to be managed. The manufacturing process, and specifically the plant layout, has to be designed taking into account this characteristic. When all products are similar, manufacturing plant layouts are relatively easy to design; difficulties come when all products are different and require specific manufacturing operations.

This paper proposes a methodology based on data mining techniques. Different steps are proposed to achieve this goal. The methodology considers: (1) identification of representative sets of products; (2) identification of representative sets of relevant manufacturing processes (for each product family); (3) categorization of new products (identification of the closest product family and the relevant layout). The focus is on data transformations that enable to extract relevant information for the manufacturing plant layout.

Key words: product families, plant layout, data mining.

1 Introduction

In nowadays markets, manufacturing companies have to be always more competitive. Manufactured products have to be customized, of the highest quality, delivered in the required place and time and at the lowest price. This is defined as mass customization [1].

Mass customization is often supported with a product platform that combines options and alternatives in order to satisfy individualized requirements. Customers select the options and alternatives they prefer to specify their own dedicated product [2]. Besides, this customization may imply a large product diversity to manage, every

S. Tichkiewitch et al. (eds.), Advances in Integrated Design and Manufacturing in Mechanical Engineering II, 373–383.

product may be different. For the suppliers it may follow small sets of variable quantities of different products to manufacture.

That huge diversity has many impacts on the manufacturing process and can therefore endanger the product quality, the lead time, the cost, etc. [3].

The manufacturing process, and specifically the plant layout, has to be designed taking into account those characteristics. When all products are similar, manufacturing plant layouts are relatively easy to design; difficulties come when almost all products are different and require some specific manufacturing operations.

In a product family all final products selected by diversified customers may be different, depending on the set on options and alternatives selected. On the manufacturing level, these differences may necessitate extra manufacturing or control operations. When a company manufactures many set of variable quantities of different products, the manufacturing plant layout may be problematic to define.

According to Phillips 30% of the cost of a manufactured part is due to handling [4]. Besides, handling is mostly impacted by the manufacturing plant layout. The layout of the manufacturing tools appears to be an essential condition for small production costs and for higher competitiveness.

In a context of the manufacturing of product families, this paper proposes to design manufacturing plant layouts with data mining techniques. Different steps are proposed to achieve this goal. The methodology considers: (1) identification of representative sets of products (using association rules and/or clustering) – products from the same set will be considered to belong to the same product family (according to the manufacturing process); (2) based on similarities on the manufacturing process, relevant manufacturing processes are identified and set in an adequate order for each product family (with association rules) – describing the plant layout; (3) classification trees categorize new products to propose the closest product family and then the relevant layout.

Section 2 contains the problem specification. The context is outlined and a case study is provided that will be exploited all over the paper. Each step of the methodology is described in Section 3. An emphasis is on data transformations necessary to complete each operation. Section 4 concludes the paper and proposes some extensions.

2 Problem Specification

Consider a company that manufactures small quantities of many different products. That company disposes of a set of machines ($M_1, M_2, \ldots, M_n$) and gives a dedicated manufacturing process for each set of parts. Each manufacturing process is different and whatever the product family the quantities to manufacture are almost equivalent (there is no “major” production).

The list of products and their manufacturing processes are presented in Table 1.

Table 1. List of products and their manufacturing processes.

	Ref #	Qty/ week	Manufacturing process						
List of products	123	5	M1	M2	M4	M7			
	234	7	M4	M7	M2	M1			
	345	4	M5	M3	M6				
	567	8	M6	M5	M3	M7			
	...	...	...						

This table represents for each product the quantity that is manufactured each week and the manufacturing process of each product. For example, there are 5 products #123 to be manufactured each week. The manufacturing process of each product #123 necessitates machines M_1, then M_2, then M_4 and finally M_7, and so on for each product.

From that data set, it is possible to extract much relevant information to define the manufacturing plant layout. Nevertheless, due to the large number of different sets of products to manufacture, Table 1 may be so large that it is impossible for a human being to have a global view on the data and to detect all relevant information for a plant layout.

In this case, data mining techniques may be advantageously employed. Data mining is defined as "the discovery of non-trivial, implicit, previously unknown, and potentially useful and understandable patterns from large data set" [5].

No data mining technique will be presented here; the emphasis is on data transformation and possible knowledge extraction for automatic design of plant layouts. For more information about data mining techniques and algorithms, the reader may refer to [6]. For applications of data mining techniques in marketing, sales and customer support, see [7]; for applications of data mining in manufacturing environments, see [5, 8].

Previous research focused on plant layout. It is possible to find methodologies [9–11] and algorithms [12–15]. This paper will follow the Systematic Layout Planning method [9], in order to propose an alternative methodology based on data mining techniques. Data mining tools are proposed to generate most of the relevant information that is needed.

3 Manufacturing Plant Layout

Several tasks of a manufacturing plant layout may be advantageously supported with data mining tools.

1/ identification of representative sets of products (using association rules and/or clustering) – products from the same set will be considered to belong to the same product family (according to the manufacturing process);
2/ based on similarities on the manufacturing process, relevant manufacturing processes are identified and set in an adequate order for each product family (with association rules) – describing the plant layout;
3/ classification trees categorize new products to propose the closest product family and then the relevant layout.

Fig. 1. Layout methodology with data mining techniques.

The proposed layout methodology with data mining techniques is represented in Figure 1. It combines three steps. Each step is described in the following subsections.

3.1 Identification of Representative Sets of Products

A global layout may be defined from the manufacturing processes described in Table 1.

In order to identify representative sets of products, it is necessary to consider relevant sets of parameters. Group technology is often supported with matrix decomposition algorithms such as [16–19]. Association rules may also be used [20]. For this application the focus is on clustering techniques. Clustering (called also segmentation) divides a population into smaller sub-populations with similar behaviour according to a predefined metric. Clustering maximizes homogeneity in each cluster and maximizes heterogeneity between clusters [21].

In order to apply clustering to data in Table 1, a first data transformation is necessary and leads to Table 2.

In Table 2, the lines represent the list of products and the columns represent the manufacturing processes involved. For example the first line means that for the manufacturing of products #123 machines M_1, M_2, M_4 and M_7 are necessary. Please note that information about order between the machines is lost, only constraints of existences appear.

Applied to Table 2, clustering methods identify the sets of products that are similar in terms of processes employed. The observed results are presented in Table 3.

Each cluster is a potential manufacturing department. For example Cluster 1 proposes a department for the manufacturing of products {#123, #234, ...} which requires machines M_1, M_2, M_4 and M_7. At this level, an expert may validate the results

Table 2. Modified data for clustering.

	Ref	Process								
		M1	M2	M3	M4	M5	M6	M7	...	Mn
List of products	123	1	1		1			1		
	234	1	1		1			1		
	345			1		1	1			
	567			1		1	1	1		
	...	...	...							

Table 3. Clusters extracted from Table 2.

Cluster 1:	products {#123, #234, ...}
Cluster 2:	products {#345, #567, ...}
...	

Table 4. Suggested departments.

Dept 1:	machines $\{M_1, M_2, M_4, M_7\}$
Dept 2:	machines $\{M_3, M_5, M_6\}$
...	

and/or propose some changes in the allocation of the machines in each department such as duplication of machines. It follows groups of machines (departments). Consider that the departments as given in Table 4 are valid.

In order to localize correctly all departments created previously next to each other, it is necessary to evaluate the circulation of products between them. Another data transformation will take advantage of former results.

For each product, movements between the departments must be identified and evaluated. Indeed one of the main objectives of a plant layout for manufacturing is to streamline the material flow. The focus must be on the minimization of handlings: the distances material and operators have to move (for production or maintenance) should therefore be evaluated.

Instead of movements of products from machine to machine such as in Table 1, movements from department to department are necessary.

By combination of Tables 1 and 4, the results as given in Table 5 are observed.

The movements between departments may be identified (see Table 6).

While it appears that, with the departments selected previously, product #123 just stays in Dept 1, products #567 moves from Dept 2 to Dept 1. Knowing the number

Table 5. Manufacturing process of each part – department level.

#123	#234	#345	#567
M1-Dept 1	M4-Dept 1	M5-Dept 2	M6-Dept 2
M2-Dept 1	M7-Dept 1	M3-Dept 2	M5-Dept 2
M4-Dept 1	M2-Dept 1	M6-Dept 2	M3-Dept 2
M7-Dept 1	M1-Dept 1		M7-Dept 1

Table 6. Manufacturing process – department to department.

	Ref #	Qty/ week	Manufacturing process		
List of products	123	5	Dept 1		
	234	7	Dept 1		
	345	4	Dept 2		
	567	8	Dept 2	Dept 1	
	…	…	…		

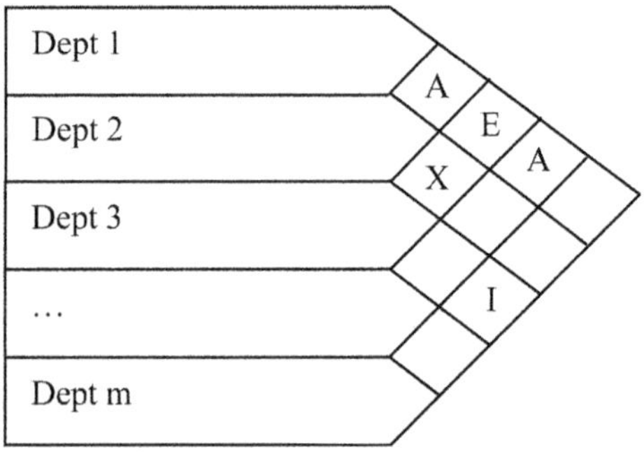

Fig. 2. Activity relationship diagram.

of products #567 to be manufactured each week and the lot size for handling; it is easy to construct an activity relationship diagram such as Figure 2.

It is then relatively easy to place all departments. Different tools exist for that task; one can cite [22] or [23] for example.

3.2 Identification of Relevant Manufacturing Processes

For the second step of the methodology, all machines belong to specific departments and the departments are situated next to each others. The objective is now to identify

Table 7. List of products and their manufacturing processes.

Dept 1	Ref #	Qty/ week	Manufacturing process				
List of products	123	5	M1	M2	M4	M7	
	234	7	M4	M7	M2	M1	
	567	8	M7				
	…	…	…				

Table 8. Developed data for each product in each department.

B_1	B_2	B_4	B_7	1_2	1_4	1_7	2_1	2_4	…	4_2	…	4_7	…	7_2	…	1_F	…	7_F
1				1				1				1						1
1				1				1				1						1
1				1				1				1						1
1				1				1				1						1
1				1				1				1						1
		1					1					1		1		1		
		1					1					1		1		1		
		…					…					…		…		…		
		1					1					1		1		1		
			1															1
			1															1
			…															…
			1															1

a relevant manufacturing process into each department so as to place the machines in an adequate order.

For each department (group of machines of Table 4) the data (Table 1) are filtered (see Table 7). The movements between departments may be identified.

It is necessary to extract sequences in the manufacturing processes. The sequence M_1–M_2 as to be identified in both products #123 and #234. The data transformation proposed in [24] makes it possible, besides it is necessary here to temperate each sequence with the number of products considered.

In Table 8 1_2 sets that machine M_1 is used before machine M_2. B_1 sets that machine M_1 is the first machine to be used and 1_F that machine M_1 is the last to be used in this department. The number of lines for each product reference is proportional to the quantity per week (it could also be divided by a common value for all references). It leads to data as given in Table 8.

Association rules may be extracted from Table 8. Association rules stress the strong associations within the data. Rules are given with two metrics (support and confidence) that makes it possible to evaluate the strength of the rule. Support consider the number of occurrence of the left side of the rule in the data, while confidence represents the ratio between the right side of the rule appearance into the data and the support of the rule. Rules with high support and confidence are strong and may be considered.

Suppose that the rule "4_7 = 1" is the strongest rule of Table 8. It means that machine M_4 is most often used just before machine M_7. Then it may be advantageous to consider the placement of machine M_4 next to machine M_7 in the order of the flux. And so on from the strongest rules to the weakest until all machines are placed. Some contradictions may appear because all sets of products are different. Some sets of products may not follow the flow.

It results in a layout for each department that is automatically generated from the data.

3.3 Classification of New Products

The last step of the methodology takes advantage of the plant layout that was proposed previously. Any new product to be manufactured will be automatically classified in the adequate department(s) by classification trees [25] (Figure 3).

Classification trees may learn from Tables 2 and 4 what are the conditions that lead to define a department. When a new product is to be manufactured, it is automatically associated to the closest cluster and then to the closest layout.

4 Conclusions and Further Work

This paper proposes a methodology to design a manufacturing plant layout automatically, with data mining techniques, from the production routing of each product.

This approach enables, when no "major" product can be identified, to define an alternative layout without over sizing the plant. The information needed to perform this definition is one that is always at disposal (the routings) and therefore does not require a consuming (in time and resources) information retrieval.

The method proposed here does not consider organizational criteria such as workers abilities or cross trading; this aspect of plant layout should be addressed in a second step. Data mining techniques could then be used on data representing the skills needed for each operation.

From time to time it is important to reconsider all the process of layout planning. The new parts may overlap the olds one and a new plant layout may be generated.

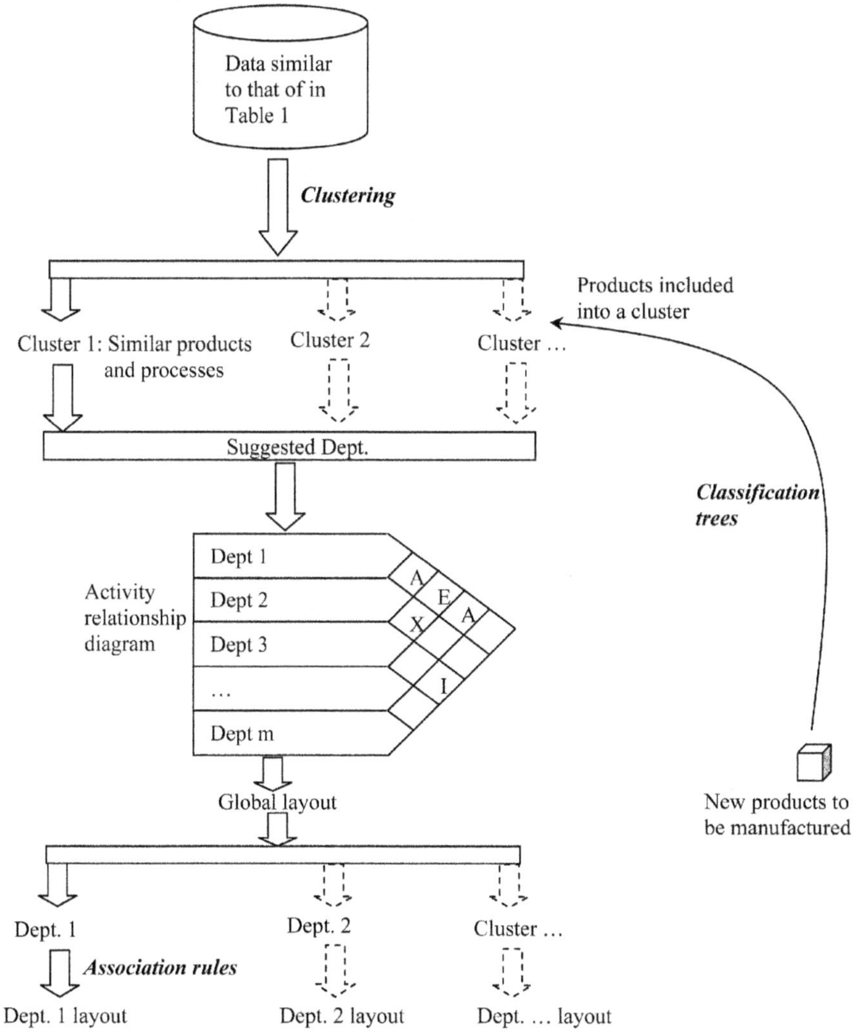

Fig. 3. Activity relationship diagram.

Further researches may include the consideration of other products' attributes such as parts' fragility or volume; the aim of the plant layout will then be the minimization of the breakages or the handling volume.

References

1. B.-J. Pine II, *Mass Customization, The New Frontier in Business Competition*, Harvard Business School Press, Boston, Massachusetts, 1993.
2. B. Agard and M. Tollenaere, Méthodologie de conception des familles de produits, *Journal Européen des Systèmes Automatisés* **37**(6), 2003, 755–777.
3. P. Child, R. Diederichs, F.-H. Sanders and S. Wisniowski, The management of complexity, *Sloan Management Review*, Fall, 1991, 73–80.
4. E.J. Phillips, *Manufacturing Plant Layout – Fundamentals and Fine Points of Optimum Facility Design*, Society of Manufacturing Engineers, 1997.
5. A.G. Büchner, S.S. Anand and J.G. Hugues, Data mining in manufacturing environments: Goals, techniques and applications, *Studies in Informatics and Control* **6**(4), 1997, 319–328.
6. U.M. Fayyad, G. Piatetsky-Shapiro, P. Smyth and R. Uthurusamy, *Advances in Knowledge Discovery and Data Mining*, AAAI Press/The MIT Press, 1996.
7. A.J.A. Berry and G. Linoff, *Data Mining Techniques: For Marketing, Sales, and Customer Support*, Wiley, New York, 1997.
8. B. Agard and A. Kusiak, *Computer Integrated Manufacturing: A Data Mining Approach*, in R.C. Dorf (Ed.), *The Engineering Handbook*, 2nd edition, CRC Press & IEEE Press, Boca Raton, FL, 2005, pp. 192.1–192.11 (Chapter 192).
9. R. Muther, *Systematic Layout Planning*, 2nd edition, CBI Publishing Company Inc., 1973.
10. J.M. Apple, *Plant Layout and Material Handling*, 3rd edition, John Wiley, New York, 1977.
11. J.A. Tompkins, J.A. White, Y.A. Bozer and J.M.A. Tanchoco, *Facilities Planning*, 3rd edition, John Wiley & Sons, 2002.
12. K.C. Lam, C.M. Tang and W.C. Lee, Application of the entropy technique and genetic algorithms to construction site layout planning of medium-size projects, *Construction Management & Economics*, **23**(2), 2005, 127–145.
13. F. Karray, E. Zaneldin, T. Hegazy, A. Shabeeb and E. Elbeltagi, Computational intelligence tools for solving the facilities layout planning problem, in *Proceedings of the American Control Conference*, 2000, pp. 3954–3958.
14. Z. Li, M. Anson and G.M. Li, A procedure for quantitatively evaluating site layout alternatives, *Construction Management and Economics* **19**, 2001, 459–467.
15. T.D. Mavridou and P. Pardalos, Simulated annealing and genetic algorithms for the facility layout problem: A survey, *Computational Optimization and Applications* **7**, 1997, 111–126.
16. A. Kusiak, *Engineering Design: Products, Processes, and Systems*, Academic Press, San Diego, CA, 1999.
17. A.M.A. Al-Ahmari, Part feature selection and family design in group technology using a fuzzy approach, *International Journal of Computer Applications in Technology (IJCAT)* **21**(3), 2004.
18. J.R. King, Machine-component grouping in production flow analysis: An approach using a rank order clustering algorithm, *International Journal of Production Research* **18**(2), 1980, 213–232.
19. Z. Albadawi, H.A. Bashir and M. Chen, A mathematical approach for the formation of manufacturing cells, *Comput. Ind. Eng.* **48**(1), 2005.

20. A. Kusiak, Non-traditional applications of data mining, in D. Braha (Ed.), *Data Mining for Design and Manufacturing*, Kluwer, Boston, MA, 2001, pp. 401–416.
21. C. Westphal and T. Blaxton, *Data Mining Solutions*, John Wiley, New York, 1998.
22. E.S. Buffa, G.C. Armour and T.E. Vollman, Allocating facilities with CRAFT, *Harvard Business Review* **42**(2), March/April 1964, 136–159.
23. C.E. Donaghey and V.F. Pire, Solving the facility layout problem with BLOCPLAN, Industrial Engineering Department, University of Houston, TX, 1990.
24. C. da Cunha, B. Agard and A. Kusiak, Improving manufacturing quality by re-sequencing assembly operations: A data-mining approach, in *18th International Conference on Production Research – ICPR 18*, University of Salerno, Fisciamo, Italy, July 31–August 4, 2005.
25. B. Agard and A. Kusiak, Data mining in selection of manufacturing processes, in O. Maimon and L. Rokach (Eds.), *The Data Mining and Knowledge Discovery Handbook*, Springer, 2005, pp. 1159–1166.

Empowering CAPP Systems with Human-Computer Interaction Study

Vincent Capponi and François Villeneuve

Laboratoire Sols Solides Structures, Ecole Nationale Supérieure d'Hydraulique et de Mécanique de Grenoble, BP 95, 1025 rue de la Piscine, 38402 St Martin d'Hères Cedex, France; E-mail: {vincent.capponi, francois.villeneuve}@hmg.inpg.fr

Abstract. Initially designed to fill the gap between CAD and CAM systems, computer aided process planning (CAPP) systems have reached very limited success and acceptance in industry. To explain this assessment, some researchers point out the fact that most existing approaches aim at complete automation, and suggest that CAPP systems should integrate the human planner into the decision-loop to improve its flexibility and efficiency. For instance, such systems should provide some relevant information about alternative plans rather than automatically computing one unique plan.

The objective of the present paper is to provide some results about the efficiency of such alternative proposal from a human-computer interaction point of view. Our work is based on an existing set-up planning method. Human decision-making is necessary because no accurate optimality metric has emerged to automatically select the set-up positions among the computed alternatives. This paper deals with the design and early evaluation of the future user-interface of a set-up planning module. A mock-up of the human-computer interface has been designed to drive the evaluation. This evaluation points out that the number of alternatives solutions should be limited.

This paper suggests this limitation is due to either of two factors: human limitation in information processing and industrial constraints in the process planning activity. To conclude, the relevancy of this proposal in an industrial context is discussed.

Key words: human-computer interaction (HCI), user-interface, CAPP, 5-axis NC machining, set-up planning.

1 Introduction

Computer-assisted process planning (CAPP) is the last gap to bridge to achieve the CAD/CAM integration challenge. Generating process plans for machining is still a difficult task for companies despite the research and industrial work that has been done for the last twenty years. Currently, most promising approaches rely on generative systems [1–3] that generate and organise the machining processes into set-ups.

S. Tichkiewitch et al. (eds.), Advances in Integrated Design and Manufacturing in Mechanical Engineering II, 385–397.

Numerous AI techniques have been used to solve the planning problem, mainly in the automotive domain. These systems mostly rely on a feature-based approach to model the part specifications and a knowledge formalisation of generic machining rules to solve the system.

A quick review shows that most developed approaches aimed for full automation. In other words, those systems could be defined as black-boxes: given geometric and technical data of the part and resources, they search for a unique solution (i.e. one process plan) that is optimal with respect to some pre-specified objective function. Most existing systems do not allow modifying directly the computed solution and the user has to run the system again with new options or new input data if the solution does not match his/her requirements.

If fully-automated CAM functions have drastically increased the efficiency of human planner in the industry, the generative planning systems which aimed at complete automation have reached very limited success and acceptance in industry. To explain this assessment, one can say that the capture and the use of knowledge for the former systems is much easier than for the latter systems, but the scientific community claims that feature-based approaches coupled with the use of artificial intelligence techniques have achieved good technical results in process planning. To explain this limited success, some researchers point out the lack of adaptation to the context and suggest that CAPP systems should integrate human actions into the decision-loop to improve their flexibility and efficiency in the workshop. Li et al. [4] developed a system called ASUPPA that proposes a framework for the human planner to improve iteratively a computed default process plan. Amara et al. [5] developed a non-deterministic multi-agent system called HICAPP that allows the human planner to modify in real-time every decision the system makes alone. For Shen and Norrie [6], such systems should give humans some relevant information about alternative plans rather than automatically computing one unique plan.

The objective of this paper is to highlight some results made in an industrial context about the efficiency of such alternative proposals to the human planner. The paper is organised as follows: Section 2 summarises our set-up planning method and the results of automated tasks. Section 3 describes the human-computer interaction study we carried out to specify the necessary interactions and the corresponding user-interface. Section 4 discusses some interesting results from the evaluation of this interface: what the number of alternatives to be displayed is. Then, conclusions about the relevance of alternative solutions in an aircraft workshop are made.

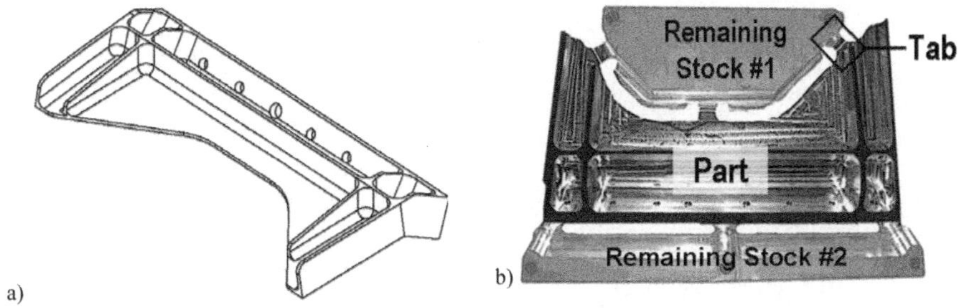

Fig. 1. (a) An aircraft structural part. (b) The corresponding workpiece.

2 A Set-up Planning Method to Machine Aircraft Structural Components

2.1 Overview

The presented work belongs in the context of a large French project developing a Computer Aided Process Planning System for the aircraft industry. This project gathered aeronautical industrialists, software developers and vendors, and research laboratories. The future CAPP system would be specifically designed for the machining of aircraft structural components with 5-axis features. Indeed, a specific approach is needed for set-up planning to take into account that structural parts are made from slab stock in two setups (front-side/back-side) on a five-axis machine tool with no intervening operations [7]. The fixture elements are included in the stock and the workpiece remains connected to the stock by tabs at the end of the machining. To post-finish the part, the tabs are manually cut by an operator. Figure 1 illustrates an aircraft structural part and its corresponding workpiece at the end of machining.

In a previous paper [8], an activity model and a method to compute alternative set-ups for such parts have been proposed. To reduce the search-space, this method relies on a restriction that mainly considers accessibility constraints. From this perspective, an ŞaccessŤ visibility-based model has been developed to handle alternative machining directions of the part. The scope of this study is limited to the generation of set-ups for 5-axis milling machines, assuming that the manufacturing-oriented model of the part is available. At the end of set up selection, some machining operations may be made on either set-up from accessibility considerations. The allocation of these operations and the sequencing of all machining operations in each set-up are achieved in another step of the proposed CAPP system. These tasks will not be considered in this paper.

The proposed method is not fully automated: it integrates the human planner at different stages of the reasoning. Obviously, the system needs the human to make decisions for which knowledge has not been captured or cannot be formalised. In the process planning of aircraft parts there are some specific tasks that cannot rely yet on

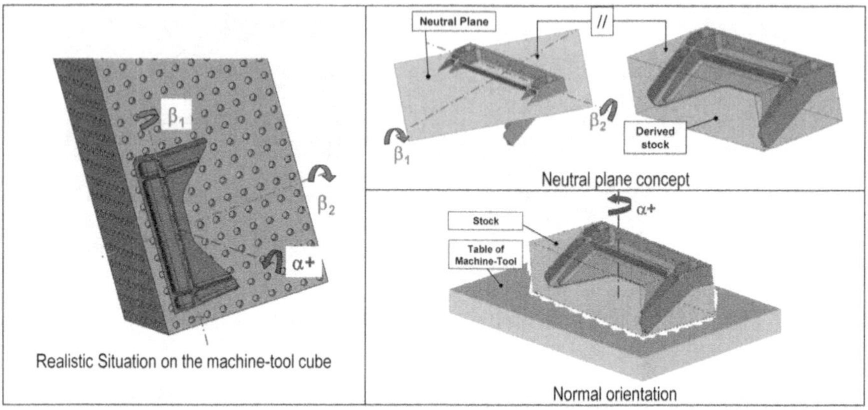

Fig. 2. Parameters modelling the orientation of the part on the machine-tool.

full automation, such as machine-tool selection or the management of the flange top (this last problem will not be tackled in this paper). However, besides these manual tasks, human decision-making is necessary because no accurate optimality metric has emerged to select automatically the set-up positions from among the computed alternatives. Indeed, one of the main objectives of our work was to study and specify how we could let the human choose between computed alternatives in order to overcome the rigidity introduced by pre-specified objective functions.

2.2 Automated Method

2.2.1 Principle of Set-up Generation

Structural parts are made from slab stock in two setups (front-side/back-side), but as the initial orientation of the part in the stock is not specified by the designer, the process planner could set it in order to ease the machining. The main objective of the process planning is to generate two opposite set-ups ensuring that all features of the part are accessible. A feature is accessible if at least one of its alternative machining accesses (i.e. machining direction or cone of direction) is reachable by the spindle of the machine-tool. Therefore, the first task of the planner is to generate set-ups considering accessibility constraints. As shown in Figure 2, the part is oriented in the stock with two parameters β_1 and β_2, within the concept of neutral plane. The neutral plane represents the main symmetry plane of the future stock and therefore the orientation of the future positioning planes as well. Once the part is oriented in the stock, the orientation of the stock (or the part) on the machine-tool is only defined with one rotation parameter α^i per set-up (Figure 2).

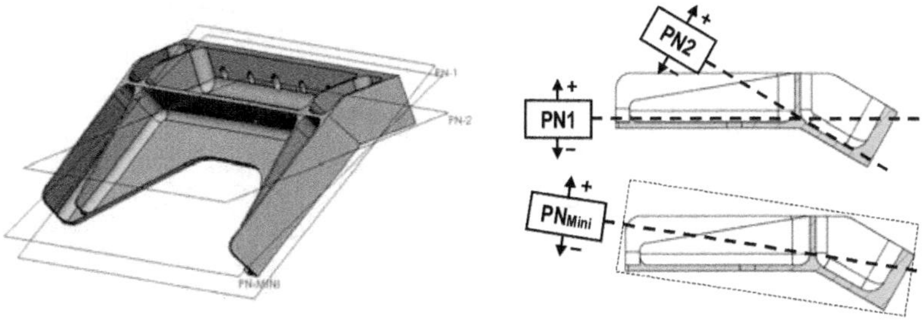

Fig. 3. Solutions of neutral planes for the part described in Figure 1.

We developed a method based on the visibility concept to compute set-up solutions that respect these accessibility constraints [IFAC 04]. This method is divided into two steps:

- *Stock orientation*. The first heuristic step determines the parameters β_1 and β_2: solutions of neutral planes are sought within a reduced space using some industrial rules. For instance, one of these rules is to keep planes that are parallel to large pocket bottoms.
- *Normal orientation*. The second deterministic step computes all viable part orientations (around the normal of the machine-tool table) for a given neutral plane and for each of the two corresponding set-ups. This step relies on an intersection algorithm that is run iteratively. This algorithm takes into account accesses in the two set-ups at the same time. First, all viable orientations are calculated independently for each set-up. Then the system retains only the combinations of orientations (one in each set-up, i.e. one on each side of the stock) that make possible the machining of the whole part.

2.2.2 Computation Results

The result of the first step is a set of neutral planes, deduced from the industrial heuristic rules. Figure 3 shows the neutral planes solutions for the part described in Figure 1.

For each neutral plane, the second step calculates the orientation solutions for the positioning of the stock (or the workpiece) on the machine-tool. The developed algorithm gives non-discrete solutions of orientation. In other words, for the same set-up, several (actually an infinite number) orientations of the part may lead to the same result from a process planning point of view: for each feature, the same milling processes (i.e. flank milling or end milling) are achievable on the considered set-up. Therefore, equivalent solutions may be defined as a range of values of the rotation parameter a, defining sectors of viable orientations. Figure 4 graphically shows the solution-space for one set-up: there are three sectors (i.e. three angular ranges)

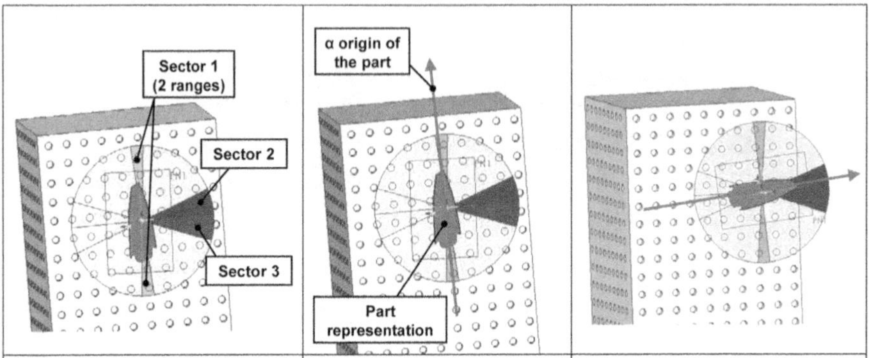

Solution of orientations (clustered in sectors) for one set-up | One orientation solution among sector 1 | One solution orientation among sector 2

Fig. 4. Solution-space and examples of viable orientations for one set-up.

that define three different process plan solutions. Within one sector, the difference between two orientations can change the machining direction of an operation, but this change is not very significant for the planner at this macro-planning stage.

3 Preliminary Design of the User-Interface

The planner needs to understand and to evaluate the different solutions to make a decision. This section presents the preliminary design of the user-interface. The preliminary design of the considered set-up planning system has followed a three stage methodology:

Technical specifications have highlighted the tasks that can be done automatically. In addition, the automated tasks have been simulated on industrial parts and their set-up solutions validated.

A human-computer interaction study has been carried out to define how the system could help the human to achieve the manual tasks and understand the computed solutions. This study specified the interactions the human should have with the computer.

An early evaluation of these proposed interactions has been done with representative user panels. Based on the specified interactions, a mock-up of the user-interface has been designed to drive the evaluation. As the technical specifications are resumed in Section 2, this section will describe the result of the two last stages. The objective of this preliminary design and early evaluation was to assess the efficiency of an alternative set-up solution proposal.

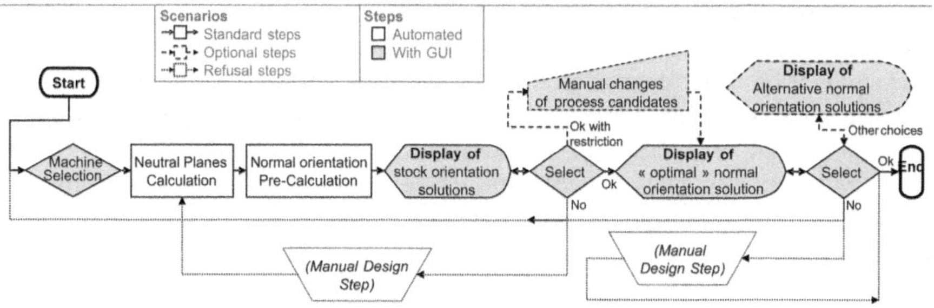

Fig. 5. Tasks of the computer aided tool for designing set-ups.

3.1 Specification of the Human-Computer Interaction

3.1.1 Scenario

Once the feasibility of the automated tasks has been proven, the next step is to define representative scenarios, that is, to define precisely which tasks have to be done manually and determine the way these tasks are successively achieved and linked with the results of the automated ones. This is context-dependent, meaning it depends on the part to machine. Four scenarios have been defined according to the machining difficulties the planner can find on the part. Working on the scenario corresponding to a part with few difficulties, the study underline the fact that the system must define milestones to let the human decide which branch of neutral plane solutions has to be developed. Indeed, the final number of set-up solutions is very high because the search-space is large and not strongly constrained. Defining a milestone in our system allows two benefits: the human planner is able to understand the few proposed solutions rapidly and he could impose his will onto the system.

The tasks of the studied scenario are illustrated in Figure 5. There are three manual tasks, and the two latter ones request the user to choose between computed solutions:

- Select the machine-tool configuration (our method did not try to automate this).
- Select the neutral plane (i.e. the orientation of the part in the stock).
- Select the couple of orientation (orientation #1 on front-side set-up; orientation #2 on back side set-up) of the part around the normal of the machine-table.

The strategy to display the final normal orientation solutions is to first present a so-called "optimal" solution (i.e. based on industrial heuristic rules that have been defined to select one discrete solution). The alternative solutions are only displayed on user demand.

3.1.2 Design Principles

The study of human-computer interaction was based on the literature guidelines [9, 10] to design interactive systems. In addition, some advice on cognitive aspects of human information processing [11] and means of displaying information [12] were used. Two main principles guided the design of specifications during our study:

The user must be aware about the state of the solution and the state of the system (what has been done/what to do next) in each task [13]. Therefore the screen must be structured to provide in an easily identifiable place some information about the current state of the solution and the current task.

The user cannot afford that a choice made in a previous step will be disputed in a subsequent step. This is a factor that could definitely end the use of the aid system if not respected. To prevent this misbehaviour in our system, we have to check if the computed neutral planes lead to viable orientations (i.e. orientations that allow the part to be completely machined) before displaying them to the user. This checking imposes a background task to run the orientation step to ensure there is at least aone viable orientation solution within the proposed neutral planes.

3.1.3 Evaluation Method

The evaluation of the human-computer interaction has been carried out with a mock-up of the future user-interface, illustrated below. This mock-up is made with successive capture of screens the user may have before his/her eyes to make his decision. The user-interface was filled with data of an industrial aircraft part (simulated output of the automated method). An industrial manager, an expert aircraft process planner and a CAD/CAM software developer composed the evaluation team. We asked the team to react about the succession of tasks and the corresponding interface.

The objective of this evaluation was to define the utility of the proposed tasks and associated interactions. The utility term defines the relevance of the tasks of the system for achieving the user objective, while the usability term defines how easy is it use the system. As a result, the utility of the proposed interactions and tasks was validated, although critical comments have been expressed. These comments will be discussed in Section 4.

3.2 Mock-up of the Future User-Interface

This mock-up has been designed considering the scenario of Figure 5. The user-interface is the physical support of the proposed interaction. It is the perceptible behaviour of the system. Efforts have been made to display the maximum amount of information graphically. Three screen captures are shown in Figure 6: (a) the display of neutral planes solution; (b) the display of optimal orientation and (c) the display of alternative orientations. In screens (a) and (b), the system provides indicators to help the user in his decision-making task between computed alternative solutions.

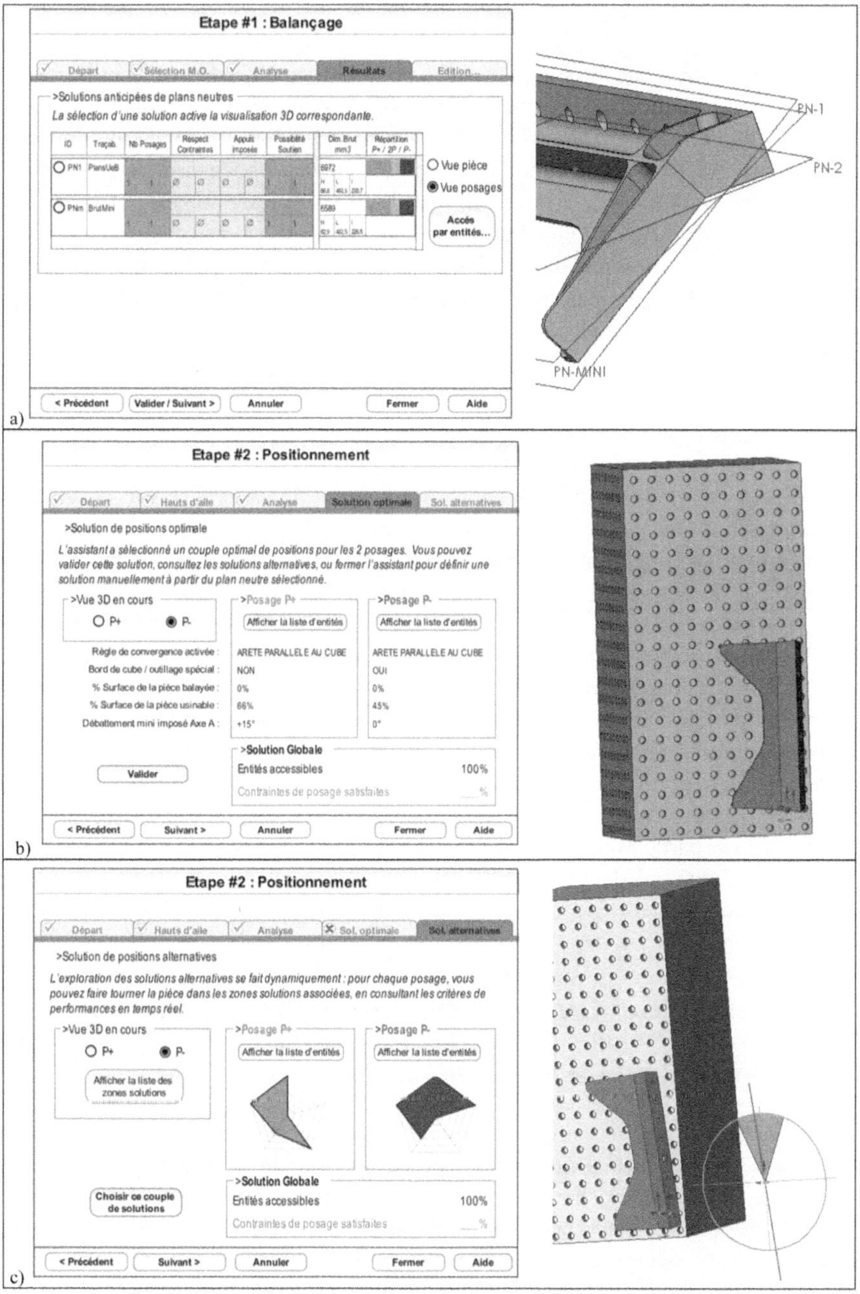

Fig. 6. Three screens of the user-interface mock-up.

3.2.1 Solution Indicators

The retained criteria are simple in order to be directly comprehensible and could be quantitative or qualitative. For the neutral planes assessment, two main indicators have been implemented: the prior ones describe the viability of the solution (does solution in two set-ups exist?) and the secondary ones have been designed to compare the solutions (initial stock dimension, distribution of the feature between the set-ups, etc.). For the orientation solution assessment, a star-shaped indicator has been proposed with five pre-specified technical criteria (number of features that can be end-milled, need of non-standard work-holder, etc.) which indicate if the solution is near the theoretical best cost solution or not. These graphical indicators are illustrated in Figure 6.

3.2.2 Displaying Process Planning Information

Besides supporting interaction, the main objective of this interface is to provide and display the relevant information on the computed alternative positions in order to aid the human planner to make his decision. To keep the technical information comprehensible, we propose mapping these data onto the solid 3D model of the part. In contrast with displaying information in a list, graphical 3D representation is able to handle many more elements and display them spatially so that the human planner does not have to remember their IDs. In addition, some tools available in 3D application such as filters or zoom could be really useful for the planner to explore the solutions. Figure 7 shows information displayed in the stock orientation step: neutral plane orientation and corresponding stock (Figure 7a); accessibility of features in the set-up (Figure 7b). Other information is displayed in the normal orientation step: situation of the part on the machine-tool table (Figure 4); available milling operation for a feature in a considered set-up (Figure 7c). This information display on the 3D model was considered to be very relevant for the planner by the evaluation team.

4 Discussion about Critical Comments from the Evaluation

4.1 The Difficult Choice of the Number of Alternatives

Some critical observations have been expressed during the evaluation. The following discussion focuses on one important problem that has been raised: the number of alternative solutions that should be displayed to the planner. Indeed, the search for the optimal solution may spend too much time in comparison to the gain earned between a viable solution and solution that is only slightly better, technologically speaking. Therefore, the number of alternatives must be limited. We suggest this limitation is due to either of two factors:

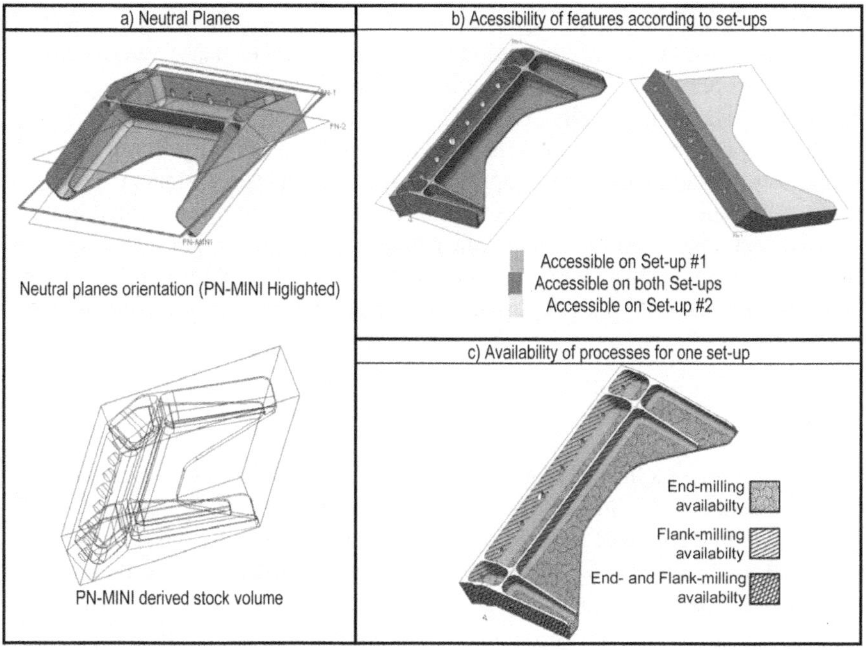

Fig. 7. Manufacturing Information displayed graphically on the 3D part model.

- The first factor is the human limited capability of information processing. From his cognitive studies, Miller [14] claims that the number of items a human can handle at the same time (short term memory capacity) is 7. An item can gather much information if it has a significance for humans, that is why we put the emphasis on graphical representation and develop simple graphic indicators.
- The second factor is the industrial context of aircraft machining, which is characterised by very short production runs and small batches. The cost of planning is very important and is split over few parts (average of ten parts per year), meaning that it is often not worth spending time on improve planning if a simplest viable solution exists.

4.2 Discussion

Thus, following the evaluation team suggestion, the proposed computer-aided tool first displays the so-called "optimal" solution of orientation and then displays alternative solutions on user demand, but according to the part geometry there might be more than three alternative solutions. To display only three solutions, the system must make a selection based on pre-specified criteria that might not match with the user's context-dependent criteria. In addition, providing the first time a unique solution may provoke an abnormal response of the user which is called "contentment behaviour": Hoc shows in his study [16] that the operator using automated complex systems prefers acceptance of a non-optimal solution rather than using complex functions to search for better solutions.

In view of these constraints, the Shen and Norrie proposition about alternative proposal seems to be difficult to apply directly in an industrial context. Even if the parts studied in this paper were specific, CAPP systems are often applied in the industrial sector with small batches rather than mass production. These contexts do not justify the extra programming time that would be needed to make the NC program as efficient as it could be. In view of this study, we suggest that the number of alternatives displayed to the user should not exceed three.

4.3 Perspectives

The proposition of a limited number of alternatives adds a problem of efficient automated selection and brings to an end the flexibility enhancement suggested by Shen and Norrie. In the light of our study, we suggest that a computer-aided process planning tool might support the generation of solution by the user, as is done in CAD systems to support the design of a product. An interesting lead to follow for future work would be to support the planner in designing the set-ups by providing contextual aids or functions which could evaluate the solution under construction.

5 Conclusion

This paper deals with a human-centred preliminary design of a CAPP system for aeronautical structural parts. The presented work carried out in a study of human-computer interaction to specify the interaction (and the corresponding user-interface) of a computer-aided set-up planning system. The following hypothesis suggested by Shen and Norrie was applied: "such systems should give the human planner some relevant information about alternative plans rather than automatically computing one unique plan".

Section 1 introduced a first assumption to explain the very limited success and acceptance of CAPP systems in industry. Section 2 summarised the set-up planning method proposed in a previous paper. Section 3 described the principles of the human-computer interaction study we carried out. The application of these principles led to the design of a mock-up of the future user-interface, which was evaluated. This evaluation points out two major results:

- the use of the 3D part model to display process plan information is relevant for the planner;
- the number of alternative solutions must be limited.

Section 4 discussed this limitation of alternatives: what the number of alternatives to be displayed according to the human information processing capability and to the industrial context of an aircraft workshop is. We suggested that this number should not exceed three and we highlighted the difficulties of applying Shen and Norrie's proposition in such a context. To overcome the quoted limitations, clues were suggested for a more efficient assistance strategy based on an interactive computation of solutions.

References

1. T.C. Chang. *Expert Process Planning for Manufacturing*, Addison-Wesley Publication, Massachusetts, USA, 1990.
2. F.J.A.M. van Houten. PART – A Computer Aided Process Planning System, Ph.D. Thesis, University of Twente, Enschede, Netherlands, 1991.
3. G. van Zeir, J.-P. Kruth, J. Detand. A conceptual framework for interactive and blackboard based CAPP, *Int. J. Production Research* **36**(6), 1998, 1453–1473.
4. X. Li, S. Kambhampati. Process planner's assistant: An interactive and iterative approach to automating process planning, in *Proceedings of DETC'97 ASME Design Engineering Technical Conference*, Sacramento, USA, 1997.
5. H. Amara, P. Dépincé, J-Y. Hascoët. A human-centered architecture for process planning, *CIRP Journal of Manufacturing Systems* **33**(4), 2004, 363–372.
6. W. Shen, D.H. Norrie. Agent based systems for intelligent manufacturing: A state of the art survey, *Knowledge and Information Systems* **1**(2), 1999, 129–156.
7. V. Capponi, O. Zirmi, D. Brissaud, F. Villeneuve. Computer Aided Process Planning, strategy and models in aircraft industry, *Journal of Engineering Manufacture* (Special Issue of the Proceedings of the Institution of Mechanical Engineers, Part B), **220**(4), 2006, 541–553.
8. V. Capponi, F. Villeneuve, H. Paris. Handling of alternative processes for machining of aeronautical parts in a CAPP system, in *Proceedings of the IFAC Conference on Manufacturing, Modelling, Management and Control* (IFAC-MiM'04), Athens, 21–22 October 2004.
9. J. Nielsen, R. Molich. Heuristic evaluation of user interfaces, in *Proceedings of CHI'90 Conference*, Seattle, ACM New York, 1990, pp. 349–356.
10. C. Bastien, D.L. Scapin. Evaluating a user interface with ergonomic criteria, *International Journal of Human-Computer Interaction* **7**(2), 1995, 105–121.
11. J. Rasmussen. The Human as a system component, in *Human Interactions with Computers* Smith and Green (Eds.), London Academic Press, 1980.
12. C. Kolski. *Interfaces homme-machine*, Editions Hermès, Paris, 1997.
13. L. Nigay. Conception assistée par ordinateur: Aménagement d'une cuisine, in *Actes de la Conférence IHM'96*, Cépaduès-Editions, 1996.
14. J.-G. Miller. The magic number Seven plus or minus two: Some limits in our capacity for processing information, *Psychological Review* **63**, 1965, 81–97.
15. J-M. Hoc. Conditions et enjeux de la coopération homme-machine dans le cadre de la fiabilité des systèmes, in *Sécurité et Cognition*, J.G. Ganascia (Ed.), Edition Hermès, Paris, 1999, pp. 147–164.

TOLERANCING METHODS

Optimization of Gear Tolerances by Statistical Analysis and Genetic Algorithm

Jérome Bruyere[1], Jean-Yves Dantan[1], Fangcai Wu[2] and Regis Bigot[1]

[1]*LGIPM, ENSAM de Metz, 4 rue A. Fresnel, 57070 Metz Cedex, France;*
E-mail: jerome.bruyere@metz.ensam.fr
[2]*Wuhan University of Technology, China*

Abstract. The purpose of functional tolerancing process is to define the geometrical specifications (tolerances) of parts ensuring functional requirements. An important distinction in tolerance process is that engineers are more commonly faced with the problem of tolerance synthesis rather than tolerance analysis. In tolerance analysis the parts tolerances are all known and the resulting geometrical requirement respect is calculated. In tolerance synthesis, on the other hand, the geometrical requirement is known from design requirements, whereas the magnitudes of the parts tolerances to meet these requirements are unknown. In this paper, we focus on the gear tolerances, and we propose an approach based statistical analysis for tolerance analysis and genetic algorithm for tolerance synthesis. Usually, statistical tolerance analysis uses a relationship between parts deviations and functional characteristics. In the case of tolerance analysis of gears, thus relationship is not available in analytic form, the determination of a functional characteristic (e.g. kinematic error) involves a numerical simulation. Therefore the Monte Carlo simulation, as the simplest and effectual method, is introduced into the frame. Moreover, to optimize the tolerance cost, genetic algorithm is improved. Indeed, this optimization problem is so complex that for traditional optimization algorithms it may be difficult or impossible to solve it because the objective function is not available in analytic form. For the evaluation of the fitness of each individual based on Monte Carlo Simulation, the number of samples is the key of precision. By a large number of samples, the precision can be improved, but the computational cost will be increased. In order to reduce the computational cost of this optimization based on Monte Carlo Simulation and Genetic Algorithms, the strategy is to adopt different precision of fitness; different numbers of samples during the optimization procedure are introduced into our algorithms.

Key words: tolerance synthesis, tolerance analysis, Monte Carlo simulation, genetic algorithm.

S. Tichkiewitch et al. (eds.), Advances in Integrated Design and Manufacturing in Mechanical Engineering II, 401–415.

1 Introduction

As technology increases and performance requirements continually tighten, the cost and required precision of assemblies increase as well. There exists a strong need for increased attention to tolerance design to enable high-precision assemblies to be manufactured at lower cost. For automotive and aircraft industries, the tolerance design activity has become an important issue in product design process, it involves allocation of parts tolerances based on the functional requirements. There is important question that would need to be looked upon: How to optimize the allocation of design tolerances? Design tolerances are allocated to achieve optimal manufacturing cost.

This topic has been under extensive research, which has resulted in several formulations and solution algorithms for systematic tolerance design considering various aspects. Tolerance design has been the focus of many techniques such as worst-case analysis [16, 19], statistical analysis [13, 18], design optimization [7–9, 17] and constraint-based reasoning. Many of these are restricted to either analysis or synthesis; only a few are applicable to both analysis and synthesis.

In this paper, we focus on the gear tolerances, and we propose an approach based statistical analysis for tolerance analysis and genetic algorithm for tolerance synthesis.

For the tolerances analysis, the evaluation of the effect of geometrical variations of individual part (gear) on functional characteristics (e.g. kinematic error) can be either worst-case or statistical. In worst-case tolerance analysis, the analysis considers the worst possible combinations of individual tolerances and examines the functional characteristic. In the case of tolerance analysis of gears, we are not sure that the worse kinematic error corresponds to the worse possible configurations of tolerances considering non-linearity of system. Statistical tolerancing is a more practical way of looking at tolerances and works on setting the tolerances so as to assure a desired yield. Usually, statistical tolerance analysis uses a relationship between parts deviations and functional characteristics. In the case of tolerance analysis of gears, thus relationship is not available in analytic form, the determination of a functional characteristic (e.g. kinematic error) involves a numerical simulation. Therefore, the Monte Carlo simulation, as the simplest and effectual method, is introduced into the frame. This aspect is detailed in Section 2.

Moreover, to optimize the tolerance cost, genetic algorithm is improved [7–9, 17]. Indeed, this optimization problem is so complex that for traditional optimization algorithms it may be difficult or impossible to solve it because the objective function is not available in analytic form. A genetic algorithm includes 3 steps: the initialization of the first population, the evaluation of the fitness of each individual of the population P and the definition of a new population by genetic operators. For the evaluation of the fitness of each individual based on Monte Carlo Simulation, the number of samples is the key of precision. By a large number of samples, the precision can be improved, but the computational cost will be increased. In order to reduce the computational cost of this optimization based on Monte Carlo Simulation and Genetic Algorithms, the strategy is to adopt different precision of fitness; different numbers of sample during the optimization procedure are introduced into our algorithms. This optimization approach is presented in Section 3.

2 Gear Tolerances Analysis

Tolerance analysis involves evaluating the effect of geometrical variations of individual part on designated functional characteristics. To do so, the best way to analyze the tolerances is to simulate the influences of parts deviations on the geometrical requirements. Usually, these geometrical requirements limit the gaps or the displacements between product surfaces. For this type of geometrical requirements, the influences of parts deviations can be analyzed by different approaches as Variational geometry [14], Geometrical behaviour law [1], Clearance space and deviation space [14, 16], Gap space [19], quantifier approach, cinematic models [15], etc. In the case of gears, their geometrical deviations impact the kinematic error, the tooth contact position, meshing interference, gap, etc. To ensure a quality level, designers limit these parameters by requirements. To simulate the influences of gears deviations on these requirements, the classical approaches cannot possibly be employed because the tolerance analysis of gears includes a kinematic aspect and a determination of contact position. Therefore, we propose an approach based on the parameterization of deviations (Section 2.1), the Tooth Contact Analysis (TCA) for the geometrical behaviour model (Section 2.2), the Monte Carlo simulation for the tolerance analysis (Sections 2.3–2.5).

2.1 Geometrical Model

Usually, geometrical behaviour description needs to be aware of the surface deviations of each part and relative displacements between parts according to gap or functional characteristic. In the case of gears, the geometrical behaviour description principally needs an exact analytical definition of tooth surface. In this application, the analytical definition of tooth surface is a parametric model which includes geometrical deviations and gaps. This model is based of the vectorial dimensioning and tolerancing (VD&T) strategy. The principle of VD&T is based on the concept of substitute surfaces. A substitute feature is an imaginary geometrical ideal surface which is represented by parametric vector.

The substitute tooth surface $\sum_t$ is described by its parametric model in the local coordinate system (S_7 first substitute surface pinion 1 or S_8 first substitute surface wheel 2) as functions of two parameters (φ, θ) or (α, γ) in a certain closed interval. The position vectors of tooth surface $\sum_t$ of pinion 1 and wheel 2 are denoted as $\mathbf{r}_k^{(i)}$ for part (i) in the coordinate system "k". The parametric model of tooth surface $\sum_t$

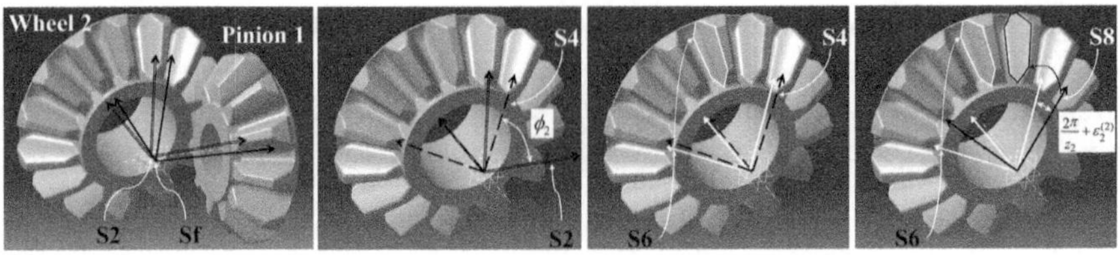

Fig. 1. Definition of wheel coordinate systems.

in the local coordinate system includes the nominal active surface model $X^{(i)}(\_,\_)$ [5], form deviations $w^{(i)}(\_,\_)$, and the crowing $\eta(\_)$. $\mathbf{n}_k^{(i)}$ is the unit normal vectors of part (i) surface in the coordinate system "k" (Equation 1 of pinion 1). This model is detailed in [2].

$$\mathbf{r}_7^{(1)}(\varphi,\theta) = \mathbf{X}^{(1)}(\varphi,\theta) + [w^{(1)}(\varphi,\theta) + \eta(\theta)] \cdot \mathbf{n}_7^{(1)}. \tag{1}$$

The model of the gear box with geometrical variations can be simulated by changing the settings and orientations of the coordinate systems. Few coordinate systems (S_7, S_5, S_3, S_1 for pinion 1 and S_8, S_6, S_4, S_2 for wheel 2) (Figure 1) [2] are introduced. They allow to parameterize geometrical deviations, gaps and displacements: error on cumulative angular pitch, position and orientation deviations between the axis of the teeth and the hole axis, rotational parameter, misalignments, etc. Therefore, we can define the parametric model of tooth surface $\sum_t$ in the global coordinate system S_f (Equation 2 where matrix M_{ij} is the homogenous coordinate transform matrix from Sj to Si).

$$\begin{cases} r_f^{(1)}(\varphi,\theta,\phi_1) = M_{f1}M_{13}(\phi_1)M_{35}M_{57}\, r_7^{(1)}(\varphi,\theta) \\ r_f^{(2)}(\alpha,\gamma,\phi_2) = M_{f2}M_{24}(\phi_2)M_{46}M_{68}\, r_8^{(2)}(\alpha,\gamma) \end{cases} \tag{2}$$

2.2 Geometrical Behaviour Simulation During the Global Meshing

The mathematical formulation of tolerance analysis is based on the expression of the geometrical behaviour of the mechanism; various equations and inequations modelling the geometrical behaviour of the mechanism are defined by the composition relations of displacements in the various topological loops [1], and the interface constraints. Interface constraints limit the geometrical behaviour of the mechanism and characterize non-interference or association between substitute surfaces, which are nominally in contact.

In the case of gears, the interface constraints are defined by the Tooth Contact determination. The principle idea of TCA [10–12] is based on simulation of tangency of tooth surfaces being in mesh. The aim of TCA is to obtain the real gear ratio at the mean contact point during the meshing. If the teeth surfaces and the relative positions

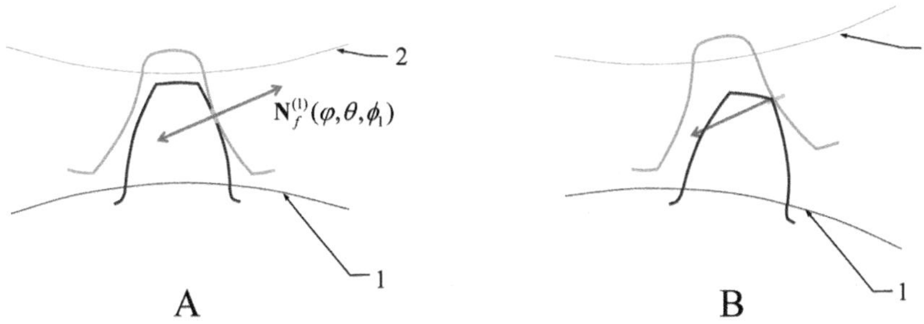

Fig. 2. Configurations of tooth contact.

are perfect, the instantaneous gear ratio would be constant. Due to misalignment and parts deviations, this instantaneous kinematic relationship is changing. The relative variations of real gear ratio are minor but accelerations induced are not negligible. Indeed, jump of angular velocity must be avoided in order to reduce noise level and vibrations [10–12].

The most difficult in TCA is to solve the system of non-linear equations that traduce contact between the two surfaces. During the meshing, the position of contact point can evolve between several configurations: contact inside the limited tooth surfaces (Figure 2a), contact on the boundary of one limited tooth surface (Figure 2b). First configuration, the surfaces are tangential, the necessary and sufficient conditions are defined by formulae 3 [10–12]. Second configuration, the contact is on the boundary of one limited tooth surface, the necessary and sufficient conditions are defined by formulae 4 ($\mathbf{N}_k^{(i)}$ is the normal vectors of part (i) surface in the coordinate system "k").

$$\begin{cases} \mathbf{r}_f^{(1)}(\varphi,\theta,\phi_1) = \mathbf{r}_f^{(2)}(\alpha,\gamma,\phi_2), \\ \mathbf{N}_f^{(1)}(\varphi,\theta,\phi_1) = c\mathbf{N}_f^{(2)}(\alpha,\lambda,\phi_2), \\ (\varphi,\theta) \in E^{(1)}, \\ (\alpha,\gamma) \in E^{(2)}, \\ (\phi_1,\phi_2) \in E^{(3)}, \\ c \in \text{Real}, \end{cases} \tag{3}$$

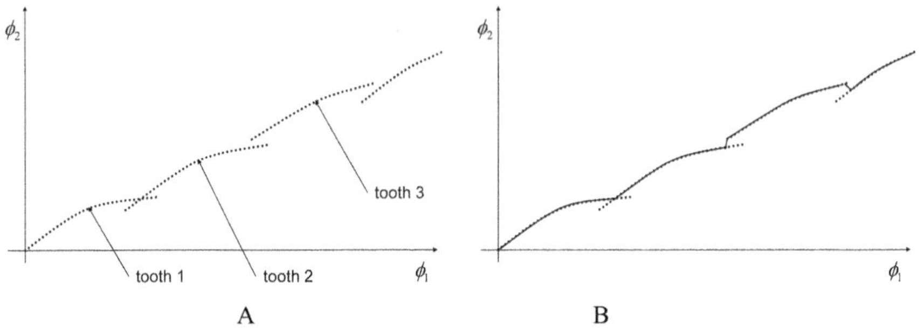

Fig. 3. Kinematic relationships during the global meshing.

$$\begin{cases} \mathbf{r}_f^{(1)}(\varphi_0, \theta, \phi_1) = \mathbf{r}_f^{(2)}(\alpha, \lambda, \phi_2), \\ \frac{\partial \mathbf{r}_f^{(1)}(\varphi_0, \theta, \phi_1)}{\partial \theta} \mathbf{N}_f^{(2)}(\alpha, \gamma, \phi_2) = 0, \\ \theta \in D \subset E^{(1)}, \\ (\alpha, \gamma) \in E^{(2)}, \\ (\phi_1, \phi_2) \in E^{(3)}, \\ \varphi_0 \text{ fixed.} \end{cases} \tag{4}$$

System of equations for meshing is known (3) or (4) and the Tooth Contact Analysis allows to determine $\phi_2, \varphi, \theta, \alpha$ and γ in function of ϕ_1. This mathematical problem has not an explicit solution in general case. We may only have an approximate numerical solution [2]. In this aim, the following method is used:

1. To choose a series of values for ϕ_1.
2. For each value of ϕ_1, to solve the system of Equations (3).
3. If one of the parameters $\{\varphi, \theta, \alpha, \gamma\}$ is not in his validity domain, to solve a new equations system like (4).
4. To analyze the instantaneous kinematics error.

This tooth contact analysis allows to define the kinematic relationship during the meshing of one tooth. To define it during the global meshing, we reproduce this analysis for each tooth which is nominally in contact, we obtain each kinematic relationship of each tooth (Figure 3a), and we calculate the superior (or inferior, that depends on direction of rotation) envelope of these relationships (Figure 3b).

2.3 Statistical Tolerances Analysis

The tolerance model is based of the vectorial dimensioning tolerancing (VD&T) model. A tolerance is an interval which limits an individual deviation of the geometrical part model [1, 14].

For statistical tolerance analysis, the possible values for an individual deviation of a component part are described by some frequency distribution. By permitting a small fraction of assemblies to not assemble or function as required, an increase in tolerances for individual dimensions may be obtained, and in turn, manufacturing costs my be reduced significantly. Statistical tolerance analysis computes the probability that the product can be assembled and will function under given individual tolerance.

Statistical tolerance analysis uses a relationship of the form (Equation 5) where Y is the response (characteristic such as gap or functional characteristics) of the assembly and $X = \{x_1, x_2, \ldots, x_n\}$ are the values of some characteristics (such as situation deviations or/and intrinsic deviations) of the individual parts or subassemblies making up the assembly [13]. The function f is the assembly response function. The relationship can exist in any form for which it is possible to compute a value for Y given values of $X = \{x_1, x_2, \ldots, x_n\}$. It could be an explicit analytic expression or an implicit analytic expression, or could involve complex engineering calculations or conducting experiments or running simulations. The input variables $X = \{x_1, x_2, \ldots, x_n\}$ are continuous random variables. In general, they could be mutually dependent.

$$Y = f(X). \tag{5}$$

A variety of methods and techniques allow to estimate the probability distribution of Y and the probability $P(T)$ of the respect of the geometrical requirement. Essentially, the methods can be categorized into four classes according to the different type of function f [13]:

- Linear Propagation (Root Sum of Squares)
- Non-linear propagation (Extended Taylor series)
- Numerical integration (Quadrature technique)
- Monte Carlo simulation

In the case of the statistical tolerance analysis of gears, the function f is not available in analytic form, the determination of the value of Y involve the running simulation. Therefore, we use a Monte Carlo simulation. Indeed, Monte Carlo technique is easily the most popular tool used in tolerancing problems. The appeal of Monte Carlo lies in its applicability under very general settings and the unlimited precision that can be achieved [2].

2.4 Algorithm of Gear Tolerances Analysis by Monte Carlo Simulation

Monte Carlo simulation is based on a very simple thinking: Large Number Law [18]. Pseudo random number generators are used to generate a sample of numbers $x_1, x_2, \ldots, x_n$, belonging to the random variables $X_1, X_2, \ldots, X_n$, respectively. The value of Y, $y_1 = f(x_1, x_2, \ldots, x_n)$, corresponding to this sample is computed. This procedure is replicated a large number of samples S. This would yield a random

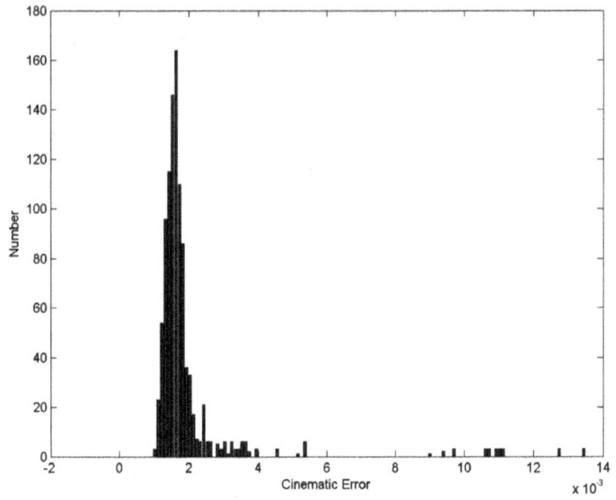

Fig. 4. Probability distribution of the maximum value of the kinematic error.

sample, $\{y_1, \ldots, y_N\}$ for Y. Standard statistical estimation methods are then used to analyze the distribution of Y (Figure 4).

The Monte Carlo simulation allows to estimate the probability $P(T)$ (Equation 6) where R is the number of the individual which respects the geometrical requirements.

$$P(T) = \frac{R}{S}. \tag{6}$$

The flow diagram of the Monte Carlo simulation and the geometrical behaviour simulation is shown in Figure 5.

2.5 Precision of the Statistical Tolerances Analysis

The precision of this statistical analysis increases proportional to $\sqrt{S}$ and therefore unlimited precision can be achieved through large number of replications. If the value of S is sufficiently large, the value of $P(T)$ will reach statistically stable results that are independent of the initial seed value. On the other hand, if the difference of the simulated results amongst various initial seed values exceeds 10%, then the number of samples, S, is not sufficient for stability to be achieved.

Monte Carlo Simulation use random sampling, the result $P(T)$ is also a random variable. The variance of $P(T)$ estimation converge with an average rate of $1/\sqrt{S}$. To characterize this random variable, 20 Monte Carlo simulations were computed for each value of S: 20, 30, 40, 50, 75, 100, 200, 300, 400, 500, 1000, and 2000. This experiment allows to estimate the range and the standard deviation of $P(T)$ (Figure 6).

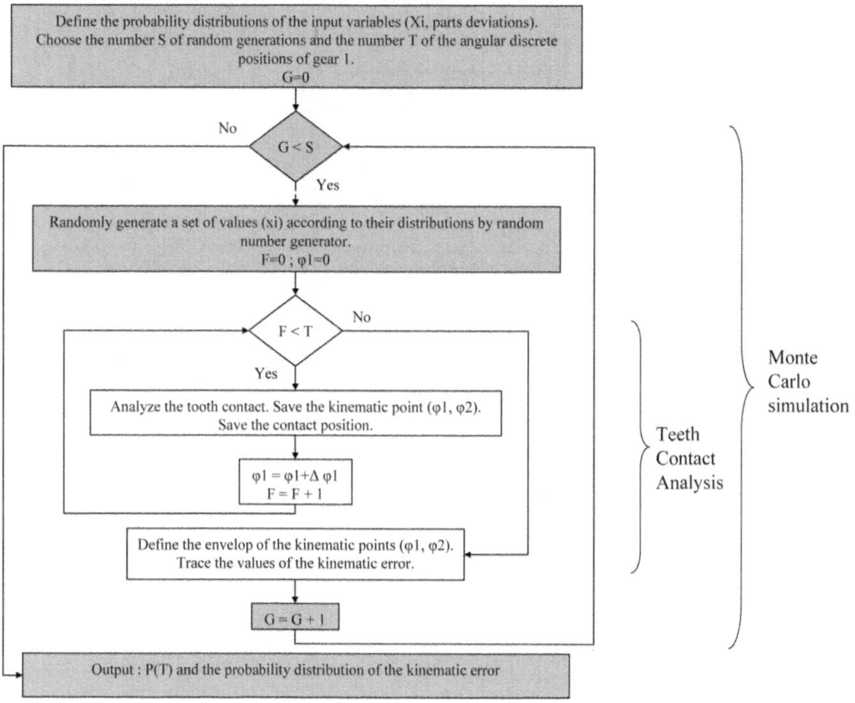

Fig. 5. The flow diagram of gear tolerances analysis.

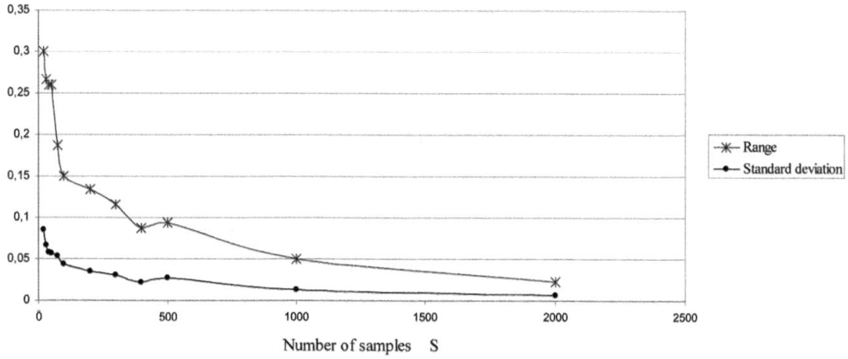

Fig. 6. Range and standard deviation of $P(T)$.

3 Tolerances Synthesis – Optimization of Gear Tolerances

Tolerances synthesis is considered an optimization problem for looking for the best balance between the risk of non-quality and cost of a mechanical product [4]. The

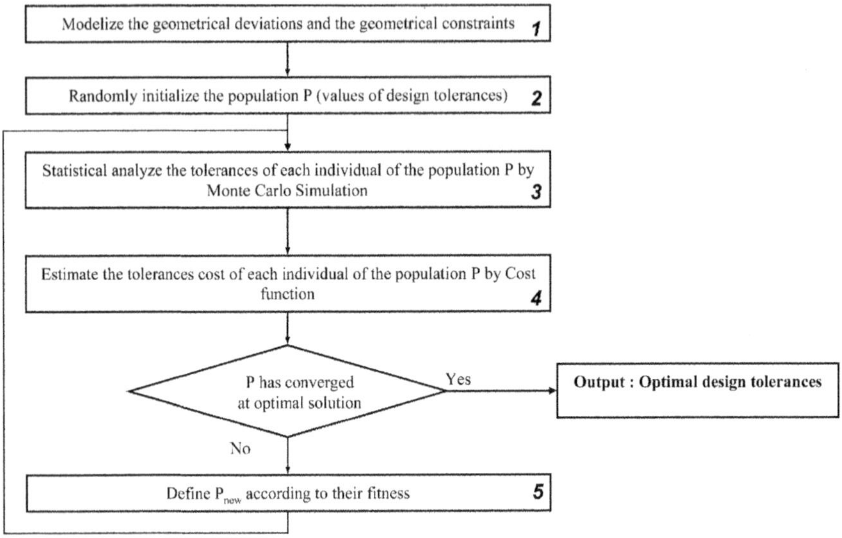

Fig. 7. The flow diagram of global approach.

optimization problem is denoted by follows:

$$\text{Minimize}\,[F(T)] \quad \text{with} \quad F(T) = C(T)/P(T), \tag{7}$$

T are the tolerance values, $C(T)$ is the cost function and $P(T)$ is the probability of the respect of requirements on kinematic error.

This optimization problem is so complex that for traditional optimization algorithms it may be difficult or impossible to solve it because the function F is not available in analytic form. Recently, due to their outstanding performance in non-linear optimization, some stochastic optimization algorithms, such as genetic algorithms and simulate annealing algorithms, have be used well to solve the problem. Therefore, this optimization is realized by genetic algorithm (Figure 7). A genetic algorithm includes 3 steps [6]: the initialization of the first population (sub-step 2), the evaluation of the fitness of each individual of the population P (sub-steps 3 & 4) and the definition of a new population by genetic operators (sub-step 5 & Section 3.3).

The fitness of an individual is the value of the function F. The evaluation of the function F includes the cost estimation (sub-step 4) which is detailed in the Sections 3.1 and 3.2, and the probability estimation (sub-step 3). This probability estimation is realized by statistical tolerance analysis (Section 2) which needs a geometrical model (sub-step 1).

3.1 Fitness Function

The fitness of an individual is the value of the function F with $F(T) = C(T)/P(T)$ [4].

For statistical tolerance synthesis based on Monte Carlo Simulation, the number of Monte Carlo samples is the key of precision of $P(T)$, which has illuminated in the Section 2.5. By a large number of samples, the precision can be improved, but the computational cost will be increased, too, especially when such as Genetic Algorithms, the iterative algorithms that need enough many times of evaluating the objective function to obtain a good result, are used solved the optimization problem [9].

In general, during the whole procedure of optimization of Genetic Algorithms, the precision requirement of fitness is not same. In initial stage of optimization, the individuals are dispersed on whole domain of solution for global optimization, and the discrepancy of fitness in the individuals is very big, therefore, the precision requirement of fitness can be wide. However, in final stage of optimization, almost individuals fall into the local domain in which the final result lied, and the discrepancy of fitness in the individuals is likely very small, so the precision of fitness must enough high that the algorithms can converge.

Thereby, in order to reduce the computational cost in statistical tolerance synthesis based on Monte Carlo Simulation and Genetic Algorithms, the strategy to adopt different precision of fitness like simulate annealing algorithms, that is, different number of samples S during the optimization procedure is introduced to this algorithm:

$$S(g) = S_0 + (S_t - S_0)\frac{1}{2}\{1 + \tanh[\eta(g/g_{\max} - \delta)]\}, \quad (8)$$

where, g is the current generation number, $g_{\max}$ is the maximal generation, S_0 and S_t are respectively the initial number of samples and final number of samples determined by the precision of the optimization solution, and η and δ are the control parameters, $\eta > 1, \delta \in (0, 1)$.

3.2 Tolerance Cost Model

A substantial amount of research has been carried out regarding optimal tolerance allocation using cost-*vs*-tolerance functions. Various functions have been proposed to describe the cost-tolerance relationship [3]:

$$C_i(T_i) = a_i + \frac{b_i}{T_i^{K_i}} \quad \text{or} \quad C_i(T_i) = \frac{b_i}{T_i^{K_i}} \quad \text{or}$$

$$C_i(T_i) = b_i e^{-k_i T_i} \quad \text{or discrete functions, etc.} \quad (9)$$

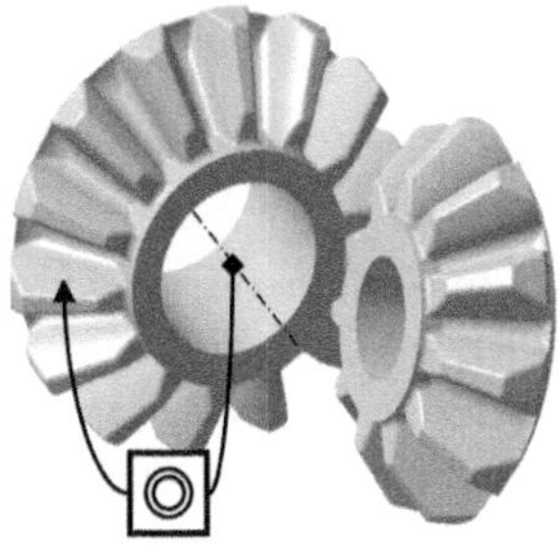

Number of teeth for pinion :	$z_1 = 11$
Number of teeth for wheel :	$z_2 = 15$
Nominal shaft angle :	$\Delta = \pi/2$
Sphere radius at small tip :	Rp = 50 mm
Sphere radius at big tip :	Rg = 100 mm
Tooth height :	h = 0.1613*(sphere radius)
Maximum kinematic error :	$ec_{max} = 2.10^{-3}$ rad

Fig. 8. Example.

Our approach can take into account continue functions or discrete functions. In the application, the first function is used. The global cost is the sum of each tolerance cost:

$$C(T) = \sum_{i=1}^{n} C_i(T_i) \tag{10}$$

3.3 Guo's Crossover

The population is refreshed continually according to following procedure (sub-step 5 – Figure 7): at each iteration, some of the chromosomes are chose to reproduce, via suitable operator (such as crossover operator, mutation, etc.), according to their value of function, as a capability of adaptation to the environment (value of the fitness). There are many genetic operators which are designed for different problems.

The Guo's crossover is improved when the genetic algorithm is coupled with Monte Carlo simulation. Guo has proposed a linear non-convex multi-parent crossover operator, used in optimization of non-linear continuous functions with non-equality constraints. It generates offspring by linear non-convex combination of m parental vectors. Its algorithm is as follows:

The chromosome is the most naturally represented by a real floating number vector, $T^{(i)} = T_1^i, T_2^i, \ldots, T_N^i)$, which T_j^i is the tolerance of x_j in the ith individual.

1. m individuals T^i, $i = 1, 2 \ldots, m$ selected from the population.
2. m random real numbers ω_i, $i = 1, 2 \ldots, m$ generated, subject to $\sum_i \omega_i = 1$ and $-\varepsilon \leq \omega_i \leq 1 + \varepsilon$ where ε is positive parameter.
3. The new individual T^{new} is generated as follows $T^{\text{new}} = \sum_i \omega_i T^i$

The interrupt criteria are maximal iteration number and convergence.

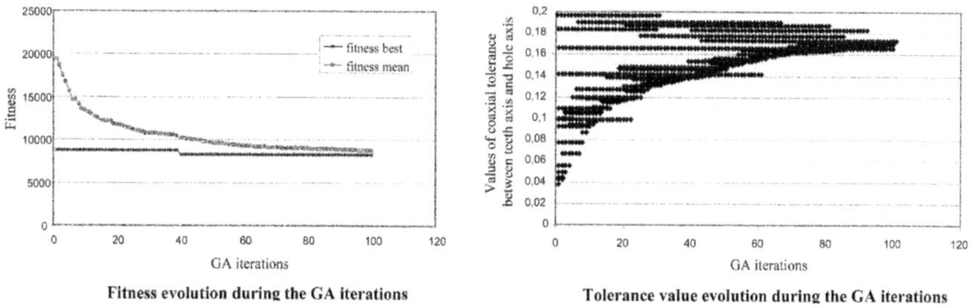

Fig. 9. Evolution of fitness and tolerance values during the GA iterations.

3.4 Application

In order to show the convenience of the proposed approach, we employ it to calculate the optimal coaxial tolerance between teeth axis and hole axis (Figure 8). The gear parameters are given in following table:

Figure 9a illustrates the convergence of the optimization by genetic algorithm: the evolution of the best fitness (smaller fitness of the population) and the mean fitness. These fitness decrease. Figure 9b shows the set of coaxial tolerance values of each population during the iterations of the genetic algorithm. We observe the convergence of the coaxial tolerance values during the GA iterations.

4 Conclusion

The problematic of this paper is "How to optimize the allocation of design tolerances of gear?" To do so, there exist some standard tables which are a set of discrete associations between tolerances and meshing quality. These tables are based on experimental results with cutting gears, they do not take into account all tolerances, and they focus on the pitch error and the misalignment. In the case of forging gear, we cannot use them to allocate the gear tolerances to achieve the optimal manufacturing cost.

Therefore, for the gear tolerance allocation to achieve the optimal manufacturing cost, we propose to use the techniques of Tooth Contact Analysis, Monte Carlo simulation and genetic algorithm which have been performed.

To optimize the tolerance cost, genetic algorithm is implemented. This optimization problem is so complex that for traditional optimization algorithms it may be difficult or impossible to solve it because the objective function is not available in analytic form. A genetic algorithm includes three steps: the initialization of the first population, the evaluation of the fitness of each individual of the population P and the definition of a new population by genetic operators.

The evaluation of the fitness needs a statistical tolerance analysis. Statistical tolerance analysis uses a relationship between parts deviations and functional characteristics. In the case of tolerance analysis of gears, thus relationship is not available in analytic form; the determination of the kinematic error involves a numerical simulation (Tooth Contact Analysis). Therefore, the Monte Carlo simulation, as the simplest and effectual method, is used. And, the main purpose of Tooth Contact Analysis is the simulation of contact of gear tooth surfaces, permitting to investigate the influence of geometrical deviations on kinematic error.

Moreover, for the evaluation of the fitness of each individual based on Monte Carlo Simulation, the number of samples is the key of precision. By a large number of samples, the precision can be improved, but the computational cost will be increased. In order to reduce the computational cost of this optimization based on Monte Carlo Simulation and Genetic Algorithms, we propose to adopt different precision of fitness; different numbers of sample during the optimization procedure are introduced into our algorithms. The complexity of this algorithm is $O(G(S_0 + S_t/2)T)\times$ complexity of TCA time where G is the maximum number of generations of the genetic algorithm, S_0 and S_t are respectively the initial number of samples and final number of samples of the Monte Carlo simulation, and T is the number of angular discrete positions of the global Teeth Contact Analysis. In a future article, we compare discrete optimization algorithms with regard to the computational cost.

References

1. Ballot E., Bourdet P.A., Computation method for the consequences of geometric errors in mechanisms, in *Proceedings of CIRP Seminar on Computer Aided Tolerancing*, Toronto, Canada, April 1997.
2. Bruyere J., Dantan J.Y., Bigot R., Martin P., Statictical tolerance analysis of gears, in *Proceedings of CIRP Seminar on Computer Aided Tolerancing*, Phoenix, USA, 2005.
3. Chase K.W., Least cost tolerance allocation for mechanical assemblies with automated process selection, *Manufacturing Review* **3**(1), March 1990, 49–59.
4. Dantan J.Y., Etienne A., Wu. F., Siadat A., Martin P., Allocation des tolérances fonctionnelles par optimisation du coût de fabrication, Journée AI2P Tolérancement, ENS Cachan, France, 2005.
5. Godet J.-C., Raffy M., Le calcul des engrenages coniques droits à développante de cercle sphérique, *Société d'Etude de l'Industrie de l'Engrenage* **53**, 1967.
6. Goldberg D., *Genetic Algorithms*, Addison Wesley, 1989.
7. Iannuzzi M.P., Sandgren E., Tolerance optimization using genetic algorithms: Benchmarking with manual analysis, in *Proceedings of the Fourth CIRP Design Seminar on Computer-Aided Tolerancing*, 1995.
8. Kanai S., Onozuka M., Takahashi H., Optimal tolerance synthesis by genetic algorithm under the machining and assembling constraints, in *Proceedings of the Fourth CIRP Design Seminar on Computer-Aided Tolerancing*, 1995.
9. Lee J., Johnson G.E., Optimal tolerance allotment using a genetic algorithm and truncated Monte Carlo simulation, *J. Computer Aided Design* **25**, 1995, 601–611.

10. Litvin F.L., *Gear Geometry and Applied Theory*, PTR Prentice Hall, Englewood Cliffs, NJ, 2004.
11. Litvin F.L., De Donno M., Peng A., Vorontsov A., Handschuh R.F., Integrated computer program for simulation of meshing and contact of gear drives, *Computer Methods in Applied Mechanics and Engineering* **181**(1–3), 2000, 71–85.
12. Litvin F.L., Theory of gearing, NASA RP-1212, Washington DC, 1989.
13. Nigam S.D., Turner J.U., Review of statistical approaches to tolerance analysis, *J. Computer Aided Design* **27**(1), 1995, 6–15.
14. Roy U., Li B., Representation and interpretation of geometric tolerances for polyhedral objects, *J. Computer Aided Design* **31**(4), 1999, 273–285.
15. Sacks E., Joskowicz L., Parametric kinematic tolerance analysis of planar mechanisms, *J. Computer Aided Design* **29**(5), 1997, 333–342.
16. Samper S., Petit J.P., Giordano M., Elastic clearance domain and use rate concept applications to ball bearing and gears, in *Proceedings of CIRP Seminar on Computer Aided Tolerancing*, USA, April 2005.
17. Shan A., Roth R.N., Genetic algorithms in statistical tolerancing, *J. Mathematical and Computer Modeling* **38**, 2003, 1427–1436.
18. Skowronski V.J., Turner J.U., Using Monte Carlo variance reduction in statistical tolerance synthesis, *J. Computer-Aided Design* **29**(1), 1997, 63–69.
19. Zou Z., Morse E.P., A gap-based approach to capture fitting conditions for mechanical assembly, *J. Computer Aided Design* **36**, 2004, 691–700.

Modal Tolerancing – Application to Gap and Flush Analyses

Pierre-Antoine Adragna, Serge Samper, Fabien Formosa and Maurice Pillet

SYMME, Université de Savoie, B.P. 806, 74016 Annecy Cedex, France; E-mail: {pierre-antoine.adragna, serge.samper, fabien.formosa, maurice.pillet}@univ-savoie.fr

Abstract. Tolerancing of geometrical forms is applied either to surfaces or to outlines of parts. We are interested, in this paper to form errors of outlines, but the presented method allows to describe all possible deviations of features as presented in [9]. That is to say that dimension, position, form, and roughness can be analysed by the modal tolerancing.

The measured feature gives a set of points that have to be filtered in order to distinguish those deviations. The decomposition of the measured geometry in a set of modal natural shapes sorts deviations from the lowest to the highest level of complexity. The set of natural modes is a theoretically unlimited basis but by using a discretisation of the feature this set will limited. This method inherits of the modal shape and finite element analysis properties:

- *Versatility*: any kind of geometry can be analysed.
- *Deviation unification*: size, position, form and roughness can be analysed.
- *Form growing complexity*: from large to short wavelengths.

In the case of surfaces, a shell element should be used in the FEA model and in the outline case, a beam (or a bar) model would be better. The behaviour of the element and boundary conditions have to be associated to the kind of deviations specified or observed. The manufacturer can see a decomposition of the actual parts in the modal shapes and look the size of the maximum error in order to control it. The metrologist who uses Fourier decomposition can easily understand this method. As this method gives a language to form errors, every actor of the product can express his needs in an unambiguous way.

In this paper we present an application of the method to an industrial example of a gap and flush analyse of a plastic assembly.

Key words: modal shapes, tolerancing, gap and flush, perceived quality.

1 Introduction

As precisions in parts are smaller, we need to look for the quality of form errors. In an other way, the perceived quality looks in the same direction. In some applications,

S. Tichkiewitch et al. (eds.), Advances in Integrated Design and Manufacturing in Mechanical Engineering II, 417–430.

the form is the major function of the product (optic, design, etc.). The metrology of form errors has to be qualified and quantified.

First, the possible deviations are sorted in a set of particular forms, and then each of them is limited in order to write the acceptable geometries. The customer can express his needs and the designer can express them. It cannot be possible to describe one by one a big number of form errors thus this set must be the smallest as possible.

The customer needs are not the same for any level of errors. In a roundness or straightness measurement, we can see that the evolution of deviation level is globally decreasing with the frequency of the form. In those particular cases, Fourier transform gives the spectrum of the form error. In the general case of a 3D-curved outline, there is no method to express and analyse. The modal shapes of the outline give to this problem a generic solution. This method can analyse a feature and truncate the shapes to the simplest ones in order to express the actual shape with the lowest number of parameters.

We first present the method, then an application to a 3D-curved outline of an industrial product in order to control gaps and flushes.

2 Actual Solutions

Definition of surfaces and specifications of form errors are given by three axis:

- Optimisation of point measurements.
- Minimisation of error by the choice of the best associating criteria between the measured and the nominal surfaces.
- The measured errors of a feature are decomposed in a basis of specific errors.

Talking about form errors measurement, in [1] the oldest works on minimisation of form deviations are presented by Carr. He refers to the decomposition of form errors with periodic functions as Pandit introduced them in [2]. The third axis supposes that a form basis is created in association with specific errors. In the optic domain, a norm (ISO 10110-5) has been written in this aim. Other solutions exist as given in [8] which presents turning errors. Capello [3] presents a harmonic association approach in order to define the number and the position of points for a plane surface measurement. This method allows the optimisation of measurement methods [4] by showing a symbolic relation linking the estimated error and the three factors (number of sampled points, their distribution, part positioning). In [5], a classification of analysed phenomenon is proposed.

In [6] Summerhays presents a method enabling the decomposition of defects in two types of basis. The first one is a set of analytic functions describing form errors whose angular dependencies are described with Fourier series and axial dependencies with second type Chebyshev polynomials. The second basis is a set of vectors of the measured surface. After comparison, the eigenvectors basis allows to model with lower order modes form errors with abrupt slope changes.

Ceglarek's team proposes a method for forms decomposition with the help of Discrete Cosine Transformation [7]. The displacements are segregated in a set of independent forms. It is possible with this method to use regular shapes as squares or rectangles (DCT is well known for image processing like JPEG compression). The use of DCT enables to build a statistical process control with form errors.

In form errors metrology, different basis has been developed in order to identify and qualify defects. There is still no general normalised method allowing this decomposition, but the need is reel.

The decomposition of a tolerance zone is not sufficient in numerous cases in order to specify a functional need for form error. When the customer needs are expressed, the normalised specifications and annotations have to explain them. We present in the following a non-ambiguous mathematical way to translate the gap and flush form functional needs.

3 Modal Deviation Decomposition

We propose a generic language enabling the mathematic expression of form errors, and giving a sense to the measures made with 3D means. The principle of our model is based on the decomposition of shapes in a natural modal basis.

3.1 Modal Analysis

In structural linear dynamics, equations of a conservative system (continued or discrete) are written under the following form:

$$M\ddot{q} + Kq = 0. \tag{1}$$

$M(n, n)$ and $K(n, n)$ are the generalised mass and stiffness matrices. $q(n, 1)$ is the vector of nodal displacements and n the degrees of freedoms number.

The free vibrations of the structure are supposed to have small magnitude and the movement equations of the structure can be linearised. The solution of (1) is written:

$$q - Q \cdot \cos \omega t. \tag{2}$$

$Q(n, 1)$ is the amplitudes vector and the natural pulsation (rad$\cdot$s^{-1}). Equation (1) has n solutions, called natural modal shapes of the structure. The natural pulsations ω_i of the modal shapes are the roots of the polynomial:

$$\det(K - \omega^2 M) = 0. \tag{3}$$

The natural modal shape Q_i are eigenvectors associated to ω_i, and are a M and K orthogonal basis in the vectorial small displacements movements of the structure. In the general case, Q_i are chosen as $Q_i^t \cdot M \cdot Q_i = 1$. Scalar product and norm of a vector A are defined with:

$$\langle A \cdot B \rangle = A^t M B \quad \text{and} \quad \|A\| = \sqrt{A_t M A}. \tag{4}$$

Orthogonality of modal shapes expresses that inertia and stiffness forces of a mode do not work in any other modal shape. There is an independence of modal shapes. Here, the orthogonality property leads to build a set of independent equations.

3.2 Modal Basis Construction

This versatile method can analyse either simple or complex geometries. First, we have to mesh the geometry in order to have each node on a measured point (we can also make an interpolation in general case).

The finite element model and the corresponding parameters can be built in order to target modal shapes to specific forms. This optimisation leads to minimise the number of modal shapes for a defined set of defects, but in general study, we use a classic model without optimisation. The boundary conditions are free type in general. The rigid body movement leads to take into account global small displacements of the measured feature. The calculus of modal solutions can be made by theories or solved by numeric tools.

3.3 Modal Basis Projection

When the natural mode shapes are set, the defect can be decomposed by a vectorial projection. Let V be the measured feature vector. It is composed of the displacements of points from the ideal to the "real" surface. Its coordinates in the modal basis is obtained as follows:

$$\lambda_i = \langle V \cdot Q_i \rangle = V^t \cdot Q_i. \tag{5}$$

In (5) λ_i is the modal coordinate (also named modal coefficient) of V. The residue vector ε is the difference between the displacement vector V and its projection in the reduced modal basis:

$$\varepsilon = V - \sum_{i=1}^{n} \lambda_i \cdot Q_i. \tag{6}$$

The scalar global residue e is the norm of ε:

$$e_i = \sqrt{\varepsilon_t \cdot \varepsilon}. \tag{7}$$

A program computes the projection of the observed defect on each mode in order to obtain the corresponding modal coefficients and residues in function of the number of selected modes. By this way, we can estimate if the truncation is efficient to represent the measured feature.

4 Gap and Flush Analysis

4.1 Industrial Case Application

The application presented in this paper is the assembly of two parts of a weighing balance. The gap and flush condition is decomposed in two cases:

- an assembly functional condition which implies to respect a gap between the two parts of the balance,
- a flush condition which implies to respect the flush between the two parts of the balance.

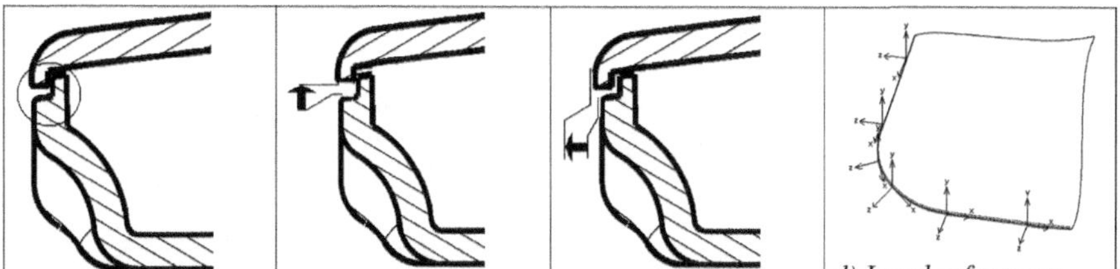

a) Gap and flush | *b) The gap* | *c) The flush* | *d) Local references placed on the gap and flush line*

Fig. 1. The gap and flush characteristics.

The gap and flush functional condition on the two parts is composed of an assembly functional condition. The parts assembly is made by translation of one part to the other on the vertical y axis, normal to the weighing plate of the balance, so it is taken as the common y axis of the local references.

4.2 Gap and Flush

Gap and flush are displacements of a surface relatively to another along a gap and flush line common to the nominal surface (no gap). Surfaces are not necessarily coplanar, there are 6 available displacements presented in the following table according to the components of the small displacements torsor associated. In the general case, it is possible to define the gap and flush problem on each point, in this case, local references will be created as much as needed. Components of the small displacement torsor are expressed relatively to a reference on the left side in the table. The analysis reference is then linked to this surface and has the following properties:

- X is collinear to the gap and flush line.
- Z is normal to the reference surface at the analysis point.
- Y is such as (X, Y, Z) is a direct orthonormal base.

Table 1. Gap and flush definition.

References	View	Displacements Torsor	Calling proposal
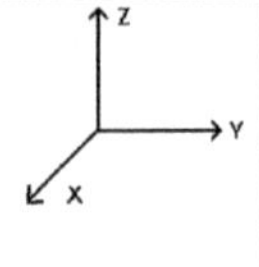	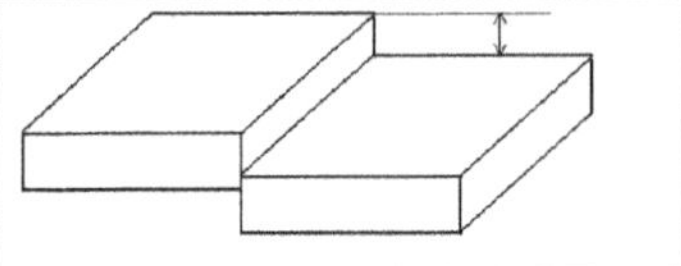	$\begin{Bmatrix} 0 & 0 \\ 0 & 0 \\ Tz & 0 \end{Bmatrix}$	Flush
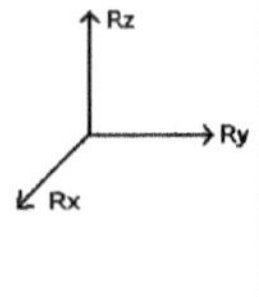	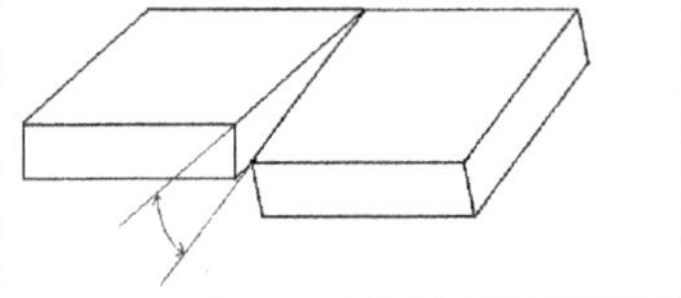	$\begin{Bmatrix} 0 & 0 \\ 0 & Ry \\ 0 & 0 \end{Bmatrix}$	Evolutionary flush
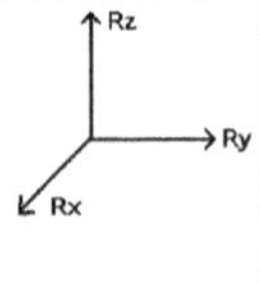	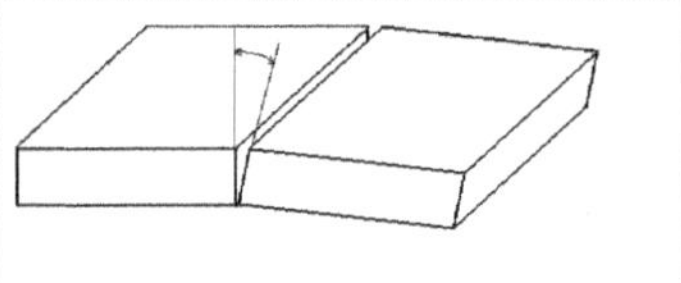	$\begin{Bmatrix} 0 & Rx \\ 0 & 0 \\ 0 & 0 \end{Bmatrix}$	Tangency flush
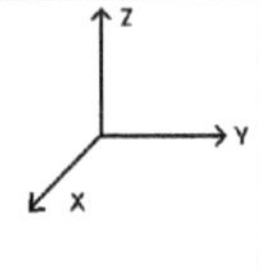	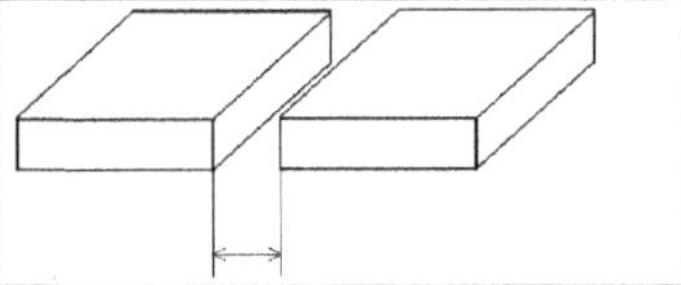	$\begin{Bmatrix} 0 & 0 \\ Ty & 0 \\ 0 & 0 \end{Bmatrix}$	Gap
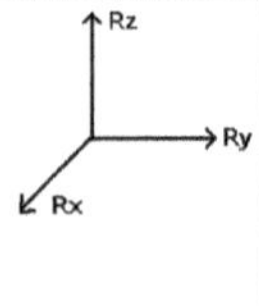	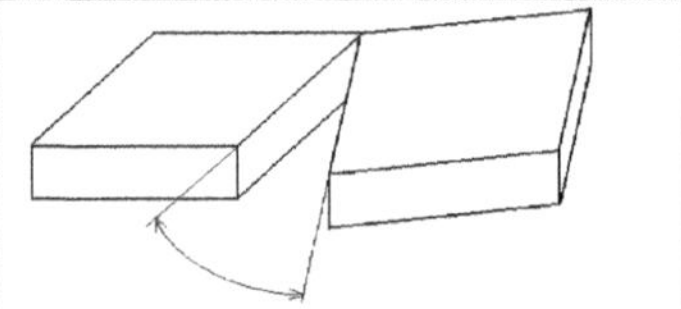	$\begin{Bmatrix} 0 & 0 \\ 0 & 0 \\ 0 & Rz \end{Bmatrix}$	Evolutionary gap
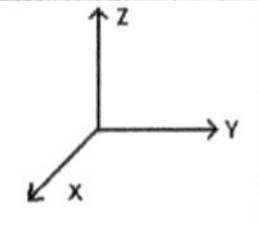	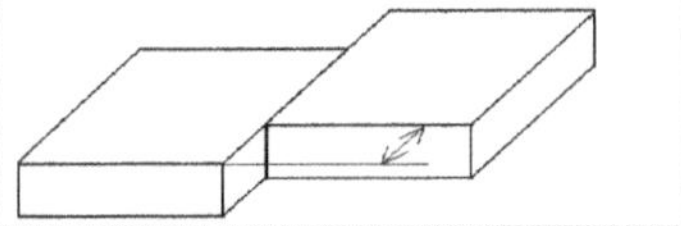	$\begin{Bmatrix} Tx & 0 \\ 0 & 0 \\ 0 & 0 \end{Bmatrix}$	Sliding gap

Remarks:

- Gaps are displacements in the (X, Y) plane.
- Flushes are displacements out of (X, Y) plane.

In the tolerancing process, we should consider that this torsor is a clearance one because it is a joint (even if the contact is not allowed), which links two parts.

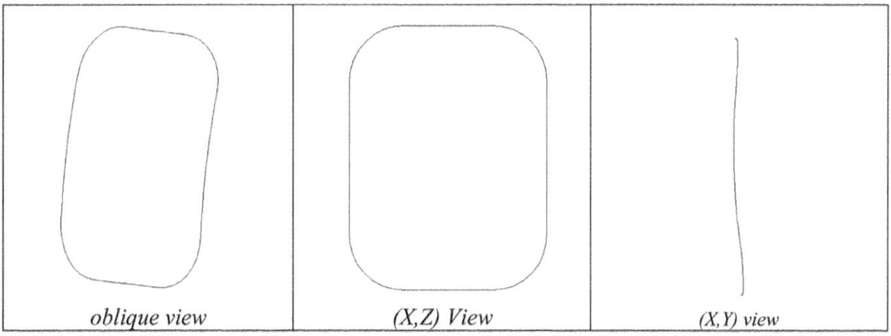

Fig. 2. The model of the gap and flush outline.

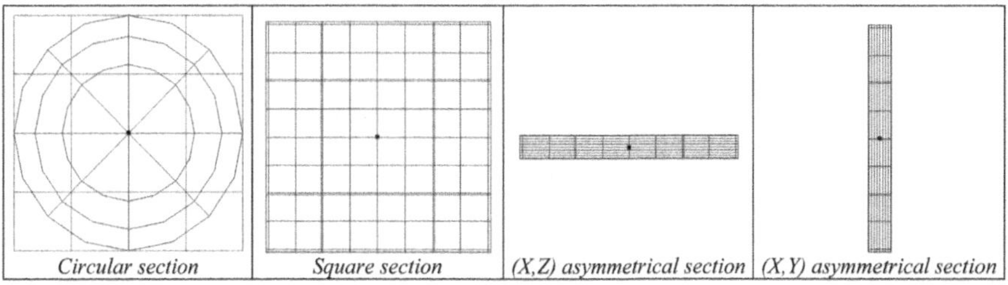

Fig. 3. Different sections used for the model.

5 Model Building

The natural modal shapes are computed by a classic finite element analysis. The model can be different form the discretised part to be analysed. It must be built in accordance to the geometric specifications. In the industrial example, we only study the outline (Figure 2) of the two parts. The finite element model fitting the geometrical specification (3D curves) is a beam one (Figure 3). The modal shapes will have to represent all the possible variations of made parts.

5.1 Good Practice

The choice of the model allows to obtain deformed geometries, which compose the modal defects basis. This basis has to best describe the measures. The gap and flush defects are expressed in a local reference linked to the gap and flush line. The good practice for the gap and flush analysis is to use a (topologic) line model based on the gap and flush line, and to define local references at each node (FEM) of the meshed model respecting the following orientation as defined in Section 4.2.

It is then possible to obtain deformed shape on each axes that allows to have three bases:

- modal deformed shape on the X axis, which describe the sliding gap,
- modal deformed shape on Y, which describe the gap,
- modal deformed shape on Z, which describe the flush.

One is interested only by those three characteristics which are translations flowing each axis of the local reference, rotations are not measured. It is then possible to analyse the measured geometry on each three bases and to characterize each component of the gap and flush, or to create a bigger basis with the union of the first three bases to directly analyse the gap and flush.

This practice can be a good one but it is necessary to define local references at each node (a program can make it by connected surfaces orientations). It is possible to shorten the method and do not use local references, but the analysis on each three components of the gap and flush will not be possible. It is possible to use a global reference and a good orientation can allow expressing one characteristic of the gap and flushing on one axis of the global reference.

In our case we choose to orientate the Y axis of the local references collinear to the Y axis of the global reference, which is the assembly axis of the parts. Thus, it is possible to keep the expression of the gap on the Y axis, the flush is expressed on the (X, Z) plane. The sliding gap is not useful here. Although it is possible to have two separate modal bases, the example uses a unique basis which characterises both gap and flush components which interest us.

5.2 Line Model

The line model is the one who seems the more appropriate to characterise gap and flush. The model is based on the geometry of the gap and flush line, which is extracted from an IGES standard file in our case.

For the computation of eigenmodes with the finite element method, the model is meshed and the element characteristics have to be defined. The model is composed of 76 elements, an element is a 2 nodes beam with 6 degree of freedom at each node, and the element size is 10 mm. A model can be considered as a "line model" when the ratio high/length equals 1/10. Material characteristics are common values, $\rho = 8000$ kg/m^3, $E = 2.1 \cdot 10^{11}$ Pa and $\nu = 0.3$. The use of a section to define the structural characteristics allows to easily change the model characteristics to obtain the searched modes.

5.2.1 Symmetrical Sections

The computation of the eigenmodes with a symmetrical section defines a basis which gap and flush modes are uniformly distributed. The orders of the modes in the obtained basis with the square or the circular sections, defined above, are identical up to the 40th mode, and then there are some inversions of the modes order. This can be explained by the difference between the inertia moments of the sections.

Remark: The following presented modes are computed with the use of a symmetrical section. One can notice that the modes characterising the gap and flush are uniformly distributed. The modes numbers printed on top of the figures shows the alternation of gap and flush modes.

5.2.1.1 Rigid modes of positioning

The computations of the eigenmodes also gives rigid modes which can be linked to positioning modes. Rigid modes are the 6 first modes (for a non-fixed model) made of pure translation and rotation of the model on each axis. Although those modes are not useful for the description of the measured deviation, they are used as positioning of the measured object and so reduce the positioning error while measuring the part.

In the following, for the application of the method, the modal basis is obtained with the circular symmetrical section giving the previous modal shapes presented.

5.2.2 Asymmetrical Sections

Using a non-symmetrical section involves differences between the bending moments on Y and Z axes. It is then possible to principally construct a basis of flexion modes on a chosen axis, and so to have a basis that describes one characteristic of the gap and flush.

5.2.2.1 Gaps

Choosing a section with a highest size in the (X, Z) plane of the global reference leads to a basis principally composed of flexion modes whose nodes displacements are principally along the Y axis. Thus the basis is gap-oriented.

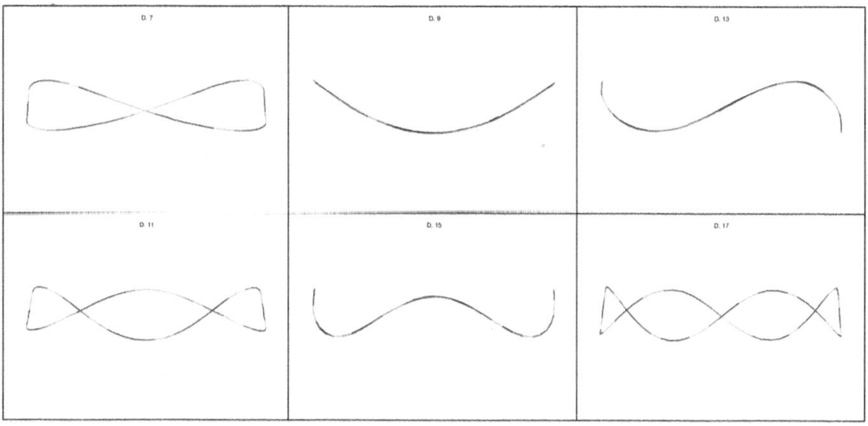

Fig. 4. Gap characterising modes in the (X, Y) plane.

5.2.2.2 *Flushes*

Choosing a section with a highest size in the Y plane leads to a basis principally composed of flexion modes whose nodes displacements are principally along the (X, Z) plane. Thus the basis is flush-oriented.

In the following, for the application of the method, the modal basis is obtained with the circular symmetrical section giving the previous modal shapes presented.

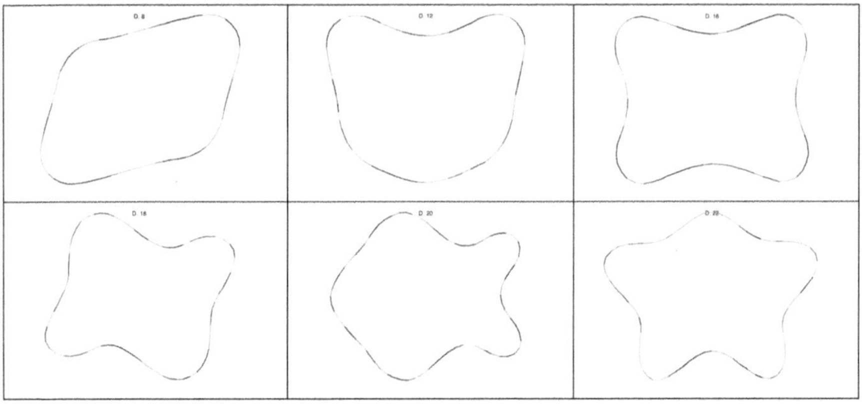

Fig. 5. Gap characterising modes in the (X, Z) plane.

5.3 Shell Model

It is possible to use another type of model and chose a shell model instead of a line model. The complexity is increased. The shell model is constructed from the line model. The gap and flush line describes a closed shape on which a surface is constructed, this surface is the geometry of the new model meshed with shell elements. The interesting nodes are those placed on the shape of the geometry, thus a filtering is necessary to only consider the nodes corresponding to the measured characteristic. Modal shapes are computed with the complete model but only the filtered nodes are kept for the modal analysis of the measured deviation.

Remark: In the case of a closed surface, it is possible to use a volume model, which is more complex. It is possible to change the dimension of the model, line.surface or surface.volume, but this increases the element number and then the computation time to obtain the eigenmodes.

It is possible to influence the eigenmodes distribution in the modal basis by increasing the thickness of the shell elements, then the membrane modes, corresponding to the flush modes, are placed earlier in the basis compared to the flexion modes, corresponding to the gap modes.

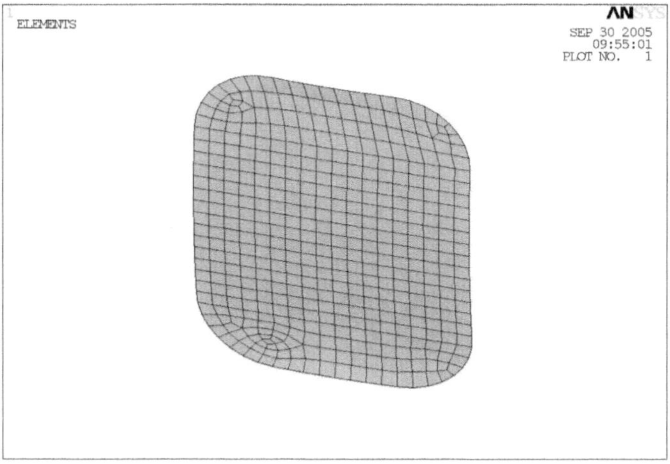

Fig. 6. The shell model of the gap and flush.

6 Part Analysis

Here is presented an application of the modal analysis method on two upper parts of the balance. The parts are being measured then, we simulated defects on balance parts in order to test the method. In the following, measurements will be simulated ones.

6.1 Deformed Shape Computation

Two geometries are calculated by defining displacement on nodes, the first deformed geometry has deviations with shorter wavelength than the second geometry. Figures 7 and 8 present the deformed shapes to analyse. From left to right: the oblique view of the deformed shape, view in the (X, Z) plane and finally in the (X, Y) plan.

6.2 Shape Analysis

The error criterion C is calculated with each node translations (3 components along the X, Y and Z axes). The deformed shape to analyse is a list of node displacements along each axis. The error criterion C is the quadratic error between the deformed and the recomposed shapes. Let us denote d_{Xmi} the displacement of the node i along the X axis of the measured geometry to analyse, stimulated here, and d_{Yri} the displacement of the node i along the Y axis of the reconstructed geometry.

$$C = \frac{\sum_{i=1}^{n}((d_{Xmi} - d_{Xri})^2 + (d_{Ymi} - d_{Yri})^2 + (d_{Zmi} - d_{Zri})^2)}{\sum_{i=1}^{n}(d_{Xmi}^2 + d_{Ymi}^2 + d_{Zmi}^2)}. \tag{8}$$

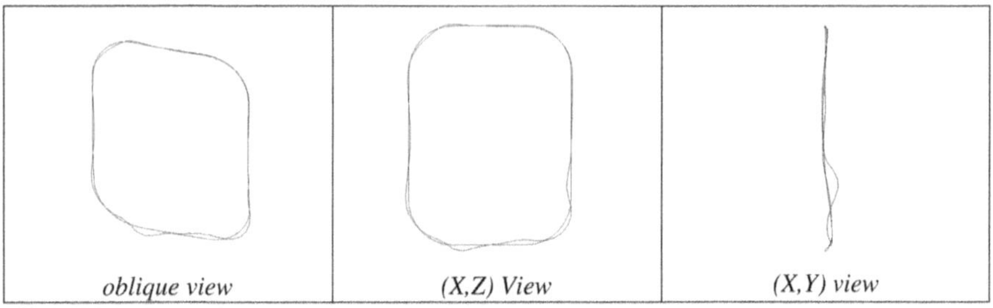

Fig. 7. First defect (short wavelength).

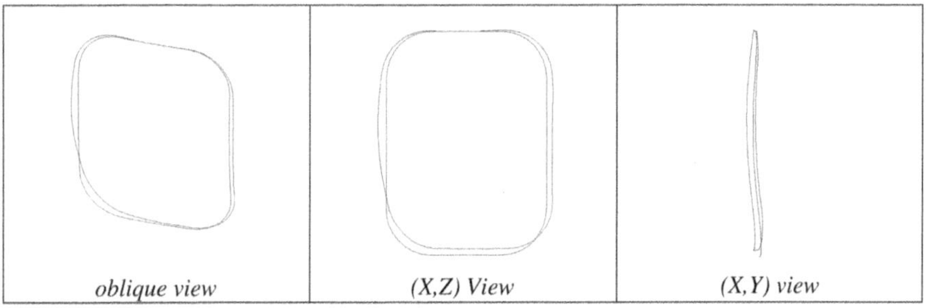

Fig. 8. Second defect (short wavelength).

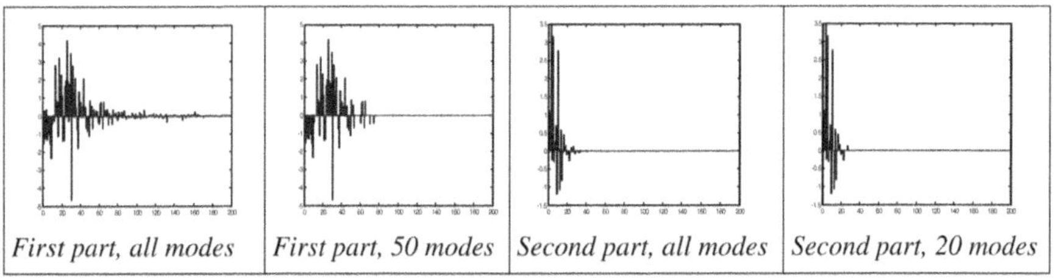

Fig. 9. Modal spectra of the two analysed geometries.

Figure 9 presents the whole and filtered modal spectra of the two analysed geometries. From left to right: the complete modal spectrum of the first deformed shape, the 50 biggest coefficients of the modal spectrum, the whole spectrum of the second deformed shape, and finally the 20 first biggest coefficient of the modal spectrum.

One notices that the use of the complete spectrum, and so the whole basis composed of 200 eigenmodes gives good results compared to the error criteria value. Both geometries have deviations with different wavelengths. The first one with shorter wavelengths deviation needs at least 50 eigenmodes to have an error value lower than 2%, this is justified by the large number high frequency eigenmodes which are not in the first 80 eigenmodes of the basis. The second geometry, which

Table 2. Error criteria for both shapes depending on the number of modal coefficients.

Number of considered modes	First deformed shape	Second deformed shape
The complete spectrum, 200	0.02%	1.9×10^{-7}%
The 50 biggest	1.7%	1.3×10^{-6}%
The 20 biggest	15.8%	0.06%

deviation has larger wavelengths, is better described with 20 modes than the previous geometry with 50. Moreover those modes are in the 30 first eigenmodes of the basis. The application shows the efficiency of the proposed method in term of geometry description. The analysed geometry is described with 76 nodes, depending on the wavelength of the deviation the quadratic error between the analyse geometry and its reconstructed geometry can be lower than 0.1% with 20 coefficients. On the other hand, more than 50 coefficients may be needed to have a quadratic error lower than 2%.

7 Conclusion

The form errors have to be modelled in a non-ambiguous language in order to perform the form quality challenge. The natural modal shape model gives interesting solutions to this issue. A feature (curve or surface) is discretised and natural mode shapes can give all the possible form errors of the corresponding real object. A vectorial projection enables to extract the modal coefficients of the measured feature with an associated residue for a given number of modes (truncation of the modal basis).

Our method is applied to an industrial example for a gap and flush analysis. The method gives a good accuracy (error less than 0.1% with 20 modes). When the numeric model is built with gap and flush orientations, the modes are associated to those behaviours and the projection of a measured feature gives the set of gap or flush form coefficients.

References

1. Carr K., Ferreira P., Verification of form tolerances. Part I: Basic issues, flatness and straightness, *Precision Engineering* **17**(2), 1995, 131–143.
2. Pandit S.M., Shaw M.C., Characteristic shapes and wave length decomposition of surfaces in machining, *Annals of CIRP*, 1981, pp. 487–492.
3. Capello E., Semeraro, Harmonic fitting approach for plane geometry measurements, *The International Journal of Advanced Manufacturing Technology* **16**, 2000, 250–258.
4. Capello E., Semeraro, The harmonic fitting method for the assessment of the substitute geometry estimate error. Part I: 2D and 3D theory, *The International Journal of Machine Tools and Manufacture* **41**, 2001, 1071–1102.

5. Capello E., Semeraro, The harmonic fitting method for the assessment of the substitute geometry estimate error. Part II: Statistical approach, machining process analysis and inspection plan optimisation, *The International Journal of Machine Tools and Manufacture* **41**, 2001, 1103–1129.
6. Summerhays K.D., Henke R.P., Baldwin J.M., Cassou R.M., Brown C.W., Optimizing discrete point sample patterns and measurement data analysis on internal cylindrical surfaces with systematic form deviations, *Journal of the International Societies for Precision Engineering* **26**(1), 2002, 105–121.
7. Huang W., Ceglarek D., Mode-based decomposition of part form error by discrete-cosine-transform with implementation to assembly and stamping system with compliant parts, *Annals of CIRP* **51**(1), 2002, 21–26.
8. Gouskov A., Tichkiewitch S. Influence of the machine-tool defects on the formation of the residual defects on a surface: Application to turning, in *Integrated Design and Manufacturing in Mechanical Engineering'98*, Kluwer Academic Publishers, 1999, pp. 341–348.
9. Adragna P.-A., Samper S., Pillet M., Favreliere H. Analysis of shape deviations of measured geometries with a modal basis, in *XVII Workshop on Supervising and Diagnostics of Machining Systems, Manufacturing Accuracy Increasing Problems*, 2006.

Evaluation of Machining Dispersions for Turning Process

Arnaud Lefebvre[1] and Valéry Wolff[2]

[1] *PRISMa, Claude Bernard University – Lyon 1, 17 rue de France, 69627 Villeurbanne, France; E-mail: arnaud.lefebvre@iutb.univ-lyon1.fr*
[2] *IUT B – GMP, 17 rue de France, 69627 Villeurbanne Cedex, France; E-mail:valery.wolff@iutb.univ-lyon1.fr*

Abstract. In order to control the design of product/process, the manufacturing companies are directed more and more towards the installation of tools of simulation and in particular simulation of machining. The simulation of machining aims to simulate the operations of machining by taking into account the dispersions of machining in order to determine the intervals of tolerances and to minimise the manufacturing costs.

In this article we propose a model of simulation of dispersions in turning based on the geometrical specifications. Our study is articulated around two trends of development: the first trend relates to the geometrical model. The geometrical model suggested must allow a follow-up of the geometry of the part during the simulation of machining. It is thus a question of carrying out a systematic treatment of the whole dimensioning and tolerancing process while being based on the principles of the ΔL method. One of the limits of the traditional ΔL method resides in the one axis exclusive study of a dimension of a part. For the turned workpieces for example the specification of coaxiality often specified by the tolerancing activity is not taken into account by this method. We also planned to integrate this type of specification in the model of simulation of machining suggested. It is more generally a question of extending the traditional model for better taking into account the multi axis specification of coaxiality and perpendicularity on the turned workpieces.

The second trend of our study relates to the widening of the field of application of the model. We propose to extend the field of application of the model by taking into account the modifications of several parameters of the manufacturing process plans, likely to involve variations of dispersions. The use of the design of experiments method makes it possible to quantify the influence of these modifications on dispersions. Integration with the initial model of parameters relating to cutting conditions or the nature of the material – parameters invariants in the initial model – allows on the one hand, to extend the field of application of the model and allows on the other hand to enrich considerably the model by simulation of dispersions of machining. This experimental study based on the model of design of experiments made it possible to confirm the assumptions of invariants retained for the construction of the initial model. Moreover, the influence of several cutting parameters on particular dispersions of machining could be revealed and quantified.

S. Tichkiewitch et al. (eds.), Advances in Integrated Design and Manufacturing in Mechanical Engineering II, 431–440.

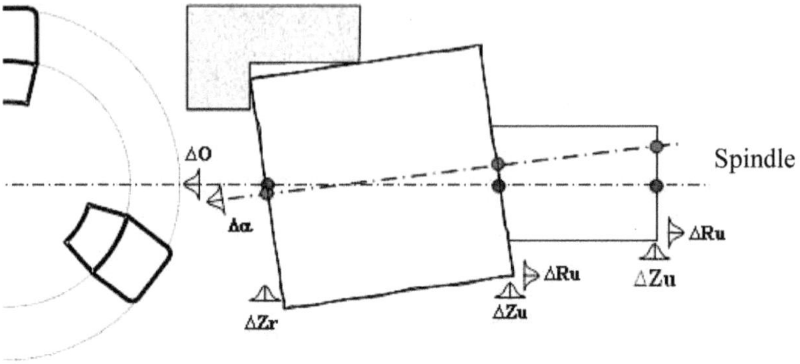

Fig. 1. Modelling of the five dispersions in turning.

Key words: Dispersions of machining, geometrical specifications, machining simulation, design of experiments.

1 Introduction

The stage of simulation of machining allows to simulate the behaviour of the operations of the process plan considering the dispersions to fix the intervals of tolerances and to produce at lower cost.

The first part of this article is devoted to modelling the dispersions in turning. We propose to extend the traditional model of ΔL by taking into account defects of coaxiality and perpendicularity on a turned part.

Experimental models resulting from design of experiments make it possible to quantify the influence of several parameters on dispersions. The detailed analysis of a model of dispersions of machining on Z axis in the second part of this article allows to quantify the variations of dispersions according to the evolution of the cutting speed and the insert type.

2 Modelling of Dispersions

We call dispersions of machining the geometrical and dimensional variations obtained on a set of real parts for a manufacturing process plan and a given machine-tool. The supposed sources of dispersions have several origins in particular related to controls with the inflexion of the tools, the cutting efforts and the geometrical defects of the machine tool. The rules which characterise each origin of dispersions can be of various nature (normal, bimodal, etc.); nevertheless we make the assumption that the resultant follows a normal distribution.

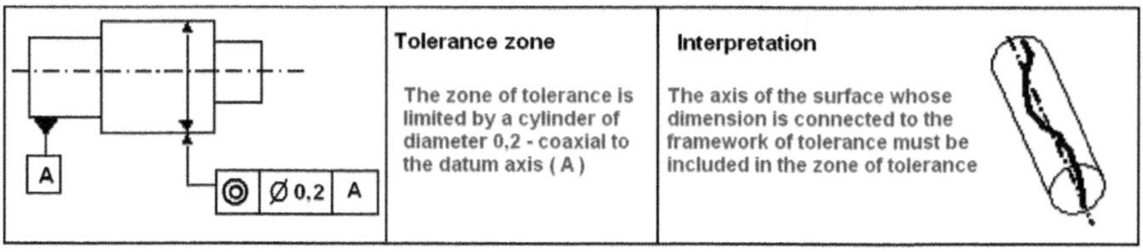

Fig. 2. GPS standards-coaxiality according to ISO 1101.

2.1 Extended Model Suggested

Considering a shouldered part the modelling of the behaviour of a machine-tool [3] is classically approached according to a study of five parameters of Δ *machine-tool* dispersions as shown in Figure 1. These dispersions are classified in two categories: the first category relates to dispersions of setting in position such as ΔO, $\Delta\alpha$, ΔZ_r. The second category includes dispersions of machining like ΔR_u (dispersion of machining according to X axis) and ΔZ_u (dispersion of machining according to the Z axis) [1].

We use the traditional method of ΔL on X and Z axes to take into account the axial dimensions and the dispersions obtained during machining. The parameters are then indicated by ΔZ_u, ΔR_u, ΔZ_r. We must use new parameters to take into account the geometrical specifications of tolerancing (coaxiality, perpendicularity, etc.): $\Delta\alpha$ and ΔO [7].

- $\Delta\alpha$ represents the angular defect of remachining of the part in the jaws in turning,
- ΔO is the defect of concentricity (between the axis of the reference surface and the spindle axis) located at the bottom of the soft jaws,
- ΔZ_r corresponds to the axial remachining error of the part in the part holder along the Z axis.

The objective of the modelling of this study of machining is to determine relationships characterising the behaviour of the machine-tool in terms of dispersions. These relations are of the type: $\Delta_{\text{machine-tool}} = f_i(p_i)$ where $\Delta_{\text{machine-tool}}$ is one of the five dispersions and p_i a set of discrete or continuous product/process parameters.

2.2 ISO Tolerancing

ISO Standards of tolerancing gathered under the term of GPS (Geometrical Products Specifications) provide a complete language to the mechanical engineers [2]. They are adopted today by the manufacturing industry. The geometrical model of dispersions in simulation of machining which we propose [9] takes into account the ISO specifications.

For example in the case of the coaxiality we defined the methods of calculation [7] necessary in order to connect the parameters of our model to ISO specifications.

Table 1. Factors and associated values.

Parameters		Type	Values	
Insert type	N	discrete	P15 (finish)	P35 (rough)
Nose radius (mm)	Rε	discrete	0.40	0.80
Cutting speed (m/min)	Vc	continuous	150	280
Material of the machined part	M	discrete	A60	XC38
Feed rate (mm/turn)	F	continuous	0.10	0.30

Table 2. Interactions retained between parameters.

Insert type – cutting speed	$N \cdot V_C$
Material – feed rate	$M \cdot f$
Cutting speed – material	$V_C \cdot M$
Insert type – feed rate	$N \cdot f$

The coaxiality relates to the relative position of the real axis of the toleranced surface and the reference axis. It never relates to surfaces but always to axes. The definition resulting from the standard and its interpretation is given in Figure 2.

This definition must be interpreted and calculated most precisely possible to make correspond the measurement obtained on three-dimensional coordinate measuring machine (CMM) with the specifications of the drawing.

2.3 Experimental Protocol

2.3.1 Design of Experiments (DOE) Parameters

We defined five product/process parameters allowing the evaluation of various dispersions. Table 1 gives the list of the parameters and their associated levels.

Particular interactions are taken into account to supplement the study (Table 2).

2.3.2 DOE Choice

We use the Taguchi's method [4–6] to limit the number of tests to be realised. The table L_{16} (2^{15}) was selected according to the criterion of orthogonality and number of degrees of freedom of the model. For each response of studied dispersion, the order of the experiments as well as the combinations of the parameters are given in Table 3.

Table 3. Table of the experiments (extract).

$N°$	Insert type	Noise radius $R\varepsilon$	Cutting speed V_c	Material M	Feed rate f
1	P15	0.4	150	A60	0.1
2	P15	0.4	150	XC38	0.3
3	P15	0.4	280	A60	0.3
...	...	...	...	...	...
16	P35	0.8	280	XC38	0.1

2.4 Experimental Results

2.4.1 Evaluation of the Answers

The responses of the experiments are obtained by measuring. The measures allow to calculate the dispersion parameters: ΔO, $\Delta\alpha$, ΔR_u, ΔZ_τ and ΔZ_u (millimetres). A reduced sample of five workpieces is enough. The used Taguchi's method is a standard L_{16} (2^{15}) experiments five times repeated.

We observe two cases. The response is calculable starting from the standard deviation obtained by the measurements of one or several dimensions:

- Case of the response obtained by the measurement of only one dimension. That relates to dispersions ΔO, $\Delta\alpha$, ΔR_u. For example, the relation relating to ΔR_u is written:

$$\Delta R_u = \frac{\left[\frac{6\times(\sigma_{\text{diametral}})_{\text{sample}}}{C_4}\right]}{2}, \tag{1}$$

  where C_4 is the weighting taken in the statistical table of the reduced samples.
- Case of the response obtained by the measurement of several dimensions d_i (standard deviation noted σ_i). The variance of required dispersion is related to the sum of the variances of dimensions concerned dimensions. That relates to ΔZ_τ dispersions (or ΔZ_u). The relations used are as follows:

$$\sigma_r = \frac{\sqrt{\sum \pm\sigma_i^2}}{\sqrt{2}} \quad \text{and thus} \quad \Delta Z_r = \frac{(\sigma_r)_{\text{sample}}}{C_4} \times 6. \tag{2}$$

The necessary values to calculate the parameters of dispersions (part diameters, points of intersection) are measured directly on the 80 parts using a coordinate measuring machine. Some of these values are only intermediates parameters.

Each batch of five parts allows to calculate the standard deviation of each answer ΔZ_τ, ΔO, $\Delta\alpha$, ΔR_u and ΔZ_u using CMM measurements. The design of experiments provides in this way 16 values for each studied response.

Table 4. Statistically significant parameters.

Dispersions responses	Statistically significant parameters
ΔR_u	Material – Cutting speed (M Vc) Nose radius (Rε) Material – Feed rate (M f)
$\Delta Z_{u\_dressage}$	Cutting speed (Vc), Insert type (N)
$\Delta Z_{u\_en\_remontant}$	Material – Feed rate (M f)

2.4.2 Analysis and Summary of the Results

We carried out the analysis of the measurements obtained on the 80 parts of the L_{16} (2^{15}) plan definite previously to determine the parameters of manufacture process plan influencing machining dispersions.

The analysis of the variance indicates that a parameter is statistically significant on the response as soon as p parameter is higher than 0.05 (level of confidence higher than 95%).

The R-squared (R^2) makes it possible to evaluate the percentage of data explained by the model. The higher the R^2 is, the more the model is usable in a predictive mode. A R^2 coefficient between $\sim$ 70 and $\sim$ 90% corresponds to an acceptable model.

3 Analysis of Dispersions Models

Table 4 presents the summary of the results of obtained dispersions. Only the answers for which parameters appear statistically significant are presented in the table. It is thus the case of dispersions ΔR_u, ΔZ_{u1} and ΔZ_{u2}.

Note: One of the characteristics of the model suggested is due to the distinction made on the level of the dispersion in ΔZ_{u1} and ΔZ_{u2} [7]. Two distinct answers will then be evaluated.

3.1 Responses Modelling

For each response of Table 4, the machining dispersion can be represented in the form of a regression function.

The obtained model by linear regression concerning ΔR_u dispersions is:

$$\begin{aligned}\Delta R_u &= 0.00876824 + 0.0182206 \times N + 0.0318719 \times R\varepsilon + 0.000080 \times V_c \\ &\quad + 0.0118769 \times M + 0.019411 \times f - 0.0000424 \times N \times V_c \\ &\quad - 0.0287376 \times N \times f - 0.0001095 \times V_c \times M \\ &\quad + 0.0615979 \times M \times f\,. \end{aligned} \tag{3}$$

The obtained model for ΔZ_{u1} dispersions is:

$$\Delta Z_{u1} = 0.0214865 - 0.00477308 \times N - 0.005125 \times R\varepsilon - 0.00004307 \times V_c + 0.0015125 \times M + 0.007 \times f + 0.000013653 \times N \times V_c. \quad (4)$$

And the model obtained for ΔZ_{u2} is:

$$\Delta Z_{u2} = -0.00899519 - 0.0151385 \times N + 0.0171875 \times R\varepsilon - 0.0000196 \times V_c + 0.0122423 \times M + 0.108 \times f + 0.00004942 \times N \times V_c + 0.069375 \times N \times f + 0.00001538 \times V_c \times M - 0.128 \times M \times f. \quad (5)$$

These mathematical models have to be used in a predictive mode to evaluate machining dispersions for the various combinations of product/process parameters defined in Table 1 [8].

3.2 Representation of ΔZ_{u1} Model

3.2.1 Graphical Chart

Contrary to the mathematical model resulting from the design of experiments (see Section 2.1), the representation of the ΔZ_{u1} dispersions model is limited. Indeed, we propose a representation which deals with statistically significant parameters (see Table 4).

In the case of ΔZ_{u1}, the two statistically significant parameters are the cutting speed (V_c) and the insert type (N). Then $\Delta Z_{u1} = f(V_c, N)$.

The surface represented in Figure 3 shows the various ranges of variation of dispersions according to the two parameters.

3.2.2 Analysis of Dispersions Variations

The graph in Figure 4 represents the evolution of ΔZ_{u1} dispersion according to the cutting speed (V_c) for each of the two modal values of the insert type (P15, P35).

(a) We can see in Figure 4a considerable influence of the cutting speed (V_c) on the values of dispersions obtained by the predictive model. Indeed the values obtained for the P15 insert type vary from 16 to 8 μm in the range of variation of the cutting speed. We also observe the same trend for the P35 insert type.
The values of dispersions are lower for the P35 insert type than for the P15 insert type. This is explained by the choice of the ranges of the cutting speed. The chosen cutting speeds for the experimental protocol must follow three use rules:

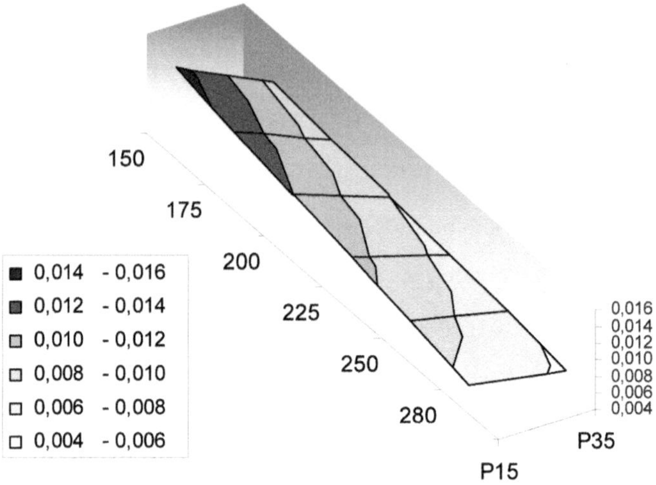

Fig. 3. $\Delta Z_{u1} = f(V_c, N)$.

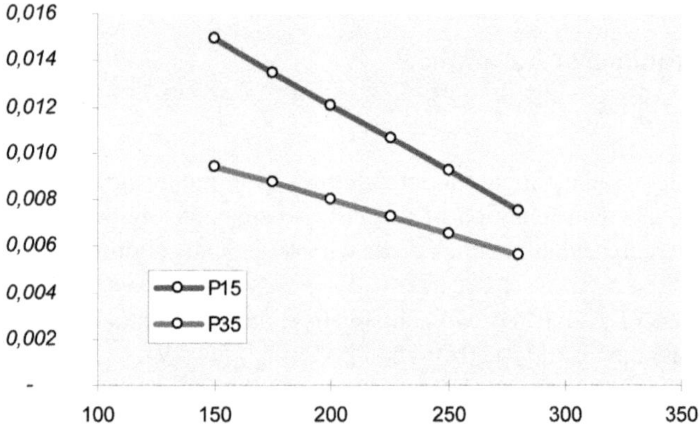

Fig. 4. Cutting speed influence on dispersions.

- The cutting speed range (V_c) must be compatible with the manufacturers insert types database.
- The limits of the machine-tool (maximum spindle speed) must be taken into account.
- The construction of a design of experiments requires to be able to use the same ranges of variation speed (V_c) whichever the other parameters values are (e.g. material, nose radius, feedrate).

In the case of the P15 insert type which is a finish insert type the used range is located partly low (minimum speed) of the possible range given by the manufacturer. Preconnized optimal speed is higher than 400 m/min. For the P35 insert type the totality of the range of the preconnized cutting speed was used. The maximum preconnized speed is close to 250 m/min.

(b) We observe a convergence of these dispersions for the two modal values P15 and P35 when it reaches the maximum values of the chosen range for V_c.

For the P15 insert type this is explained by the fact that at the convergence point the used speed (280 m/min) approaches the optimal speed.

In the next future, we will simulate the process planning of a turned part using the predictive models according to the different cutting parameters. We will then validate this predictive models by using a set of machined parts (real parts).

4 Conclusion

The simulation of machining and the characterisation of the machine-tools in terms of dispersions of machining are a necessary stage in the control of the costs. In the first part of this article we presented an extended model of dispersions in turning based on multiaxes defects of coaxiality or perpendicularity. It allows thus to take into account a more complete definition of the geometrical model – as well on the dimensional level and on the geometrical level. Moreover, for the five types of selected dispersions, angular dispersions, remachining dispersions, an experimental model was elaborated and gave us the possibility to quantify the influence of several parameters like the type of material or the cutting speed.

In the second part of the article, we clarify an example of dispersions model. We showed in experiments the influence of the cutting speed as well as the insert type. The model which is graphically represented allows to quantify the values of dispersions according to the statistically significant parameters. These results allow on the one hand to enrich considerably the initial model of dispersions in which the process plan parameters were fixed. In addition to the aspect of formalisation of a know-how, they allow to better characterise the machine-tool and thus to improve the quality of the machined parts.

References

1. Anselmetti B., Villeneuve F., Computer aided process planning: Two ways for including a user's know-how, in *Proceedings of the CIRP Seminars, Manufacturing Systems: 23rd Int. CIRP Conference on Production Systems*, Nancy, June 1991, *CIRP Annals* **21**(3), 1992, 167–172.
2. Anselmetti B., *Cotation de fabrication et métrologie*, Volume 3, Editions Hermès, 2003.
3. Dessein G., Redonnet J-M., Lagarrigue P., Rubio W., Correction des trajectoires d'une machine outil à commande numérique par une qualification des dispersions selon l'usinage, in *IDMME'98, 2ème Conférence Internationale sur la Conception et la Fabrication*, Compiègne, 27–29 May 1998, p. 759.

4. Goupy J., *La méthode des plans d'expériences Optimisation du choix des essais et de l'interprétation des résultats*, Editions Dunod/Bordas, 1988.
5. Pillet M., Regnier G., Les plans d'expériences: Outil de la démarche expérimentale, Institut pour la Promotion des Sciences de l'Ingénieur, Paris, 1995.
6. Pillet M., Appliquer la maîtrise statistique des procédés MSP/SPC, Les éditions d'organisation, 1996.
7. Wolff V., Le suivi de la cotation des pièces fabriquées pour la conception coopérante en mécanique, Thèse de doctorat, INSA de Lyon, 2000.
8. Lefebvre A., Renaud J., Wolff V., Représentation des dispersions d'usinage à partir de cartographie de connaissances métier, Chapter 18, Hermès Science Publication, ISBN 2-7462-0952-7, 2004.
9. Wolff V., Lefebvre A., Renaud J., Cartographie d'un modèle de dispersions en tournage, 9ème Colloque AIP Primeca, La Plagne, 2005.

METAL CUTTING AND HIGH SPEED MACHINING

Fundamental Understanding of the Segmented Chip Genesis for Smart Machining. A Contribution in Hard Material Turning

T. Mabrouki[1], S. Belhadi[2] and J.-F. Rigal[1]

[1]*LaMCoS, INSA-Lyon, CNRS UMR5259, Bât. J. Jacquard, Domaine scientifique de la Doua, 69621 Villeurbanne Cedex, France; E-mail: tarek-mabrouki@insa-lyon.fr*
[2]*Laboratoire (LMS), Université du 8 mai 1945, BP 401, Guelma 24000, Algérie*

Abstract. The main idea in this paper deals with the importance of smart machining improvement by introducing physical based understanding. The case of chip formation, in hard turning, is treated via the existence of a direct relationship between serrated-chip morphology simultaneously with force component signals (resulting from a high frequency acquisition) and with the width of facets detected on a workpiece machined surface. The forces component variations can reach several tens of Newton. The shear segmentation of the chip is a high frequency phenomenon. Because it can excites machine-tool, induces insert wear and affects surface integrity, this phenomenon has to be understood and controlled by the smart machining devices. The paper is a contribution by considering an experimental and a numerical approach.

Key words: segmented chip, smart machining, cutting forces, cutting numerical model.

1 Introduction

Nowadays, the realization of autonomous smart machining systems is a goal fixed to industrial institutions [1]. The economical challenges facing the manufacturing sector that produces metal parts and their fabrications correspond to an urgent need to re-examine the basis of manufacturing environment. The aim is to enable dramatic improvements in the productivity and cost of designing, planning, producing, and delivering high-quality products within short cycle times. Five essential thrust areas, which are in relationship with the smart machining concept, can be identified:

1. Process definition and design,
2. Smart equipment operation and process control,
3. Fundamental process and equipment understanding,
4. Health security monitoring and assurance,
5. Integration framework.

S. Tichkiewitch et al. (eds.), Advances in Integrated Design and Manufacturing in Mechanical Engineering II, 443–459.

The objective of researches in this smart machining domain is to develop and validate the metrology infrastructural tools that enable industry to characterize, monitor, and improve the accuracy, reliability and productivity of machining operations. This requires a science-based understanding and unambiguous representation of material removal processes and machining system performance. The present paper focuses on third area defined above. It concerns a fundamental understanding of the multi-physical phenomena occurring during machining at mesoscopic scale (tool-tip/chip scale). Mechanical dynamic behaviour of the manufacturing process robustly defines the production's limits of high speed cutting (HSC) and hard material turning (HT) [2]. Smart equipments (spindle, transducers, etc.) and performing control devices have to be developed closely considering the dynamic excitation generated by the cutting process. In bibliography, many contributions are available for the high speed milling process considering the intermittent cutting in the cases of a flexible workpiece [3], the dynamic and thermo-mechanical behaviour of the spindle [4], and the machine tool structure [5]. This contribution is centred on the dynamic of continuous cutting in hard turning. Traditional mesoscopic approach considers that components of turning cutting forces are constant versus time. But, by considering nonlinear and multi-physical approaches of the cutting process, it is shown that cutting forces oscillate due to a periodic shear band apparition in the chip section especially during the turning of hard material [6]. Because of its geometry, the chip obtained is commonly named a saw-tooth chip. So, in the aim to improve smart machining concept, it is important to develop some physical comprehensions related to this chip formation. Indeed, hard turning process has been applied in many industrial cases. Frequently, it is used in manufacturing bearings, shafts, gears, cams and other power transmission mechanical components. Although there are many significant papers and different methodologies adopted in studying saw-tooth chip formation, it is noted that there are few studies treating their effect on cutting force components and machined surfaces especially in hard turning. In this framework, this paper is focused on genesis comprehension of saw-tooth chip resulting from the turning of hardened AISI 4340 steel, treated at 47 HRC. Two methodologies are adopted which are experimental tests and numerical simulations. The experimental methodology is based on the acquisition of chip segmentation frequencies according to different cutting speeds and feed rates. The measurement of chip segmentation frequencies is realized by three methods:

- High frequency signal acquisition of cutting forces,
- Chip geometric measurements based on microscopic observations,
- Measurement of surface roughness in the cutting direction.

The second methodology concerns numerical simulations which are based on an adiabatic calculation achieved with Abaqus/Explicit software.*

2 Experimental Study of Chip Formation and Smart Machining Concept

During this study, two working parameters were considered; cutting speed ($V_C = 60$ to 120 m/min) and feed rate ($f = 0.2$ to 0.47 mm/rev). Cutting depth was kept equal to a constant value ($a_P = 1$ mm). In the following, it is proposed to study the frequency of the shearing plane formation accompanying saw-tooth. For that, firstly, we performed measurements, at high frequency sampling, of cutting force signals. Secondly, geometrical measurements on chip saw-tooth were made. Finally, the frequency related to facet appearing on the machined surface was calculated.

2.1 Experimental Device

The experiments concern straight turning of an AISI 4340 workpiece, which was hardened by heat treatment to 47 HRC. The machine tool and the tool-insert used are respectively a universal lathe Gallic 20 (Muller and Pesant Company*) and coated carbide insert (AC700G grade), from Sumitomo Electric company.* This insert is referenced (CNMA 12 04 04), and is fixed on a tool-holder (PCLNR 25 25 M 12) from Sandvik company.*

The measuring equipment is composed of a standard dynamometer (Kistler 9257B*), a charge amplifier and a high frequency data acquisition device (NI 4472) of National Instrument.* The signal acquisition and data treatment were carried out with the LabView software.*

2.2 Frequency Measurements of the Formation of Chip Shearing Planes

2.2.1 Sampling of the Cutting Force Signal

Figure 1a represents a temporal signal made after machining and acquisition, in real time, at high frequency of cutting forces. It shows that the magnitudes of these forces are not constant in time. They are characterized by a force bandwidth ΔF_i. Index i denotes (a: axial force, r: radial force and c: cutting or tangential force).

The frequential representation of force temporal signals is treated by using a Fast Fourier Transform (FFT) presented in Figure 1b. The latter shows the presence of an important peak at frequency, F_{HZSCF} (Spectrum of Cutting Forces) equals to 1750 Hz and with a higher power level. This is clearly the most powerful frequency identified.

2.2.2 Measurements of Saw-Tooth Shapes

The aim of this section is to propose a calculus procedure of the saw-tooth genesis frequency during machining. This calculus is based on measurements realized on the chip longitudinal section.

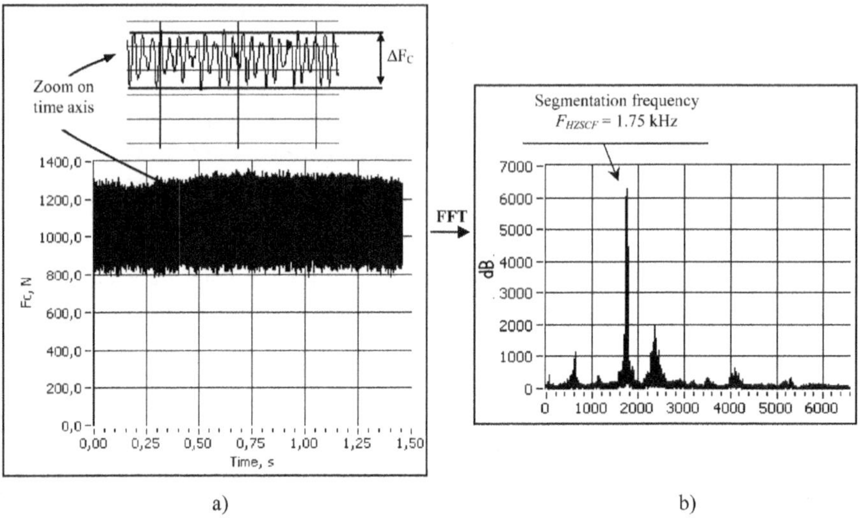

Fig. 1. (a) Evolution of the force, F_C, according to time and (b) FFT representation of the force F_C signal ($V_C = 80$ m/min, $f = 0.4$ mm/rev, $a_P = 1$ mm).

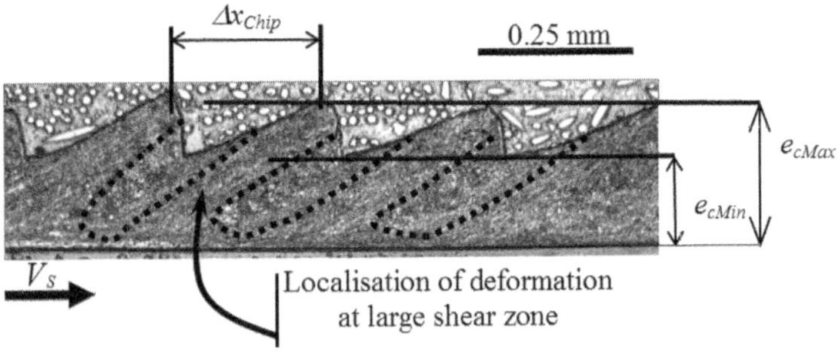

Fig. 2. Saw-tooth chip longitudinal section of machined AISI 4340 (47 HRC) for (V_C = 120 m/min, $f = 0.2$ mm/rev) (This section is polished and attacked with the nital (acid water solution).)

Measurements are made by using a microscope measuring device. By considering a mean chip-slip speed (chip evacuation speed), V_S (m/min), on tool rake face and the distance Δx_{Chip} measured between two consecutive shearing planes (in the direction of the tool rake face, mm), (Figure 2). Consequently, the frequency corresponding to shearing planes formation determined from chip geometry, F_{HZCG}, can be established:

$$F_{\text{HZCG}} = \frac{1000 V_s}{60 \Delta x_{\text{Chip}}}. \tag{1}$$

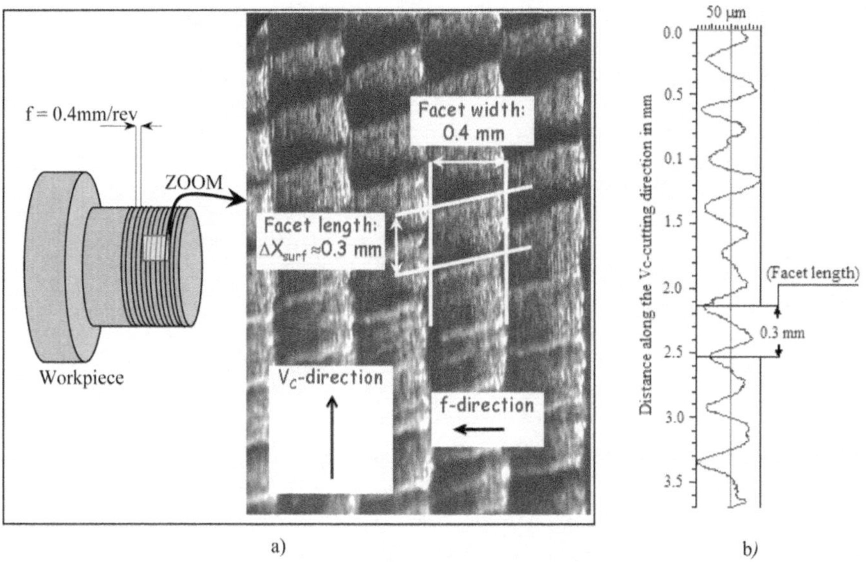

Fig. 3. (a) Texture of the machined surface and (b) profile plotting in the cutting direction (V_C = 120 m/min, f = 0.4 mm/rev and a_P = 1mm).

By assuming that the mass of metal deformed during machining is constant, it can be written that:

$$\rho_1 V_C f a_p = \rho_2 V_S\, e_C\, l_c \tag{2}$$

ρ_1 and ρ_2 are metal densities just before and just after cutting (i.e. in the chip) in kg/cm^3, V_C is cutting speed in m/min, f is feed rate in mm/rev, a_P is depth of cut in mm, e_C means chip thickness in mm, l_C is the width of the chip in mm.

By neglecting material compressibility, it is assumed that the ratio ρ_1/ρ_2 is equal to the unit. Consequently, the chip-slip speed V_S on tool rake face is shown by the following equation:

$$V_S = V_C \frac{f\, a_p}{e_C\, l_C}. \tag{3}$$

2.2.3 Measurements on Machined Surface

Figure 3a presents a microscopic photo concerning the workpiece machined surface (in the case of V_C = 120 m/min and f = 0.4 mm/rev). This photo shows the presence of lays resulting from machining cinematic which are parallel to cutting speed direction. Grooves are characterized by a width equals to the feed rate. It is important to underline the existence of facets along these lays.

In order to quantify this texture, a series of roughness measurements was carried out (Figure 3b) on machined surface cutting direction where we are interested in the

Table 1. Cutting force variation according to feed rate and cutting speed.

V_C, (m/min)	f, (mm/rev)	Fa, (N)	ΔFa, (N)	Fr, (N)	ΔFr, N	F_C, (N)	ΔF_C, (N)
60		609	150	397	200	1122	350
80	0.40	586	285	394	400	1102	550
100		552	350	380	430	1035	450
120		527	450	362	585	0991	600
	0.20	427	90	233	90	0624	130
80	0.30	494	190	297	190	0817	225
	0.40	586	285	394	400	1102	550
	0.47	602	325	445	425	1207	535
	0.20	427	135	231	120	611	160
100	0.30	501	320	302	280	840	310
	0.40	556	650	386	650	1033	650
	0.47	572	445	415	475	1165	450

measurement of facet length ΔX_{surf} shown on roughness profile (Figure 3a). This profile is aperiodic curve, whose wavelength corresponds to ΔX_{surf}. The frequency corresponding to the appearance of these facets is given by:

$$F_{\text{HZSrf}} = \frac{1000 V_C}{60 \Delta x_{\text{Srf}}}. \tag{4}$$

2.3 Result and Discussion

In Figure 2, which shows a segmented chip obtained at $V_C = 120$ m/min and $f = 0.2$ mm/rev, the presence of a localized deformation zone due to a high shear strain is underlined. These deformations cause chip thickness irregularity from ec_{min} to ec_{max}. Consequently, chip evacuation speed can not be supposed constant during time. Irregularities, in the form of saw-tooth chips, can be correlated with force measurements. Table 1 shows the variation, with cutting conditions (V_C, f), of axial force F_a, radial force F_r and tangential force F_C components and the variations of corresponding bandwidth forces ΔF_i. It is noted that mean values of cutting force components (F_a, F_r and F_C) increase considerably with the feed rate yielding to chip section increase (Figure 4). This same tendency of variation, according to considered cutting parameters, concerns the magnitude variations (ΔF_a, ΔF_r and F_C) of each force component. The force components decrease slightly when the cutting speed increases, and the magnitude variations increase weakly when compared to the feed rate effect (Figure 5).

In what follows, we intend to establish a relationship between the experimental detected frequency, F_{HZSCF}, the frequency F_{HZCG} calculated by performing geometrical measurements on collected chips after machining and the frequency, F_{HZSrf}, which is calculated based on the facet length ΔX_{surf} measurements (Table 2).

The obtained results presented in Figure 6a show that the feed rate variation, for a fixed cutting speed, does not have a great influence on the frequencies of the

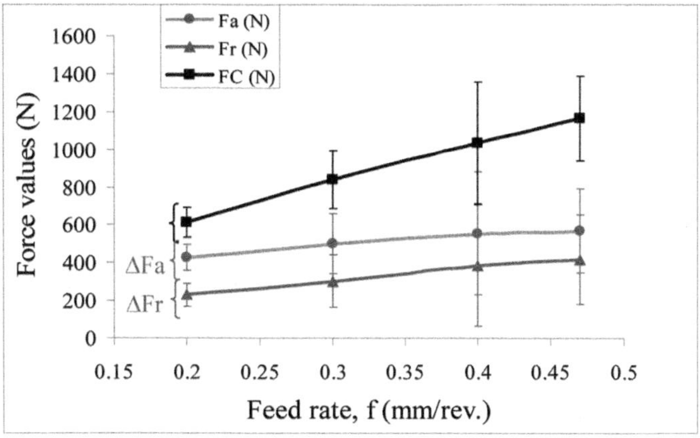

Fig. 4. Evolution of force magnitudes versus feed rate $V_C = 100$ m/min.

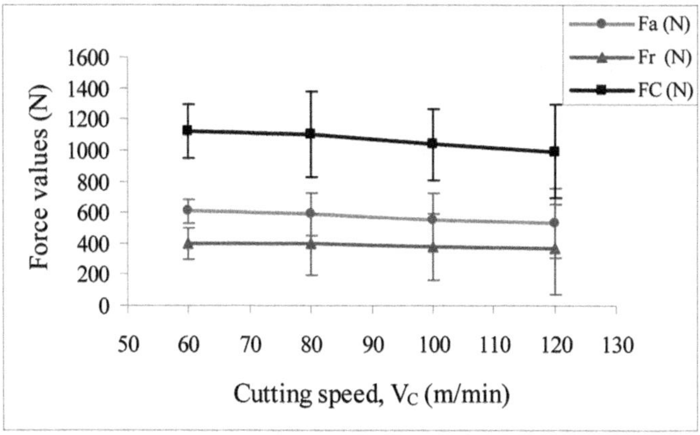

Fig. 5. Evolution of force magnitudes versus cutting feed for $f = 0.4$ mm/rev.

shearing bands. This influence is larger when the cutting speed increases Figure 6b. Nevertheless, in Figure 7 it is shown that cutting speed has a considerable effect on the shearing band frequency value. Indeed, frequency segmentation increases as cutting speed increases because of the direct effect of the cutting speed on the chip evacuation speed (Equation (2)).

In addition, Figures 6 and 7 make possible to compare the two frequency values F_{HZCG}, F_{HZSCF} and F_{HZSrf}. It is observed that they vary in the same way with the considered cutting parameters. Nevertheless, a difference between these frequencies can be observed. This is due to some phenomena accompanying the saw-tooth chip formation, such as elastic deformation and chip-tooth slipping, which are not taken into account in formulae (1–3). Indeed, chip genesis is accompanied by some inter-

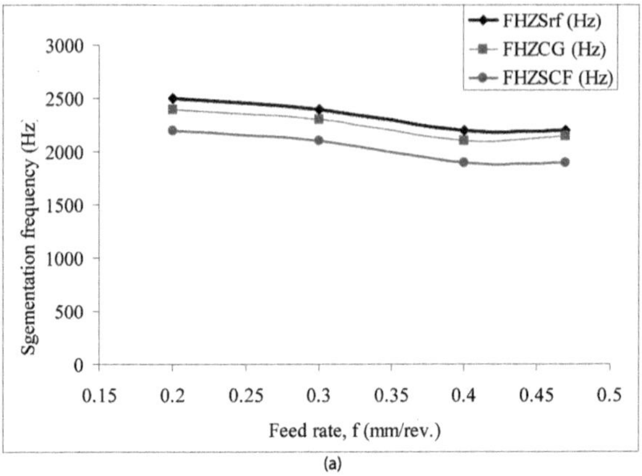

Fig. 6. Segmentation frequency versus feed rate for (a) $V_C = 80$ m/min and (b) $V_C =$ 100m/min.

related mechanisms inside the chip such as localized shear, adiabatic shear and also catastrophic shear zone in the form of extensive cracks [7]. The results of all these mechanisms can explain the differences between these different frequency results.

The third column of Table 2 presents the facet length ΔX_{surf} needed to make a comparison with the distance ΔX_{chip} (fourth column). For a good comprehension, these two distances are drawn at the form of curves evolving with cutting speed and feed rate as mentioned in Figures 8 and 9, respectively. We note that, the values of ΔX_{surf} are lower than ΔX_{chip}. This can be explained by the large displacement and deformation in chip. Consequently, is is expected that the F_{HZSrf} frequency will be higher than the F_{HZCG} frequency as is shown in Figures 4 and 5. Additionally, it is underlined on the one hand that as cutting speed increases as the length ΔX_{surf} and

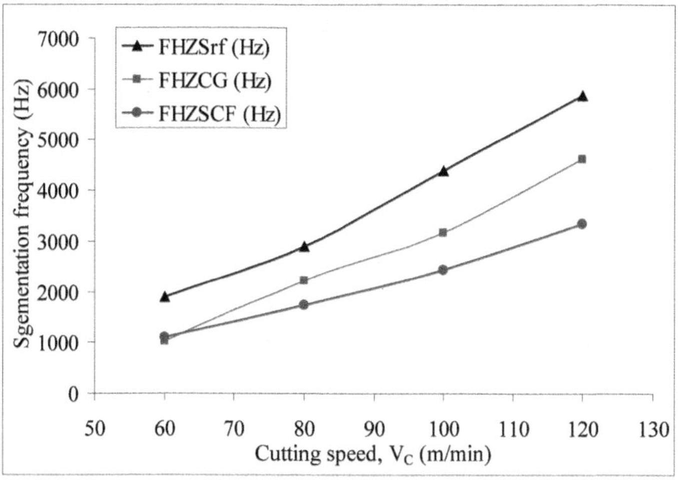

Fig. 7. Segmentation frequency versus cutting speed for $f = 0.4$ mm/rev.

Table 2. Summary of roughness and frequency results ($a_P = 1$ mm).

V_C, (m/min)	f, (mm/rev)	Fa, (N)	ΔFa, (N)	Fr, (N)	ΔFr, N	F_C, (N)	ΔF_C, (N)
60	0.40	609	150	397	200	1122	350
80		586	285	394	400	1102	550
100		552	350	380	430	1035	450
120		527	450	362	585	0991	600
80	0.20	427	90	233	90	0624	130
	0.30	494	190	297	190	0817	225
	0.40	586	285	394	400	1102	550
	0.47	602	325	445	425	1207	535
100	0.20	427	135	231	120	611	160
	0.30	501	320	302	280	840	310
	0.40	556	650	386	650	1033	650
	0.47	572	445	415	475	1165	450

the distance ΔX_{chip} decrease. On the other hand, as the feed rate increases as the mentioned distances increase.

3 Numerical Simulation of Chip Formation and Smart Machining Concept

3.1 Data and Methodology

In Section 2, the complexity of the chip genesis has been experimentally shown. The difference between experimental results and simple geometrical models underlines this complexity. More physical explanations must be brought. In this framework,

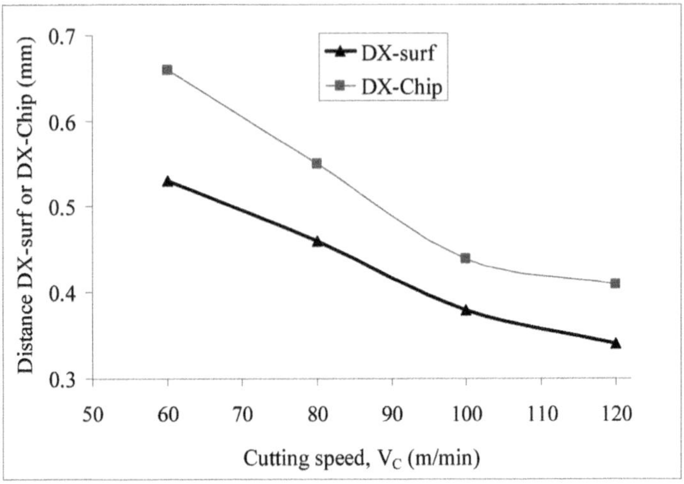

Fig. 8. Facet width of machined surface or distance between two consecutive shearing planes versus cutting speed for $f = 0.4$ mm/rev.

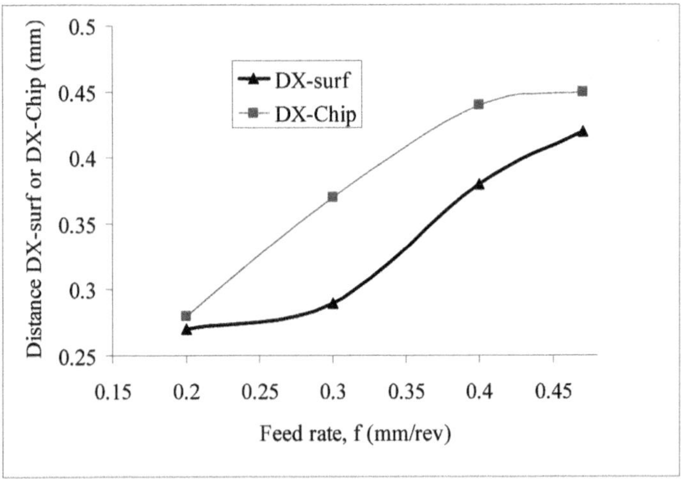

Fig. 9. Facet width of machined surface or distance between two consecutive shearing planes versus feed rate for $V_C = 100$ m/min.

we have developed a numerical model based on Abaqus/Explicit software. It treats the physical phenomena during tool/workpiece interaction. The aim is to improve physical comprehension about the observations made during previous experimental tests. The workpiece mechanical and thermal properties, contact conditions between the tool and machined workpiece, and boundary conditions were defined. An adia-

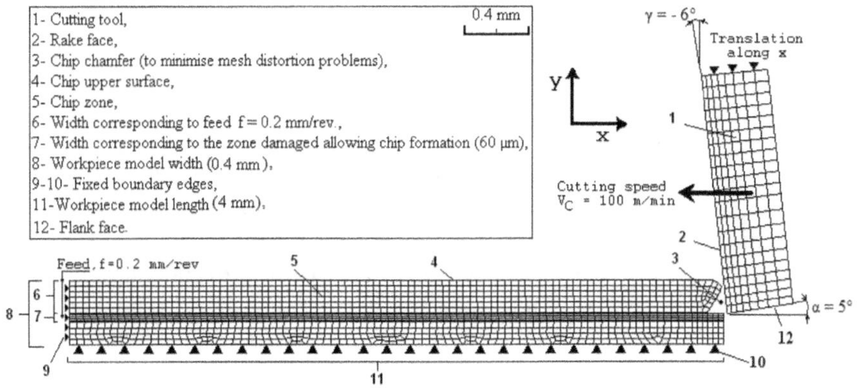

Fig. 10. Mesh used for cutting simulations [8].

batic calculation was performed. The cutting modelling used quadrilateral elements with a plane strain assumption in both the workpiece and the cutting tool. Because machining is an extremely dynamic event with considerable changes in geometry it seems important to improve the mesh aspect by adopting an adaptive remeshing technique. The model mesh used during simulations is shown by Figure 10, a representation of the studied orthogonal cutting model. The tool has the following geometrical angles: entering angle $\chi = 90°$, rake angle $\gamma = -6°$ and flank (or clearance) angle $\alpha = 5°$. Working parameters tested are feed rate f(0.2 mm/rev) and cutting speed V_C(100 m/min). In the case of orthogonal cutting conditions, feed rate corresponds to the material thickness which will be cut (Figure 10, feature 6). It is supposed that workpiece edges (Figure 10, features 9 and 10) are fixed. The cutting operation is realized by the tool movement with a cutting speed V_C. For mesh efficiency reasons, allowing the minimization of mesh distortion at the beginning of the cutting operation, the workpiece is chamfered as shown in Figure 6 (feature 3) [8].

For the material behaviour of the machined workpiece (AISI 4340), a Johnson–Cook model [9] was used for Abaqus/Explicit software. The Johnson-Cook model is a strain-rate and temperature dependent visco-plastic material model suitable for problems where strain rates vary over a large range (10^2 s^{-1}–10^6 s^{-1}), and temperature changes due to plastic deformation caused by thermal softening. This model represents the equivalent flow stress as the following equation:

$$\bar{\sigma} = \underbrace{(A + B\bar{\varepsilon}^n)}_{\text{Elasto-plastic term}} \underbrace{\left[1 + C \ln\left(\frac{\dot{\bar{\varepsilon}}}{\dot{\bar{\varepsilon}}_0}\right)\right]}_{\text{Viscosity term}} \underbrace{\left[1 - \left(\frac{T - T_{\text{room}}}{T_{\text{melt}} - T_{\text{room}}}\right)^m\right]}_{\text{Softening term}}, \tag{5}$$

where $\bar{\sigma}$ is equivalent stress, $\bar{\varepsilon}$ equivalent strain, $\dot{\bar{\varepsilon}}$ equivalent plastic strain rate, $\dot{\bar{\varepsilon}}_0$ reference plastic strain rate (1.0 s^{-1}), T_{room} room temperature, T_m and melting temperature. A is the initial yield stress [MPa], B is the hardening modulus, n is the

Table 3. Johnson–Cook and cumulative damage law parameters [9].

A (MPa)	B (MPa)	n	C	m	D_1	D_2	D_3	D_4	D_5
792	510	0.26	0.014	1.03	0.05	3.44	−2.12	0.002	0.61

work-hardening exponent, C is the strain rate dependency coefficient [MPa], and m is the thermal softening coefficient. The Johnson–Cook parameters used to simulate the behaviour of AISI 4340 workpiece are specified in Table 3.

In order to simulate the appearance of chip formation a fracture damage model evokes chip detachment behaviour is adopted. It is a separation criterion allowing the tool tip movement. This fracture damage is modelled according to a cumulative damage law.

$$D = \sum \left(\frac{\Delta \bar{\varepsilon}}{\bar{\varepsilon}_f} \right). \tag{6}$$

$\Delta \bar{\varepsilon}$ represents the equivalent strain increment. According to the Johnson–Cook model [9], the cumulative strain $\Delta \bar{\varepsilon}$ is updated at every analysis increment. $\bar{\varepsilon}_f$ is the equivalent failure strain and is expressed as the following equation:

$$\bar{\varepsilon}_f = \left[D_1 + D_2 \exp \left(D_3 \frac{P}{\bar{\sigma}} \right) \right] \times \left[1 + D_4 \ln \left(\frac{\dot{\bar{\varepsilon}}}{\dot{\bar{\varepsilon}}_0} \right) \right] \times \left[1 + D_5 \left(\frac{T - T_{\text{room}}}{T_{\text{melt}} - T_{\text{room}}} \right) \right], \tag{7}$$

where $\bar{\varepsilon}_f$ depends on the ratio $\dot{\bar{\varepsilon}}/\dot{\bar{\varepsilon}}_0$, the equivalent plastic strain $\dot{\bar{\varepsilon}}$, the ratio of hydrostatic pressure to equivalent stress $P/\bar{\sigma}$, and the temperature. It also depends on the damage constants (D_i, $1 \leq i \leq 5$), which are determined experimentally. This shear failure model is used to carry out the chip detachment. It is based on the value of the equivalent plastic strain at element integration points; failure is assumed to occur when the damage parameter D exceeds 1. So, the deviatoric stress components will be set to zero for those points and will remain zero for the rest of the calculation. The pressure stress is then required to remain compressive; that is, if a negative pressure stress is computed in a failed material point in an increment, it is reset to zero. In the proposed model, in order to save calculation time, this damage behaviour concerns only a zone of 60 μm in width as shown in Figure 10 (feature 7). The damage constants of AISI 4340 standard alloy steel [10] are presented in Table 3.

As already said, an adiabatic calculation was adopted for this application. Heat generated during cutting (Figure 11) is the result of plastic deformation and friction. The latter is based on Coulomb's friction law, which has been used in several previous publications [11, 12]. So, if we consider a local temperature rise of ΔT_p in the workpiece during a period of time Δt, the heat generation rate $\dot{q}_P$ due to inelastic work is given by Equation (8)

$$\dot{q}_P = \rho C_P \frac{\Delta T_P}{\Delta t} = \eta_P \sigma_{\text{eff}} \dot{\bar{\varepsilon}}, \tag{8}$$

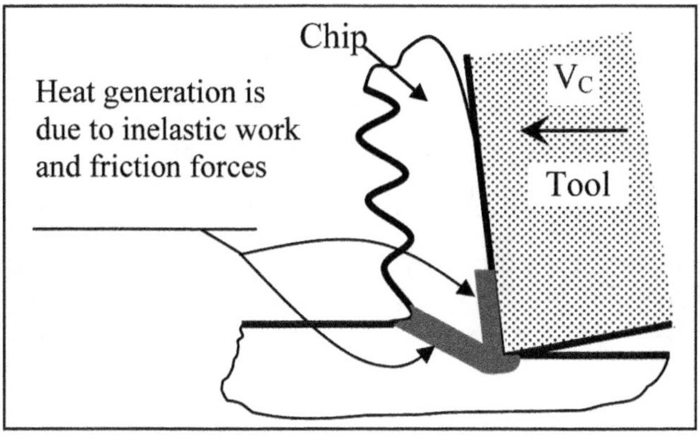

Fig. 11. A schematic representation of heat generation sources.

Table 4. Workpiece and tool thermal parameters.

Mechanical and Thermal Parameter	Density ρ (Kg/m^3)	Elastic modulus E (GPa)	Poisson's ratio ν	Specific heat C_p (Jkg^{-1}°C^{-1})	Thermal conductivity λ (W m^{-1}C^{-1})	Expansion (µm.m^{-1}°C^{-1})	T_{melt} (°C)	T_{room} (°C)
Workpiece	7850	205	0.3	475	44.5	13.7	1520	25
Tool	11900	534	0.22	400	50	×	×	25

where η_P is the fraction of inelastic heat (it is assumed that $\eta_P = 0.90$ [11, 12]). The heat generated by friction is due to a rise in temperature ΔT_f during a period of time of Δt due to friction forces. According to literature [11, 12], the fraction of dissipated energy η_f caused by friction is equal to 1. From η_f an amount of heat, J, remains in chip. The volumetric heat flux $\dot{q}_f$ corresponding to friction is calculated according to the following equation:

$$\dot{q}_f = \rho C_P \frac{\Delta T_f}{\Delta t} = \eta_f J \bar{\tau} \dot{\gamma}. \tag{9}$$

The shear stress $\bar{\tau}$ is given by Coulomb's law, and $\dot{\gamma}$ is the slip strain rate. Consequently, Equation (10) gives the energy equation defining the temperature field.

$$\lambda \nabla^2 T - \rho C_P \frac{\Delta T}{\Delta t} + \dot{q} = 0, \quad \text{where} \quad \dot{q} = \dot{q}_P + \dot{q}_f. \tag{10}$$

For this simulation, workpiece and tool thermal parameters values are given by Table 4.

3.2 Numerical Results and Discussion

In this section, numerical results dealing with temperature evolution, equivalent plastic strains and cutting force F_C during saw-tooth chip formation are presented. In

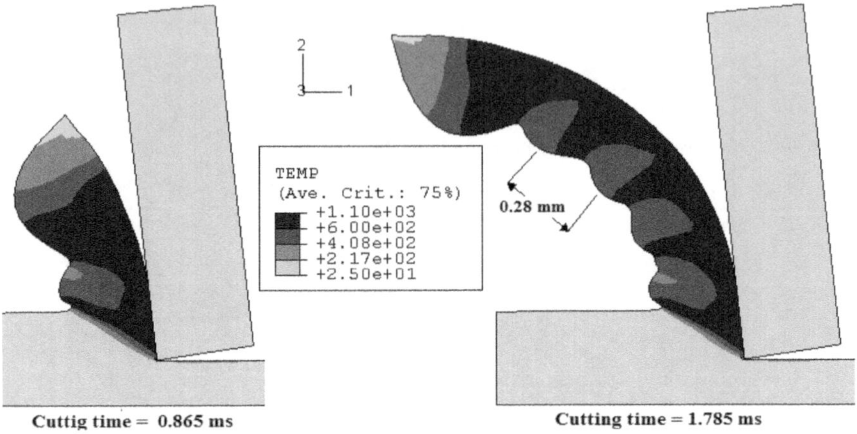

Fig. 12. Temperature evolution (Temp in °C), (f = 0.20 mm/rev, V_C = 100 m/min).

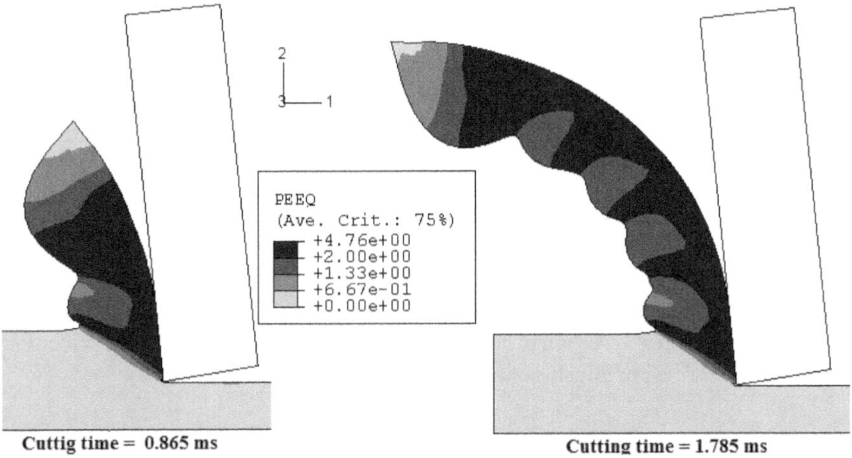

Fig. 13. Distribution of the equivalent plastic strains (PEEQ) in saw-tooth chip (f = 0.20 mm/rev, V_C = 100 m/min).

simulations using Abaqus/Explicit software, the Johnson–Cook damage model sets the deviatoric stress components at zero for the remainder of the simulation. This yields to an advancement of the tool-tip along the cutting zone.

In Figure 12, we observe the evolution of temperature inducing a localized thermal softening. This causes an increase in the ductility of material characterized by high values (2 to 4.76) of equivalent plastic strains (Figure 13) similar to the strain state observed in experimental tests (Figure 2). This state is known as a shear-localized deformation region [13, 14]. It is underlined that during the cutting oper-

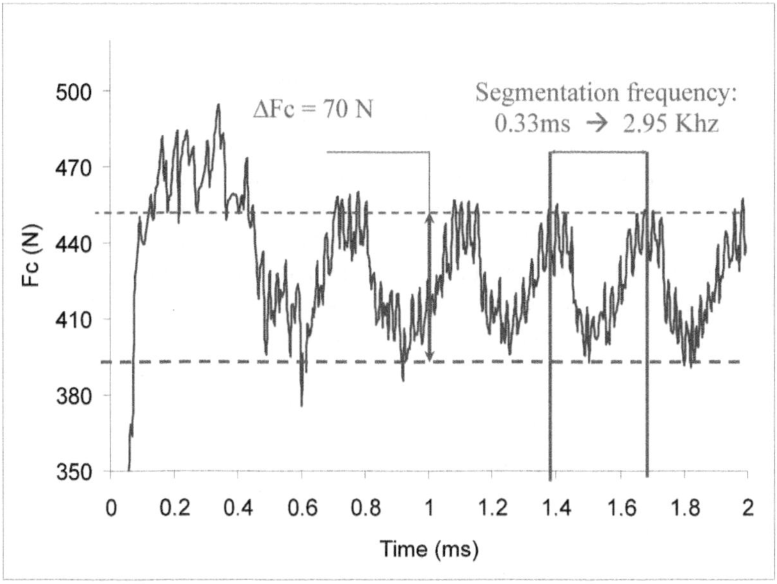

Fig. 14. Variation of cutting force magnitude ($f = 0.20$ mm/rev, $V_C = 100$ m/min).

ation, temperature rise occurs in the separation zone (near the tool-tip) and moves towards the chip curl-up side. This will induce a waved periodic chip known as a saw-tooth chip.

Figure 14 presents the variation of cutting force F_C magnitude given by simulation as a function of cutting time. It is shown that the F_C force oscillates in two manners. The first relates to oscillations of small periods corresponding to a numerical disturbance. The second one corresponds to a relatively high period (almost 0.33 ms). It is the result of chip segmentation. For working parameters of ($V_C = 100$ m/min and $f = 0.2$ mm/rev), we note on one hand that, the value of frequency found numerically is approximately 2950 Hz. This value is near that found experimentally (3151 Hz) (Table 2). On the other hand, the inner-to-peak oscillations observed on numerical results are almost equal to 70 N (Figure 14). This value is near the half of what it was recorded experimentally (Table 1) at the same working parameters. The non-considering of the dynamic behaviour of all machining system can explain the precited difference. Moreover, the cutting force periodicity can be at the origin of machining system excitation. In particular F_C variation can induce, as we have shown in Figure 3, the presence of facets on the machined surface along the cutting speed direction.

4 Concluding Remarks

In order to improve smart machining systems, by introducing physical based understanding we have proposed an experimental and numerical contribution concerning segmented chip genesis in hard material turning. In this paper it is shown that the high frequency variation of the cutting force components reaches several tens of Newton. This cyclical variation involves fatigue behavior of the tool-insert and causes damages (fracture, wear, etc.). Consequently, it seems important to control, during machining, the segmentation frequency. This can improve the accuracy, reliability and productivity of machining operations. So, thanks to segmentation frequency prediction added to some tools and machine constraints (spindle supported torque, power limitation, etc.), it could be possible to make a best choice of working parameters. As a result, an improvement of some quality criteria such as finished surface roughness and integrity will be realized. The comprehension of chip segmentation events could bring more ideas concerning the tool working conditions such as thinking to monitor the tool displacement by making movement compensation, when it seems significant, to avoid an undesirable excitation due to chip segmentation. Also, cyclic chip segmentation can allow a prediction of tool damage. The presented ideas can participate to the improvement of machining optimizations and leading to the realization of autonomous smart machining systems.

The smart machining concept presents many strategies related to productivity improvement through innovative technology. So, the comprehension of some physical phenomena corresponding to material removal by cutting tool seems to be an important key to overcome difficulties related to machining control and monitoring. An objective, among others, of many companies and national institutes is to build a smart machining system which will be able to monitor and optimize its operations autonomously. For that, the study of phenomena accompanying chip segmentation and predicting its frequency will permit a best choice of working parameters. As a result, an improvement of some quality criteria such as finished surface roughness and integrity will be realized. Potentially, this comprehension of chip segmentation can be presented by mathematical and informational framework to set up optimization models that can handle the different scenarios related to a concerned machined operation.

Note

*Specific commercial equipment is identified to fully describe the experimental procedures. This identification does not imply endorsement by the authors.

References

1. Research's programs, NIST, Manufacturing Engineering Lab., http://www.mel.nist.gov/proj/
2. D.E. Dimla Sr, The impact of cutting conditions on cutting forces and vibration signals in turning with plane face geometry inserts, *J. Mat. Proc. Techn.* **155–156**, 2004, 1708–1715.
3. S. Ratchev, S. Liu and A.A. Becker, Error compensation strategy in milling flexible thin-wall parts, *J. Mat. Proc. Techn.* **162–163**, 15 May 2005, 673–681.
4. C.-W. Lin, F. Jay Tu and J. Kamman, An integrated thermo-mechanical-dynamic model to characterize motorized machine tool spindles during very high speed rotation,ă *Int. J. of Mach. Tools and Manuf.* **43**(10), August 2003, 1035–1050.
5. P. Krajnik and J. Kopač, Modern machining of die and mold tools, *J. of Mat. Proc. Techn.* **157–158**, 20 December 2004, 543–552.
6. L. Deshayes, Méthodologie d'étude de la coupe, liaison entre Couple Outil Matière et Pièce Outil Machine, PhD Thesis, INSA-Lyon, 2003, 271 pp.
7. G. Poulachon and A.L. Moisan, Hard-turning: Chip formation mechanisms and metallurgical aspects, *Trans. ASME J. Manuf. Sci. Eng.* **122**, 2000, 406–412.
8. T. Mabrouki and J.-F. Rigal, A contribution to a qualitative understanding of thermo-mechanical effects during chip formation in hard turning, *J. Mat. Proc. Techn.* **176**(1–3), 2006, 214–221.
9. G.R. Johnson and W.H. Cook, Fracture characteristics of three metals subjected to various strains, strain rates, temperatures and pressures, *Eng. Fract. Mech.* **21**(1), 1985, 31–48.
10. D.J. Benson, A mixture theory for contact in multi-material Eulerian formulations, *J. Comput. Methods Appl. Mech. Engrg.* **140**, 1997, 59–86.
11. K. Li, X.-L. Gao and J.W., Sutherland, Finite element simulation of the orthogonal metal cutting process for qualitative understanding of the effects of crater wear on the chip formation process, *J. Mat. Proc. Techn.* **127**(3), 2002, 309–324.
12. E. G. Nga, D.K. Aspinwall, D. Brazil and J. Monaghan, Modelling of temperature and forces when orthogonally machining hardened steel, *Int. J. Mach. Tools & Manuf.* **39**, 1999, 885–903.
13. T. Mabrouki, L. Deshayes, R. Ivester, J.-F. Rigal and K. Jurrens, Material modelling and experimental study of serrated chip morphology, in *7th CIRP Int. Workshop on Modelling of Machining Operations*,ăăENSAM Clunny,ăFrance,ă4/5/2004 0–5/5/2004 0-4.
14. S. Belhadi, T. Mabrouki, J.-F. Rigal and L. Boulanouar, Experimental and numerical study of chip formation during straight turning of hardened AISI 4340 steel, *J. Engin. Manuf., Part B, Proc. IMechE* **219**, 2005, 515–524.

Identification of Plunge Milling Parameters to Compare with Conventional Milling

Mohamad Al-Ahmad, Alain d'Acunto and Patrick Martin

Laboratoire de Génie Industriel et Production Mécanique (LGIPM – EA 3096), CER ENSAM, Metz Technopôle, 4 rue Augustin Fresnel, 57078 Metz Cedex 3, France; E-mail: {mohamad.al-ahmad, alain.dacunto, patrick.martin}@metz.ensam.fr

Abstract. The operation of axial or plunge milling is one of the most effective methods, there is for removing metal in rough and semi rough machining where only the Z-axis is used. It has been proposed especially for cutting deep cavities or vertical walls of moulds and dies. This new operation of milling greatly contributes to shortening the entire production process, and therefore, saving significant production time. However, it remains confidential and needs additional investigations for controlling and showing its advantages.

In this paper, we first introduce and identify the operation of plunge milling. Also, we define the technical data allowing the choice of a tool and the determination of using conditions in this operation. We specify the characteristics and the cutting parameters of this operation and those of the tool. We then evaluate the performance and the effectiveness of this operation by drawing a comparison between plunge and conventional milling operations. This comparison is realised according to different criteria: geometrical configurations and cutting parameters, machining strategies, cutter edge trajectory, power and cutting forces, aspects of productivity (material removal rate), and time of machining. Finally, an example of deep cavity has been achieved for showing the effectiveness of this operation compared to conventional milling. The results of machining show that the values of the volume of metal removed and the power are more important in plunge milling, however the time of machining is less important.

Key words: plunge milling, cutting forces, productivity, machining time.

1 Introduction

Plunge milling operation is used for jobs such as roughing cavities and walls in moulds and dies. It has been proposed lately for cutting vertical deep walls. Figure 1 shows the diagram of plunge milling in which a milling cutter with insert tips is used. The advantage of this operation is that stable cutting can be performed because the cutting is carried out by single point tool, so that the cutting forces are relatively small and do not change during cutting. Furthermore, the surface machined by plunge milling has greater accuracy than the one cut by conventional milling [1].

S. Tichkiewitch et al. (eds.), Advances in Integrated Design and Manufacturing in Mechanical Engineering II, 461–474.

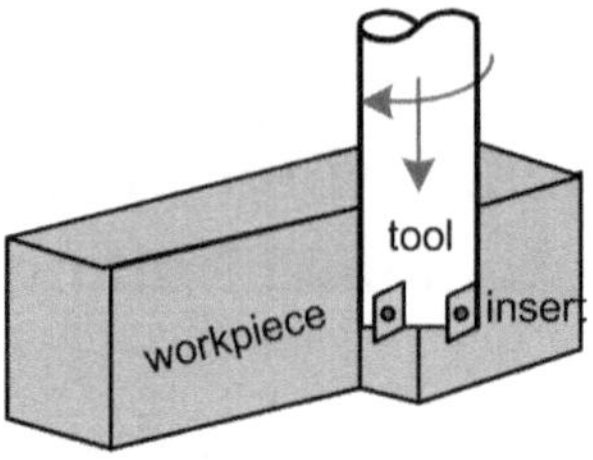

Fig. 1. Plunge milling operation.

Manufacturing industry faces one requirement in the technical data for choosing and determining the operating conditions of a cutting tool for a plunge milling operation (machine tool, material, workpiece/operation, intermediate systems, etc.). These data are used both to identify tool geometry and for their use.

In this study, we specify the characteristics and the operating parameters of this operation. Furthermore, we identify the tool geometry, the cutting conditions, and the cut geometry [2, 3]. An object model for plunge milling operation is presented by using the formalism of United Modelling Language (UML) [4]. Later, plunge milling was compared with conventional milling to clarify the accuracy and efficiency of plunge milling. In this comparison, we take into account: the geometrical configuration, the machining strategy, the cutting edge trajectory, the power and cutting forces. Finally, an example of deep cavity has been achieved for showing the effectiveness of this operation compared to conventional milling.

2 Plunge Milling Operation

2.1 Plunge Milling Cutter

Plunge milling cutter developed by the tool manufacturers consist of the cutter body and the inserts (Figure 2). The cutter is assembled on an extension whose length is often 3 to 4 times higher than the tool diameter. Moreover, the connection with the machine tool must take into account the types of mechanical interface between the holder tool and the spindle of the machine tool. However, to define the tool geometry, it is necessary to distinguish three types of parameters:

- Parameters of the body such as the overall length, the number of teeth, the tooth spacing, and the direction of helix.
- Parameters of the inserts such as the corner shape (radius, chamfer), and the chip breakers.
- Parameters of the assembly such as the fixing method, the cutter diameter, and the cutting length.

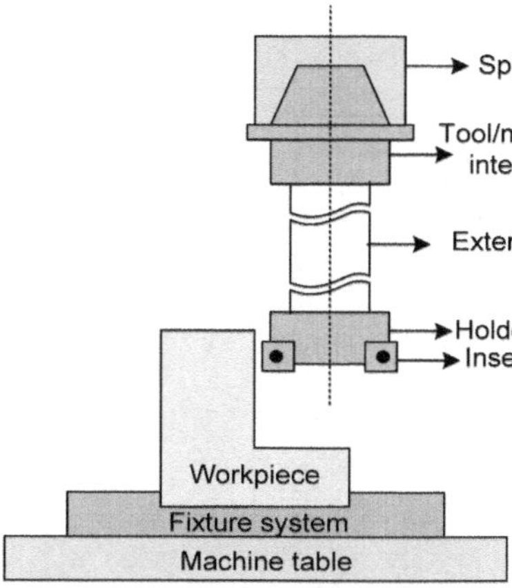

Fig. 2. Plunge milling cutter.

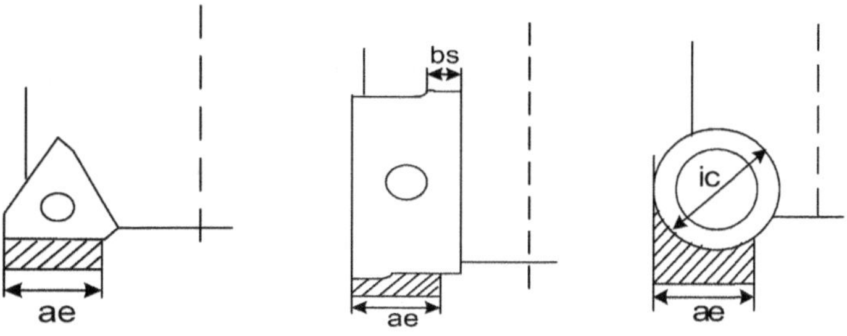

Fig. 3. Types of the inserts.

The geometry of the insert in plunge milling cutter is very important, were, we can defined various inserts whose choice is determined by the angle of edge direction, the cutter geometry and the type of operation (Figure 3). Three types of inserts geometry are used in plunge milling whose forms are round, rectangular and triangular and the chip geometry is different for each type [5].

2.2 Cutting Parameters

The working zone of cutting tools in plunge milling operation is defined by the cutting parameters (Figure 4). The definition of this operation includes the following characteristics:

- Radial engagement ae, corresponds to the distance of tool engagement in the workpiece. It is function of the insert dimension.

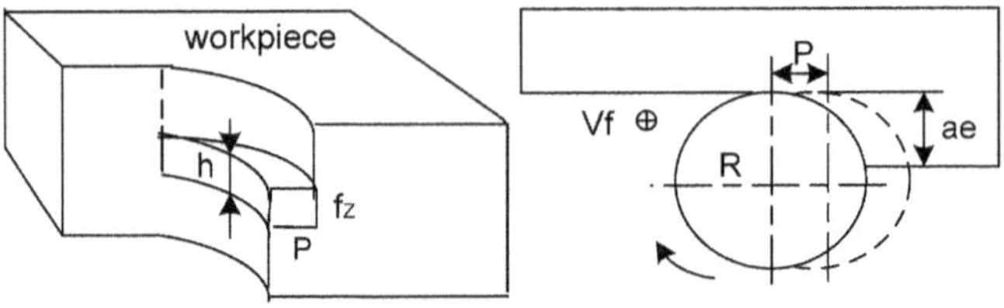

Fig. 4. Cutting parameters.

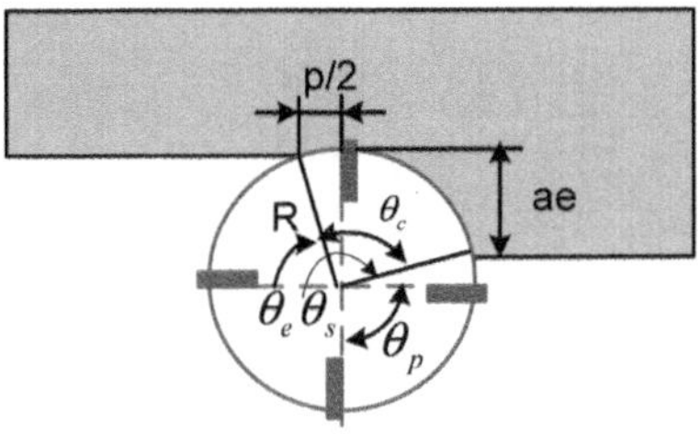

Fig. 5. Teeth engaged in the workpiece.

- Thickness of cut h, it is evaluated through the cutter rotation at the entry and the exit of the workpiece (first and last contact with the workpiece). On the other hand, the thickness of cut is stable in the medium (between the first and last contact with the workpiece) [6].
- Pick feed P, corresponds to the distance between two passages of the cutter in the workpiece. The pick feed value has an influence on the quality of the final surface [7].
- Axial engagement ap, it does not exist in this geometrical configuration but the feed rate per tooth can be similar to an axial engagement for one cutter rotation.
- Feed rate per tooth f_z, corresponds to the linear distance crossed by one tooth.
- Cutting speed V_c, corresponds to the speed when the cutting edge machines the surface of the workpiece.
- Number of teeth engaged in the workpiece Z, it is a function of various parameters such as cutter diameter R, radial engagement ae, and pick feed P (Figure 5). To calculate the numbers of inserts participate in cutting simultaneously, two angles should be defined initially: spacing angle θ_p and swept angle θ_c, were $\theta_p = 2\pi/z$, and $\theta_c = \theta_s - \theta_e$. The number of cutting inserts simultaneously engaged in the workpiece Z_c can be determined by $Z = Z \cdot \theta_c/2\pi$.

As for the other machining operations, we also find the following parameters:

- Instantaneous cutting section A_D, where the maximum cutting section is defined by the following formulation: $A_{D_{\max}} = P \cdot h_{\max}$.

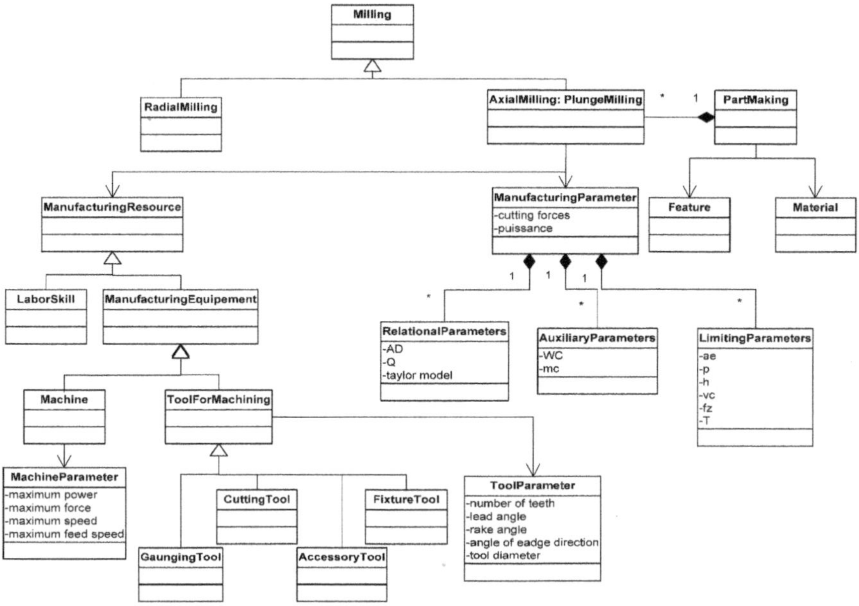

Fig. 6. Milling (plunge milling) object model.

- Material removal rate Q, it is defined by multiplying the cutting section with the feed movement: $Q = A_D \cdot V_f$.

3 Modelling of Plunge Milling Data

By taking into account the cutter geometry and the cutting parameters, we can present an initial object model for plunge milling operation. This type of modelling is important for designers to evaluate manufacturability and manufacturing cost and time in the early product development stage. The model is represented in the class diagrams in the format of the Unified Modelling Language (UML) [4].

In this study, we propose to develop an object model of milling by introducing the plunge milling operation in this model. Figure 6 shows the milling operation classes. Milling is an activity of machining by using multiple points cutting method. There are two derived classes: radial milling and axial or plunge milling. Part marking has a sequence of plunge milling and represents a part fabrication activity. Manufacturing resource is a class that represents a physical object or a labor skill that is used in a plunge milling. Manufacturing equipment is a class that represents a piece of equipment (machine, tool for machining) that is used in plunge milling. The derived classes from manufacturing equipment are machine and tool for machining. Man-

Table 1.

Parameters	Plunge milling	Conventional milling
Feed rate per tooth (f_z)	Parallel with tool axis	Perpendicular with tool axis
Axial engagement (ap)	–	Limited with fixed value
Pick feed (P)	Specific	–

ufacturing parameter is a class that represents three types of parameters (relational parameters, auxiliary parameters, and limiting parameters). The relational parameters include the cutting section, the material removal rate, and the constants of the generalised Taylor model. The auxiliary parameters include the specific cutting energy W_c and its coefficient m_c. The limiting parameters include radial engagement, pick feed, thickness of cut, feed, cutting speed, and tool life.

4 Determination of the Limits and the Performances of Plunge Milling Operation by Comparing with Conventional Milling

To ensure an equitable comparison between plunge and conventional milling, we position ourselves in the range of machining a pocket [8]. Moreover, we used the same types of machining strategies and cutting conditions for each operation. The steps of comparison define the parameters and the formulations which control the milling operations. This comparison is realised according to different criteria: geometrical configurations and cutting parameters, machining strategies, cutter edge trajectory, power and cutting forces, productivity aspects (material removal rate), and machining time.

4.1 Geometrical Configurations

In plunge milling, the relative movement of the cutter with respect to the workpiece consists of a rotation around the tool axis and a feed movement along the Z-axis (Figure 7a). On the other hand, in conventional milling, the movement of the cutter consists of a rotation around the tool axis and a feed movement perpendicular to the Z-axis (Figure 7b).

The common cutting parameters of both machining operations are: the cutting speed (V_c) and radial engagement (ae). The specific parameters of each machining operation are classified in Table 1.

4.2 Cutting Edge Trajectory

The trajectory of a tooth in the matter follows a cycloid curve determined by the movement of a point fixed on the cutting edge (Figure 8).

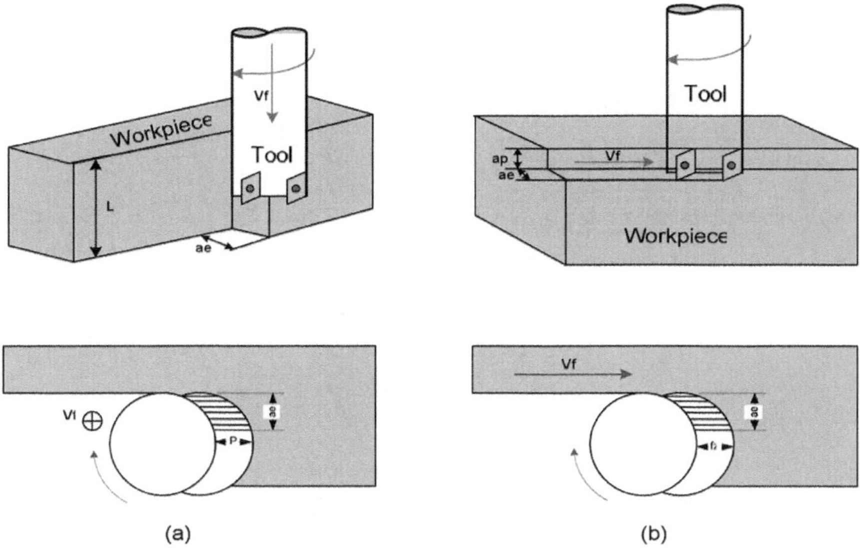

Fig. 7. Geometrical configurations and cutting parameters of plunge milling and conventional milling operations.

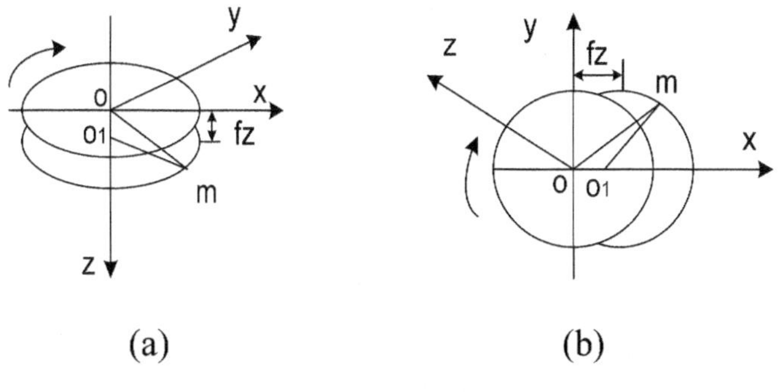

Fig. 8. Cutting edge trajectory in plunge milling (a) and conventional milling (b).

Table 2 defines the equations of the movement of a point m on the cutting edge. We can note that the cycloid trajectory in conventional milling is defined in the plane (x, y), while, for plunge milling, it is defined in the space (x, y, z).

4.3 Machining Strategy

In the framework of machining a pocket, we applied the technique of rough machining for realising a simple and deep pocket (Figure 9). The strategies suggested to

Table 2.

Plunge milling	Conventional milling
$\mathbf{Om} = \mathbf{OO}_1 + \mathbf{O}_1\mathbf{m}$	$\mathbf{Om} = \mathbf{OO}_1 + \mathbf{O}_1\mathbf{m}$
$\mathbf{OO}_1 = Nf_z t\mathbf{z}$	$\mathbf{OO}_1 = Nf_z t\mathbf{x}$
$O_1m = R\cos\theta\mathbf{x} + R\sin\theta\mathbf{y}$	$O_1m = R\cos\theta\mathbf{x} + R\sin\theta\mathbf{y}$
$\theta = \pi Nt$	$\theta = \pi Nt$
$\mathbf{Om} = R\cos(\pi Nt)\mathbf{x} + R\sin(\pi Nt)y + \mathbf{N}f_z t\mathbf{z}$	$\mathbf{Om} = [R\cos(\pi Nt) + Nf_z t]\mathbf{x} + R\sin(\pi Nt)\mathbf{y}$

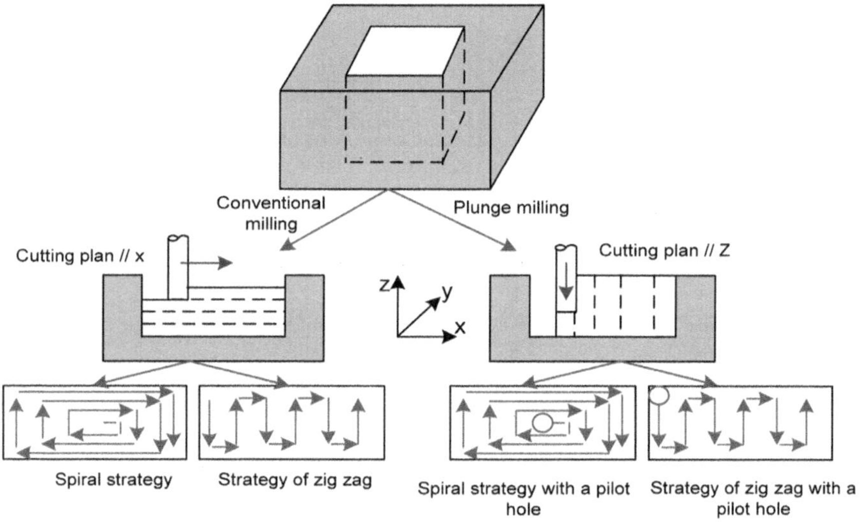

Fig. 9. Machining strategies.

achieve this pocket by using plunge and conventional milling are: (i) spiral strategy, where the direction of tool movement is homothetic with the contour of the pocket. (ii) Strategy of zigzag, where the tool movement follows a rectilinear direction. The choice of a machining strategy takes into account the type of tool, the type of machining, and also the machining constraints (e.g., accessibility, visibility). In the first time, the machining of the pocket is realised in the case of conventional milling, by passages of tool along the x-axis with different altitudes. Then, in the case of plunge milling, by passages of the tool along the z-axis with pilot hole if that is necessary.

4.4 Power and Cutting Forces

Based on the simple mechanistic approach, we determine the power and the cutting forces by using simple formulations defined in the technical documents of Sandvik [5]. According to the chip geometry, we determined the cutting section of each operation to determine the power and the cutting forces (Table 3). In this case, only the

Table 3.

Plunge milling	Conventional milling
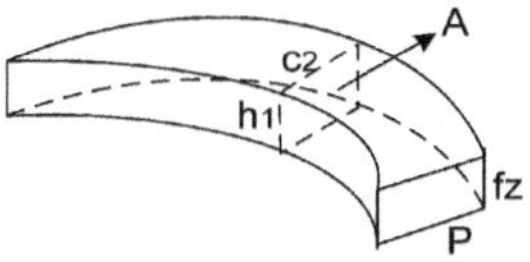 $A = c_1 * h_1$ $A_{\max} = f_z.P.\sin\theta$ $F_t = A * K_c$ $P_{ct} = Q_t * K_c$	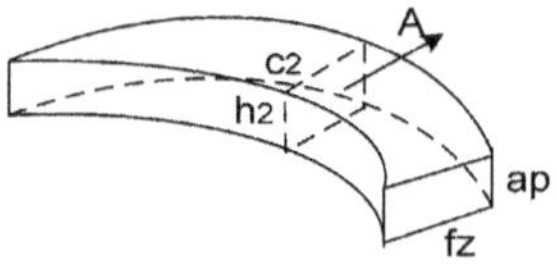 $A = c_2 * h_2$ $A_{\max} = ap * f_z.\sin\theta$ $F_f = A * K_c$ $P_{cf} = Q_f * K_c$

cutting section is different between the two operations. The power is given according to the material removal rate and the cutting specific coefficient.

In plunge milling, the feed movement is achieved in the axial direction that minimises the radial forces and the tool deflection [9]. In conventional milling, the radial force is important and generates a tool deflection [10, 11].

4.5 Instantaneous Material Removal Rate

The productivity of the machining operations is characterised by the material removal rate. The material removal rate is obtained by multiplying the chip area A by the feed movement V_f. The feed movement is a parameter imposed by the operating conditions.

Regarding our work, we only determined the chip area for each operation (Figure 10). The following formulation determines the maximum chip area for plunge and conventional milling:

$$Q = A \cdot V_f \quad \text{where} \quad \begin{cases} A_{t\max} = ae \cdot p \text{ in milling plunge} \\ A_{f\max} - ae \cdot ap \text{ in conventional milling} \end{cases} \tag{1}$$

4.6 Machining Time

In this study, we are interested in determining the machining time for plunge and conventional milling in order to calculate the machining cost. The machining time (T_u) consists of the time of cut (t_c), the extra time of machining (t_m) and the time

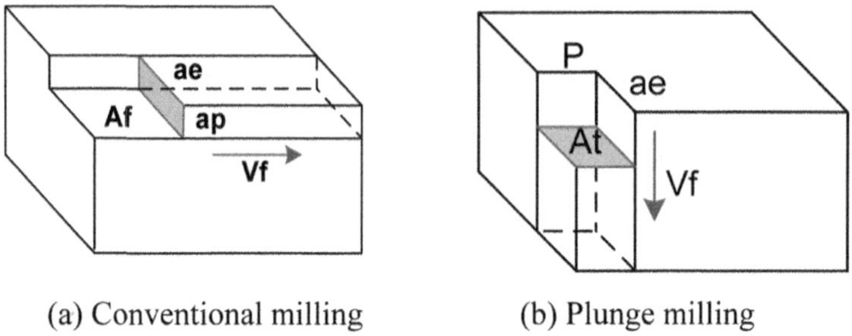

Fig. 10. Material removal rate.

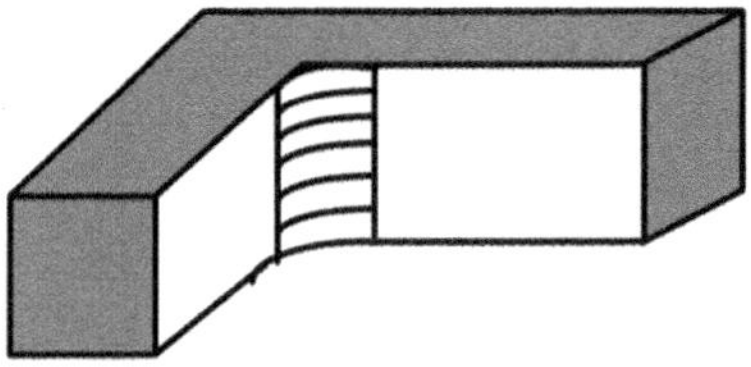

Fig. 11. Example of remaining material.

of the machining correction ($_{\text{rep}}$) [12]. The machining time can be expressed by the following relation: $T_u = t_c + t_{\text{rep}} + t_{hm}$.

Sometimes, an excess of material may remain in the machining, due on one hand to the weak local curvature of the surface and on the other hand, to the radius of the cutting tool (Figure 11). The time of the machining correction (t_{rep}) is the time necessary to remove the material that remains after the machining. The extra machining time (t_{hm}) correspond to the hm time of rapid movement of the cutting tool. The determination of extra machining time is made by analytical method [13]. This time is calculated by using the rapid speed and the total distance of rapid return of the cutting tool. We are going to determine the time of cut in the following section.

4.6.1 Cutting Time

The cutting time needed for realising a cavity in rough machining is determined from the length of cut and the advance speed. The cutting time in plunge milling (t_{ct}) and conventional milling (t_{cf}) for only one passage of the tool are expressed by the following relations:

$$t_{ct} == \frac{L_t}{V_{ft}} = \frac{L_t \cdot \pi \cdot D}{Z \cdot f_z \cdot V_c},$$

$$t_{cf} == \frac{L_f}{V_{ff}} = \frac{L_f \cdot \pi \cdot D}{Z \cdot f_z \cdot V_c}. \tag{2}$$

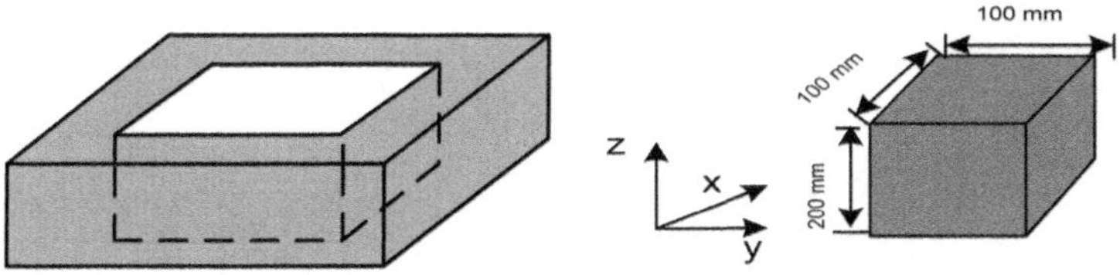

Fig. 12. Simple and deep pocket.

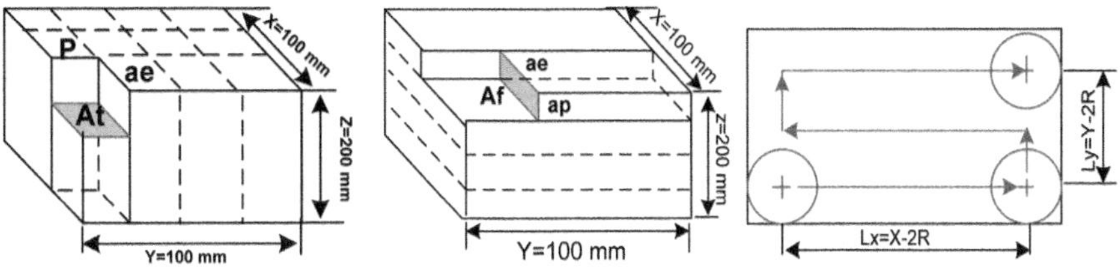

(a) *(b)* *(c)*

Fig. 13. Machining of the pocket.

In the case of conventional milling, when the height of the cavity is important, the principle consists in analysing the volume to remove with the height of the cavity in several slices. For each level, the advance speed must be changed [12]. A coefficient of reduction k (between 0.1 and 0.5) is adopted for the modification of the advance speed.

$$V_{f(i+1)} = k \cdot V_{f(i)}; \quad i = 1, 2, \ldots, \frac{ap}{z}. \tag{3}$$

This result shows that the advance speed decreases with the increase of the depth of the cavity and it requires an additional cutting time.

5 Example

In order to establish a real comparison between plunge and conventional milling, we present an example of machining a simple and deep pocket by using the strategy of zigzag. The volume of matter removed by this technique is a cube (Figure 12).

The machining of this cube by plunge and conventional milling is defined by elementary volumes. Figure 13 shows the various elementary passages and the lengths of cut for each operation.

Using the technical document of Sandvik [5], we used compatible cutting conditions (tool, material, radial engagement, feed rate per tooth, cutting speed, specific cutting parameters, etc.). Concerning the specific cutting parameters, we respect the limiting values in this document. According to equation (1), we draw the evaluation

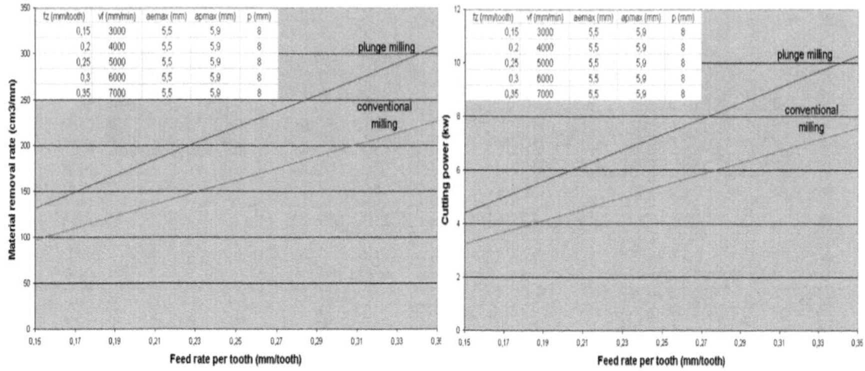

(a) (b)

Fig. 14. (a) Material removal. (b) Cutting power.

of the material removal rate in function of feed rate per tooth for each operation (Figure 14a). The values in the table respect the limits of each parameter.

According to the material removal rate which is previously determined and the specific cutting coefficient [5], we trace the evaluation of power (Table 3) in function to the feed rate per tooth (Figure 14b).

The following equations determined the cutting time necessary to realised the pocket defined previously in both operations.

$$t_{ct} = = \frac{L_t}{V_{ft}} \cdot \left(\frac{x}{ae} \cdot \frac{y}{p}\right) = \frac{L_t \cdot \pi \cdot D}{Z \cdot f_z \cdot V_c} \left(\frac{x}{ae} \cdot \frac{y}{p}\right),$$

$$t_{cf} = = \frac{L_f}{V_{ff}} \cdot \left(\frac{x}{ae} \cdot \frac{z}{ap}\right) = \frac{L_f \cdot \pi \cdot D}{Z \cdot f_z \cdot V_c} \left(\frac{x}{ae} \cdot \frac{z}{ap}\right). \tag{4}$$

According to equation (4), we trace the evaluation of the cutting time in function of the depth of the cavity for the two operations (Figure 15).

According to this example, the material removal rate and the cutting power in plunge milling is more important than the one of conventional milling. On the other hand, the cutting time in plunge milling is less important than the one of conventional milling.

These results are not very precise due to the fact that the cutting forces model (cutting specific coefficient are not the same for the two operations) and the determination of the power must be more realistic. Furthermore, the determination of cutting times considered the extra time is the same for the two operations.

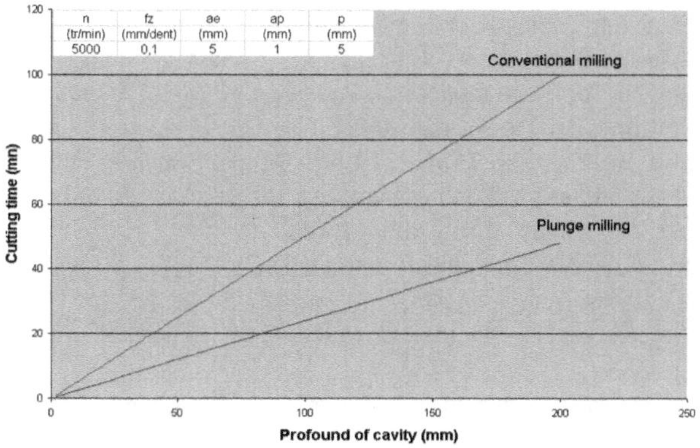

Fig. 15. Cutting time.

6 Conclusion

In this work, we specify the characteristics of plunge milling operation (technical data, tool characteristics, and operating conditions). Then, a comparison between plunge and conventional milling is achieved to clarify the accuracy and efficiency of this operation. In this comparison, we take into account: the geometrical configuration, the machining strategy, the cutting edge trajectory, the power and cutting forces. Finally, an example of deep cavity has been achieved for showing the effectiveness of this operation compared to conventional milling. We evaluated the performances of plunge milling operation (important material removal rate, short cutting time) and one of its limits (power) through the machining of a simple pocket. However, the operation of plunge milling requires additional investigations. In order to ensure an optimal use of this operation, it is necessary to determine the working zones of cutting tools (couple tool material).

References

1. S. Wakaoka, Y. Yamaneb, K. Sekiyab, N. Narutaki. High-speed and high-accuracy plunge cutting for vertical walls. *Journal of Materials Processing Technology* **127**(2), 2002, 246–250.
2. H.J. Fu, R.E. DeVor, S.G. Kapoor. A mechanistic model for the prediction of the force system in face milling operations. *Journal of Engineering for Industry* **106**, 1984, 81–88.
3. R.E. De Vor, W.A. Kline, W.J. Zdeblick. A mechanistic model for the force system in end milling with application to machining airframe structures, in *Proceedings of 8th North American Manufacturing Research Conference*, Vol. 8, 1980, pp. 297–303.
4. S.C. Feng, E.Y. Song. Information modeling of conceptual process planning integrated with conceptual design, in *5th Design for Manufacturing Conference*, Baltimore, Maryland, 10–13 September 2000.

5. Sandvik Coromant, Technical Document, 2004.
6. M. Al-Ahmad, A. D'Acunto, C. Lescalier, O. Bomont. Modelling of cutting forces in plunge milling, in *7th International Conference on Advanced Manufacturing Systems and Technology*, Udine, Italy, 9–10 June 2005.
7. M. Al-Ahmad, A. D'Acunto, P. Martin. Etude des performances et limites de l'opération de tréflage – Appliquer à lŠentité poche, in *17ème Congrès Français de Mécanique*, Troyes, France, 29 August–2 Septembre 2005.
8. A. Hatna, R.J. Grieve, P. Broomhead. Automatic CNC milling of pockets: Geometric and technological issues, *Computer Integrated Manufacturing Systems* **11**(4), 1998, 309–330.
9. Copyright Sofetec. De la technologie du tréflage, article appeared in *Machines Production 743*, 2001, p. 27.
10. E.J.A. Armarego, N.P. Deshpande. Computerized end-milling force predictions with cutting models allowing for eccentricity and cutter deflections, *Annals of the CIRP* **40**, 1991, 25–29.
11. J. Tlusty, S. Smith, W.R. Winfough. Techniques for the use of long slender end mills in high-speed milling, *Annals of the CIRP* **45**, 1996, 393–397.
12. Z. Bouaziz, J. Ben Younes, A. Zghal. Methodology of machining costs evaluation for die and mould manufacturing, *Journal of Materials Processing Technology* **152**(2), 2004, 237–245.
13. M. Field, R. Kegg, S. Suescher. Computerized cost analysis of grinding operations, in *Annals of the CIRP STC G*, 1980, pp. 233–239.

Agility in High Speed Machining: Optimization between Tool and Holder

Abdelkarim Ben Mhenni, Christian Mascle and Marek Balazinski

Manufacturing Department, Ecole Polytechnique de Montréal, C.P. 6079 – Succ. Centre-Ville, Montréal (Québec), Canada H3C 3A7;
E-mail: {abdelkarim.ben-mhenni, christian.mascle, marek.balazinski}@polymtl.ca

Abstract. In high speed machining (HSM), forces are smaller than in conventional machining. Shrink-fit tools have better behaviour in HSM than other types. However their damping capacity remains the major disadvantage. We prove in this research that there is possibility of having a satisfying maximum torque for high speed machining with a better damping ratio and repeated assembly if one optimizes contact surface roughness and other shrink fit characteristics. We show that roughness has main effect on friction and therefore torque capacity. The relation between maximum torque and surface roughness is parabolic and shows a minimum for average asperities sizes. Our approach consists in optimizing, by means of experimental designs, the parameters of the shrink fit which are contact surface roughness and interference.

Key words: high speed machining, shrink fitting, tool-holder, roughness, repeated assembly.

1 Introduction

1.1 Problem Statement

High speed machining (HSM) is into full expansion, the era of machining at low speeds (up to 10,000 rpm) is lapped and the profits in time and money carried out by the technique encourage the industrialists to choose HSM. Technological progress gave us spindles which can reach the 100,000 rpm and materials which can cut at speed as high as 7000 m/min, however these performances are still far reaching, indeed the weak link between the spindle and the cutting tool prevents us from fully benefiting from HSM. The tool is usually maintained by the holder which is maintained in its turn by the spindle. The problem with HSM is centrifugal force resulted by high speeds can cause the holder to retract when the spindle and holder touch only a long taper. Two-face contact at the spindle interface achieved with HSK (hollow shank taper) can solve this problem. At the beginning of the 1990s the holders HSK were invented in order to meet the new needs for the interface holder-spindle, since,

S. Tichkiewitch et al. (eds.), Advances in Integrated Design and Manufacturing in Mechanical Engineering II, 475–488.

many research are made on the HSK to understand their behaviour, to compare them with other technology and to improve them.

Concerning the tool-holder interface, a solution adopted was the shrink-fit assembly which consists to heat the holder by means of a specialized heater. At normal temperature, the holder bore is slightly undersize compared to the tool shank. Heating the holder open this bore, allowing the tool to be inserted. As the holder cools, the bore shrinks around the tool to create a concentric and rigid clamp. Shrink-fit presents no connection element between the two parts to assemble such as bolts, keys or others. It allows easy achievements, without unbalance or overall eccentricity.

There are alternatives to shrink-fit holder, hydraulic and mechanical deformation holder. Hydraulic holder equalizes clamping pressure around the tool using a reservoir of oil. The pressure on oil is increased by turning a screw causing an expanding sleeve to grip the tool shank. The fluid reservoir may offer some capacity to damp vibration, but clamping mechanism is easy to damage. Mechanical deformation holder features a bore that not round but instead has a three-lobed shape suggesting a triangle. A clamping device mechanically distorts this bore to make it a round, allowing the tool to be inserted. When the force is released, the holder snaps back to initial shape, clamping the tool with out heating or cooling.

The qualities sought for the interface tool-holder are minimum run-out, rigidity, balance, vibration damping of use and less cost. The shrink-fit solution is considered as valuating the investment for very fast HSM process and very committed HSM users, it offers superior concentricity, balance and ease of use. Its main default is the low capacity of vibration damping. The behaviour of the shrink-fit assembly in machining at high speed still remains to be studied and we are convinced that there are potentials to arise at end to better control this method

1.2 Related Work and Contribution

To our knowledge, no study was published on the shrink fit assembly used in machining. All the documents on shrinking which we found covers subjects for specific applications others that the HSM. Most of previous studies focus on the load or torque capacity of shrink fit, while trying to model the effect of some parameters or treatments on these loads, the damping aspect and repeatability of assembly are almost neglected.

Thornley and Elewa [9] investigated the effect of size, surface roughness and the interference value on the static and dynamic stiffness of shrink-fitted joints. They focus on dynamic behaviour and conclude that the tangential displacement and micro-slip increase with increasing roughness and decrease with increasing interference [9].

Kato et al. [15] studied in 1989 the effect of surface roughness on fitting strength of shrink fit between ceramic and metal element. They take into account elastic-plastic deformation of asperities and found theoretical estimations showing good agreement with experimental values in various interferences and temperatures.

Radhakrishnan, Ventakeswara and Ramamoorthy have published a series of papers on shrink fitted assemblies. All their studies are experimental and their objective is to improve the performances of the shrink assembly by surface strengthening or plating. They compare also the deformations of surface in press and shrink fitted assemblies and found it to be more pronounced in press-fit case [13]. El Hadidy [14] did an experimental study of the geometrical parameters, surface roughness and interference of shrink-fitted joints. He uses R_a, R_t and R_z to describe roughness. He concluded that the load carrying capacity can be improved if one uses the right parameters [14].

Yang [3] and Fontaine et al. [2–4] studied the effect of surface roughness, form defect and surface shape on shrink-fit characteristics. They criticized the French standard AFNOR NF E22–620 which does not take into account surface roughness and fix friction coefficient to 0.15.

Truman et al. [6–8] used the micro-mechanical approach to better understand the stick-slip phenomena and its influence on the holding torque. Later Booker and Truman [5] did a comparison of shrink-fit holding torque using probabilistic, micro-mechanical and experimental approaches. The probabilistic approach is found to be the closer to experimental results. Metallic mating surfaces in machine-joints offer a good source of energy dissipation, resulting in damped dynamic structural response.

Murty and Padmanabhan [1] investigated the effect of surface topography on damping in machine joints. The damping was found to increase with an increase in surface roughness up to a certain roughness value. Thus, in general, the energy loss per cycle is found to be influenced by tangential force, normal force, joint material and surface conditions. An expression which leads to evaluation of energy dissipation in the form of hysterisis loop area is given in which the influence of material and surface topography parameters are shown directly. Concerning repeated assembly, Radhakrishnan [16] affirmed that where interference fits are to be dismantled and reassembled, finer surfaces are capable of maintaining their load carrying capacity.

2 Torque Capacity

2.1 Lamé's Formulae

Lamé and Clapeyron's theory of the thick walled cylinder created in 1833 is the model used currently to design the interference-fit assemblies. In this theory, assumptions are that the two assembled parts are ideal solids: small elastic strain, isotropic materials, uniform applied forces, perfectly polished surfaces, etc.

The contact pressure between sleeve and shaft made of different materials is given by

$$p \frac{\delta}{d_c\left[\left\{\frac{(1-\gamma_{sl})+(1+\gamma_{sl})\lambda_{bc}^2}{E_{sl}(\lambda_{bc}^2-1)}\right\}+\left\{\frac{(1-\gamma_{sh})+(1+\gamma_{sh})\lambda_{lc}^2}{E_{sh}(\lambda_{lc}^2-1)}\right\}\right]}. \quad (1)$$

The torque capacity of the assembly, doing the assumption that the contact pressure p in the interface joint is uniformly distributed, can be estimated using the following relation:

$$T = \frac{\pi}{2} f d_C^2 \, pL, \quad (2)$$

where f is the friction coefficient between mating surfaces [18, 19].

2.2 Standards

All standards on interference fits use Lamé's theory of thick walled cylinders with the above assumptions. For ANSI B4.1-1967 (R87) standard, shrink or force fits are characterized by approximately constant bore pressures throughout the range of sizes. Interferences vary almost directly as the diameter, and the differences between maximum and minimum values of interferences are small. These are divided into five classes depending on applications. For metric system ANSI B4.2-1978 (R94) and ANSI B4.3-19787 (R94) define limits and fits for cylindrical parts, and provide tables listing preferred values. The standard ANSI B4.2-1978 (R94) is essentially in accord with ISO R286. Preferred interference fits for locational interference, medium-drive and force fits are H7/p6, s6 or u6 if hole basis or P7, S7 or U7/h6 if shaft basis. The NF E 22 620 and NF E 22 621 standards is dedicated to shrink fit assembly. According to this standard many limits are set:

- Surfaces are very polished ($R_a = 0.2$–0.8 µm for diameters $d < 180$ mm , $R_a = 0.8$–1.6 µm for $d > 180$ mm).
- Limit of interference: $\delta/d < 1.6 \times 10^{-3}$.
- Limit of nominal diameters $40 < d < 650$ mm.
- Smoothing $= 3(R_{aA} + R_{aM})$.
- The friction coefficient $f = 0.15$.

2.3 Empirical Formulae

Many handbooks provide a graphical interpretation of these relationships for shrink-fits to save time, although some interpolation is required and accuracy lost. The charts developed show the relationship between basic geometry ratios for the hub and shaft, interference value and the radial pressure at the interface. The following semi-empirical relationship to determine holding torque has also been proposed [7], and includes the effects of surface roughness at the interface of the shaft and hub:

$$T_{H+SR} = \frac{\pi}{4} f L_H \, E(\delta - 0.8\left[R_{ZS} + R_{ZH}]d_s[1 - \left(\frac{d_s}{D_H}\right)^2\right], \quad (3)$$

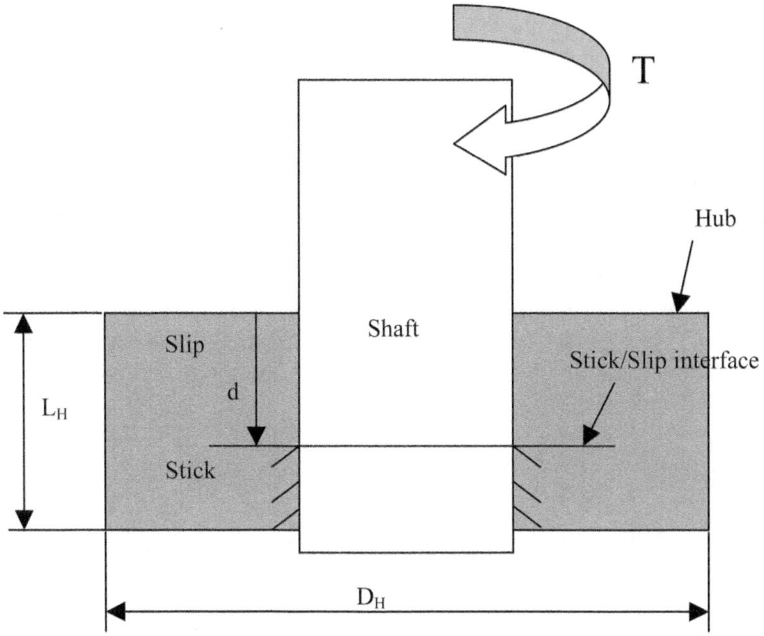

Fig. 1. Schematic of shrink-fit assembly.

where R_{ZH} and R_{ZS} are the surface peak-to-valley average roughness in an axial direction of the shaft and the hub respectively [17].

2.4 Other Approaches

Micro-mechanical approach consists to formulate the problem with regards to the local stresses along the shaft/hub interface. Truman et al. [8] affirm that residual stresses which develop along the interface under low cyclic loading can enhance the performance of the assembly. It is postulated that a zone of slip develops at the surface of the shaft/hub interface and advances monotonically with increasing torque (Figure 1). Also, under cyclic loading conditions, frictional shakedown could occur in some materials. This means the assembly can support more torque before slip takes place at the surface than originally thought as a result of the generation of beneficial residual stresses. As described above, Lamé's theory predicts failure of a shrink-fit when the applied torque exceeds the holding torque of the assembly. The holding torque is determined from knowledge of the interference pressure and the geometry of the shrink-fit assembly. Implicit in this treatment is the fact that the complete length of the embedded shaft either does or does not slip. For micro-mechanical theory the position of the stick/slip interface is determined by applying the Coulomb friction law on a point wise basis, slipping occurring if the shear traction, $\sigma_{r\theta}$, exceeds the normal pressure, given by fP, where f is the coefficient of friction and P is the radial pressure. The stresses then redistribute in order to satisfy the frictional inequalities along the interface. The stick/slip boundary may thus be found as a function of applied torque.

In probabilistic design, statistical and computational methods are combined to investigate the interaction of the independents and dependents parameters, having characterized distributions, to determine the probability of failure or distribution of loading stress through the objective functions or relationships that model the problem. The statistical variability of each parameter should be estimated. The results of the probabilistic approach can be found by using the analysis of variance. Booker et al. Use this approach and found giving the closest result to experimentation [5].

3 Damping

The effect of damping for the vibratory process is completed by the transformation (dissipation) of the mechanical energy of the vibratory movement into another form of energy, generally the heat, which can be evacuated by the system. If the vibratory process presents self-excited vibrations (ex: chattering), the vibrations can be prevented by an adequate damping capacity of the system. Dependent on vibration parameters (amplitudes, frequencies and especially directivity), the vibrations can increase or reduce the friction.

The principal sources of damping in the mechanical systems are the dissipation of energy in structures material, in the components interface (both stationary and moving joints) and in special components of damping. The second one, structural damping is the result of the mechanical dissipation of energy generated by friction which results from the micro or macro-movement between the components. Structural damping is difficult to model and the common method of its determination is the experimentation. Generally, internal damping is negligible in front of structural damping [1].

Designing an optimum structural joint giving rise to desired damping, which is involved in our case of study, is possible if all surface topography factors are understood. Damping in structural joints is associated with frictional losses caused by slip or small relative interfacial movement of component parts on one another. When unlubricated joint surface are involved, this damping mechanism is called frictional damping. Damping in unlubricated joints might be due to three of interactions: macro- and micro-slip and cyclic deformation of the asperities. For our research which is a case of machine tool joints, cyclic plastic deformation of asperities would appear to be the most significant when the magnitude of the interface pressures involved is considered.

4 Repeatability

There are few researches on repeated assemblies, the results indicate that for turned surfaces there is generally a fall of load capacity if the number of assembly increases. For rectified surfaces, the load capacity remains constant and in some cases there it increases, this can be explained by smoothing of the irregularities and with the increase in the bearing pressure. The studies in this field are no relevant and no complete study was devoted to this topic in spite of its importance. Radhakrishnan predict better behaviour of repeated assembly for finer surface [16]. This agrees with his results which predict that finer surface can transmit bigger loads because the real area of contact is bigger, but other authors affirm that if roughness exceeds a certain value, the load capacity can increase due to the interlocking of asperities.

5 Statistical Design of Experiments

The (statistical) design of experiments (DOE) is an efficient procedure for planning experiments so that the data obtained can be analyzed to yield valid and objective conclusions. In an experiment, we deliberately change one or more process variables (or factors) in order to observe the effect on one or more response variables. DOE begins with determining the objectives of an experiment and selecting the process factors for the study. An Experimental Design is the laying out of a detailed experimental plan in advance of doing the experiment. Well chosen experimental designs maximize the amount of "information" that can be obtained for a given amount of experimental effort.

The three main characteristics of a good tool/holder interface in HSM are the holding torque capacity, the damping ratio and the repeatability of assembly. According to bibliographic research on shrink fits, it us proven that there is possibility of having a satisfying maximum torque for high speed machining with a better damping ratio and repeated assembly if one optimize contact surface roughness and other shrink fit interferences value. We had fixed the aim to optimize the link between tool and holder in order to meet the HSM requirements. Our approach consists in optimising by the means of experimental designs the parameters of the shrink fit which are contact surface roughness and interference.

The holding torque capacity depends mainly on the friction coefficient and the interface pressure between mating surfaces. These two parameters are influence by interference and roughness. The first step to understand this relation is to know how they influence friction coefficient and then the other assembly characteristics. The choice of an experimental design depends on the objectives of the experiment and the number of factors to be investigated. For the Response Surface Method (RSM), the experiment is designed to allow us to estimate interaction and even quadratic effects, and therefore give us an idea of the local shape of the response surface we are investigating. We will use RSM designs in order to find improved or optimal

Fig. 2. Friction testing apparatus.

process settings. For the friction tests we used the L_{27} Taguchi's table with roughness interference us independent variable and only friction as dependent variable. For torque, damping ratio and repeatability we used central composite design (CCD).

6 Experimental Approach

Our study comprises a theoretical modelling which was validated by an experimental application. The characterization of surface is completed using a profilo-roughometer Mitutoyo SV4000, dimensions are measured with precision of 1 μm with the CMM, the torques are given with a machine especially designed in our laboratory and damping is measured using the method of impulse by using the Cutpro software.

6.1 Friction Experiments

The first step of our experiments is the coefficient of friction measurement, generally, the main factors that affect friction in a shrink-fit are the smoothness of the contacting surfaces, lubrication, temperature and interface pressure, the tendency being a decrease in friction with higher pressures generally. In our application, there is no lubrication and the temperature is not varying so much. There has been a limited amount of work in determining the coefficient of friction for various interface pressures and surface conditions that may exist in shrink-fit assemblies.

The friction testing apparatus is shown in Figure 2. It comprises a block of steel with a clearance hole at the centre where a shaft of the same diameter as those for the shrink-fit assembly can be rotated against two profiled pins. The pins and shaft materials are respectively H13 and O1, this choice is done to meet the most common industrial cases. The pins, presented in Figure 3, are positioned at right angles to the shaft

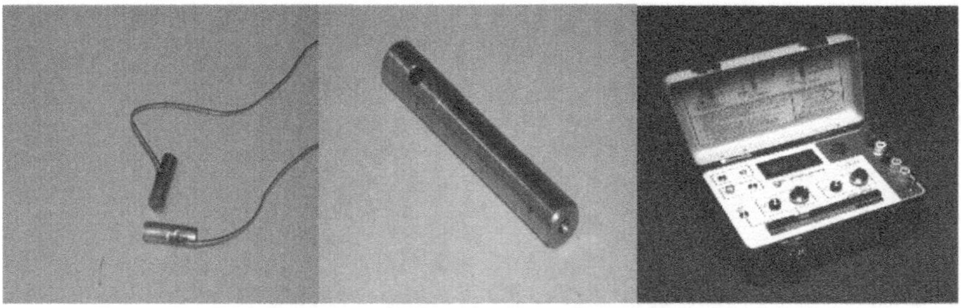

Fig. 3. Pins with strain gages, shaft and P3500 instruments used to measure the strain.

Fig. 4. Special piece used to machine pin on lathe.

axis and made out of the same material as the hub. They have a square cross-section and a concave surface matching as closely as possible the diameter of the shaft, the special piece used to machine pin on lathe is showed in Figure 4, various speeds and feeds are used to obtain desirable roughness. All roughness parameters such us R_a, R_z, R_t, ..., are measured using the profilo-roughometer Mitutoyo SV4000 presented in Figure 5. The Mitutoyo is used too to measure the diameters and roundness of pin end. Bolts accurately calibrated with strain gauges on each pin give the required pre-load and the ability to control the interface pressure with the shaft. Strain gages are mounted in the longitudinal direction and there signals are amplified with an P3500 amplifier (Figure 3) in order to obtain the strain in μm/m. The measurement apparatus provides a very good approximation to subsequent shrink-fit assembly test conditions, and the interface pressure, contact area and profile and the manner of manufacture (including surface roughness) were closely simulated in the apparatus. The torque capacity is finally measured using a simple machine specially designed and manufactured in our labs. The torque is increased gradually until reaching the

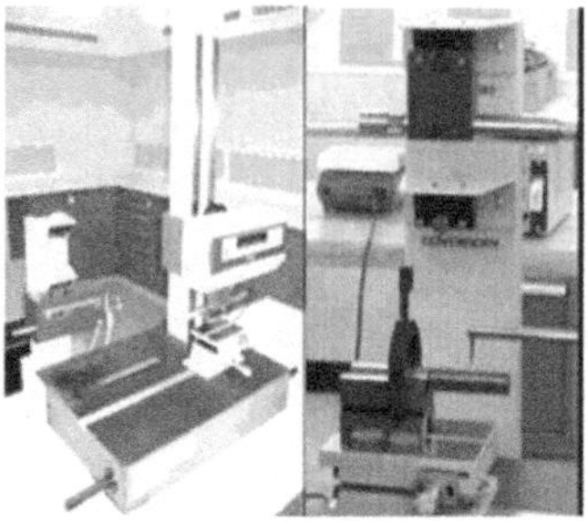

Fig. 5. Profilo-roughometer Mitutoyo SV4000.

assembly capacity. The measurement of the loosening torque is repeated five times in order to obtain the behaviour to the assembly repeatability.

6.2 Torque, Damping and Repeatability Experiments

The shaft and the hub characteristics, such us the materials, the length of contact and the diameter, are designed to be closer to the real case of tool and holder. Surfaces are machined with different speed and feeds in order to obtain the desired roughness. The roughness is measured in the longitudinal direction, it was considered the worst-case situation as surface roughness is generally higher than in any other direction because of the radial turning and reaming process used to generate the shaft and hub bore respectively. Dimensional measurements are achieved with precision as small as 1 μm with the Mitutoyo Coordinate measuring Machine (CMM). Before the specimens were assembled, the shaft and hub were cleaned thoroughly with acetone to eliminate contaminants. Once all precondition measurements are done, the assembly of the shaft and the hub is achieved by the means of thermal expansion of the hub using the Shrink-Master machine shown in Figure 6.

Once assembled the damping ratio is determined using an impulse hummer kit (Figure 7), and the Cutpro software to analyze the outputs signal.

The holding torque was measured up to the point where total slip along the length of the shaft/hub interface occurred using the same system as friction experiment shown in Figure 2. Then, the torque was released, the assembly was dismantled, the damping ration roughness was measured again, and after this the torque was applied again to measure the holding torque for slip on the next loading cycle. This process was repeated 5 times for each shrink-fit assembly, thereby subjecting each to both monotonie and cyclic torques.

Fig. 6. Shrink-master machine.

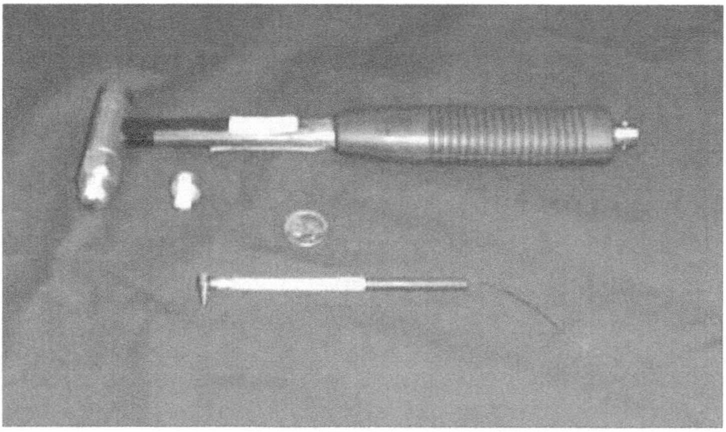

Fig. 7. Impulse hummer kit.

7 Results and Discussion

Friction tests are carried out taking care to their randomization, the histogram and the normal probability plot of friction results follows a normal distribution (Figure 8). Scatter plots show the responses independence of time order and factor levels.

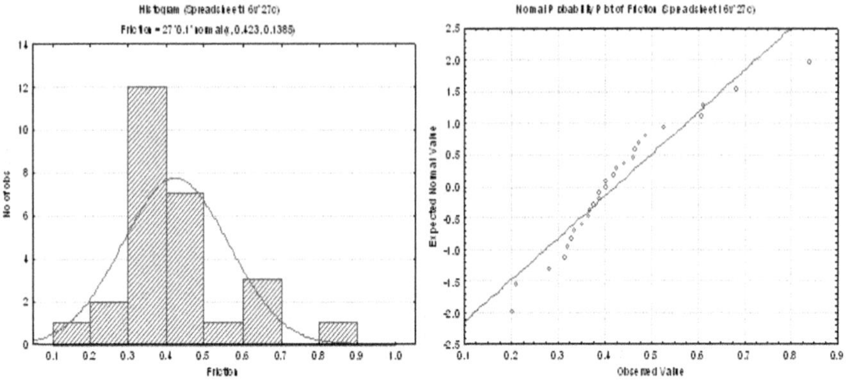

Fig. 8. Friction histogram and normal probability plot.

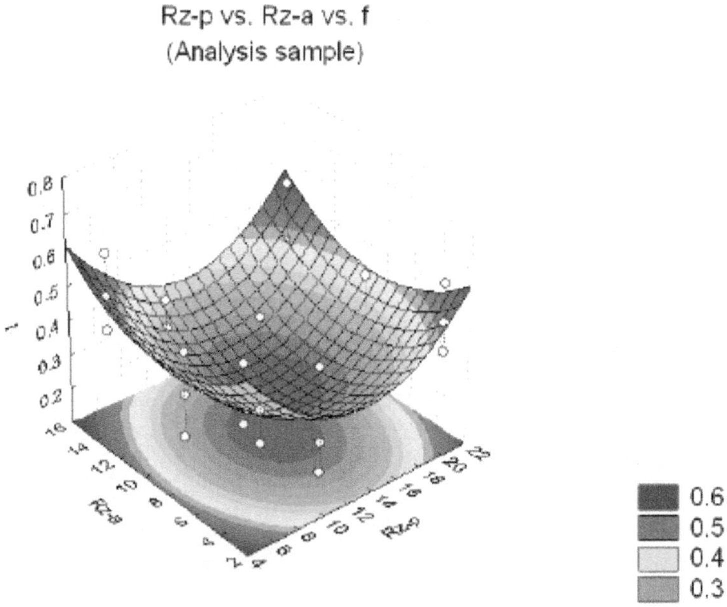

Fig. 9. Friction fitted surface plot.

A quadratic polynomial is used to model the relation between friction and independent parameters. The fitted surface of response between friction (*z* axis), pin roughness (*x* axis) and shaft roughness (*y* axis) is presented in Figure 9. The friction coefficient is high for fine roughness, it decrease with roughness increasing until cer-

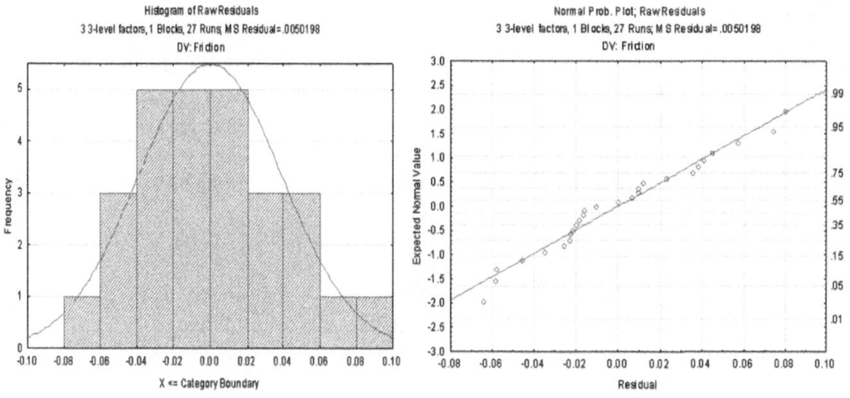

Fig. 10. Residual histogram and normal probability plot.

tain value (around 10 μm), and then it get high again. The variability decomposition shows that main effects are pin and shaft roughness, their interaction and interference.

Basis hypothesis of normal residual distribution is satisfied for our model. Figure 10 shows the histogram of raw residuals and the normal probability plot.

8 Conclusion

A better understanding of the effects of roughness and interference on friction and shrink fits characteristics is achieved. A statistical design of experiments is used. Results show the existence of correlation between friction and roughness and interference. The interaction between roughness of mating surfaces has a main effect on friction coefficient. The application of the results for tool and holder interface confirm for us that one can optimize the surface roughness and interference in order to obtain desired torque without loosing any damping or repeatability capacity. Results can be usefull for similar shrink fit applications such us machines joints.

The Lamé formula for interference fits, largely used, does not consider the roughness of contact surface. We show that roughness has main effect on friction and therefore torque capacity. The relation between maximum torque and surface roughness is parabolic and show a minimum for average asperities sizes. Indeed, for smooth surfaces the bearing area is larger and so the contact pressure. This pressure decreases when asperities sizes increase until a certain minimum where a second phenomenon starts to dominate which is the overlapping of the asperities. This phenomenon makes possible to obtain higher maximum torque with lower contact pressures.

References

1. A.S.R. Murty, K.K. Padmanabhan. Effect of surface topography on damping in machine joints, *Precision Engineering* **4**, 1982, 185–190.
2. J.F. Fontaine, I.E. Siala. Form defect influence on the shrinkage fit characteristics, *European Journal of Mechanics, A/Solids* **1**, 1998, 107–119.
3. G.M. Yang et al. Contact between two rough surfaces of a cylindrical fit, *Journal of Materials Processing Technology* **123**, 2002, 490–497.
4. J.F. Fontaine, I.E. Siala. Optimisation of the contact surface shape of a shrinkage fit, *Journal of Materials Processing Technology* **74**, 1998, 96–103.
5. J.D. Booker, C.E. Truman, S. Wittig, Z. Mohammed. A comparison of shrink-fit holding torque using probabilistic, micromechanical and experimental approaches, *Proceedings of the Institution of Mechanical Engineers, Wilson Applied Science & Technology Abstracts* **2**, 2004, 175–187.
6. A. Sackfield, J.R. Barber, C.E. Truman. A shrink fit subject to torsion, *European Journal of Mechanics, A/Solids* **21**, 2002, 73–84.
7. C.E. Truman, A. Sackfield, D.A. Hills. Torsional loading of a finite shrink-fit, *Proceedings of the Institution of Mechanical Engineers, Wilson Applied Science & Technology Abstracts* **216**, 2002, 1109–1115.
8. C.E. Truman, J.D. Booker. Failure analysis of shrink-fitted shafts, *Materials Science Forum, Transaction Tech. Publications* **440–441**, 2004, 447–454.
9. R.H. Thornley, I. Elewa. The static and dynamic stiffness of interference shrink-fitted joints, *International Journal of Machine Tool Manufacturing* **28**(2), 1988, 141–155.
10. B. Ramamoorthy, V. Radhakrishnan. A study of the surface deformations in press and shrink-fitted assemblies, *Wear* **173**, 1994, 75–83.
11. P. Venta, K. Rao, B. Ramamoorthy, V. Radhakrishnan. Role of interacting surfaces in the performance enhancement of interference fits, *International Journal of Machine Tool Manufacturing* **35**(10), 1995, 1375–1384.
12. P. Venta, K. Rao, B. Ramamoorthy, V. Radhakrishnan. Effect of plating and temperature on the strength of shrink fitted assemblies, *International Journal of Machine Tool Manufacturing* **33**(3), 1993, 475–481.
13. B. Ramamoorthy, V. Radhakrishnan. Performance improvement of shrink-fitted assemblies by surface strengthening, *Proceedings Institution of Mechanical Engineers* **206**, 1992, 207–213.
14. A.F. El-Hadidy. Influences of geometry, fitting and surface roughness on load carrying capacity of shrink fitted joints, *Metallwissenschaft und Technik* **51**(7), 1997, 251–257.
15. K. Kato, T. Kayato, I. Nitta, S. Shimoda. The effect of surface roughness on fitting strength of shrink fit between ceramic and metal element, *Transactions of the ASME* **111**(7), 1989, 318–325.
16. R.V. Ramachandran, V. Radhakrishnan. Influence of surface finish on interference fits, *International Journal of Production Research* **12**(6), 1974, 705–719.
17. W. Beitz, K.-H. Kuttner. *Dubbel Handbook of Mechanical Engineering*, Springer-Verlag, London, 1993.
18. A. Leluan. Assemblages fretté, Techniques de l'ingénieur, B5500.
19. J.S. Becker. *Shrink and Press-Fitted Assemblies*, Mechanical Design Handbook, McGraw-Hill, ISBN 0070540381, 1996.

Kinematic Performances in 5-Axis Machining

Sylvain Lavernhe[1], Christophe Tournier[1] and Claire Lartigue[2]

[1]*Laboratoire Universitaire de Recherche en Production Automatisée, ENS de Cachan – Université Paris Sud 11, 61 avenue du Président Wilson, 94235 Cachan Cedex, France; E-mail: {sylvain.lavernhe, christophe.tournier}@lurpa.ens-cachan.fr*
[2]*IUT de Cachan, Université Paris Sud 11, 9 avenue de la Division Leclerc, 94234 Cachan Cedex, France*

Abstract. This article presents a predictive model of the kinematic behaviour during 5-axis machining. This model highlights differences between the programmed tool-path and the actual follow-up of the trajectory. Within the High Speed Machining context, kinematic limits of the couple CNC-machine-tool have to be taken into account in the model. The originality of the model is the use of the inverse-time method to coordinate machine-tool axes, whatever their nature (translation or rotation). The model reconstructs the actual relative velocity tool-surface from each axis velocity profile highlighting trajectory portions for which cutting conditions are not respected.

Key words: 5-axis machining, high speed machining, predictive model, kinematic behaviour.

1 Introduction

Today, machining of aeronautic free form surfaces (turbine blades, etc.) is carried out using multi-axis milling centre. The use of rotational axes authorizes the choice of the tool orientation relatively to the surface. Hence, these supplementary degrees of freedom allow reduction costs, improving the surface quality and decreasing machining time. In the context of High Speed Machining (HSM), axis and structure solicitations are especially high. The machining process is modified and may alter the surface quality: velocity drops cause marks, vibrations and high solicitations of each axis generate bad cutting conditions.

As the machined surface results from each stage of the process, numerous parameters have to be taken into account to control the final result (Figure 1). The Computer Aided Machining (CAM) stage computes the tool-path, which constitutes and approximation of the CAD model according to the interpolation format (linear or polynomial), the driving tool direction and the CAM parameters (machining tolerance, scallop height). The choice of the parameter values directly influences the

S. Tichkiewitch et al. (eds.), Advances in Integrated Design and Manufacturing in Mechanical Engineering II, 489–503.

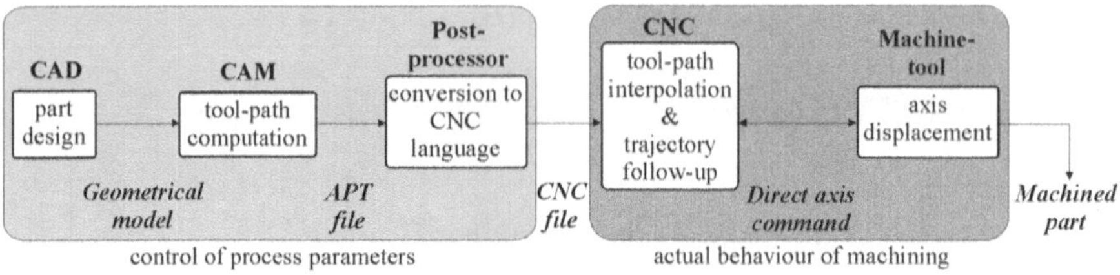

Fig. 1. Structure of the process.

surface quality. Following, the post-processing stage converts the calculated tool-path into an adapted file for the numerical controller (NC), called the CNC file. The CNC file contains the set of tool positions and axis orientations and the corresponding feed rates. In 5-axis machining, the post-processor may also solve the Inverse kinematic Transformation (IKT) to express the tool-path into direct axis commands. The tool-path interpolation and the trajectory follow-up are performed by the CNC. As this stage, the follow-up strongly depends on the CNC parameters such as time cycles, velocity limitations and special functions (look-ahead). The kinematic behaviour during machining is also affected by the machine-tool architecture and axis capacities. Therefore, from the CAM stage to the actual machining, numerous parameters influence kinematic performances in 5-axis machining which may alter the surface quality and machining time. In particular, differences exist between the calculated trajectory and associated feed rates defining the CNC file and the actual tool movement relative to the surface.

This paper deals with a predictive model of the machining behaviour with the objective of evaluating the actual relative velocity tool-surface during machining from the CNC file. For a couple machine-tool – CNC, the model predicts the kinematic behaviour through axis velocities. The final purpose is to find an optimal strategy which allows respecting programmed feed rate as well as possible. The methodology proposed consists in three main steps (Figure 2). First, the programmed tool-path $(X_{pr}, Y_{pr}, Z_{pr}, i, j, k)$ is transformed into a trajectory in the articular space (X_m, Y_m, Z_m, A, C) by solving the IKT. Once axes are coordinated, the post-processor generates the velocity profiles for each axis: velocity limitations along the trajectory resulting from axis and CNC capacities are computed; a model of the CNC treatments and the special functions is established. The originality of this stage is the use of the inverse-time method, allowing a similar treatment for translation axes and rotation axes. Finally, the relative velocity tool-surface is reconstructed considering the machine-tool architecture allowing the prediction of bad cutting condition zones. In a first approach the linear interpolation only is modelled. The application is carried out on a 5-axis milling centre Mikron UCP 710 with an industrial numerical controller Siemens 840D.

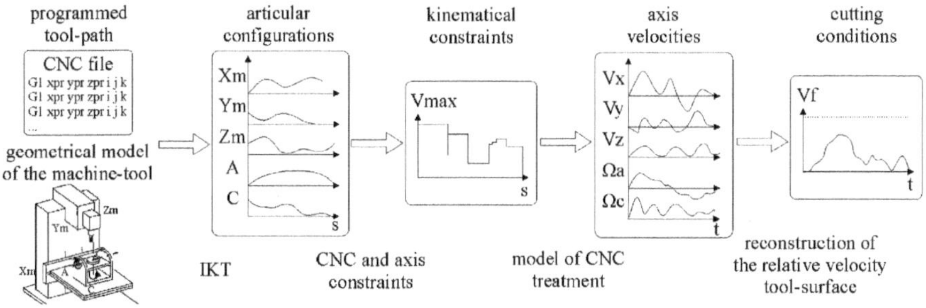

Fig. 2. Structure of the predictive model.

The next section deals with the IKT for a $CAXYZ$ serial architecture. It underlines the difficulty in the choice of solution to match the CNC behaviour. Generation of axis profiles is detailed in Section 3. The inverse-time method used to express velocity constraints is exposed. Limitations from CNC and axis capacities are formulated to construct kinematic profiles. An illustration is exposed in Section 4. In Section 5, the relative velocity tool-surface is recalculated from each axis velocity.

2 Trajectory into the Articular Space

The tool trajectory results from the axis displacements of the machine-tool. For each Cutter location (Cl point) computed by the CAM software, the tool position and its axis orientation relative to the surface are achieved by combining axis articular positions. From the programmed tool-path, the IKT calculates corresponding articular configurations. Two calculation modes are possible, the IKT can be performed in real time by the NC unit during machining or it can be computed off-line by a dedicated post-processor [1]. As our objective is to predict articular configurations computed in real time by the NC unit, a post-processor is created to simulate the IKT by the CNC. Therefore, the role of this post-processor is to compute the articular configurations (X_m, Y_m, Z_m, A, C) for each Cl point $(X_{pr}, Y_{pr}, Z_{pr}, i, j, k)$ programmed in the part frame (Figure 3).

Given the machine-tool architecture (Figure 13), the IKT leads to equations (1) and (2). Details on the geometrical modelling of the machine-tool and the IKT formulation are exposed in Appendix.

$$\begin{bmatrix} 0 \\ 0 \\ 1 \\ 0 \end{bmatrix}_{(X_m, Y_m, Z_m)} = P_{mb} \times P_{bp} \times P_{ppr} \begin{bmatrix} i \\ j \\ k \\ 0 \end{bmatrix}_{(X_{pr}, Y_{pr}, Z_{pr})} , \qquad (1)$$

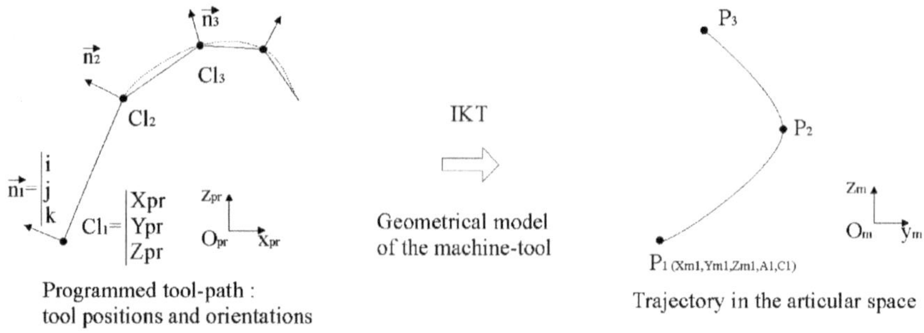

Fig. 3. Inverse kinematic transformation.

Table 1. Domains of solutions (A_1, C_1) and (A_2, C_2).

	$i<0$	$i=0$	$i>0$
$j<0$	$A_1 = \text{acos}(k)$ $C_1 = -\text{atan}(i/j)$ $A_2 = -\text{acos}(k)$ $C_2 = -\text{atan}(i/j)+\pi$		
$j=0$	$A_1 = \text{acos}(k)$ $C_1 = -\pi/2$ $A_2 = -\text{acos}(k)$ $C_2 = \pi/2$	$A=0$ C undefined	$A_1 = \text{acos}(k)$ $C_1 = \pi/2$ $A_2 = -\text{acos}(k)$ $C_2 = -\pi/2$
$j>0$	$A_1 = \text{acos}(k)$ $C_1 = -\text{atan}(i/j)+\pi$ $A_2 = -\text{acos}(k)$ $C_2 = -\text{atan}(i/j)$		

$$\begin{bmatrix} X_m \\ Y_m \\ Z_m \\ 1 \end{bmatrix}_{(O_m, X_m, Y_m, Z_m)} = P_{mb} \times P_{bp} \times P_{ppr} \begin{bmatrix} X_{pr} \\ Y_{pr} \\ Z_{pr} \\ 1 \end{bmatrix}_{(O_{pr}, X_{pr}, Y_{pr}, Z_{pr})} . \quad (2)$$

Equation (2) directly gives the axis commands (X_m, Y_m, Z_m). If we consider that the axes of the programming basis and the table basis are parallel, equation (1) leads to the following system of equations:

$$\begin{cases} i = \sin(C) \times \sin(A), \\ j = -\cos(C) \times \sin(A), \\ k = \cos(A). \end{cases} \quad (3)$$

System (3) has two domains of solutions ($A_1 > 0$ or $A_2 < 0$). In function of the (i, j, k) values, solutions vary (Table 1).

Due to physical limitations, the range of angle A is limited to $[-30°; +120°]$. So, not all solutions are practically possible. Different possible cases are summarized in Table 2.

Table 2. Sets of solutions for the IKT.

values of k	[-1 ; -0.5 [	[-0.5 ; 0.866 [	[0.866 ; 1 [	1
number of	0	1	2	∞
solutions	no solution	(A_2,C_2)	(A_1,C_1) or (A_2,C_2)	A=0 and C=unspecified

When several solutions exist, the choice of one of them influences the axis behaviour. Indeed, switching from domain 1 ($A_1 > 0$) to domain 2 ($A_2 < 0$) makes the table C rotates about 180°. This rapid movement can cause rear gouging, especially when using toroïdal tool [2, 3]. Moreover, during this movement, the displacement of the cutter contact on the surface is small; the relative velocity tool-surface is almost zero. As cutting conditions are not satisfied, velocity drops will cause marks on the surface. The discussion on the best choice that can be done according to the kinematic behaviour is the subject of current work and will not be discussed here. The objective of the post-processor is to approach the actual behaviour of IKT performed by the CNC. During tests on our system, incoherent movements can be noticed. The choice of solution done in real time may be arbitrary. Indeed, for two similar trajectories, the CNC can choose two different solutions; thus an inversion on the A axis appears whereas it is not necessary. Moreover, for certain cases, chosen solution leads to axis displacements out of range. So, predict what will be the solution chosen by the CNC is difficult. After having tested several cases on the CNC, the post-processor tries to choose the solution that matches the one done by the CNC.

At this stage, axis configurations corresponding to Cl points are computed.

3 Prediction of Axis Velocities

Prediction of the relative velocity tool-surface requires the estimation of each axis velocity. Axis kinematic behaviour during the follow-up of the trajectory depends on the tool-path geometry and the performance of the components (CNC, machine-tool axis, etc.). Several limitations linked to the HSM context have to be taken into account to generate kinematic profiles.

3.1 Programming Method

Usually, in 3-axis machining, axis displacements of the machine-tool correspond to the displacements of the tool relatively to the surface. For multi-axis machining, due to the rotational axes, axis displacements are totally different. During the follow-up, the main difficulty is to coordinate translational and rotational axes. This difficulty is emphasised by the differences between the kinematic capacities of rotational axes and translational axes. The proposed method consists in programming velocities by

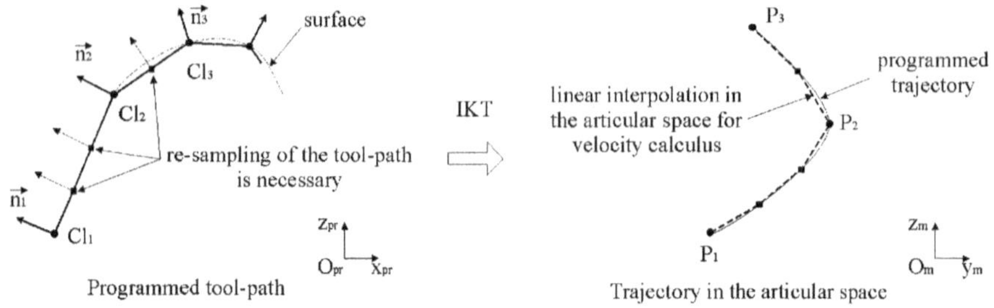

Fig. 4. Articular interpolation for velocity calculus.

the inverse-time. It defines axis velocities by specifying the time to go from one point to the following one. As a result it is easier to coordinate axes with different natures of movements (translation, rotation). As many axes as necessary can be interpolated. For instance, when programming the tool-path in the part frame, the feed rate V_f is defined as constant between two Cl points (Cl_1, Cl_2). If the length of the segment is L, the tool is supposed to move during Δt from Cl_1 to Cl_2 (equation (4)).

$$\Delta t = \frac{L}{V_f} \quad \text{with} \quad L = \|\overline{\text{Cl}_1\text{Cl}_2}\|. \tag{4}$$

To each couple Cl_i-n_i programmed in the part frame corresponds one axis configuration P_i ($P_{i\text{ axis }1}$, $P_{i\text{ axis }2}$, $P_{i\text{ axis }3}$, ...). In the articular space, the time allocated to each axis to go from one position $P_{1\text{ axis }j}$ to the following one $P_{2\text{ axis }j}$ should be Δt. Thus, for the segment (Cl_1, Cl_2) the axis speeds to apply to the machine-tool are:

$$V_{12\text{ axis }j} = \Delta P_{12\text{ axis }j} \times V_{\text{time inverse}}$$

$$\text{with} \quad \Delta P_{12\text{ axis }j} = P_{2\text{ axis }j} - P_{1\text{ axis }j} \quad \text{and} \quad V_{\text{time inverse}} = \frac{1}{\Delta t}. \tag{5}$$

In this way, the calculus of axis velocities is adapted for linear interpolation in the articular space. However, the programmed tool-path is theoretically interpolated in the part frame. The trajectory could be curved in the articular space between two axis configurations. Hence, differences can appear, especially if distances between successive configurations are high (Figure 4).

If a best approximation of the actual velocity is required for the velocity calculus, it can be increased by re-sampling the tool-path [4]. New articular configurations are computed by the IKT and the estimation of the curve length in the articular space is improved.

To summarise, this method enables to coordinate and define axis velocities from the programmed feed rate expressed by the inverse-time.

3.2 Expression of Kinematic Constraints

Within the context of high speed machining, CNC parameters and components capacities are also critical for the trajectory follow-up. As programmed velocities are very high, axis dynamics and CNC treatment limit kinematic performances. During tests on our system, several constraints are noticed. Three of them seem to be most critical; they are identified and implemented in the model:

- the maximal kinematic capacities;
- the velocity limitation at block transition;
- the velocity limitation due to the cycle time of CNC.

For a serial multi-axis machine, kinematic capacities of each axis are different. They are limited by CNC parameters according their inertia and position into the chain. For each axis, maximum available velocity, acceleration and jerk are respectively $V_{\text{max axis } j}$, $A_{\text{max axis } j}$ and $J_{\text{max axis } j}$. Moreover, during trajectory interpolation, these capacities have to be coordinated to follow-up the trajectory. Hence, given the current position along the trajectory, the feed rate is limited by the less powerful axis. These constraints are integrated in the model by computing the maximum available velocity, acceleration and jerk ($V_{\text{max inverse-time}}$, $A_{\text{max inverse-time}}$, $J_{\text{max inverse-time}}$) for each articular segment (equations (6) to (8)).

$$V_{\text{max inverse-time}} = \min\left(\frac{V_{\text{max axis } j}}{\Delta P_{\text{axis } j}}\right) \quad \text{for} \quad \Delta P_{\text{axis } j} \neq 0, \tag{6}$$

$$A_{\text{max inverse-time}} = \min\left(\frac{A_{\text{max axis } j}}{\Delta P_{\text{axis } j}}\right) \quad \text{for} \quad \Delta P_{\text{axis } j} \neq 0, \tag{7}$$

$$J_{\text{max inverse-time}} = \min\left(\frac{J_{\text{max axis } j}}{\Delta P_{\text{axis } j}}\right) \quad \text{for} \quad \Delta P_{\text{axis } j} \neq 0. \tag{8}$$

The follow-up also depends on the geometry of the trajectory. In 5-axis machining, discontinuities may appear along the trajectory in the articular space. Tangency and curvature discontinuities involve slowdowns during the follow-up. To prevent velocity drops, the CNC smoothes these discontinuities by inserting arcs of circle on block transition (Figure 5).

Therefore, velocity is non null but limited by CNC parameters. The post processor models this rounding, taking into account the rounding tolerance value which defines the radius of rounding (equation (10)) [5, 6]. The maximum velocity to cross this rounding can thus be computed with the maximum centripetal acceleration allowed. On the CNC used, the parameter Curv_Effect_On_Path_Accel distributes the acceleration capacities between the centripetal acceleration and the tangential acceleration along curved contours (equation (9)) [6, 7]. This velocity limitation is applied at block transition considering that rounding length is short enough to consider that the velocity remains constant on the arc of circle.

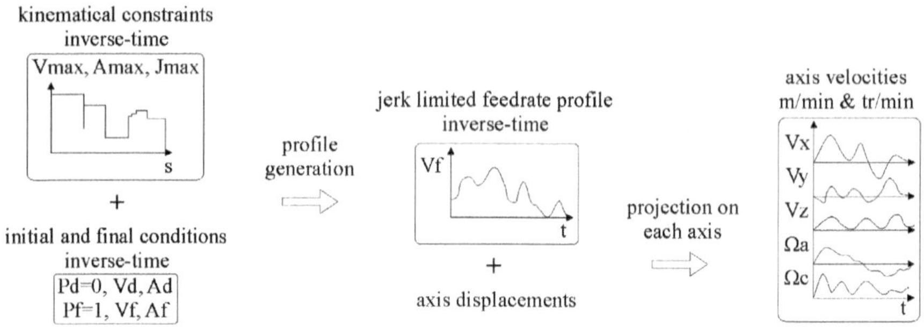

Fig. 5. Rounding of tangency discontinuity.

$$A_{\text{centripetal allowed}} = A_{\max} \times (1 - \text{Curv\_Effect\_On\_The\_Path\_Accel}), \tag{9}$$

$$V_{\max} = \sqrt{R \times A_{\text{centripetal allowed}}} \quad \text{with} \quad R = e \times \left(\frac{\sin\left(\frac{\alpha}{2}\right)}{1 - \sin\left(\frac{\alpha}{2}\right)} \right). \tag{10}$$

The last constraint implemented in the post-processor is the maximum velocity relative to the CNC time cycle. As feed rates are high, cycle times have to be short enough to compute new instructions before the previous ones are reached. Otherwise, the axes stop and wait for the next cycle. Indeed, for short segments, the interpolator cycle time can be longer than the duration of axis displacements. For example, if the feed rate is equal to 20 m/min and the interpolator cycle time is 3ms, for all segments with a length less than 1 mm axes stop at the end of segment by lack of information. To avoid this problem, the maximum axis velocity is limited in order to respect cycle times (equation (11)).

$$V_{\text{max time inverse}} \leq \frac{1}{\text{interpolar cycle time}}. \tag{11}$$

Constraints presented above are expressed by the inverse-time in the post-processor. So, all trajectory long, maximal velocity, acceleration and jerk limit the kinematic performance of each axis. At this stage, the model generates kinematic profiles taking into account these limitations.

3.3 Generation of the Axis Kinematic Profiles

Due to HSM, structures of machine-tools are more and more solicited. High axis velocities and high curvatures require high accelerations. Trajectory discontinuities in the articular space create jerking. Such machine behaviour is harmful for the machine and the part quality. Consequently, jerk limited velocity profile is used to drive axes. This way of piloting reduces jerking and smoother transitions are achieved [8, 9]. Actual velocities generated by the CNC are modelled by jerk limited velocity profiles.

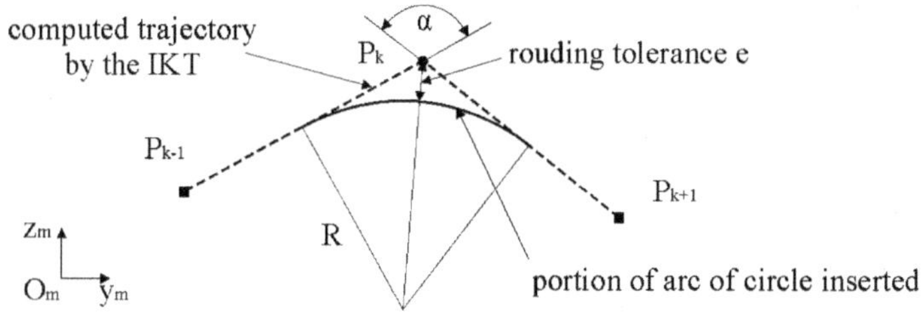

Fig. 6. Generation method of axis velocity profiles.

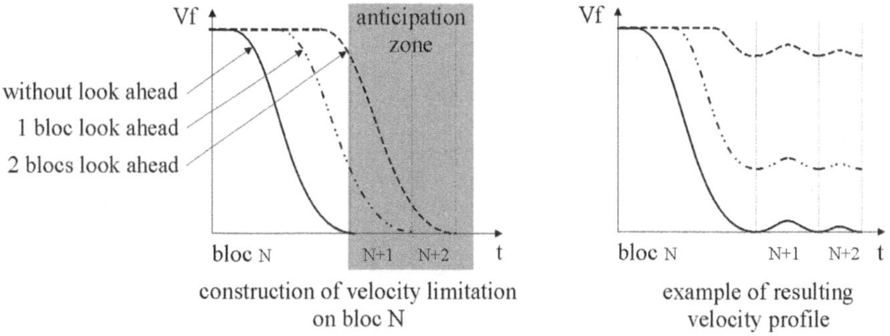

Fig. 7. Influence of look ahead function on velocity profile.

For each segment of the articular trajectory, a inverse-time feed rate profile is computed from kinematic constraints, initial and final conditions. Finally, axis velocities are reconstructed by multiplying the inverse-time profile by the axis displacements (Figure 6).

Henceforth, CNC has special functions to improve the follow-up in HSM. As one of them, the dynamic anticipation, is of major influence for high velocities, it is modelled to fit the interpolator treatment. This function, called "look ahead" is used to prevent overshoots. For example, if the anticipation is realised on 10 blocks; in order to generate the profile on block N, the geometry of the trajectory until block $N + 10$ is taken into account. Hence, the velocity deceleration is delayed and overshoots are avoided. The use of such function enables to reach higher velocities (Figure 7).

4 Application

To illustrate our approach, the machining behaviour on a blending radius of 5 mm is studied (Figure 8). The programmed machining strategy is parallel to plane with toroïdal mill ($D = 10$ mm, $R_c = 1$ mm). Chordal deviation (machining tolerance) is set to 0.01 mm; tool inclination is set to 5° and feed rate to 5 m/min. 5-axis machining

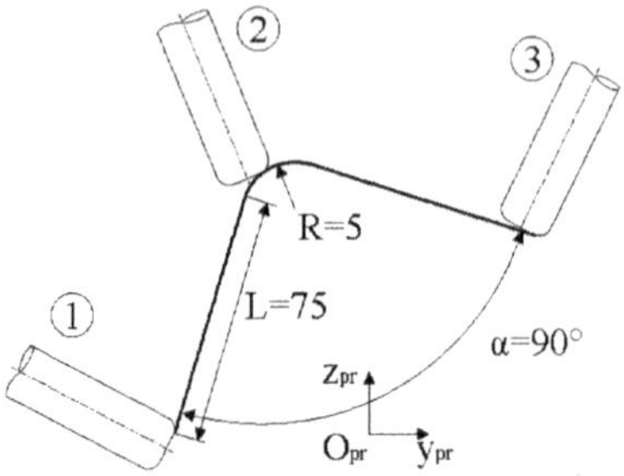

Fig. 8. Machining of the blending radius.

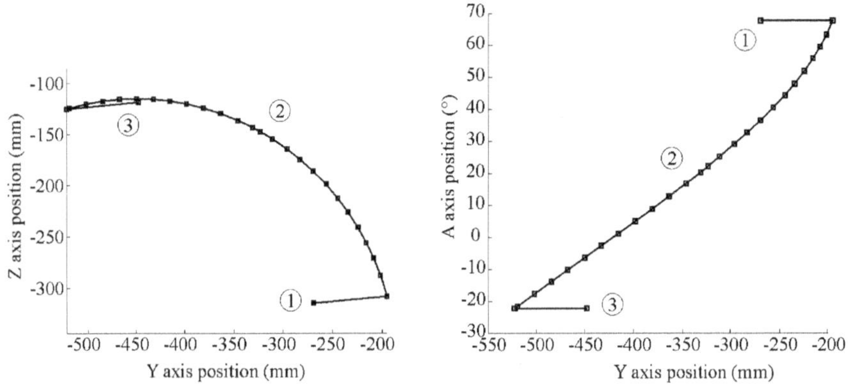

Fig. 9. Articular configurations for Cl points.

is carried out on the Mikron milling centre. The programming frame is oriented on the rotate table such as only the YZA axes are solicited.

Results of predictive model are presented above. First, the IKT is performed. Figure 9 shows the computed axis configurations corresponding to the programmed Cl points.

Given the articular configurations, axis capacities and CNC parameters velocity limitations are predicted along the trajectory (Figure 10). They are expressed in the inverse-time method because profile generation is not yet done.

The final stage consists in computing the jerk velocity profile taking into account previous constraints. Then, for each segment, the inverse-time velocity is multiplied by axis displacements to obtain each axis velocity. Figure 11 compares the predicted velocities to records done in real time through the CNC during machining.

For the first and the last blocks, where tool axis orientation does not change, predicted axis velocities are equal to the measured one. Differences appear when rotational axis moves. The CNC treatment seems to change when commuting from

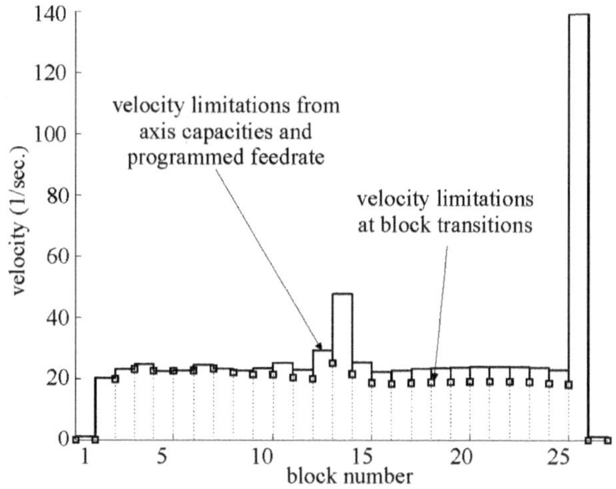

Fig. 10. Inverse-time velocity limitations for each block.

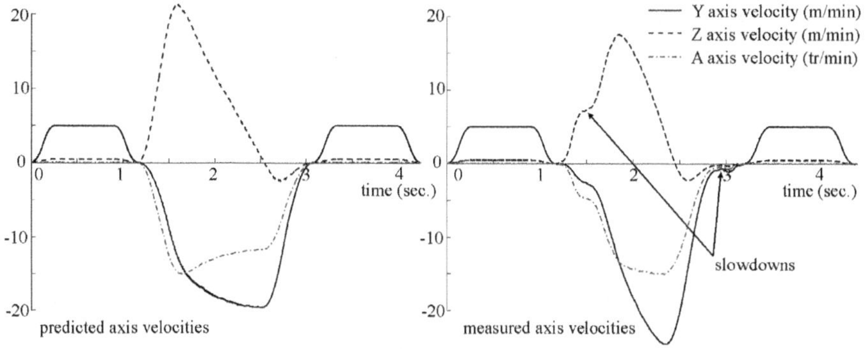

Fig. 11. Comparison between predicted and measured velocities.

two axis interpolation (YZ) to three axis interpolation (YZA). Indeed, on block 2, 3, 25 and 26 ($t \approx 1.5$ and 3 sec), initial and final accelerations are set to zero; slowdowns appear, whereas, between blocks 4 and 24, profile is optimal. This constraint limits axes to reach higher feed rate. So, predicted axis behaviours are locally more dynamic than the actual ones. These treatments modifications by the CNC are difficulties in predicting machining behaviour.

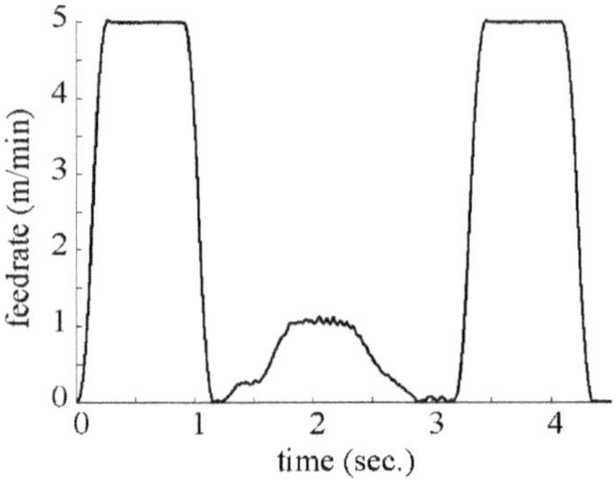

Fig. 12. Actual feed rate along the tool-path.

5 Evaluation of Cutting Conditions

To evaluate the cutting conditions, actual feed rate of the tool relatively to the surface is rebuilt from each axis velocity. Cutting conditions have to be expressed on the Cutter contact point (Cc point). Equation (12) expresses the tool feed rate in the machining direction. It can be approximated by the velocity on Cl point as Cc and Cl points are close enough and rotational axis velocities are low compared to the translational ones (equations (13) and (14)).

$$\mathbf{V}_{\mathrm{Cc,tool/surface}} = \mathbf{V}_{\mathrm{Cc,tool/machine}} - \mathbf{V}_{\mathrm{Cc,surface/machine}}, \tag{12}$$

$$\mathbf{V}_{\mathrm{Cc,tool/machine}} = \mathbf{V}_{\mathrm{Cl,tool/machine}}, \tag{13}$$

$$\begin{aligned}\mathbf{V}_{\mathrm{Cc,surface/machine}} &= \mathbf{V}_{\mathrm{Cl,surface/machine}} + \mathbf{CcCl} \wedge \mathbf{\Omega}_{\mathrm{surface/machine}} \\ &\approx \mathbf{V}_{\mathrm{Cl,surface/machine}}.\end{aligned} \tag{14}$$

Thus, combining Cl positions, axes velocities and machining parameters, cutting conditions are evaluated along the tool-path. Figure 12 shows that programmed velocity is not respected on the bending radius. Indeed, the feed rate reaches 5 m/min along first and last blocks since they are long enough. When the tool axis orientation varies, discontinuities created in the articular space make the feed rate drop to 0 m/min. Along the radius, the maximal reached velocity is about 1 m/min.

Given the $CAXYZ$ architecture, the actual feed rate depends on axes capacities, chordal deviation, rouding tolerance, etc. It depends also on the position and orientation of the part in the machine frame. According to the distance between part and rotational axes, variations in tool axis orientation may induce large displacements. These displacements cause high solicitations of translational axes.

To summarize, the reconstruction on the relative feed rate tool-surface gives a criterion to qualify cutting conditions, and so, the quality of the machined part.

6 Conclusion

Multi-axis machining and high velocities create significant solicitations on machine-tool axes, reducing the machined part quality. This is due for the main part to the kinematic performances of the couple CNC-machine-tool. In the paper we have proposed a predictive model of the kinematic behaviour during 5-axis machining to highlight differences between the programmed tool-path and the actual execution of the tool trajectory. For this purpose, a post-processor is developed to approach the inverse kinematic transformation done in real time by the CNC. Limits of the CNC and the machine-tool are exposed. Therefore, axis capacities and CNC special functions are modelled. The inverse-time method used to coordinate axes and to generate jerk limited kinematic profiles constitutes the originality of this model. It allows interpolating axes, whatever their number and their movement (translation or rotation). The results of the predictive model are compared with tests carried out on an industrial milling centre. The global behaviour is similar showing this efficiency of our model. Nevertheless, small differences can be noticed when machining short length segments and for transition between 2 and 3-axis interpolation. Finally, the model builds the relative velocity tool-surface highlighting trajectory portions for which cutting conditions are not respected. Works in progress will now take advantage of our predictive model to find an optimal machining strategy, computing the tool position and the tool axis orientation, to respect the programmed feed rate as well as possible.

Appendix: Formulation of the IKT

The architecture of the milling centre Mikron UCP 710 is $CAXYZ$. Two rotations are applied on the part, and the tool orientation is fixed in the machine frame (Figure 13).

The different frames are:

- the machine frame (O_m, x_m, y_m, z_m) is linked to the structure of the machine-tool. Its axes are parallel to the XYZ axis; the z_m axis of this frame is parallel to the axis of the tool and is pointing to the tool tip;
- the tilt frame (S, x_b, y_b, z_b) is linked to the tilt table: x_b is parallel to x_m, S is located on the A axis and is given by equation (15):

$$\mathbf{OmS} = m_x \cdot \mathbf{x}_m + m_y \cdot \mathbf{y}_m + m_z \cdot \mathbf{z}_m. \tag{15}$$

- the table frame (R, x_p, y_p, z_p) is linked to the rotary table: z_p is parallel to z_b, R is defined as the intersection between the C axis and the upper face of the table (equation (16));

$$\mathbf{SR} = b_y \cdot \mathbf{y}_b + b_z \cdot \mathbf{z}_b. \tag{16}$$

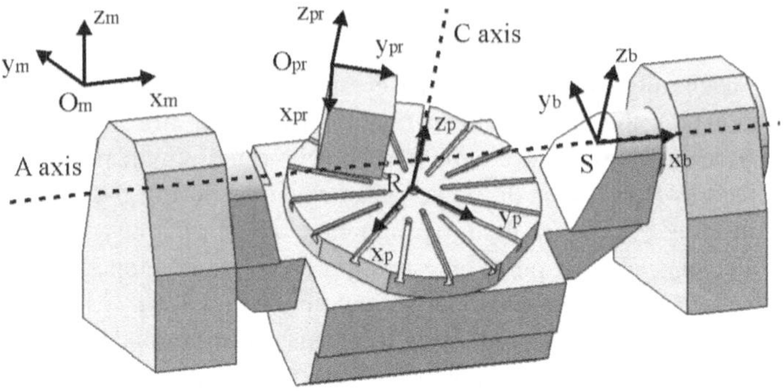

Fig. 13. Definition of different frames.

- the programming frame $(O_{pr}, x_{pr}, y_{pr}, z_{pr})$ is linked to the part: it represents the frame used for CAM computation. Its origin O_{pr} is given by equation (17).

$$\mathbf{RO}_{pr} = p_x \cdot \mathbf{x}_p + p_y \cdot \mathbf{y}_p + p_z \cdot \mathbf{z}_p. \tag{17}$$

The parameters m_x, m_y, m_z, b_y and b_z are fixed values, identified on the machine-tool. The position of the part on the table define the parameters p_x, p_y and p_z. To switch easily between different frames, we generally define P_{ij} the matrix that converts a vector initially expressed in the frame j into the frame i: $\mathbf{V}_{\text{frame } i} = P_{ij} \cdot \mathbf{V}_{\text{frame } j}$ (equation (18)):

$$P_{mb} = \begin{bmatrix} 1 & 0 & 0 & m_x \\ 0 & \cos(A) & \sin(A) & m_y \\ 0 & -\sin(A) & \cos(A) & m_z \\ 0 & 0 & 0 & 1 \end{bmatrix}, \quad P_{bp} = \begin{bmatrix} \cos(C) & \sin(C) & 0 & 0 \\ -\sin(C) & \cos(C) & 0 & b_y \\ 0 & 0 & 1 & b_z \\ 0 & 0 & 0 & 1 \end{bmatrix},$$

$$P_{ppr} = \begin{bmatrix} a & d & g & p_x \\ b & e & h & p_y \\ c & f & i & p_z \\ 0 & 0 & 0 & 1 \end{bmatrix}. \tag{18}$$

A and C are the angular values to command rotational axis. The parameters $a, b, c, d, e, f, g, h, i$ define the orientation of the part on the table.

Equation (19) expresses that the tool orientation is fixed in the machine frame; the tool orientation on the surface is given by the two rotational axes on the part.

$$\begin{bmatrix} 0 \\ 0 \\ 1 \\ 0 \end{bmatrix}_{(x_m, y_m, z_m)} = P_{mb} \times P_{bp} \times P_{ppr} \times \begin{bmatrix} i \\ j \\ k \\ 0 \end{bmatrix}_{(x_{pr}, y_{pr}, z_{pr})}. \tag{19}$$

Then, once the values of angles A and C are computed, the X, Y and Z axis correct the displacements induced on the part and move the tool to reach the CL point. The commands to apply to axis of translation are given by equation (20).

$$\begin{bmatrix} X_m \\ Y_m \\ Z_m \\ 1 \end{bmatrix}_{(O_m,x_m,y_m,z_m)} = P_{mb} \times P_{bp} \times P_{ppr} \times \begin{bmatrix} X_{pr} \\ Y_{pr} \\ Z_{pr} \\ 1 \end{bmatrix}_{(O_{pr},x_{pr},y_{pr},z_{pr})} . \quad (20)$$

Acknowledgments

This work was carried out within the context of the working group Manufacturing 21 which gathers 11 French research laboratories. The topics approached are: the modelling of the manufacturing process, the virtual machining and the ermerging of new manufacturing methods.

References

1. A. Affouard, E. Duc, C. Lartigue, J.-M. Langeron, P. Bourdet. Avoiding 5-axis singularities using tool-path deformation, *International Journal of Machine Tools & Manufacture* **44**, 2004, 415–425.
2. Y.H. Jung, D.W. Lee, J.S. Kim, H.S. Mok. NC post-processor for 5-axis milling machine of table-rotating/tilting type, *Journal of Materials Processing Technology* **130–131**, 2002, 641–646.
3. M. Munlin, S.S. Makhanov, E.L.J. Bohez. Optimization of rotations of a five-axis milling machine near stationary points, *Computer-Aided Design* **36**, 2004, 1117–1128.
4. D. Vouillon, G. Poulachon, L. Gasne. De la modélisation à l'usinage 5 axes d'un impeller, in *Assises MO & UGV*, Clermont Ferrand, 2004, pp. 141–150.
5. A. Dugas. Simulation d'usinage de formes complexes, PhD Thesis, Ecole Centrale Nantes, December, 2002.
6. V. Pateloup. Amélioration du comportement cinĆmatique des machines outils UGV. Application au calcul des trajets d'évidement de poches, PhD Thesis, Blaise Pascal University-Clermont II, July, 2005.
7. SINUMERIK 840D/840Di/810D Basic Machine, Description of Functions (FB1), 10.04 Edition, 2001.
8. K. Erkorkmaz, Y. Altintas. High speed CNC system design. Part I: Jerk limited trajectory generation and quintic spline interpolation, *International Journal of Machine Tools & Manufacture* **41**, 2001, 1323–1345.
9. P.-J. Barre. Commande et entrainement des machines-outils à dynamique elevée – Formalisme et applications, Habilitation à Diriger des Recherches, Université des Sciences et Technologie de Lille, December, 2004.

Dynamic Analyses and Design Optimization of High-Speed Spindle-Bearing System

Vincent Gagnol[1], Belhassen C. Bouzgarrou[1], Pascal Ray[1] and Christian Barra[2]

[1]*LAMI, Mechanical Engineering Research Group, Université Blaise Pascal – Clermont II Institut Français de Mécanique Avancée, BP 265, 63175 Aubière Cedex, France; E-mail: {vincent.gagnol, belhassen-chedli.bouzgarrou, pascal.ray}@ifma.fr*
[2]*PCI-SCEMM, rue Copernic, 42000 Saint-Étienne, France; E-mail: christian.barra@pci.fr*

Abstract. High-Speed Machining (HSM) is widely used in the manufacturing industry. However premature failure without alarming signs leads the development of spindle technology to be strategically critical for the HSM implementation. This paper presents a software package using Matlab for dynamic modelling and simulation of HSM spindle. It has been developed for industrial use and is composed of several modules. First a design module allows elaborating a Finite Element Model(FEM) on the basis of a dedicated rotor-beam element. A readjustment module allows tuning specific material properties such as Young modulus, shear modulus and damping coefficients in order to fit model results to the experimental frequency response function. A simulation package allows to perform dynamic analyses such as transient analyses, mass unbalance analyses and synchronous excitation analyses representative of the milling tool teeth impact. Specific dynamic effects that are underestimated in previous studies are highlighted. Spindle dynamics properties are shown to be strongly dependent on rotation frequency and have to be evaluated for accurate stability prediction. Finally, optimization studies are performed on spindle design and on cutting condition by integrating the predicted FRF of the spindle to the Budak–Altintas chatter stability approach.

Key words: spindle-bearing system, dynamics prediction, design optimization.

1 Introduction

High Speed Machining (HSM) is a technology used to increase productivity and reduce production costs. Sudden premature spindle failures at high speeds make the development of this technology strategically critical for its implementation. The evolution of HSM spindle design consists of integrating numerous mechanical functions to increase system performance at constant volumes. Consequently, interactions and couplings between the various functions become significant, and spindle behavior

S. Tichkiewitch et al. (eds.), Advances in Integrated Design and Manufacturing in Mechanical Engineering II, 505–518.

becomes very difficult to evaluate effectively. Designers, limited primarily to their own experiments, cannot guarantee a sufficient level of reliability, which severely penalizes the industrial integration of modern HSM spindles.

Spindle dynamics is shown to be influenced by a large number of factors, including holder characteristics [1], spindle shaft geometry and drawbar force [2] and the stiffness and damping provided by the bearings. Most of these factors are independent on the spindle speed in opposition to bearing stiffness that changes according to preload and spindle speed. Wang and Chang [3] performed a Finite Element Model (FEM) analysis on a two-bearing spindle using a Rayleigh beam, without taking high-speed effects into account. Results show the importance of internal bearing contact angle variation on the high vibration modes. Alfares and Elsharkawy [4] showed the effects of axial preloading of angular contact ball bearings on the dynamics of a spindle. In a high-speed spindle the bearing axial preload is kept essentially constant during thermally-generated changes to spindle dimensions by spring or hydraulic arrangements. Typically, these spindles are directly driven by brushless motors and the spindle shaft is supported by hybrid angular contact bearing. In this paper, the bearing model formulation proposed in the work of Lim [5] and Jorgensen [6] is integrated into the spindle FEM.

For accurate dynamics prediction the spindle speed-dependent characteristics must be evaluated. Many works showed the causal link of rotational speed on the thermal aspects in a spindle [6–9]. Stein and Tu [7], Bossmanns and Tu [8] represent this phenomenon by a closed loop linking the preload to the thermal dilatation of a bearing. An algorithm of preload estimation under cutting operation is developed by Spiewak and Nickel [9] based on an experimental vibration study. These works show the importance of having an accurate evaluation of rolling bearing characteristics but do not take into account rotating spindle shaft dynamic variations.

Faassen [10] and Schmitz [11] proposed considering such dependencies on the basis of experimental transfer function identification at different spindle speeds. However this method cannot be implemented in a manufacturing environment due to the excessive machine immobilization time needed to characterize even a few rotating transfer functions. Furthermore, experimental modal characterization using a classic hammer test cannot supply reliable information about the dynamic properties of a rotating shaft.

This paper presents a software package using Matlab for dynamic modelling and simulation of HSM spindle. In the first part, a brief geometrical and technological description of the considered spindle-bearing system is given. A software design module allows elaborating a finite beam element model on the basis of a simplified spindle geometry.

In the second section the software processor is introduced. Spindle FEM consists in assembling the rotating entity and the rolling bearing models. A special rotor-beam element has been developed [12] and is implemented. The rolling bearing stiffness

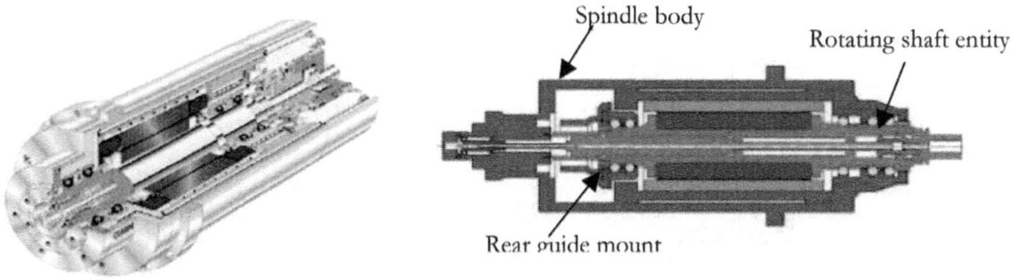

Fig. 1. The HSM spindle bearing system.

matrix $\mathbf{K}_i$ is calculated around a static functioning point on the basis of Lim's formulation [5].

While detailed knowledge of the spindle is in general not available in the manufacturing environment, a readjustment algorithm allows material properties such as the Young modulus, shear modulus or damping variable to be tuned to fit model results to the experimental Frequency Response Function (FRF).

In Section 3, simulation results are presented. Specific rotating dynamic effects are highlighted and give rise to additional features, which makes this model depart from the usual non-rotating spindle beam models. Dedicated experiments are carried out on a motorized high-speed milling spindle of 25 kW at 15000 rpm. They show a good agreement between experimental and model results and confirm the validity of model predictions.

In Section 4, optimization studies are performed on spindle design and on cutting condition by integrating the predicted FRF of the spindle to the Budak–Altintas chatter stability approach [13]. A concurrent spindle benchmarking is carried out on the basis of their stability lobes diagram. Then, bearings arrangement are optimized in order to obtain a chatter vibration free cutting operation at the desired speed and depth of cut for a given cutter.

2 Model Building

A technological analysis makes it possible to identify the structural elements that contribute to the total dynamic behavior of the spindle. The structural entity (spindle body, rotating shaft entity, rear guide mount) and its connections are presented in Figure 1.

A modelling approach based on FEM analysis coupled to experimental modal identification was carried out on the different spindle entities [14]. Results showed that the dynamic properties of the rotating entity are strongly dominant in overall dynamic spindle behaviour. The rotating entity is composed of five subentities, namely the shaft, rotor, tool, tool holder and tool holder mount (Figure 2). The geometrical

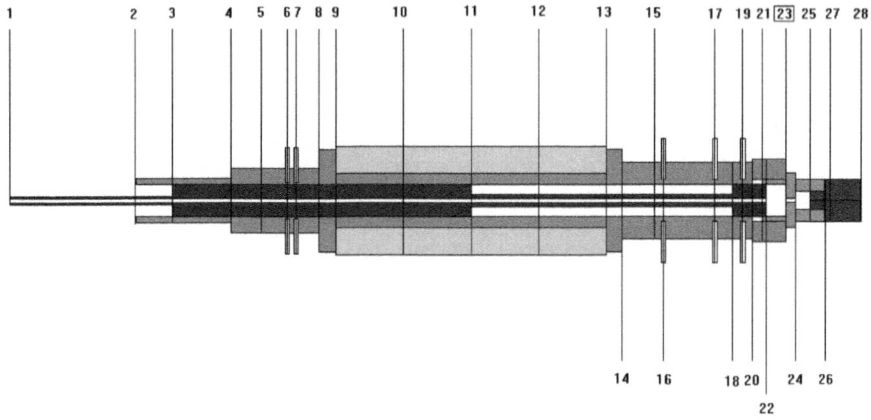

Fig. 2. Finite element model of the spindle bearing system.

rotating entity model has been simplified in order to be meshed by a few number of beam elements. Node numbers are indicated in Figure 2.

The proposed spindle software focus on the rotating entity dynamics.

All modules of software operation require a description of the spindle, which should be provided by means of a model file. This file allows the main Finite-element program to construct the model from: geometry, material properties, boundary conditions, actuators and sensor node locations, unbalance, misalignment.

3 Processor Description

Rotating entity motion is considered as the superposition of rigid and elastic body displacements. Dynamic equations are obtained using Lagrange formulation associated with a finite element method. Due to the size of the rotor sections, shear deformations must be taken into account. Then, the rotating entity is derived using Timoshenko beam theory [15]. The relevant shape functions are cubic in order to avoid shear-locking [16]. A special three-dimensional rotor-beam element with two nodes and six degrees of freedom per node was developed. The rotation effect gives rise to additional features, making this model depart from the usual non-rotating beam model. *Lagrange* equations and the additive properties of kinetic and elastic energy enable rotor dynamics equations to be obtained [12].

As in the classic finite element procedure, dynamic rotor equations are obtained by assembling element matrices. The rotor entity model is assembled with the rolling bearing model. The interfaces between the different subentities which composed the rotating entity are assumed to be rigid. The damping model used draws on *Rayleigh* viscous equivalent damping, which makes it possible to regard the damping matrix

$\mathbf{D}$ as a linear combination of the mass matrix $\mathbf{M}$, the spindle rigidity matrix $\mathbf{K}_{\text{spindle}}$ and the rolling bearing rigidity matrix $\mathbf{K}_{\text{bearing}}$. The set of differential equations can be written as:

$$\mathbf{M}(\mathbf{q}_N)\ddot{\mathbf{q}}_N + (\mathbf{C}(\mathbf{q}_N, \dot{\mathbf{q}}_N) + \mathbf{D})\dot{\mathbf{q}}_N + \mathbf{K}\mathbf{q}_N = \mathbf{F}(t), \quad (1)$$

where $\mathbf{M}(\mathbf{q}_N)$ and $\mathbf{K}$ are the mass and stiffness matrices, $\mathbf{C}(\mathbf{q}_N, \dot{\mathbf{q}}_N)$ matrix contains the rotational dynamics effect, $\mathbf{D}$ is the viscous equivalent damping matrix, $\mathbf{q}_N$ and $\mathbf{F}$ are the nodal displacement and force vectors.

An accelerating rotor gives rise to time-variant matrices since $\mathbf{M}(\mathbf{q}_N)$ and $\mathbf{C}(\mathbf{q}_N, \dot{\mathbf{q}}_N)$ are time-dependent matrices. The treatment of the rotor dynamics in a pseudo-constant rotational speed manner allows establishing that:

$$\mathbf{C}(\mathbf{q}_N, \dot{\mathbf{q}}_N)\dot{\mathbf{q}}_N = 2\Omega\mathbf{G}\dot{\mathbf{q}}_N - \Omega^2\mathbf{N}\mathbf{q}_N.$$

$\mathbf{G}$ and $\mathbf{N}$ proceed from the matrix $\mathbf{C}$ decomposition and are respectively representative of gyroscopic and spin softening effects. Ω is the rotor's angular pulsation.

Hence, considering the assumption of pseudo-constant rotational speed, rotor dynamics is described by means of linear time-invariant matrices.

$$\mathbf{M}_0\ddot{\mathbf{q}}_N + (2\Omega\mathbf{G} + \mathbf{D})\dot{\mathbf{q}}_N + (\mathbf{K} - \Omega^2\mathbf{N})\mathbf{q}_N = \mathbf{F}(t), \quad (2)$$

where $\mathbf{M}_0$ is the constant part of matrix.

3.1 Experimental Characterization and Model Readjustment

The material properties of the rotor and the tool holder mount cannot be precisely known due to their geometrical and physical complexity. The modelling of damping in order to represent the dissipative properties of spindle components also represents a real problem. A readjustment procedure based on a two-step algorithm is proposed to tune the previous parameters in order to fit model results to experimental ones. These parameters are readjusted by minimizing the gap between the measured and the modelled tool tip node frequency response function for a non-rotating spindle. Step one adjusts the frequency properties of the model by tuning the rigidity parameters, such as the *Young modulus E* or the *Shear coefficient G*, and the material volume mass parameter ρ. The parameter δ allows rigidity variations caused by the rear bearings floating mount to be taken into account. Step two tunes the damping ratio of the model tool tip FRF by adjusting the α, β and γ coefficients of the damping matrix. The damping model used draws on *Rayleigh* viscous equivalent damping, which makes it possible to regard the damping matrix $\mathbf{D}$ as a linear combination of the mass matrix $\mathbf{M}$, the spindle rigidity matrix $\mathbf{K}_{\text{spindle}}$ and the rolling bearing rigidity matrix $\mathbf{K}_{\text{bearing}}$. A Matlab optimization routine was used with a least squares type objective function (Figure 3). Experimental determination of the dynamic characteristics of the entity was carried out by an impact hammer test and a Bruel and Kjaër acquisition rack and FFT analyzer.

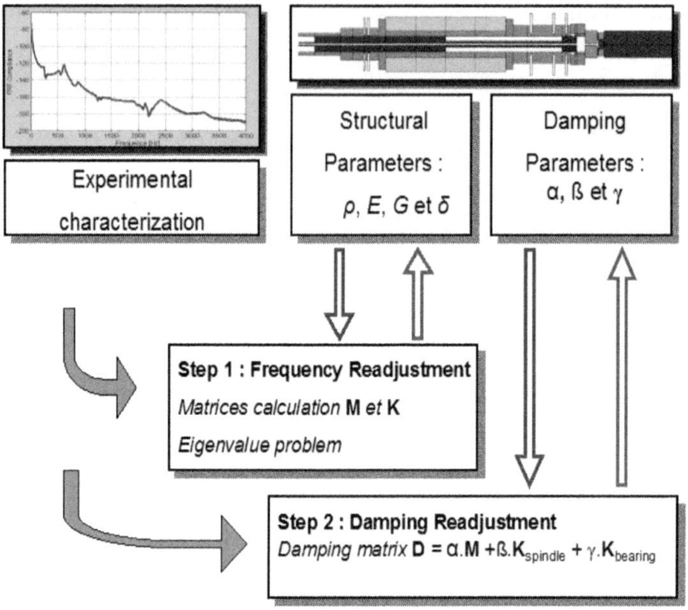

Fig. 3. Algorithm of numeric model readjustment.

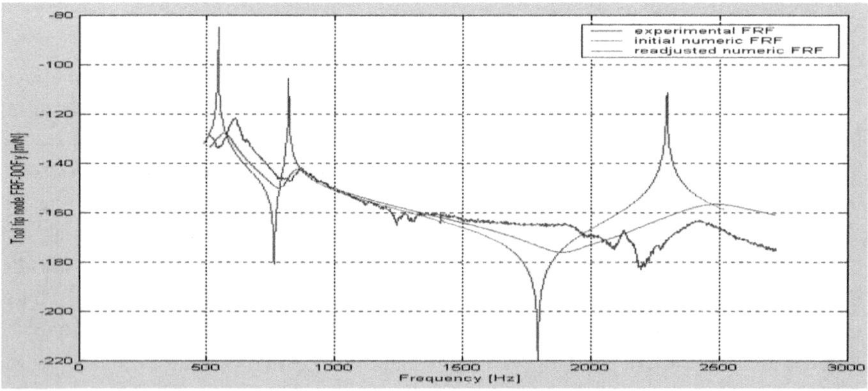

Fig. 4. Experimental, initial and readjusted FRF.

Finally the spindle database is updated to contain the geometrical and the final readjusted material parameters.

The readjustment procedure yields a good correspondence in a bandwidth of 500 to 1500 Hz (Figure 4). The first and second bending modes are modelled at a realistic frequency value and with close damping ratio.

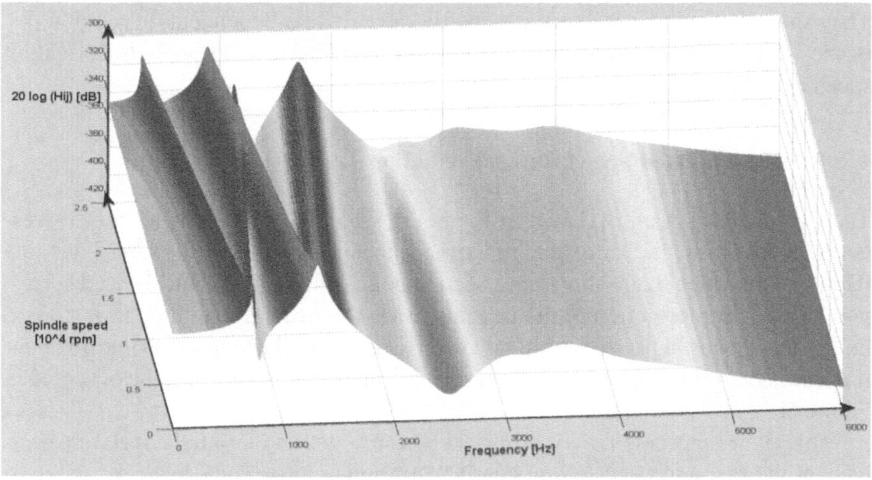

Fig. 5. 3D speed-dependent Frequency Response Function.

4 Dynamic Analyses Package and Model Results

The main features of dynamic analyses package are the computation of natural frequencies, modes shapes, response to mass unbalance excitation or to synchronous excitation representative of a milling tool excitation and finally computation of stability lobes diagram.

4.1 Spindle Dynamic Analysis

Readjusted model can be used to investigate high-speed rotational effects on spindle dynamics. Natural frequencies are determined as a function of the speed of rotation. Two computational loops on both frequency and spindle speed ranges allow us to generate a 3D surface (Figure 5) which represents tool tip node FRF variations according to spindle speed. The top view can be interpreted as a Campbell diagram and each constant-speed vertical layer represents the tool tip node FRF at this rotation speed.

Dynamic effects due to high rotational speed and elastic deformations, such as gyroscopic coupling and spin softening, have a significant influence on spindle behaviour. For the non-rotating spindle, two orthogonal bending mode shapes are associated to equal bending frequencies, due to the rotor axial-symmetry. When the spindle rotates, the frequencies associated with these two mode shapes move apart, one above and one below the rest value, as a function of spindle speed. These two different modes are called forward and backward whirls in rotor dynamics literature [15]. Dedicated experiments allowed validating such results with good agreements.

They show that spindle dynamics properties are strongly dependent on rotation frequency (Figure 6a). Consequently, for accurate spindle dynamics prediction, rotation speed must be taken into account.

4.2 Reliability Based Speed Range Prediction

The Campbell diagram is shown in Figure 6a. The two black straight lines represent respectively an excitation due to a mass unbalance and a milling forces due to a five-tooth tool. The figure shows intersection points of potential critical speed corresponding to the intersection of the bending mode frequencies with the black straight line of the synchronous excitation due to a five-tooth tool excitation. Vibration consequences on rolling bearing nodes around the intersection points can be observed in Figure 6b.

This figure presents the vibration level as a function of spindle speed respectively at the rear and front rolling bearing. Amplified vibration levels correspond to the rotation speed values predicted by the Campbell Diagram. Rear and front rolling bearing loads can reach an amplification ratio of respectively 1200 and 1800. This ratio is dependent of the damping matrix **D** and special care has to be taken when introducing it. These results are of great interest when evaluating spindle rolling bearing life expectancy. Some spindle speed ranges should be avoided during machining operations in order to avoid decreasing spindle life expectancy. Typically, in the case of a five-tooth tool it is thus preferable to avoid rotating speed around 6500, 8500 and 13000 rpm because of the consequent lowering of the spindle rolling bearing lifespan.

4.3 Spindle Selection for a Milling Operation

Application of the software allows performing spindle benchmarking. Dynamic stiffness of the modes which induce the allowable depths of cut in stability limits theory of Budak–Altintas [13] represents an important issue for machine tools since it may lead to spindle, cutter and part damages. When the critical depth of cut of the stability lobes at the cutting spindle speed is maximized, the cutting is the most stable. The resulting stability relationships are shown in Equations (3) to (5) where a_{lim} is the limiting chip width or axial depth of cut in peripheral end milling operations, Z is the number of cutter teeth, and K_r the specific cutting energy coefficient, which relates the radial cutting force F_r to the tangential cutting force, F_t (Figure 7). K_t is a second specific cutting energy coefficient that relates F_t to the chip area, n is the spindle speed in rpm, f_c is the chatter frequency, and k is an integer that corresponds to the individual lobe number. $H_{pq}(j\omega)$ is the speed-dependent receptance of the tool tip node. The negative real part R_{pq} and the phase φ of this function are used to determine optimal cutting parameters that maximize the chatter-free material removal rate.

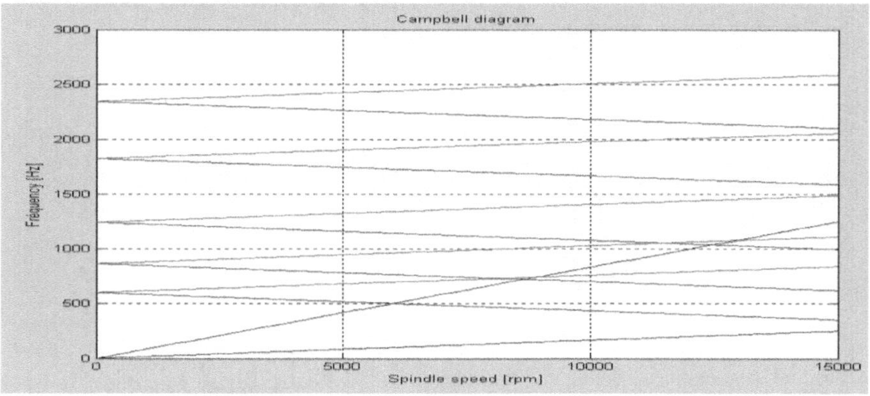

a) Campbell diagram

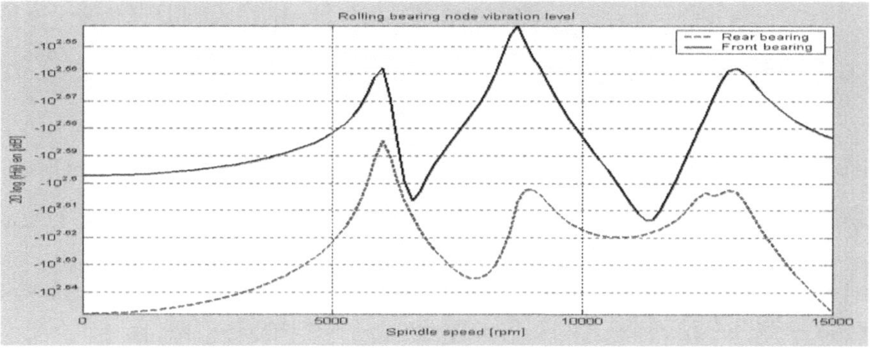

Fig. 6. Speed-dependent spindle dynamics.

In practice the minimum value of the real part of the transfer function determine the value of $a_{\lim}$. The speed-dependent receptance at the tool tip node can be deducted from Equation (2).

$$\mathbf{H}(j\omega) = (-\mathbf{M}_0\omega^2 + (2\mathbf{\Omega}\mathbf{G} + \mathbf{D})\omega + (\mathbf{K} - \mathbf{\Omega}^2\mathbf{N}))^{-1}, \tag{3}$$

$H_{pq}(j\omega) = R_{pq}(\omega) + jI_{pq}(\omega)$ where $j^2 = -1$. R_{pq} and I_{pq} are respectively the real part and the imaginary part of the frequency response function. p and q are respectively the sensor and excitation node numbers.

Resolving the cutting forces in the Y and Z directions in time domain,

$$\{F(t)\} = \begin{Bmatrix} F_y \\ F_z \end{Bmatrix} = \frac{1}{2} a K_t \mathbf{A}_0 \mathbf{\Delta}(t),$$

where

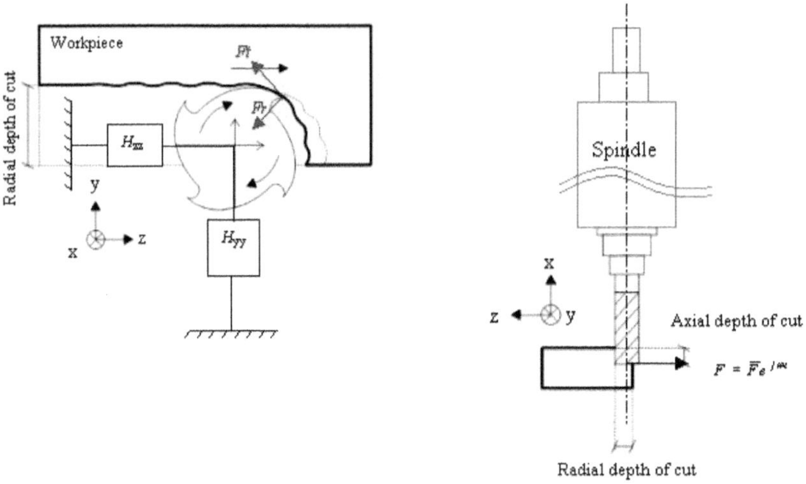

Fig. 7. Self-excited vibrations of the spindle milling system.

$$\mathbf{A}_0 = \frac{Z}{2\pi}\begin{bmatrix} \alpha_{yy} & \alpha_{yz} \\ \alpha_{zy} & \alpha_{zz} \end{bmatrix}$$

is a time-invariant but immersion-dependent directional cutting coefficient matrix.

$$\mathbf{\Delta}(t) = \begin{bmatrix} y(t) - y(t-\tau) \\ z(t) - z(t-\tau) \end{bmatrix},$$

where $(y(t), z(t))$ and $(y(t-\tau), z(t-\tau))$ represent the dynamic displacements of the spindle-tool set at the present and previous tooth period, respectively.

$$a_{\mathrm{lim}} = \frac{-2\pi K}{K_t Z}\frac{1}{R(\omega)}, \tag{4}$$

where $K = f(\alpha_{xx}, \alpha_{xz}, \alpha_{zx}, \alpha_{zz})$,

$$n = \frac{60}{T} = \frac{60 \times 2\pi f_c}{Z(\pi + 2k\pi - 2\varphi} \quad \text{with} \quad \varphi = -\tan^{-1}\left|\frac{I(\omega)}{R(\omega)}\right|. \tag{5}$$

Three concurrent spindles are evaluated through their comparative stability lobes diagram (Figure 8). Comparison are made using a two-tooth, 16 mm Jabro tool JH420 cutter mounted on a Kelch shrink-fit tool holder, cutting an aluminium AS7 workpiece.

It can be observed that the locations of stability lobes are significantly different. The stability limits min calculated, using the model, allows choosing the most adequate spindle for a machining process given. In our case, the choice of spindle (a) is relevant for milling operation in cutting speed around 9000 and 13000 rpm. Spindle (b) and (c) have very closed stability lobes and could be use for spindle speed around 10000 and 15000 rpm.

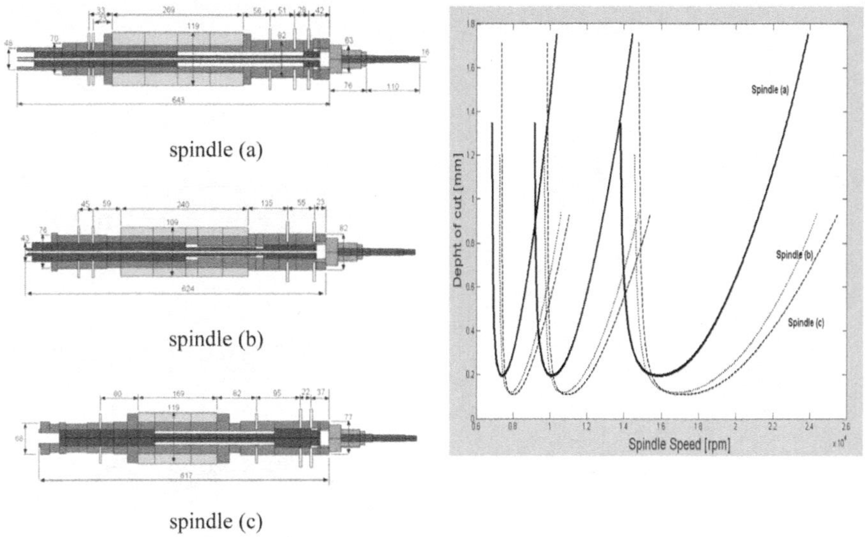

Fig. 8. Comparative spindle stability lobes diagram.

4.4 Optimization of Bearing Locations

In this section an optimization study on bearing and rotor location is carried out, at the spindle design stage, in order to adjust spindle behavior to a given milling operation. This kind of application is a useful tool for the designer. Objective function and design variables are defined as follows for our present design optimization problem:

Objective function. The cutting conditions, the depth of cut and spindle speed, must be under the stability lobes in order to avoid chatter vibrations in metal cutting. The location of lobes is dependent on the natural frequencies of the spindle system, and the allowable depth of cut depends on the dynamic stiffness of the modes. We propose automated tuning of the spindle modes in order to create stability lobes at the desired spindle speed and depth of cut. The tuning of the spindle dynamics is achieved by optimizing the distribution of bearings and rotor along the spindle shaft.

Design variables. Here the bearing spans and rotor position are selected as spindle design parameters. The tuning of the spindle dynamics is achieved by optimizing the distribution of bearings and rotor part along the spindle shaft. The bearing spans' constraints are set by considering the spaces of lubrication devices.

The proposed optimization study is performed for a commercially existing spindle. The initial spindle design is represented in Figure 9 as spindle (a). The design variables of the motorized spindle with five bearings (three front and two rear bearing) are represented on the spindle (b) in Figure 9. The X(1) to X(6) are the design variables which define the bearing and the rotor locations. Spindle design configuration and associated stability lobe diagram are represented at different stages

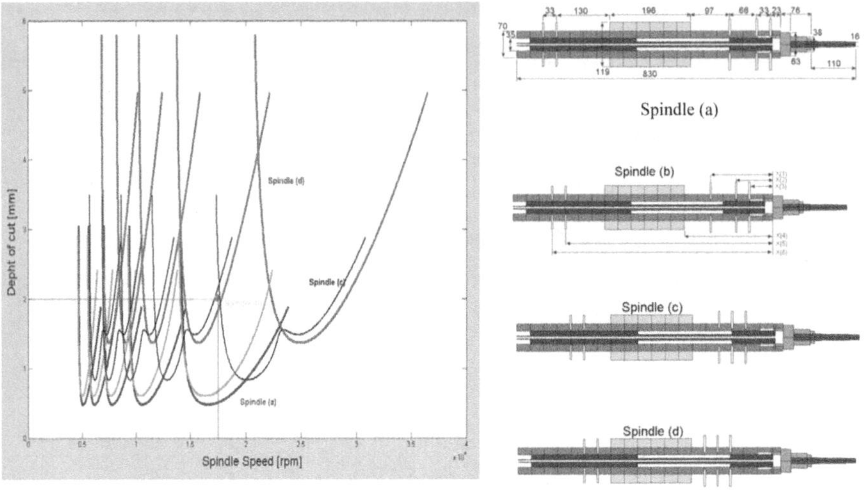

Fig. 9. Stability lobes for initial, intermediate and optimized spindle design.

of the optimization routine. It is assumed that the user wishes to use the machine in aluminium AS7 workpiece with a two-fluted end mill at a desired depth of cut of 2 mm and 17500 rpm spindle speed. These cutting conditions are predominantly used in the considered automotive part cutting operations. Hence, they represent the target point to attain through the optimization procedure.

Spindle geometry and their associated stability lobes diagram are represented at different stage of the optimization routine (Figure 9).

Cutting is not stable for the initial spindle (a) and for spindle (b). It becomes stable for the design configuration of spindle (c). Then cutting again becomes unstable for the spindle (d) design. The optimized spindle configuration is spindle (c).

5 Conclusion

This paper presents a software packages for spindle dynamic predictions. The software has been developed for industrial use. The processor implements a classic finite element procedure based on specific dynamic rotor equations. The rotating entity is derived using the *Timoshenko* beam theory where the effect of rotation gives rise to a few additional features, which make this model depart from the usual non-rotating beam model. Results show that spindle dynamics properties are strongly dependent on rotation speed and have to be evaluated for accurate spindle dynamics prediction.

Dynamic analyses package allows providing rules of expertise on HSM spindle. Specific spindle speed ranges can be predicted to avoid excessive vibration levels on spindle component and to limit the risk of shortening spindle life expectancy. The choice of adequate spindle speed can be realized on static or dynamic characteristics. By integrating the predicted model in the chatter vibration stability law, which indicates whether the design would lead to chatter vibration free cutting operation, accurate spindle cutting condition can be predicted. Optimization studies can be performed on design variable to improve criteria such as reliability or production rates. A bearing spacing optimization is realized in order to create stability lobes at the desired spindle speed and depth of cut.

The model, by its modularity, is easily adaptable to other types of spindles and makes it possible to qualify a spindle for a process of machining given. It can also provide a guideline for machine tool designers to estimate correct system dynamics for different operating conditions.

Acknowledgements

This work was achieved within the TIMS Research Group, granted by the Regional Council of Auvergne, the French Ministry of Research, the CNRS and the Cemagref.

References

1. J. Agapiou, E. Rivin, C. Xie, Toolholder/spindle interfaces for CNC machine tools, *Annals of the CIRP* **44**(1), 1995, 383–387.
2. S. Smith, P. Jacobs, J. Halley, The effects of drawbar force on metal removal rate in milling, *Annals of the CIRP* **48**(1), 1999, 293–296.
3. W.R. Wang, C.N. Chang, Dynamic analysis and design of a machine tool spindle-bearing system, *Journal of Vibration and Acoustics* **116**, 1994, 280–285.
4. M.A Alfares, A. Elsharkawy, Effects of axial preloading of angular contact ball bearings on the dynamics of a grinding machine spindle system, *Journal of Materials Processing Technology* **136**(1–3), 2003, 48–59.
5. T.C. Lim, Vibration transmission through rolling element bearing, Parts I to IV, *Journal of Sound and Vibration* **139**(2), 1990, 179–199.
6. B.R. Jorgensen, Y.C. Shin, Dynamics of spindle-bearing systems at high speeds including cutting load effects, *Journal of Manufacturing Science and Engineering* **120**, 1998, 387–394.
7. J.L. Stein, J.F. Tu, Active thermal preload regulation for machine tool spindles with rolling element, *Journal of Manufacturing Science and Engineering* **118**, 1996, 499–505.
8. B. Bossmanns, J.F. Tu, A thermal model for high speed motorized spindles, *International Journal of Machine Tools and Manufacture* **39**(9), 1999, 1345–1366.
9. S.A. Spiewak, T. Nickel, Vibration based preload estimation in machine tool spindles, *International Journal of Machine Tools and Manufacture* **41**(4), 2001, 567–588.
10. R.P.H. Faassen, N. van de Wouw, J.A.J. Oosterling, H. Nijmeijer, Prediction of regenerative chatter by modelling and analysis of high-speed milling, *International Journal of Machine Tools and Manufacture* **43**, 2003, 1437–1446.

11. T. Schmitz, J. Ziegert, C. Stanislaus, A method for predicting chatter stability for systems with speed-dependent spindle dynamics, Society of Manufacturing Engineers, Technical Paper TPO4PUB182, *Transactions of NAMRI/SME* **32**, 2004, 17–24.
12. V. Gagnol, C.B. Bouzgarrou, P. Ray, C. Barra, Dynamic analysis of a high speed machine tool spindle-bearing system, in *CD-ROM Proceedings of the ECCOMAS Conferences*, Madrid, Spain, June 21–24, 2005.
13. Y. Altintas, E. Budak, Analytical prediction of stability lobes in milling, *CIRP Annals* **44**(1), 1995, 357–362.
14. V. Gagnol, C.B. Bouzgarrou, P. Ray, C. Barra, Modelling approach for a high speed machine tool spindle-bearing system, in *CD-ROM Proceedings of the International Design Engineering Technical Conferences*, Long Beach, USA, September 24–28, 2005.
15. S. Timoshenko, *Vibration Problems in Engineering*, D. Van Nostrand Company, New York, 1928.
16. G. Genta, *Vibration of Structures and Machines. Practical Aspects*, Springer-Verlag, 1995.

Identification of Cutting Relations in High Speed Milling

Stéphanie Bissey-Breton[1], Gérard Poulachon[2] and François Lapujoulade[3]

[1]*STFU/LUP, CEA Valduc, 21120 Is sur Tille, France; E-mail: stephanie.breton@cea.fr*
[2]*LaBoMaP, ENSAM, Rue Porte de Paris, 71250 Cluny, France;*
E-mail: gerard.poulachon@cluny.ensam.fr
[3]*LMSP, ENSAM, 151, Bd. de l'Hôpital, 75013 Paris, France;*
E-mail: francois.lapujoulade@paris.ensam.fr

Abstract. The knowledge of the cutting forces and their evolution along time are very interesting for the optimisation and the monitoring of the machining process. The prediction of the time evolution of the cutting forces could allow optimizing tool geometries, or better adapting the cutting conditions and the machine tool to the considered machining operation, in order to minimize the level of cutting forces. It could also be helpful to optimize tool trajectories in CAM software for example. The objective of this study is to improve the qualification procedure of a tool for milling a given material, considering first the local geometry of the cutting edge, and going back then to the global geometry of the considered tool or tool family. This methodology can be adapted for all global shapes of tools, even in the case of complex geometries. Only the local definition of the cutting edge must be constant so that the methodology could be applicated to several tools of the family. This study has also been possible thanks to the use of a dynamometer with accelerometric compensation allowing to obtain good quality signals even at high frequencies of tool rotation.

Key words: cutting forces, milling, predictive model, rake angle, helix angle, identification, cutting relation.

Notation

a_p, a_e	=	axial, radial depth of cut
b	=	width of cut
f	=	feed rate
$(\mathbf{g}, \mathbf{n}, \mathbf{a})$	=	reference linked with the rake face
γ_n	=	normal rake angle
h	=	undeformed chip thickness
λ_s	=	helix angle
λ_e	=	chip flow angle

S. Tichkiewitch et al. (eds.), Advances in Integrated Design and Manufacturing in Mechanical Engineering II, 519–530.

λ_n = angle between PQ and PR (Figure 1)
N = spindle rotation speed
$(\mathbf{r}, \mathbf{t}, \mathbf{a}_x)$ = radial, tangential, axial directions
V_c = cutting speed

1 Introduction

The calculation of cutting forces in milling is important at different steps of the machining process preparation: choice or dimensioning of tools and machines, prediction of deformations and vibrations occurrence, optimization of cutting parameters, etc. Numerous papers have been published about this field and the basic principles are well established. Models allowing predicting cutting forces in milling with a good precision can be set up. The precision can be improved by sharpening the model but the necessary determination of an important number of constants linked with the couple machining material/tool is very penalizing in terms of time and costs of calculation. The simplest models only take into account one or two coefficients, for example the specific cutting pressure that appear in data bases, but do not allow to calculate the time evolution of cutting forces in the cases of wide machining configurations such as 5 axes milling. More complex models need to use experimentation in order to determine the cutting constants, and the cost of this experimentation will be all the higher since the results are given for a fixed tool/material pair.

The objective of this study is to show that it is possible to obtain models of average complexity valid for a wide family of tools in a given material. The experimentation is then reduced to test a short number of tools representative of the considered family of tools. The objective is reached by introducing two coefficients in the cutting relations: the rake and helix angles, which complete the representation of the interaction between the material and the cutting edge, in addition with the classical variables such as the cutting speed, the undeformed chip thickness and the width of cut. Thus, the constants of the model are characteristic of the action of a cutting edge in a given material. The definition of the cutting edge is dependant of the tool material, the coating and the edge preparation. Each tool with the same definition of cutting edge belongs to the same family, even if its global shape is different (e.g. cylindrical, ball end mill, taper end mill, bull nose end mill). It is also possible to include tools with variable helix or rake angles along the edge, as it is often the case for ball end mills, but with the condition that the edge remains with the same definition everywhere. The preferred application field is solid end mills.

2 Model Formulation of Forces in Oblique Cutting

The basic methodology is classical and described in numerous articles or books of the literature [1, 2]. The segmentation principle consists into dividing the tool into disks

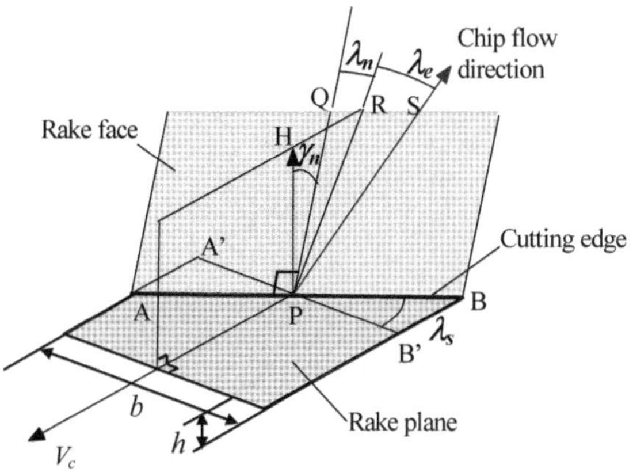

Fig. 1. Principles of oblique cutting.

elements of constant thickness perpendicularly to its axial direction, determining the force acting on each segment by application of the cutting relation and finally, a summation all over the flutes engaged in the cut yields to the total forces acting on the milling cutter at a given instant time. This methodology relies on the hypothesis of independence of the segments because the calculation is made without taking into account what is happening on the neighbouring segment. This hypothesis is realistic in the case of low bended edges but does not work in zones with high curvature, with variable morphology, complex chip breaker, or in zones where the chip formation is particular such as in parts near the tool axis in the case of front milling for example.

The segmentation and the determination of the working conditions of each segment need all information about the definition of the tool shape, the kinematics between the tool and the workpiece, and also the global shape of the workpiece (limiting surfaces). All this information is geometrical and kinematics calculations. The action of each segment is characterised by five parameters, which are the cutting speed (V_c), the undeformed chip thickness (h), the width of cut (b), the helix angle (λ_s) and the normal rake angle (γ_n). The helix and the rake angle only depend of the tool; the cutting speed and the undeformed chip thickness are linked with the tool geometry, its kinematics and the machining conditions. The width of cut depends on the segmentation. The action of each segment takes place in a configuration of oblique cutting as shown in Figure 1.

The width of cut (b) corresponding to the edge segment AB is the distance between point A′ and point B′, perpendicularly to the cutting speed direction. The undeformed chip thickness (h) is the distance between point P and the free surface of the workpiece, perpendicularly to the rake plane. The different methodologies used to determine h in function of the position of point P during the rotation of the tool

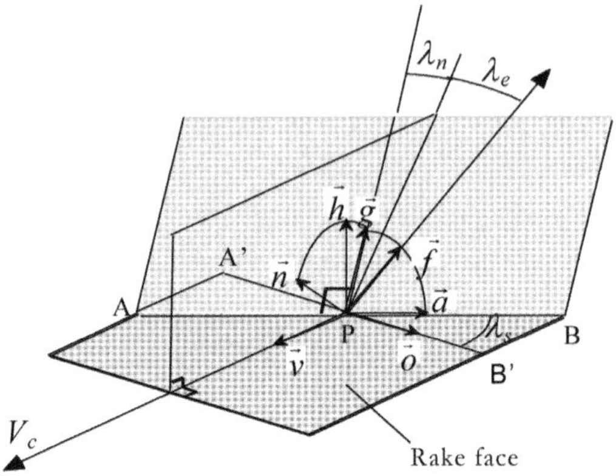

Fig. 2. Coordinate systems of the oblique cutting.

would not be detailed here. It can be noticed that this could be done with more or less precision.

Different coordinate systems can be used to express the cutting forces. The reference usually used to describe the cutting forces is {**h**, **v**, **o**}, as defined in Figure 2. In the case of a cylindrical milling cutter, this reference corresponds to the coordinate system $\{\mathbf{r}, \mathbf{t}, \mathbf{a}_x\}$ where **r**, **t** and $\mathbf{a}_x$ are respectively the radial, tangential and axial directions. In the {**h**, **v**, **o**} reference, the cutting relation used by Altintas [2] gives:

$$Fh = Kh_1 \cdot b + Kh_2 \cdot h \cdot b, \tag{1}$$

$$Fv = Kv_1 \cdot b + Kv_2 \cdot h \cdot b, \tag{2}$$

$$Fo = Ko_1 \cdot b + Kp_2 \cdot h \cdot b. \tag{3}$$

The application of this cutting relation to all the segments of a tool implies that the rake angle γ_n and the helix angle λ_s are the same all along the tool. The six coefficients K_{h1}, K_{h2}, K_{v1}, K_{v2}, K_{o1} and K_{o2} depend on these angles. It could be possible to write these six coefficients as new functions of two parameters λ_s and γ_n but it would introduce a great number of new coefficients, for example: $K_{h1} = K_{h0} + K_{h11} \times \lambda_s$. Therefore, in this study, it was decided to obtain simpler relations, by using the fact that the chip flows out along the rake face in the chip flow direction defined in Figure 2 and characterised by the chip flow angle λ_e. The new coordinate system that is used to describe the cutting forces is {**g**, **n**, **a**}, linked with the cutting edge and the rake face of the tool. The cutting forces can then be expressed by:

$$\mathbf{F} = F_n \cdot \mathbf{n} + F_f \cdot \mathbf{f} = F_n \cdot \mathbf{n} + F_g \cdot \mathbf{g} + F_a \cdot \mathbf{a}, \tag{4}$$

with

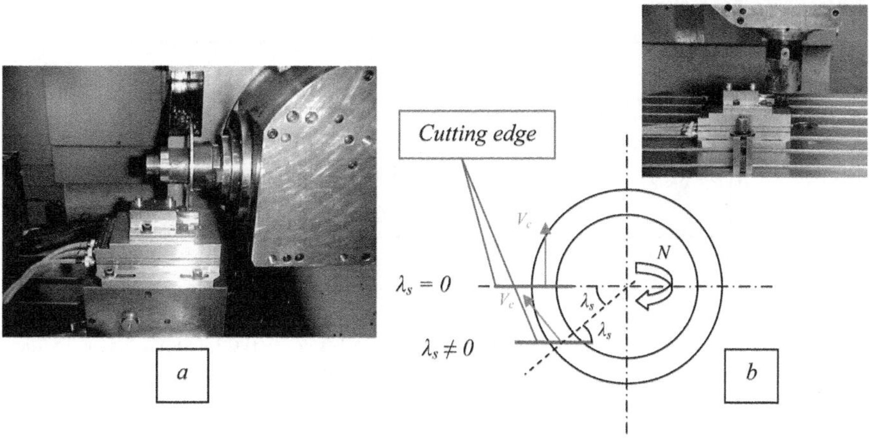

Fig. 3. Test set-up to see the influence of the rake angle (a) and the helix angle (b).

$$F_g = F_f \times \cos(\lambda_e + \lambda_n) \quad \text{and} \quad F_a = F_f \times \sin(\lambda_e + \lambda_n). \tag{5}$$

The following relation gives the angle λ_n:

$$\tan(\lambda_n) = \sin(\gamma_n) \times \tan(\lambda_s). \tag{6}$$

The usual components F_h, F_v and F_o are then obtained thanks to F_g, F_n and F_a. The cutting relation can be written as follows:

$$F_n = f_n(h, b, \lambda_s, \gamma_n, V_c), \tag{7}$$

$$F_f = f_f(h, b, \lambda_s, \gamma_n, V_c), \tag{8}$$

$$\lambda_e = f_\lambda(h, b, \lambda_s, \gamma_n, V_c). \tag{9}$$

The question is then to determine the shape to give to these three functions. The measurements in milling cannot bring a direct answer to this question, because the cutting relation must be identified. The problem supposes that the cutting relation shape is known. In order to easily separate the influences of the different parameters of the cutting relation on the cutting forces, it has been decided to proceed to tests in turning. Disks or tubes were machined with a straight flute milling cutter. The cutting edge is defined as the one of the other tools of the considered family of tools. The cutter is fixed on the dynamometer (Figure 3).

In order to create obliquity and "introduce" a helix angle on straight flute milling cutters in the case of Figure 3b, the tool was radially moved relative to the tube. Consequently, an obliquity angle appears between the cutting speed vector $\mathbf{V}_c$ and the cutting edge. Results for cdifferent materials have already been presented in [3, 4]. The main results are the following:

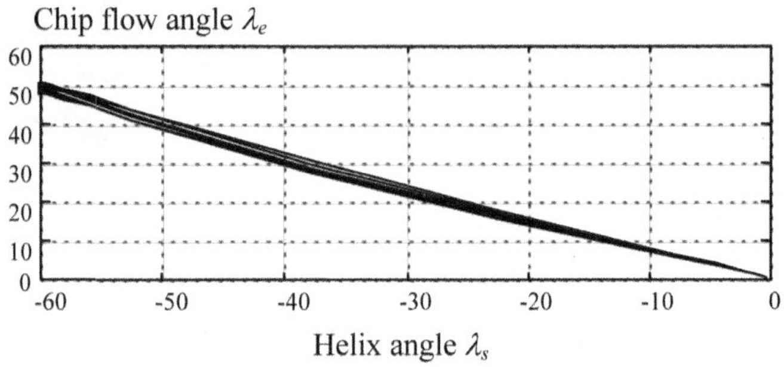

Fig. 4. Chip flow angle in function of the helix angle for $h = 0.025, 0.05, 0.1, 0.15$ mm. (Z35cn1702 steel – Non-coated solid carbide end mill)

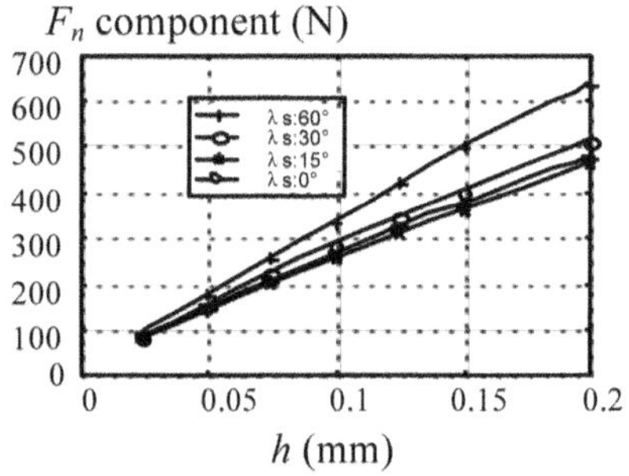

Fig. 5. Evolution of the F_n component in function of h for different values of λ_s.

- The chip flow angle λ_e is nearly proportional to the helix angle λ_s (Figure 4).
- The influences of the undeformed chip thickness and the width of cut well respect a relation such as $F_n = (F_0 + A(\lambda_s) \cdot h)b$, for the low values of h, where A is a function of the helix angle λ_s. Concerning values higher than 0.2 mm (Figure 5), the cutting relation must be expressed by a non-linear function.
- The influence of the rake angle γ_n is limited but linear. It is possible to introduce it into the cutting relation thanks to a corrective and multiplicative term: $(1 + K_\gamma(\gamma_n - \gamma_{n0}))$. The angle γ_{n0} is the most representative rake angle of the family of tools.

Finally, the shape of the cutting relation that has been obtained is given by equations (9), (10) and (11):

$$\lambda_e = K_{\lambda e} \times \lambda_s, \tag{10}$$

$$F_n = (K_{n0} + A_n \cdot h)b(1 + K_{ng}(\gamma_n - \gamma_{n0})) \quad \text{with} \quad A_n = K_n \left(\frac{1}{\cos \lambda_s}\right)^{K_{n\lambda_s}}, \tag{11}$$

$$F_f = (K_{f0} + A_f \cdot h) b (1 + K_{fg}(\gamma_n - \gamma_{n0}) \quad \text{with} \quad A_f = K_f \left(\frac{1}{\cos \lambda_s} \right)^{K_{f\lambda_s}} . \quad (12)$$

This relation is expressed with seven principal coefficients and two secondary coefficients in order to introduce the correction due to the rake angle γ_n.

3 Identification of the Cutting Relation in Cylindrical Milling

In order to identify the cutting relation linked with a family of tools, for a given material, it is necessary to realise a reduced number of tests with a limited number of tools of the considered family and to choose the appropriate representative tools.

As the variation of the undeformed chip thickness is automatically created by the rotation of the milling cutter, only three tools are necessary for the tests. Then, one test per tool, with measurements of the three components of the cutting forces, is enough to identify the nine coefficients of the model. However, this is not realistic regarding the defaults of the modelisation and the experimental errors. The cutting relation allows predicting the cutting forces in every machining configuration imaginable with the considered family of tools. However, it is preferable to be in an "interpolation" situation and not an "extrapolation" one. That is to say that the tested tools and configurations must be chosen to include all the possible configurations. For example, it is better to choose two tools with 10° and 60° helix angles than two cutters with 20° and 30° helix angle to predict the action of a tool with a 45° helix angle.

Even if it is possible to use ball end mills or bull nose end mills, the geometry of those tools is not always perfectly known. This could introduce important factors of errors. It is then better to use simple cylindrical tools.

A very important point is the great influence of geometrical and kinematics defaults on the cutting forces. For example, an excentration of the tool can significantly change the final values of the cutting forces. It is thus necessary to consider these defaults during the calculation of the undeformed chip thickness. Therefore, their values must be known, but the direct measure is difficult and gives low precision. It is preferable to identify them and to consider them as other parameters to determine, all the more since their identification is easy: these factors are the only ones that allow making the difference between the actions of the different flutes of the tool, if the definition and the wear of the edges remain the same. For a considered tool, tests realised with very different conditions give very close results in terms of identification of the excentration.

Tests with cylindrical end mills are yet difficult to set up because of the action of the extremity of the tool. Several effects can be superposed during the rotation of the cutter: presence of a little corner radius, non-validity of the basis hypothesis of the segmentation principle, action of the front edges due to a tool deformation, etc. It is clear that these effects are not taken into account in the cutting relation: the relation

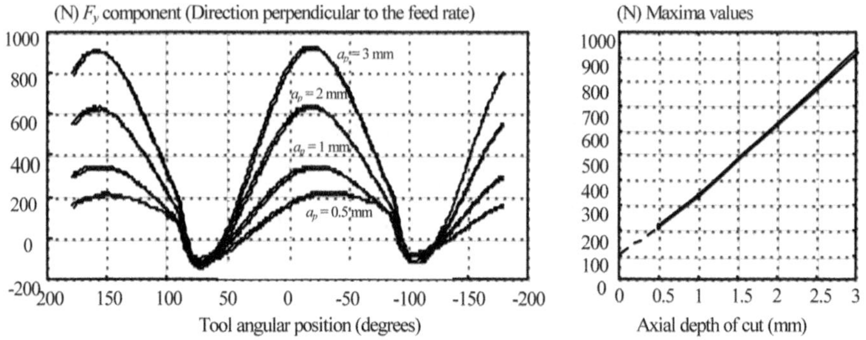

Fig. 6. F_y component in the direction perpendicular to feed rate in function of the angular position of the tool. Material: X38 CrMoV 5 (AISI H11). Tool: PWZ solid non-coated carbide end mill with $\lambda_s = 40°$. Slotting operation, $N = 2020$ rpm, $f_z = 0.1$ mm/rev.

is only valid for the "lateral" edges. The action of the lateral edges implies that forces are proportional to the axial depth of cut a_p. It can be noticed in Figure 6 that it is not always the time. For cylindrical end mills, it is necessary to add a model describing the action of the extremity of the tool to the one relative to the action of the lateral edges. In the opposite, if it is the "lateral" cutting relation that must be identified from tests with cylindrical end mills, the action of the extremity of the cutter must be eliminated. A good methodology consists in systematically realising two tests with two values of the axial depth of cut a_{p1} and a_{p2}. If the action of the extremity is considered the same in both cases, the force $F = F_1 - F_2$ obtained by subtraction of the results of both tests is considered to correspond to the action of only lateral flutes during a test with a depth of cut $a_p = a_{p1} - a_{p2}$.

4 Experimental Results

4.1 Experimental Achievement

The family of tools that is considered is composed of non-coated solid carbide end mills from PWZ. There are cylindrical tools, bull nose end mills or ball end mills with different diameters and different numbers of flutes. Tests were carried out in X38 CrMoV 5 (AISI H11) steel which is a die and moulds steel, heat treated at 43 Hrc hardness value. During milling, cutting forces were measured with a dynamometer called "Dyna TK" and set up by the laboratory. The principle of this dynamometer is based on the use of piezoelectric sensors associated with an inertial compensation giving it a flat pass-band up to 8 kHz. This dynamometer is described more precisely in [5, 6]. The good quality of the signals obtained after measurements allows to well

Table 1. Summation of identificated coefficient in the cutting relation.

λ_s	K_{λ_e}	K_{n0}	A_n	K_{f0}	A_f
0°	–	18	2411	34	743
20°	0.45	23	2482	42	738
40°	0.51	23	2869	30	916
60°	0.45	17	3633	38	1526

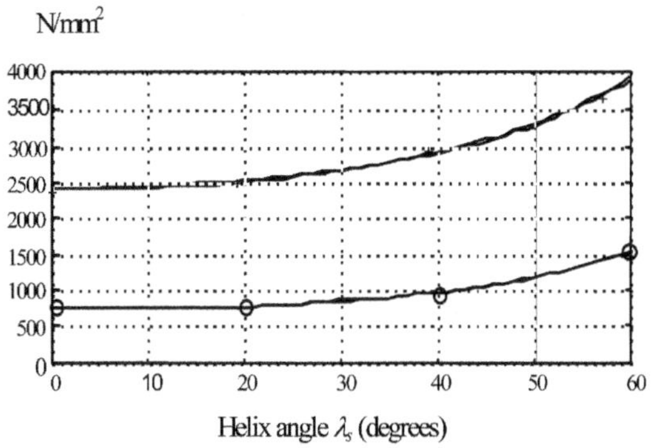

Fig. 7. Evolution of A_n and A_f coefficients in function of the helix angle λ_s.

analysing them and to make easier the identification of the cutting relations. Actually, the dynamometer introduces very low distortion, especially on straight fronts.

4.2 Results

The identification of the cutting relation has been realised with cylindrical end mills of 20 mm diameter, with helix angles of 0°, 20°, 40° and 60°, and a rake angle of 4°. Numerous tests (about 150) have been set up with different radial depths of cut and various feed rates. For the identification, only the tests in slotting were used. They better cover all the values of the undeformed chip thickness and give better quality experimental results than tests with a lower radial engagement, because of the non-appearance of high frequency excitation (no shocks).

Table 1 sums up the identification results thanks to the series of tests realised in slotting with different cylindrical tools of the considered family.

Figure 7 shows that the coefficients A_n and A_f well follow the relations described by equations (9), (10) and (11).

For the other coefficients, the mean values are considered for all the tests. The cutting relation can thus be obtained with the following values, thanks to equations (9), (10) and (11):

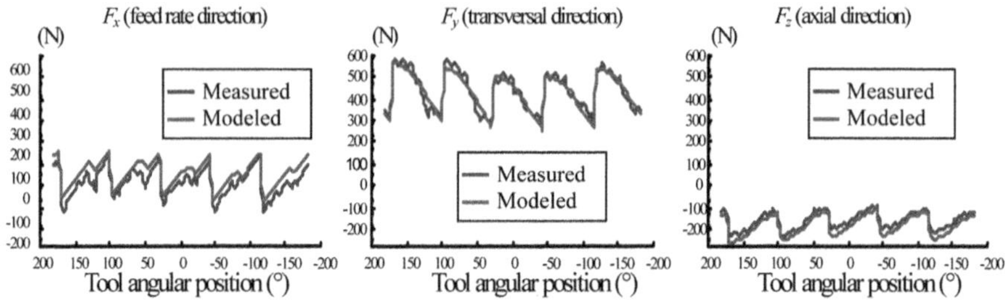

Fig. 8. Cutting forces components for a cylindrical end mill ϕ 20, 5 flutes, $\lambda_s = 30°$, $\gamma_n = 12°$, $a_p = 1.5$ mm, $a_e = 15$ mm, $f_z = 0.1$ mm/rev, $N = 2020$ rpm.

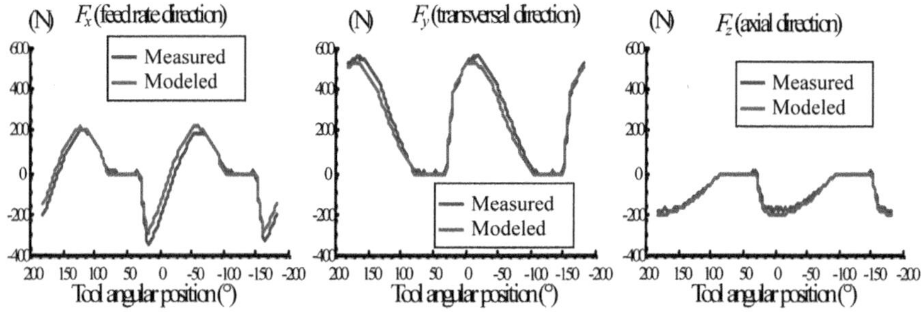

Fig. 9. Cutting forces components for a cylindrical end mill ϕ 12, 2 flutes, $\lambda_s = 27°$, $\gamma_n = 10°$, $a_p = 2$ mm, $a_e = 9$ mm, $f_z = 0.1$ mm/rev, $N = 3270$ rpm.

$$K_{\lambda_e} = 0.47, \ K_{n_0} = 20.3, \ K_n = 2407, \ K_{n\lambda_s} = 0.708,$$

$$K_{f_0} = 36, \ K_f = 710, \ K_{f\lambda_s} = 1.08.$$

Complementary tests realised with tools with a 10° rake angle allow obtaining the values of the last correcting coefficients: $K_{ng} = -0.012$ and $K_{fg} = -0.015$.

The application of the cutting relation to other tools of the family but with significant differences from the tools used for the identification, gives good results: they are presented on the following cases. Figures 8 and 9 correspond to cylindrical end mills with different number of teeth and diameters. Figure 10 shows the case of a milling cutter with a different global shape (ball end mill) and with an inclination of 30°. A great concordance can be noticed between experimental values and calculated ones. This concordance is not as good in the case of the ball end mill (Figure 10), but this one has been used with a mean value of the undeformed chip thickness of 0.02 mm instead of 0.08 mm for the identification tests in slotting, with 0.1 mm feed rate.

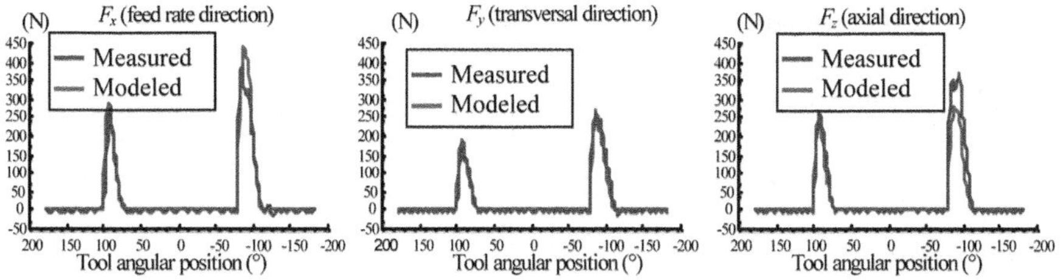

Fig. 10. Cutting forces components for a ball end mill ϕ 16, 2 flutes, $\lambda_s = 30°$, $\gamma_n = 10°$, $a_p = 2$ mm, $a_e = 1$ mm, $f_z = 0.1$ mm/rev, $N = 3790$ rpm, inclination $= 30°$.

5 Conclusions and Perspectives

The tests that have been realised show the possibility to use a unique cutting relation, including influences of the rake and helix angles, for a wide family of tools, yet the same type of edge defines them. The cutting relation is obtained from a limited number of tests with the same tools of the family, preferably with a cylindrical geometry. It is thus possible to characterize with a good precision the behaviour of the entire family of tools machining a given material. The resulting cutting relation allows predicting the cutting forces supported by a tool of the family in various using conditions.

The cutting relation and the procedure of identification could be improved in order, for example, to study the case of low undeformed chip thickness. A model should also be developed to take into account the effects due to the extremity of the tool on cutting forces, which becomes predominant with low axial depths of cut. A similar methodology is being developed for operations of oblique grooving or axial milling operations.

Other developments of the cutting relation could also take place in the field of the machining material: studies are realised in order to verify if the shape of the cutting relation is valid when the family of tools is used for machining another material than the one considered for its identification.

A software has also been set up to sum up the information in a data basis: coefficients of the cutting relation linked with each edge-material pair that could be considered in a machining operation are saved in this data basis and the software allows to calculate the cutting forces that could be developed in the machining conditions entered for the prediction. This software could be a great help for tool designers for example or to choose the adapted a milling cutter for a given operation on a given machine.

Acknowledgements

The PWZ Company and the CEA are gratefully acknowledged for their support to this work.

References

1. M.E. Merchant, Basic mechanics of the metal process, *Trans. ASME Journal of Applied Mechanics* 1944, 168–175.
2. Y. Altintas, *Metal Cutting Mechanics, Machine Tool Vibrations and CNC Design*, Cambridge University Press, ISBN 0-521-65973-6, 2000.
3. F. Lapujoulade, A. Khellouki, Identification de lois de coupe en fraisage latéral à partir d'essais de tournage, in *Integrated Design and Production CPI'2003*, Meknès (Morocco), 2003.
4. S. Bissey, G. Poulachon, F. Lapujoulade, Integration of tool geometry in prediction of cutting forces during milling of hard materials, in *5th IDMME (Int. Conf. on Integrated Design and Manufacturing in Mechanical Engineering)*, Bath (England), 2004.
5. F. Lapujoulade, Measuring of cutting forces during fast transient periods, in *First French and German Conference on High Speed Machining*, Metz (France), 1997, pp. 372–376,
6. F. Lapujoulade, G. Coffignal, J. Pimont, Evaluation des forces de coupe en fraisage à grande vitesse, in *2nd IDMME*, UTC Compiègne (France), May 1998.

Using 3D CAD Models for Programming High Speed Machining

Zenobia Weiss, Roman Konieczny and Radoslaw Paszkiewicz

Institute of Mechanical Technology, Poznan University of Technology, Poznan, Poland; E-mail: {zenobia.weiss, roman.konieczny, radoslaw.paszkiewicz}@put.poznan.pl

Abstract. The paper analyses the factors that have an influence on the quality of the HSM process, as experienced within the industrial process. The discussed factors include the limitations of machine tools and the problem of design producibility that has an impact on process programming for HSM technology. The mistakes made in the stages of product design, when 3D CAD model of product geometry is prepared have the considerable effect on process programming, machining and final surface quality. The paper describes also the method of NC part programming in CAM systems that can eliminate disadvantages of machine tools. It shows the options of machining cycles designed for HSM.

Key words: high speed machining, CAD/CAM, CNC programming.

1 Introduction

High speed machining technology (HSM) introducing cutting speeds that are five to ten times higher than in conventional machining offers new possibilities for increasing process efficiency and for obtaining a good workpiece surface quality. In comparison to conventional cutting, HSM provides reduced cutting forces, a reduced heating of the workpiece and higher machining accuracy by higher metal removal rate. These advantages of HSM have been reported in many scientific papers and many application notes [1, 2].

However HSM is still used on a lower scale than might be expected, taking into account all the above-mentioned advantages. There are many problems that should be solved to the successful implementation of HSM. The technology has the potential to have an impact on cutting tools, spindles, tool holders, and controls. It is essential to be aware of the fact that the task of preparing a CNC program for high speed machining is more than a matter of introducing new values for spindle rotary speed, feed rate and depth of cut. NC programmers must be aware of all the constrains imposed by the machine tool and take them into consideration in the design process.

S. Tichkiewitch et al. (eds.), Advances in Integrated Design and Manufacturing in Mechanical Engineering II, 531–541.

These programmers will also have to adapt their existing machining strategies to increase the efficiency, accuracy and safety of the cutting tool. As the feed rate and spindle speed are considerably higher than those used in conventional machining the programmers must foresee the manner in which the tool will sink into the raw material and the avoidance of rapid changes in the direction of movement is essential. A wide-ranging cooperation between the designer and the process engineer, that allows for an analysis of the CAD part model in respect of the its producibility through HSM, is also required.

2 Limitations of HSM Process Design

The analysis of a machine tool revealed that, despite being assigned for HSM, it has many limitations in terms of the dynamic movements and computational properties of the control system and that these limitations have a significant impact on the possibility of obtaining a high standard surface quality. In HSM technology the machine tool control must process in a shorter time period a large amount of NC data. The block processing time indicates how long the control unit processes one block of data and this parameter is used for an evaluation of the properties of the CNC control systems. The shorter the time period, the more efficient the control system. The typical values of the block cycle time T are in the range of 0.5 ms to 10 ms [3]. The block processing time and the length of the interpolated section l directly effects the maximal feed rate $v_{f_{\max}}$.

$$V_{f_{\max}} = \frac{l}{T} \cdot 60. \quad (1)$$

Consequently, NC part programs for complex 3D shapes, which because of quality requirements consist of short line segments, not only produce a large amount of data, but also limit the possible feed rate. This is a significant disadvantage for HSM technology that requires a high feed rate per cutting tooth. For example, during described in this paper, machining tests, where the CNC control processed the NC block in time $T = 4$ ms and where the track segment was $l = 0.01$ mm, the maximal feed was limited to the value $v_{f_{\max}} = 0.15$ m/min. When the control unit is unable to process the data in the required time, the feed movement may stop temporary and can cause tool standstill tracks to appear on the working surface.

There is another problem, which is a consequence of the limited acceleration ability over individual axes. Very often, during the sudden change in the direction of movement, the accelerations possible from the machine axis drivers are not sufficient to achieve the programmed feed rate values. For example, during the tool movement along one axis (e.g. X axis) the speed of the other axis drive (e.g. Y axis) is equal to zero. In order to maintain the constant feed speed during a rapid change of direction in movement along another axis, the axis drive speed should, within an infinitesimal time period, change from zero to the programmed value [1, 3]. The rapid acceleration

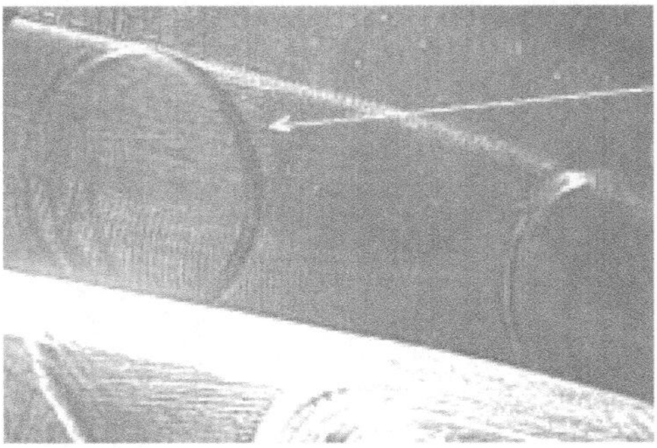

Fig. 1. The tracks made on the working surface by stopped tool [3].

variations can cause vibrations in the mechanical system of machine tool and the reduction of the feed rate. It means that the value of the feed rate per cutting edge decreases dramatically and as a result the tool reduces its life period and the surface texture deteriorates. The characteristic tracks appear in those locations where the feed rate comes to a stop because of the changes of movement direction [3].

3 CAD Model Requirements

As the limitations of the machine tool are difficult to eliminate, another way to reduce the deterioration effect has to be searched for. The opportunity for this can be found in the proper leading of the tool paths. To avoid the tool path sharp angles that causes the rapid changes in movement directions and feed rate reduction the CAD models provided by designers should ensure the producibility properties. The main problem in this field is the lack of tangency in the transition between the two surface plains. This problem was not significant with conventional machining, but in HSM technology it has considerable importance.

Figure 2 illustrates the effects after building surfaces by lofting, which has thrown the profiles without tangency condition (a) and with given tangency condition (b). Surface E was tacked on the cross-sections A-A, B-B and C-C. In the next case (Figure 2b), tangential passes between surfaces E and D and also E and F were also specified.

The lack of a tangency condition in the CAD model has an effect on the tool path quality obtained in the CAM system. At the point where the transition between surfaces does not have tangency conditions, sharp passes in the tool path segments (Figure 3a) may appear instead of the required smooth passes (Figure 3b). In combination with the above-described machine tool limitations this can cause the deterioration of the surface quality.

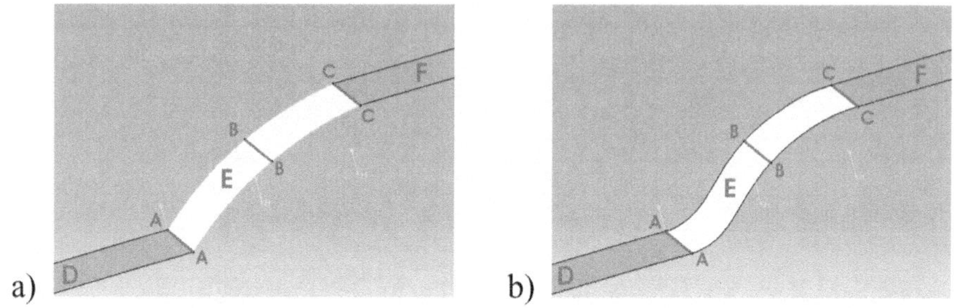

Fig. 2. The result of loft operation: (a) without tangency condition, (b) with specified tangency between surfaces.

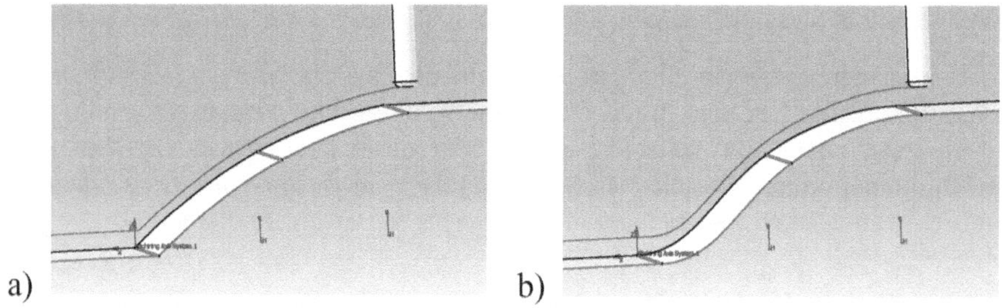

Fig. 3. Tool path: (a) machining the surfaces without tangential transitions, (b) machining the tangential surfaces.

In the case of pockets machining (removing material from within a closed geometry), it is recommended that the largest possible fillets are used for the corners (Figure 4). Machining the corners, with the fillet radius equal to the tool radius, leads to the situation shown in Figure 4a, where a rapid change in the direction of movement is required. To avoid this effect, a tool with a smaller radius, or a bigger fillet radius in the corners, could be applied. This modification provides tool paths with the advisable rounded transitions between the segments in the pocked corners.

In pockets machining it is recommended that the pockets are designed as open pockets. This approach allows for a longer tool life by providing better cutting conditions in the tool lead in stage. There are many advantages, especially for cutting hard materials, when the tool approaches the part in the manner illustrated in Figure 5a. In this case, the tool works with the teeth placed in its circumference. In the 'Helix' or 'Ramping' strategies the tool life period is reduced by the considerable axial force which is generated.

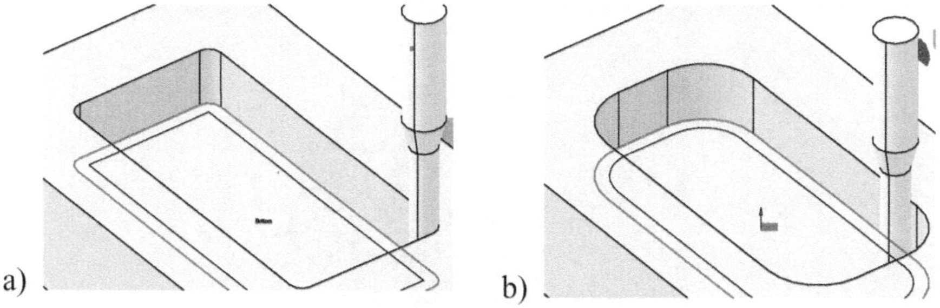

Fig. 4. Machining the pocket corners: (a) with corner fillet equal to tool radius, (b) with corner fillet bigger than tool radius.

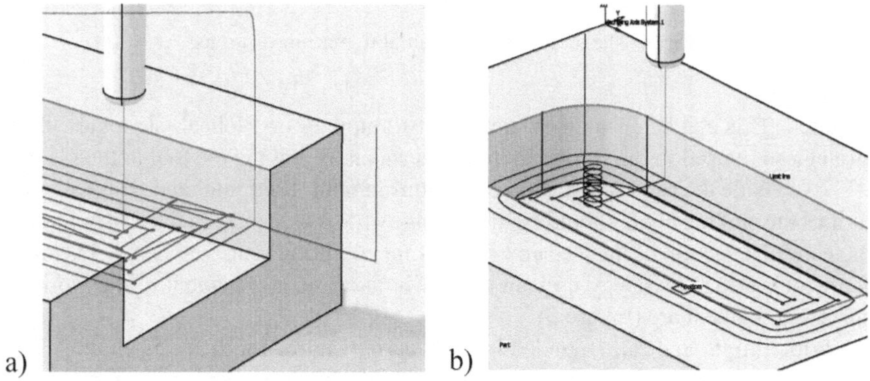

Fig. 5. The methods of tool lead-in: (a) the case of open pocked, (b) the case of closed pocked with the 'Helix' strategy.

4 Machining Strategies Adopted to HSM

The efficient tool paths, in the case of high speed machining, excepting the small distance between consecutive passes, should ensure tool protection and a high feed rate. Where possible, it is essential to maintain a constant tool load as any load changes may cause variations in tool deflections and this, in turn, can lead to loss of accuracy, surface texture and tool life period.

Upon analysis of all tool paths generated by particular machining cycles, the points that should be changed to adopt the paths to HSM requirements became evident. Special attention was paid to the method in which the tool lead-in to and lead-out of the part. The approach to the blank by the tool directly at full cutting depth with a working feed rate can lead to excessive tool load and therefore a shorten life period. Consequently, in HSM technology, a sinking of the tool using a skew movement is

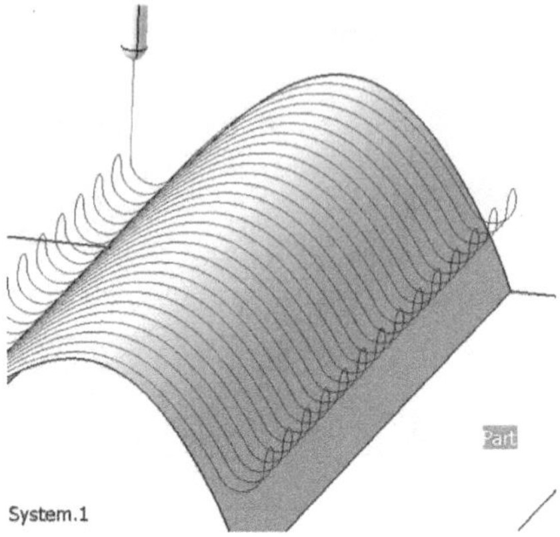

Fig. 6. The tool lead-in tangential to machined surface.

required. This can be ensured by applying ŚRampingŠ or 'Helical' strategies. Both strategies are used to minimize the tool load shock by sinking the tool in the blank.

To provide the right tool lead-in conditions in both 'Ramping' and 'Helical' conditions the sinking angle should be in the range of 1–3 grade [2]. In the finishing state the careful selection of the lead-in and lead-out condition is also essential. The most frequently utilized strategy uses an approach on a line at a tangent to the working surface or tangent arc (Figure 6).

With a high feed rate, rounded internal corners allow for faster machining. The inconvenience of the solution is that the rest of the material needs to be removed in an additional machining cycle (re-machining operation).

The efficiency obtained within HSM technology is, however, so high that the process is very efficient even when an additional machining operation is added. With the very common 'Along Lines in XY Plane' strategy, the traditional rectangular reverses between successive parallel tool paths are replaced by curvilinear loops which allow the maintenance of a high feed rate. With the 'Constant Z' machining strategy, the tool path follows the workpiece contour on a specified level of the Z axis and then descends to the next Z level. This strategy provides a constant tool load as the tool maintains contact with the workpiece along the whole path. Conventionally, the alternative for this strategy is milling with 'Zig Zag' shape paths but this does cause a decrease in the life of the tool as it must leave the material and return at the end of each stroke.

These, and other options connected with the smoothing of the tool paths, can be found in several machining cycles in the CAM module of the CATIA V5 system.

The pocketing operation in the strategy definition chart allows for the choice of HSM dedicated options. In the case of the 'Zig Zag' shape path, rounded corners are applied and in 'Concentric milling' or 'Contour parallel milling' an additional transition is generated between tool paths (Figures 7 and 8).

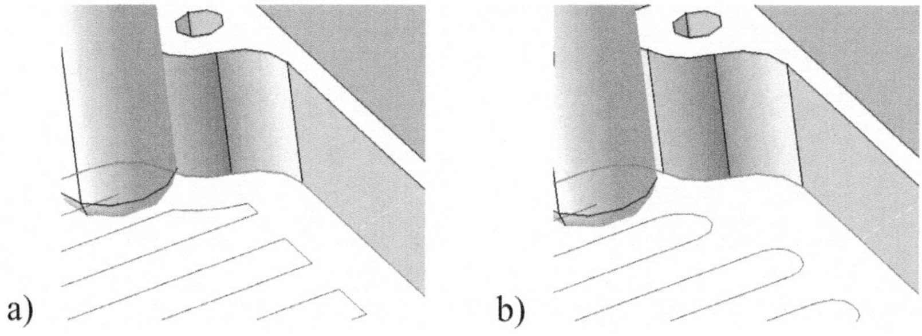

Fig. 7. Pocketing operation with 'Zig Zag' milling strategy (a) traditional rectangular reverses, (b) with rounded corners – HSM option.

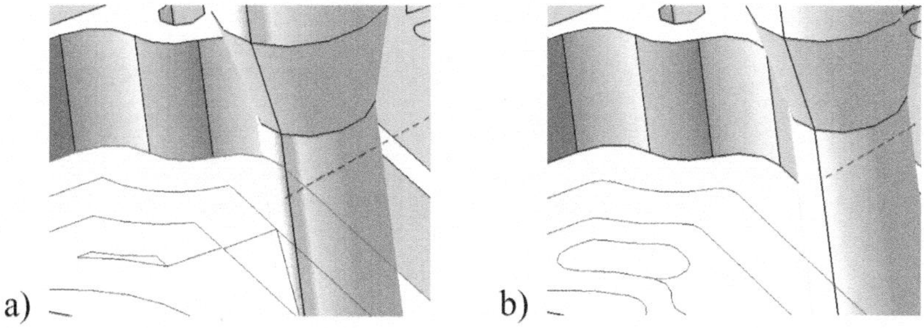

Fig. 8. Pocketing operation with concentric milling strategy (a) without rounded corners, (b) with rounded corners – HSM option.

The same options – inserting fillets in sharp corners, inserting fillets in the finishing cut, and defining the passes between tool paths – can be found by planning the facing milling operation using both concentric and spiral strategies (Figures 9 and 10).

Pencil milling moves the cutter along the internal corners and it is used for removing the remains of material left by the previous cycles. In this strategy it is not possible to interfere in the tool paths because the finishing cycle is executed in one pass. Inserting, for example, the filets in the sharp corners of the toll path will result in workpiece shape errors. This operation does not have any options dedicated to HSM. However, by defining the lead-in and lead-out macros, the tool paths can be adapted to high speed machining needs.

Beside the wide range of dedicated to HSM machining strategy options that can be found in CAD/CAM systems the feed rate optimization function is also very useful. It allows for the adaptation of the feed rate to the changes of tool load. When the tool load is reduced the feed rate could increase automatically and a shorter machining time would occur.

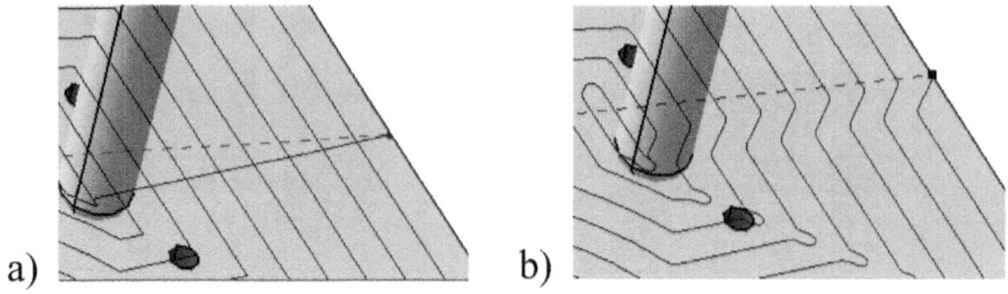

Fig. 9. Facing milling with concentric strategy (a) without soft passes between tool paths, (b) with soft passes between tool paths.

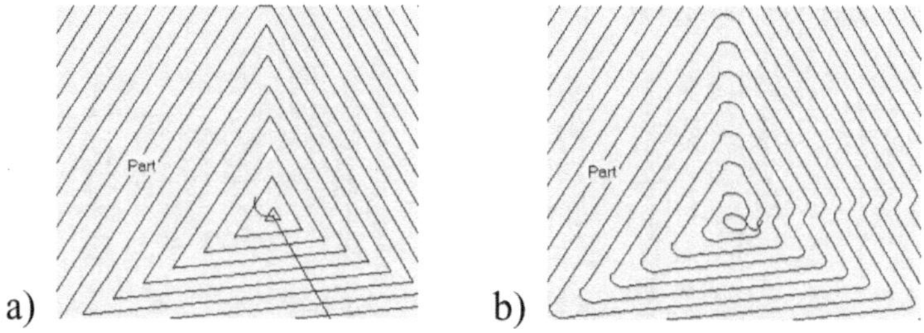

Fig. 10. Spiral milling (a) without HSM option, (b) with HSM option.

5 Programming the Machining of Forging Dies

Work to put into practice the high speed machining technology for making drop forging dies commenced in 2003 at the Poznan University of Technology and the tool shop of the company Aesculap Chifa. The aims were to machine the die by implementing the new technology, which would shorten the working time, reduce the costs and eliminate the handwork. Initially, the company bore great expense by purchasing the machining centre and tooling system (milling tools and contraction holders). The company expected to reduce the costs by making the die about 20% cheaper and decreasing the working time by about 15%. A five year investment return was assumed. The base for preparing the new technology and NC programming were two computer stands with CATIA V4 and CATIA V5 systems.

The first attempts at machining the dies with HSM technology did not bring the expected results. The surface texture was far from adequate and handwork finishing was necessary. Additionally the tools' life was unsatisfactory and significantly lower than the producer's data indicated.

To eliminate these problems, all elements of the processes that may have influenced the machining quality were examined.

Machining operations, strategies and parameters were defined using the dialog method in the CATIA V5 CAM module. In the rough milling of the die container the 'Spiral' strategy was applied (Figure 11). The die draft was roughly machined

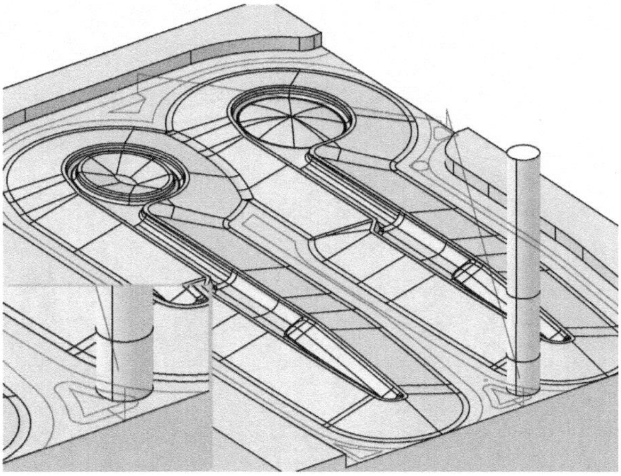

Fig. 11. Rough milling of the die container.

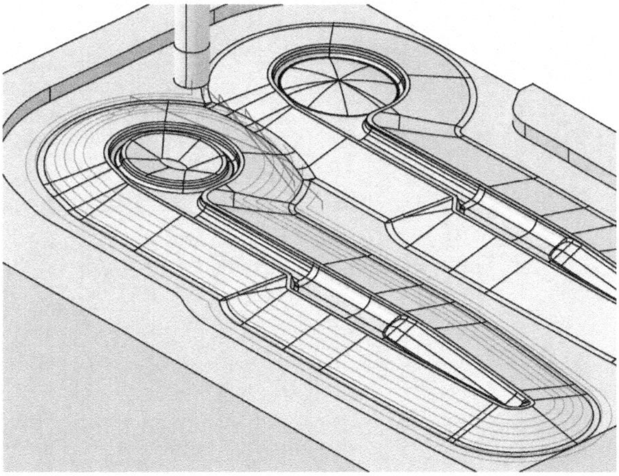

Fig. 12. Rough milling of the die draft.

with the 'Contour Parallel' strategy (Figure 12) and then finally with the 'Constant *Z*' operation, in which the HSM feature was applied.

The finishing operation was carried out by 'Along Line in *XY* Plane' strategy with use of HSM features (Figure 14).

The forging dies were machined on a Deckel Maho DMC 70V hi-dyn vertical milling centre equipped with a electrospindle 30 000 RMP. The machine was controlled by a Heidenhein TNC 426 CNC control with a 'Look Ahead' function.

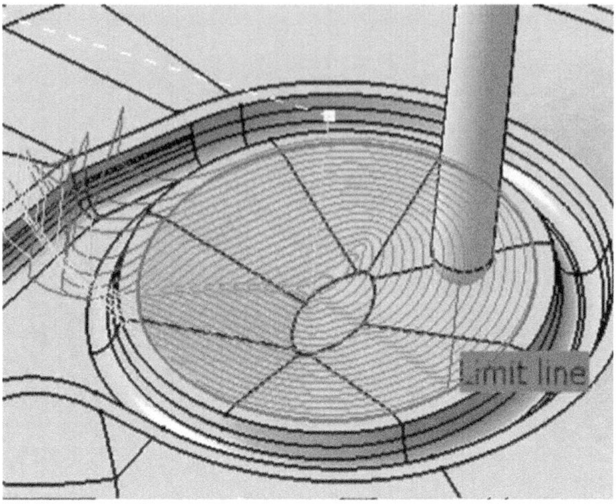

Fig. 13. The rough machining of the forming cavity; 'spiral' milling strategy – option of rounding sharp passes for HSM.

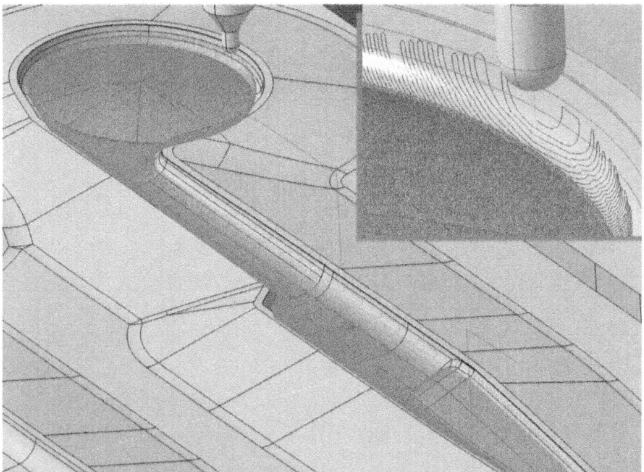

Fig. 14. The finishing operation of machining the forming cavity; strategy 'along line in XY plane' – with HSM features.

6 Conclusion

The results obtained indicate the advisability of using HSM technology for manufacturing the forging dies. The possibilities for improvement in the particular elements of the machining process and the positive impact upon the final process results were evident. The designers and process engineers realized that all the aspects of HSM

Fig. 15. The forming die after rough milling.

technology that can affect the machining process are of great importance in product and process development and can eliminate the large majority of process faults. As a result of the new work organization, the machining with HSM technology allowed a 35% saving in process time and a 25% saving in process costs compared to traditional production technology.

The basis for the successful application of the HSM technology were the advances in machine tool technology, controls, cutting tools and CAD/CAM systems. Machining strategies in CAD/CAM systems are now adopted to the specific requirements of HSM machining and are able to ensure the uniform cutting conditions relative to speed, tool load, and radial and axial feed.

References

1. S. Altmueller, S. Knodt, T. Hedwig. Issues, trends and developments in HSM, Online, World Wide Web page of Modern Machine Shop, http://www.mmsonline.com/articles.
2. R. Field. High speed machining of dies and molds, Online, World Wide Web page of Tooling University, http://www.toolingu.com.
3. M. Loftus. Causes of high-speed machining irregularities, in *Proceedings of MoldMaking 2002 Conference*, Communication Technologies Inc, Chicago, USA, 2002, pp. 126–142.
4. U. Heisel, A. Feinauer. Dynamic influence on workpiece quality in high speed milling, *Annals of the CIRP* **48**(1), 1999, 321–324.
5. D. Marinac. Smart toolpath for HSM, *Manufacturing Engineering* **125**(5), 2000.
6. V. Pateloup, E. Ducand, C. Lartigue, P. Ray. Pocketing optimization for HSM geometry tool path and interpolation mode influence on dynamic machine tool behaviour, *Machine Engineering* **3**(1), 2003, 375–382.

Author Index

Subject Index

GPSR Compliance

The European Union's (EU) General Product Safety Regulation (GPSR) is a set of rules that requires consumer products to be safe and our obligations to ensure this.

If you have any concerns about our products, you can contact us on ProductSafety@springernature.com

In case Publisher is established outside the EU, the EU authorized representative is:

Springer Nature Customer Service Center GmbH
Europaplatz 3
69115 Heidelberg, Germany

Batch number: 10809436

Printed by Printforce, the Netherlands